Acta Physica Austriaca
Supplementum XXI

Proceedings of the
XVIII. Internationale Universitätswochen für Kernphysik 1979
der Karl-Franzens-Universität Graz
at Schladming (Steiermark, Austria)
28th February—10th March 1979

Sponsored by
Bundesministerium für Wissenschaft und Forschung
Steiermärkische Landesregierung
CERN (European Organization for Nuclear Research)
ICTP (International Centre for Theoretical Physics)
Sektion Industrie der Kammer der
Gewerblichen Wirtschaft für Steiermark

1979

Springer-Verlag
Wien New York

Quarks and Leptons as Fundamental Particles

Edited by Paul Urban, Graz

With 184 Figures

1979

Springer-Verlag

Wien New York

Organizing Committee

Chairman

Prof. Dr. H. Mitter
Ordinarius am Institut für Theoretische Physik
der Universität Graz

Committee Members

Dr. L. Mathelitsch
Dr. W. Plessas
Dr. H. Zankel

Secretary

M. Krautilik

Library of Congress Cataloging in Publication Data. Internationale Universitätswochen für
Kernphysik der Karl-Franzens-Universität Graz, 18th, Schladming, Austria, 1979. Quarks and
leptons as fundamental particles. (Acta physica Austriaca : Supplementum ; 21.) "Proceedings of the
XVIII. Internationale Universitätswochen für Kernphysik 1979 der Karl-Franzens-Universität Graz
at Schladming (Steiermark, Austria), 28th February—10th March 1979; sponsored by
Bundesministerium für Wissenschaft und Forschung ... [et al.]." 1. Quarks—Congresses. 2. Leptons
(Nuclear physics)—Congresses. I. Urban, Paul Oskar, 1905—. II. Austria. Bundesministerium für
Wissenschaft und Forschung. III. Title. IV. Series. QC793.5.Q2522I57. 1979. 539.7'211. 79-23328

ISSN 0065-1559
ISBN-13:978-3-7091-8576-6   e-ISBN-13:978-3-7091-8574-2
DOI: 10.1007/978-3-7091-8574-2

CONTENTS

Acta Physica Austriaca, Suppl. XXI, 1--4 (1979)
© by Springer-Verlag 1979

OPENING ADDRESS

by

H. MITTER
Institut für Theoretische Physik
Universität Graz

This year's subject is "Quarks and Leptons as Fundamental Particles". You are all familiar with the enthusiasm, with which theorists working in high-energy physics follow the recent development in unifying the world of particles and their interactions. So it was probably no surprise to you that we have chosen this title for our school. Since I personally have become more sceptical against enthusiasm with increasing age (one could also say, the burnt child abhors the fire), I had even liked to put a question mark at the end of the title: then, however, the majority of the theorists would probably have disagreed. Enthusiasm is human, maybe some amount of it is even necessary for any progress. So let us be enthusiastic until our friends in the experimental department get us down to earth again. In the past they did that, at least. In the last two years they seem to work in the opposite direction. Perhaps it is worthwhile to remember, how the whole development started. At the beginning there was probably the PEN club: of

course, not the wellknown group of more or less well-
known poets, but a more silent, almost secret group of
physicists, who believed, that three spin-$\frac{1}{2}$ fields
(Proton, Electron, Neutrino) should suffice for a
description of weak and electromagnetic interactions.
The bosonic glue fields could always be imagined as a
kind of "bound-state field". During the further
development these ideas took concrete form (through
the Salam-Weinberg model, for instance) the gauge
approach showed to be essential, generalizations were
found (the number of constituents was increased), new
leptons were discovered in due course. At the same time
the idea of quarks as fundamental constituents of the
hadronic world has gained ground slowly but steadily.
The colour degree of freedom was added, QCD came up
and opened the way to further unification, in which we
consider now quarks and leptons as building blocks. Of
course, we have now to live with at least 12 (maybe
with 18) quarks and a considerable number of leptons:
in order to give a name to our club we would need a
good part of the whole alphabet and it seems really
worth-while to look, how good the basis for our enthusiasm
is and where we stand.

In order to do this, we have tried to select a
number of subjects, which should give an account both
of the speculative and the phenomenological side of
the recent development. We were lucky to find a considerable number of excellent and expert speakers so
that the presentation will be a balanced one. The only
slight disbalance, which I can see, is that quarks will
appear in the lectures perhaps more frequently than
leptons. Since most of the latter objects are with us

for a long time, it seems, however, not quite unjustified to devote less attention to old friends and
more to foreigners, to these elusive quarks, which we
first were all anxious to see, whereas now many of us
would be happy if they were not found at all. (An
Austrian expression on confined quarks would be: Sachen
gibts, die gibts gar nicht = things exist which don't
even exist!) Two or three years ago it was just to say
that weak and electromagnetic interactions are "ahead"
of the strong ones as far as basic understanding is
concerned. Meanwhile the situation has started to change.
By the end of our school we shall see, how this race
stands now.

As always we have also included lectures on
experimental results in our field, in order to keep
track of the solid ground of all speculations. This
time we shall be confronted with the most recent
results obtained in Hamburg with the new storage
rings. If our experimental friends have worked as
hard as usual, these will be very interesting results.

A number of seminars will complete the program.
They will give some of the participants the possibility
to discuss their recent results and ideas with their
colleagues. A new achievement, which we try out for the
first time this year, are the Stammtische, where our
speakers and other interested people may meet in the
evening in order to discuss special problems inside
(and outside) physics.

This is all what I want to say about the program.
You have already started to try out your ability in

solving problems of classical mechanics, I mean motion
with more or less friction on curved trajectories in a
gravitational field, and I want to encourage you to
continue, even if you have the impression that the
Planai has to be treated as a statistical ensemble.
All the organizers of the school wish you a pleasant
stay here in Schladming. We hope that you enjoy both
the school and the environment.

Acta Physica Austriaca, Suppl. XXI, 5—80 (1979)
© by Springer-Verlag 1979

LECTURES ON QUARKS[+]

by

G. MORPURGO
Istituto di Fisica dell'Università di Genova
Istituto Nazionale di Fisica Nucleare,
Sezione di Genova
Viale Benedetto XV, I-16132 Genova

CONTENTS

[+] Lectures given at the
XVIII. Internationale Universitätswochen für Kernphysik, Schladming, Austria, February 28 - March 10,1979.

LECTURE I: SOME RECENT RESULTS OF THE
STATIC QUARK MODEL

1. INTRODUCTION

The contents of these three lectures will be as
follows: today I discuss some recent results in the
naive quark model for the "old" pre-new flavour states.
In the second and third lectures I deal with the search
for quarks.

Clearly there is not much connection between the
contents of today's lecture and that of the subsequent
ones; indeed the only connection is a personal one: my
interest in the search for quarks was stimulated long
time ago, (in 1965), by the successful predictions of
the non-relativistic quark model [1,2]. Since then many
things have changed; the figure 1 tries to summarize the
evolution of the quark model and is, I hope, self-
explanatory. But the "nuclear physics" or naive treat-

ment of the old (pre-new flavours) hadrons still appears
to me productive in the sense of establishing connections
between different facts and showing the interest of
measuring new quantities.

The following discussion is based on the work of
Isgur and Karl [3], of Le Youanc et al. [4] and of
De Rujula, Georgi, and Glashow [5]; I have limited myself to take a few of the many problems dealt in the
above work and summarize them here.

2. THE MAGNETIC MOMENTS OF THE BARYONS
AND THE V→Pγ TRANSITIONS

Part of my stimulus to reconsider these problems –
that have been with us since the beginning [1,2] – is
due to a recent very precise measurement of the magnetic
moment of the Λ hyperon [6]:

$$\mu_\Lambda = -0.6138 \pm 0.0047 \text{ n.m.}$$

This is an important result because it allows to
determine the magnetic moment μ_λ of the strange quark
(I will indicate quarks as $n, p, \lambda, \ldots$. Neutron, Proton
and Λ-particle will be N,P,Λ). If we write

$$\mu = \sum_i \mu_i \ell_i$$

as it was done originally [1] the result $\mu_p/\mu_N = -\frac{3}{2}$ is
obtained assuming:

$$\mu_p : \mu_n = 2: -1$$

(and, of course, assuming a $56,0^+$ representation of the nucleons). But we never had a good value for μ_λ. The result I mentioned above provides the good value:

$$\mu_p : \mu_n : \mu_\lambda = 2 : -1 : -0.65 \ .$$

Indeed the measured value of μ_Λ of -0.6138 n.m. is the value $-\frac{1}{3}\mu_p$ one would predict from the canonical SU_3-ratio multiplied by 0.65. Thus

$$\mu_\lambda / \mu_n = 0.65 \ .$$

There are some other values of magnetic moments that one can deduce with this result, namely:

1) $\mu_{\Sigma^+} = \{1-\frac{1}{9}(1-\frac{\mu_\lambda}{\mu_n})\}\mu_p = 0.96\mu_p = 2.68$ n.m.

$\qquad\qquad\qquad\qquad$ (Exp. $= 2.83 \pm 0.25$ n.m.)

2) $\mu_{\Sigma\Lambda} = -\frac{1}{\sqrt{3}}\mu_p = -1.63$ n.m. ($|\text{Exp.}| = 1.82 \ ^{+0.25}_{-0.18}$ n.m.)

3) $\mu_{\Sigma^-} = -1.05$ n.m. (Exp. $= -1.48 \pm 0.37$ n.m.)

4) $\mu_{\Xi^-} = -0.46$ n.m. (Exp. $= -1.85 \pm 0.75$ n.m.)

3) and 4) above have large experimental errors, and there are so far no data on $\mu(\Omega^-)$ (predicted to be -1.84 n.m.) and $\mu(\Xi^0)$ (predicted to be -1.39 n.m.). I stress the importance of further precise measurements of the magnetic moments of baryons.

Since we are speaking of magnetic moments, I
digress briefly to consider the vector meson → Ps-
meson + γ transitions. This was an early successful
check of the naive quark model [7]. Afterwards the
measurements on K^{*0} → K^0 + γ and ρ → π + γ showed
considerable discrepancies with the predictions of the
model. Now the situation is this: if we a) use the
value of μ_λ = 0.65 μ_n given previously and b) insert
(Isgur [8]) a common (form factor)^2 of 0.72 (thus
reducing the ω → π + γ rate from the theoretical
prediction - for a point ω - of 1200 keV to the observed value of 870 keV) then all the V → Pγ decays -
except the ρ → πγ decay - are predicted correctly
within the experimental error. This is shown in the
table below taken from ref.[8] with minor modifications.

I conclude this digression on the V → Pγ decays
remarking that a good measurement of the ratio:

$$\frac{K^{*0} \to K^0 \gamma}{K^{*+} \to K^+ \gamma} = \frac{(1 + X)^2}{(2 - X)^2}$$

would provide a determination of X = $\frac{\mu_\lambda}{\mu_n}$ additional to
that obtained from μ_Λ and would, therefore, be of great
interest.

V → Pγ decays (calculated with $\mu_\lambda = \mu_n \cdot 0.65$ and (form factor)^2 = 0.72)

Decay	Theory (Γ in keV)	Exp. (Γ in keV)
$\omega \to \pi\gamma$	870 (*)	870
$\rho \to \pi\gamma$	90	35 ± 10
$\rho \to \eta\gamma$	55	50 ± 13
$\omega \to \eta\gamma$	5.8	$3.0 \begin{smallmatrix}+2.5\\-1.8\end{smallmatrix}$
$\eta' \to \rho\gamma$	114	< 300
$\eta' \to \omega\gamma$	12	< 50
$\phi \to \pi\gamma$	8.2	5.9 ± 2.1
$\phi \to \eta\gamma$	75	65 ± 15
$\phi \to \eta'\gamma$	0.27	–
$K^{*0} \to K^0\gamma$	135	75 ± 35
$K^{*+} \to K^+\gamma$	90	< 90

(*) A point-ω would give 1200 keV.

3. THE MAGNETIC MOMENTS AND THE MASSES OF THE CONSTITUENT QUARKS

Are the masses of the constituent quarks inversely proportional to the magnetic moments? That is, having established that $\mu_\lambda/\mu_n = 0.65$, does this mean that $m_n/m_\lambda = 0.65$? (Here m_n and m_λ are the masses of the n and constituent quarks). Although almost always in the literature the answer is positive, this is not necessarily so. Assume, for instance, that the mass of

a bound <u>constituent quark</u> is very heavy, so heavy that
its Dirac magnetic moment is completely irrelevant.
Then the magnetic moment of this quark is not determined by its mass: it is determined by the mass of
some mesons around the quark and, <u>in such case, μ_λ/μ_n</u>
<u>does not measure the ratio between the masses of the</u>
<u>quarks but determines the ratio between the average</u>
<u>masses of these mesons.</u> If, therefore, we state that:

$$\frac{\mu_\lambda}{\mu_n} \simeq \frac{m_n}{m_\lambda}$$

we do implicitly assume that the masses of the quarks,
and not those of the mesons in the quark clouds,
determine the magnetic moments. This assumption has,
I repeat, to be underlined.

4. THE CHROMOSTATIC POTENTIAL AND THE DETERMINATION
OF m_n/m_λ FROM THE $\Sigma - \Lambda$ MASS DIFFERENCE

If we had an independent way to determine m_n/m_λ
we could check if $\mu_\lambda/\mu_n = m_n/m_\lambda$. Do we have this
independent way? A clean independent way does not
really exist at the moment, unfortunately. Here, by
clean we mean something that is independent of disput-
able assumptions. What one can do, and was done, is to
take the nonrelativistic reduction of the chromostatic
potential between quarks discussed by De Rujula, Georgi,
and Glashow (D.G.G.) [5] and to calculate with it the
$\Sigma - \Lambda$ mass difference to the m_n/m_λ ratio. In view of
the importance of the determination of the quantity

$$C = \frac{\mu_\lambda m_\lambda}{\mu_n m_n} - 1$$

(the deviations of C from zero giving the non-point Dirac character of the bound constituent quark) this procedure is certainly not satisfactory, because it introduces another assumption - the validity of the D.G.G. potential - that is not well established. On the other hand the D.G.G. potential can be used to produce several other results - we refer here particularly to the work of Isgur and Karl [3] - that so far appear to be in reasonable agreement with the facts. The analysis below appears therefore of some interest.

As a matter of fact the only part of the D.G.G. potential of interest here (and to which we shall confine our attention) is a term having the form:

$$V = \frac{16}{9}\pi\alpha_s \sum_{i>k} \frac{\vec{s}_i \cdot \vec{s}_k}{m_i \cdot m_k} \delta(r_{ik}) \tag{1}$$

where m_i and m_k are the masses of the quarks, α_s is the chromo-fine-structure constant and the other symbols are obvious. From this expression D.G.G. calculate (the derivation is simple and I refer to their paper):

$$\Sigma - \Lambda = \frac{2}{3} (1 - \frac{m_p}{m_\lambda}) (\Delta - N) \tag{2}$$

where $\Sigma - \Lambda$ stands for $M_\Sigma - M_\Lambda$ and similarly for $\Delta - N$. Thus

$$\frac{m_p}{m_\lambda} = 0.62$$

(note that $m_p \overset{\sim}{=} m_n$ so we use the two interchangeably).
If this result was unobjectionable, it would have been
very interesting because it would have meant that -
assuming the D.G.G. potential to be correct - the
equation C = 0 was correct to a great precision: a
striking conclusion. However, even if we assume that
the D.G.G. potential is correct, the result given
above - as remarked by Le Youanc et al.[4] - is not
justified. Indeed, if m_λ is different from m_p, the
space part of the Λ (or Σ) wave function is different
from the space part of the P (or Δ) wave function.
Taking this difference into account we do not get the
eq. (2) anymore. We get instead:

$$\Sigma - \Lambda = \frac{2}{3}(\Delta-N) \ \{1 - \frac{m_p}{m_\lambda} \ [\frac{1}{4} + \frac{3}{4} \ (\frac{2m_p+m_\lambda}{3m_\lambda})^{\frac{1}{2}}]^{-3/2}\} \tag{3}$$

giving $\frac{m_p}{m_\lambda} = 0.48$ rather different from 0.65.
Thus C appears to be substantially different from zero
(C = 1.35) in agreement perhaps with naive expectations,
but in disagreement with the very use of the D.G.G.
potential the derivation of which is based on essentially
point-Dirac quarks.

However, the conclusion $\frac{m_p}{m_\lambda} = 0.48$ - obtained from
the eq. (3) - is once more questionable. Indeed - even
leaving aside the non-relativistic character of the
calculation - the same D.G.G. potential used to calculate
the $\Sigma-\Lambda$ mass difference introduces also SU_6-configuration

mixings - to be discussed in a moment - whereas the equations (2) and (3) have been obtained for pure ($\underline{56}$, 0^+) states. According to G. Karl [9], if these configuration mixings are considered, the value m_p/m_λ is again raised from 0.48 to $\simeq$ 0.6.

Any stronger conclusion seems, at the moment, unjustified.

5. CHECKS OF THE CONFIGURATION MIXING INTRODUCED BY THE CONTACT SPIN-SPIN PART OF THE D.G.G. POTENTIAL

To form an idea if the contact spin-spin part of the D.G.G. potential is correct (at least for the case $m_i = m_k$ in (1)) we can calculate, following the ref.[3], the SU_6 mixing produced in the proton by the interaction (1). Calling A the factor $\frac{16}{9}\pi\alpha_s m_p^{-2}$ and determining A from the Δ -N mass difference, it is:

$$\Delta -N = \frac{3A(3km_p)^{3/4}}{4\sqrt{2}\ \pi^{3/2}} \tag{4}$$

where in the calculation use has been made of a harmonic-oscillator model with spring constant k. We do not need to specify the constant k because this is eliminated in the computation of the SU_6 mixing. (Note, incidentally, that the equation (4) allows to compute A and - if the spring constant k is known - α_s can be determined. We do not attach much significance to this deduction of α_s because the harmonic oscillator suffers from a serious defect

that is often forgotten: due to the presence of one
parameter only one cannot fix independently the
radius and the distance between levels. Once you fix
the distance between levels the radius is fixed and
viceversa; and, experimentally, the two are not very
well reproduced).

But let us go on. Writing the state of the neutron
as

$$|\text{Neutron}\rangle = |\underset{\sim}{56},0^+\rangle' + \beta|70,0^+\rangle_{N=2} \qquad (5)$$

(where the dash on $|\underset{\sim}{56},0^+\rangle'$ is there to recall that we
have in fact a certain combination of $|\underset{\sim}{56},0^+\rangle$ involved -
compare the ref.[3] -) the mixing coefficient β turns
out to be - in the harmonic-oscillator model -

$$\beta = -\frac{\Delta - N}{M_2 - M_0} \frac{\sqrt{3}}{2\sqrt{2}} \qquad (6)$$

where $M_2 - M_0$ is here the average mass-difference between the $N = 2$ and $N = 0$ bands: ~ 660 MeV. From (6) it
is:

$$\beta = -\frac{300}{660} \cdot 0.61 = -0.27 \ . \qquad (7)$$

It is then possible to calculate the average square
radius of the neutron: as well known this is zero if a
pure $\underset{\sim}{56}$ representation for the nucleon is used. The
reason is that the space and the spin-unitary spin part
of the wave function factorize in that case. With the
impure SU_6 function (5) we get instead:

$$\frac{\sum_i <e_i \; r_i^2> \text{ neutron}}{\sum_i <e_i \; r_i^2> \text{ proton}} = -0.16$$

in good agreement with the experiment (Exp. $\overset{\sim}{=}$ -0.15).
Isgur, Karl, and Konjuk also point out in their paper
[3] that the same mixing accounts also for the
violations of the Moorhouse rule [10] in the e.m.
transitions $D_{15} \rightarrow$ Pγ and $D_{15} \rightarrow$ Nγ, and that the
Becchi-Morpurgo E2 selection rule remains practically
unchanged due to the very small effect of tensor
mixing on the nucleon state. I refer for this to
their paper.

We now may ask: are the "classical" ratios
μ_P/μ_N, D/F, and g_A/g_V altered by the SU_6 mixing just
considered? The answer is that they are substantially
unchanged. Thus [11]:

$$\mu_P/\mu_N = -\frac{3}{2}(1+ \frac{1}{3} \; tg^2\theta) = -\frac{3}{2}(1 + 0.026)$$

$$(D/F)_{\text{Axial}} = \frac{3}{2}(1-\frac{1}{2}tg^2\theta) = \frac{3}{2}(1 - 0.03)$$

$$g_A/g_V = \frac{5}{8} \; (1 - 0.06) \; .$$

There is almost nothing to say on μ_P/μ_N or D/F. (D/F
derived only by the neutron β-decay and $\Sigma^- \rightarrow \Lambda\beta$ - decay
is experimentally in good agreement with the $\frac{3}{2}$ value of
the pure 56). Instead g_A/g_V deserves a comment. The experimental value of g_A/g_V is 1.25 and it is seen that
the SU_6 mixing, although acting in the right direction,

does not reduce sufficiently the $\frac{5}{3}$ result.

In my opinion we are here in presence of a "renormalization" effect of the axial coupling constant - a non-point constituent quark. The effect is analogous to that - in the prequark period - of the factor 1.23 that appeared multiplying γ_5 in the N → P weak current $<N|\tau_-\gamma_\mu(1+1.23\gamma_5)|P>$. Here, at the quark level, we have something of the order 0.75 in the quark current: $<n|\tau_-\gamma_\mu(1+0.75\gamma_5)|P>$. That the constituent quarks should be non-point seems to me fairly clear, but much confusion exists on this in the literature. As a matter of fact the situation, as **far** as g_A/g_V is concerned, is complicated by the relativistic corrections; and so it is difficult at the moment to decide which part of the discrepancy can be attributed to the renormalization effect and which to the relativistic corrections. To derive a reliable value for both these quantities appears to be an important question. It is not easy, however.

To summarize: the prescription (5) with $\beta = -0.27$ appears a reasonable mixing prescription for the nucleon, and to have related the Δ-N mass difference to the neutron radius is satisfactory. Of course there are many other quantities to be discussed like the $<\Delta|$Axial Current$|N>$ or $<\Delta|$E.m. current$|N>$ transition matrix elements, but I refer for this to the literature [11].

To end this section I would like to make a remark on a problem of signs that is somewhat confusing. We have adopted here the sign convention of ref.[3] for the $|\underline{7}0,0^+>_{N=2}$ wave functions. The Paris group [4] adopts an opposite sign and therefore should have for β

a positive sign rather than a negative one like in (7).
Indeed, it obtains a positive sign; but it instead of
determining β in the way we did follow, determines it
by trying to fit the ratio $F_2^{en}(x)/F_2^{ep}(x)$ in the parton
model - this was the original intention of the Paris
group [4] - then gets a value of β opposite in sign,
and therefore a wrong sign for the neutron-charge
radius.

In my presentation here I have purposedly not
tried to discuss the question of $F_2^{en}(x)/F_2^{ep}(x)$, because
I feel that the connection of the quark model to the
parton model - that is how to perform the appropriate
Lorentz transformation - is not entirely clear. This
is in fact still one of the important open problems
with which we are confronted.

6. EXCITED P HYPERONS AND EFFECTS OF THE $m_\lambda - m_p$ MASS DIFFERENCE

In concluding this lecture I would like to present
another set of results obtained by Isgur and Karl [3,12]
on the excited baryon states. As a matter of fact I will
describe only a small fraction of the calculations by the
above authors, that fraction which is most directly
relevant to the question of the m_λ/m_p ratio discussed
previously. I will not enter at all on a much more wide
set of calculations performed by Isgur and Karl and
dealing with a systematic application of the Q.C.D.
chromostatic potential to the excited states of baryons.
The point is this: if m_λ and m_p are different and if we
call

$$\underset{\sim}{\varrho} = \sqrt{\tfrac{2}{3}}(\underset{\sim}{r}_\lambda - \frac{\underset{\sim}{r}_p + \underset{\sim}{r}_n}{2}) \quad \text{and} \quad \underset{\sim}{r} = \frac{1}{\sqrt{2}} (\underset{\sim}{r}_p - \underset{\sim}{r}_n)$$

the $|\underset{\sim}{70,1}^-\rangle$ excited states of strangeness -1 baryons will have different excitation energies depending on whether it is the $\underset{\sim}{\varrho}$ coordinate or the $\underset{\sim}{r}$ coordinate to be excited.

Indeed the Hamiltonian is, in an harmonic oscillator description:

$$H = \frac{\underset{\sim}{P}^2}{2M} + \frac{\underset{\sim}{P}_r^2}{2m_P} + \frac{\underset{\sim}{P}_\varrho^2}{2\underset{\sim}{m}_\lambda} + \tfrac{3}{2}k(r^2 + \varrho^2) \tag{7}$$

where $\underset{\sim}{P}$ is the total momentum of the hyperon (= 0 at rest) and

$$\underset{\sim}{m}_\lambda = \frac{3}{2x+1}\, m_P \qquad \text{where } x = \frac{m_P}{m_\lambda} .$$

Clearly from (7) we have:

$$\omega_\varrho = \sqrt{\frac{3k}{m_P}} \qquad \omega_\lambda = \sqrt{\frac{3k}{\underset{\sim}{m}_\lambda}} = \sqrt{\frac{3k}{m_P}} \cdot \sqrt{\frac{3x+1}{3}} .$$

Thus

$$\hbar\omega_\varrho - \hbar\omega_\lambda = \hbar\omega_\varrho \left(1 - \sqrt{\frac{2x+1}{3}}\right) . \tag{8}$$

If, typically, we take $\hbar\omega_\varrho$ = 500 MeV, the right-hand side of (8) is 65 MeV or 100 MeV depending on whether

we choose $x = 0.65$ or $x = 0.48$. A simple example of
application is the (zero-order) mass difference of
the $\Lambda(1830|\frac{5}{2}^-|70,1^-)$ and the $\Sigma(1765|\frac{5}{2}^-|70,1^-)$. Both
these states have $J^P = \frac{5}{2}$ and because they have an
orbital angular momentum 1^-, they have necessarily
spin $S = \frac{3}{2}$. Thus the wave function of the $\Lambda(1830)$
having isospin zero is necessarily of the form:

$$\Lambda(1830) = \alpha_p{}^\alpha n^\alpha{}_\lambda \frac{(\xi_p - \xi_n)}{\sqrt{2}} \psi_o \quad \text{and is an } \xi \text{ excitation.}$$

On the other hand $\Sigma(1765)$ has isospin 1 and its wave
function is necessarily:

$$\Sigma(1765) = \alpha_p{}^\alpha n^\alpha{}_\lambda \sqrt{\frac{2}{3}}(\xi_\lambda - \frac{\xi_p + \xi_n}{2})\psi_o \quad (\text{a } \rho \text{ excitation}).$$

We thus have the energy difference stated above. Isgur
and Karl then go on computing systematically the
corrections to the zero order using essentially the
D.G.G. chromostatic potential. We shall not do so but
I find rewarding that this simple consequence of a
different λ versus p quark mass explains qualitatively
the order of the above levels.

To conclude I would like to simply mention two
questions that appear to me important for different
reasons, but are not often discussed in the literature,
hoping to stimulate some discussion:

1) Are the effective masses of the bound (constituent)
   quarks the same for all phenomena, that is independent of the particular transition or level
   under study? I feel that it would be of importance

to study as many situations as possible in which a mass dependence appears and see how consistent are the mass values obtained.

2) Can one have, in the frame of the static quark model, <u>direct</u> evidence for the momentum and angular momentum carried by the gluons?

With these questions I end this first lecture; in the two ensuing ones we shall survey the experimental searches for quarks as announced at the beginning.

REFERENCES FOR LECTURE I

[1]   G. Morpurgo, Physics 2 (1965) 95.
[2]   G. Morpurgo,
      a) Rapporteur talk in Proc. of the XIV. Conference on High-Energy Physics, Vienna 1968, ed.J.Prentki and J.Steinberger, CERN Geneva 1968, p. 225.
      b) Lectures on the quark model in: Theory and Phenomenology in Particle Physics, part A, ed. A. Zichichi, Academic (1969) p. 83 .
[3]   N. Isgur and G. Karl, Phys. Rev. 18 (1978) 4187; N. Isgur, G. Karl,and R. Koniuk, Phys. Rev. Lett. 19 (1978) 1269.
[4]   A. Le Youanc, L. Oliver, O. Pène,and J. Raynal, Phys. Rev. D18 (1978) 1591.
[5]   A. De Rujula, H. Georgi,and S.L. Glashow, Phys. Rev. D12 (1975) 147.
[6]   L. Schachinger et al., Phys. Rev. Lett. 41 (1978) 1348.
[7]   C. Becchi and G. Morpurgo, Phys. Rev. 140 (1965) B687.

[8] N. Isgur, Phys. Rev. Lett. 36 (1976) 1262.

[9] G. Karl (private communication 1979).

[10] R.G. Moorhouse, Phys. Rev. Lett. 16 (1966) 772, 968.

[11] A. Le Youanc, L. Oliver, O. Pène, J.C. Raynal, Phys. Rev. D15 (1977) 844.

[12] Compare the first ref.[3] and also: N. Isgur and G. Karl, Phys. Lett. 74B (1978) 353.

FIGURE CAPTION FOR LECTURE I

Fig. 1. A short summary of the evolution of the quark model.

The standard picture

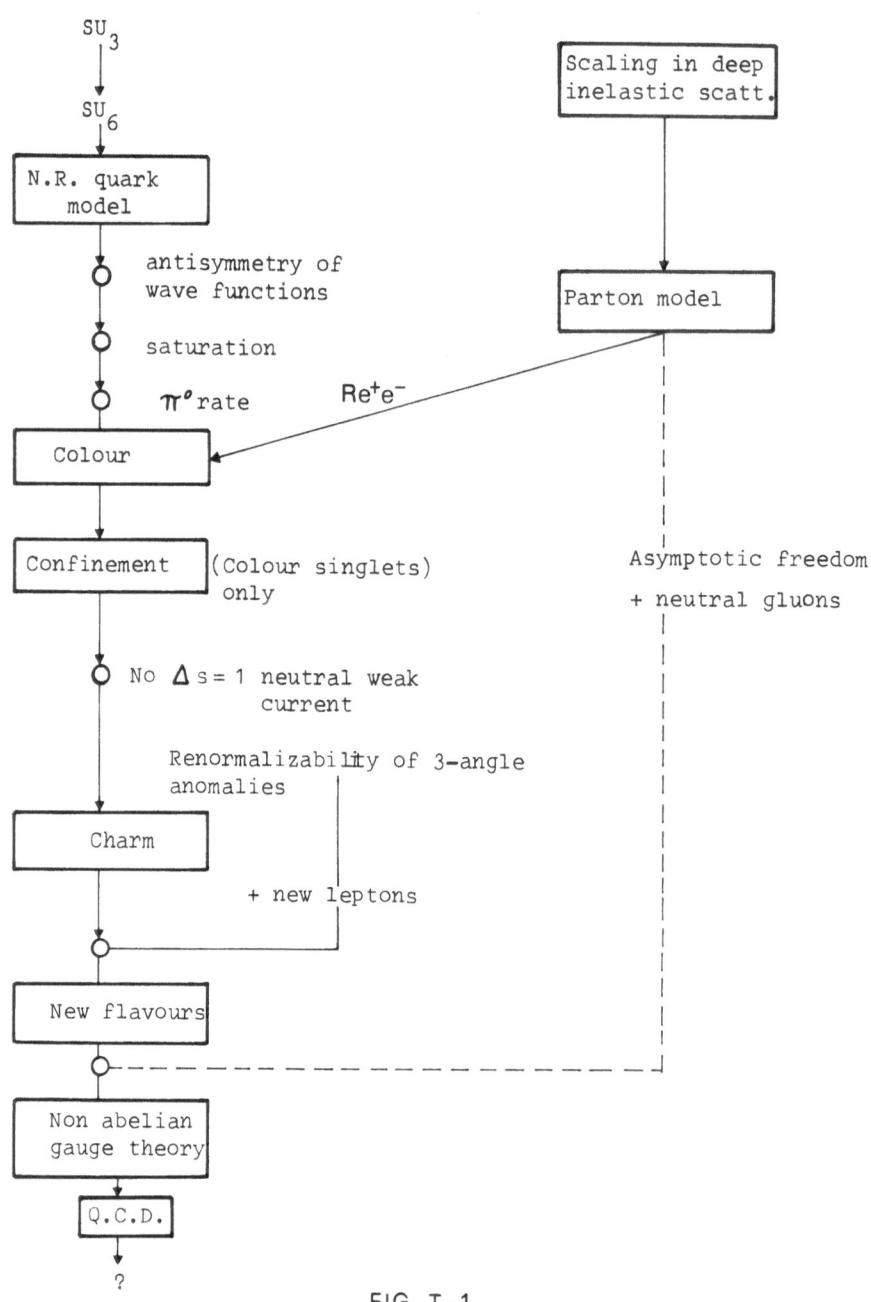

FIG. I-1

LECTURE II: SEARCH FOR QUARKS. A GENERAL SURVEY

1. INTRODUCTION

In this lecture and in the ensuing one I will speak of the experiments to detect isolated quarks. Today's lecture contains a broad survey of all the methods, except for the experiments based on the magnetic levitation electrometers, to be described in detail in the next lecture. In what follows the word quark will refer, except when otherwise noted, to fractionally charged objects.

A recent, very extensive survey has been written by L. Jones [1]; another detailed review of the terrestrial experiments is due to Kim [2]. I will rely heavily on these papers and, except for minor details, the contents of today's lecture will be a short guide to the above excellent surveys [3].

In my opinion the search for isolated quarks should go on independently of the theoretical prejudices or fashions concerning confinement. Indeed at the moment the general success of the quark model and the lack of evidence for quarks is the paradox of our days. Also one should be open-minded on the masses of the isolated quarks. Certainly - as it was noted long ago - if the isolated quarks are very massive [4], the difficulty of not seeing them produced in accelerator or cosmic-rays experiments is immediately accounted; so this is perhaps a reasonable assumption; but different possibilities cannot be discarded, e.g. not so heavy quarks but necessity of a tunnel effect to liberate them; or forces between quarks that strongly decrease

with increasing relative energy so that when one increases
the energy to go beyond the production threshold the
force acting on the quark to be liberated becomes very
weak [5], etc.

2. METHODS OF SEARCH FOR QUARKS

The methods of search for quarks can be classified
as follows: a) production and detection of quarks in
accelerator experiments, b) in cosmic rays, c) search
for stable quarks in matter. In connection with the
use of the word stable it should be noted that at
least one - the lightest - among the fractionally
charged quarks should be stable if the charge conservation holds. We are going to deal with the above
subjects in the order a), b), c).

3. ACCELERATOR EXPERIMENTS

By far the largest number of accelerator experiments have been done at the CERN and FNAL proton
accelerators in collisions of protons against nuclei
(also some experiments at the ISR of CERN have been
done). If the quark mass m_q is much larger than the
proton mass m_p the threshold energy in the lab. frame
in a reaction like:

$$N + N \rightarrow q + \bar{q} + N + N \tag{1}$$

(where N is a nucleon), is:

$$E_{Thr.Lab.} \cong \frac{2m_q^2}{m_p} \quad (m_q >> m_p) \ .$$

For m_q = 10 GeV the threshold energy is 200 GeV, and the largest quark mass that can be produced with a proton beam of 400 GeV (the FNAL energy) in the reaction (1) is $m_q \stackrel{\sim}{} 14$ GeV (in fact the Fermi motion in the target nucleus somewhat increases the maximum m_q). At the ISR one can reach m_q = 28 GeV, with, however, a much smaller luminosity.

The detection of the quarks in these experiments is based on the measurement of the ionization at minimum (related to the square of the quark charge). Because no quark signal has been ever found, these experiments only give an upper limit to the <u>differential</u> cross section for quark production in the conditions of the experiment (that is in the solid angle $\Delta \Omega$ and momentum interval Δp accepted by the detection system). To obtain an upper limit to the <u>total</u> σ_q - such as the upper limits plotted in the ensuing Fig. 1 - one usually makes some kind of extrapolation calculating the phase space available - a function of the assumed mass of the quark m_q. One can assume that the phase-space distribution is uniform; or more appropriately one can extrapolate assuming some p dependence of the production cross-section such as - in standard notation - exp$-6p_\perp$ (GeV) or exp$-Ax$ exp$-\frac{p_\perp^2}{p_o^2}$. The result of a typical experiment is therefore represented in the Fig. 1 by a curve. Consider, for instance, in the Fig. 1 the lower curve labelled c referring to the experiment by Nash et al. indicated in the figure caption. From this curve we read that for an assumed quark mass of 7 GeV the cross section σ_q for quark production is less than $\sim 5.10^{-39}$ cm^2. For an assumed m_q = 9 GeV the upper limit to σ_q is $\sim 10^{-39}$ cm^2. As soon as we go beyond m_q = 10 GeV, which

is the threshold for quark production at the energy
used in the experiment, the curve raises steeply
(indicating that the experiment does not give any
more information on σ_q). Incidentally, the curve c
just mentioned is the result of a constrained phase-
space analysis of the experiment by Nash et al. just
mentioned; whereas the upper curve c in the Fig. 1
is obtained analyzing the data from the same experiment
in terms of an isotropical phase-space extrapolation.
The other curves (a, d, e) refer to other experiments
as indicated in the figure caption. In particular the
curve e (from an ISR experiment) arrives to a value of
$m_q \sim 28$ GeV, with, however, a much smaller luminosity.
We have not completed the Fig. 1 with a few very recent
experiments [6] because these do not change the
qualitative conclusions. Note, finally, that the dashed
line in the upper right corner of the Fig. 1 gives the
upper limit to the quark production cross-section obtained in the cosmic-rays experiments as we shall say
later on. One sees that, although the cosmic rays allow
to explore the quark-mass region from 30 GeV up, in
practice their very small intensity makes this exploration not very rewarding in terms of an upper limit
to σ_q. But, of course, the scope of these experiments
is not that of finding an upper limit, but to find the
quarks, and any possible way to increase the energy of
the projectiles must be exploited. It must be added
that the Fig. 1 refers to quarks with charge (1/3)e;
a similar graph could be presented, but we omit it,
for (2/3)e quarks; the upper limits to σ_q for (2/3)e
quarks are some 5 times worse than those for (1/3)e
quarks.

4. THE INTERPRETATION OF THE UPPER LIMIT TO σ_q

What is the meaning of the limits given above for σ_q? To answer this question one should know what is the expected cross section for an assumed mass m_q. However, we are totally in the dark here, for two reasons: a) we do not have a theory of quarks at large distances (we may or may not have in QCD a theory of quarks at small distances); b) even if we had such a theory it would not be easy to compute reliably the production cross-section due to the familiar difficulties of calculating with strong interactions.

This being the situation we can only mention what has become traditional in this field.

1) A typical unit for σ_q that is often adopted in this kind of discussions (call it σ_o) was introduced long time ago by Adair and Price [7] in connection with one of the early experiments in cosmic rays. It is:

$$\sigma_o = \pi \left(\frac{\hbar}{m_q c}\right)^2 \quad \text{(at values of } E_{Lab} \gtrsim 3E_{Thr} \text{)}$$

This is a huge unit, σ_o being equal to $\sim 1.25 \cdot 10^{-29}$ cm^2 for $m_q = 10$ GeV.

2) Detailed models by Chilton et al. [8] or more recently by Halzen and Gaisser [9] - we refer to the original papers for a discussion of the assumptions underlying these models - give σ_q approximately 20 times less than σ_o (for $m_q \sim 10$ GeV).

3) A new point of some interest has emerged recently,

namely that the statistical model is not at all reliable
for the prediction of rare events. This conclusion is a
by-product of the recent discovery by the Columbia-FNAL
group of the Y meson (at a mass of $\sim$ 10 GeV) in proton-
nucleus collisions. According to the statistical model
the dependence of the production cross-section on the
mass of the produced particle is governed by the factor
$\exp(-M(\text{in GeV})/0.16)$; therefore according to the
statistical model the ratio between the production
cross sections of $Y(10)$ and $\psi(3.1)$ should be:

$$\left.\frac{\sigma(Y(10))}{\sigma(\psi(3.1))}\right|_{\text{stat.}} = \exp\left(-\frac{(m_Y - m_\psi)}{0.16}\right) \simeq 1.7 \cdot 10^{-19}$$

whereas the experimentally observed ratio is:

$$\left.\frac{(Y(10))}{(\psi(3.1))}\right|_{\exp} \simeq \frac{2.10^{-35}}{3.10^{-32}} \simeq 10^{-3} .$$

Clearly this discrepancy makes the statistical model
totally untenable for this kind of predictions (note
that the quark-production cross-section according to
the statistical model would have been:

$$\sigma_q \simeq 10^{-27} \exp\left[-\frac{m_q(\text{GeV})}{0.16}\right]\text{cm}^2 \quad \text{implying } \sigma_q \simeq 7.10^{-55} \text{ cm}^2$$

for $m_q = 10$ GeV).

5. COSMIC-RAYS EXPERIMENTS [1]

Cosmic-rays experiments can be classified as follows: a) Minimum-ionization experiments, b) delay experiments. The minimum-ionization experiments (either with electronic or with visual (cloud-chamber) detection) try to detect the quarks on the basis of their ionization. The delay experiments try to reveal the quarks exploiting their (presumably) high mass: if massive particles are produced in a shower high up in the atmosphere together with many light particles, there will be a delay between the arrival times of the heavy and light particles on a detector placed few kilometers down; this delay can be used as a first signature for a quark produced in a shower; an additional signature can then be requested through a measurement of ionization. (The delay experiments are of interest, of course, in the discovery of any massive, sufficiently stable particle with or without fractional charge). Both the minimum ionization experiments with electronic detection and the delay experiments have established an upper limit to the flux of quarks at or near to the sea level of the same general order of magnitude (though the type of events they reveal is different):

$$\phi_q < 10^{-11} \; (cm^2 \cdot sr \cdot sec)^{-1} \; . \tag{2}$$

Concerning the minimum-ionization experiments (with electronic techniques) it must be stressed that if two particles hit the detector simultaneously, the event does not appear as a quark signal, even if one of the particles is a quark; in fact a typical apparatus for these experiments (an example is given in Fig. 2) measures the total energy deposited in each counter.

Therefore, if quarks often travel accompanied by other particles, experiments of this type have a bias against their detection. Note that it will be difficult that future experiments   push down, even by a factor four or five, the number on the right hand side of the in-equality (2). To appreciate the effort implied in establishing the inequality (2) note that the flux of muons at the sea level is approximately one per $cm^2 \cdot sr \cdot minute$. The upper limit (2) implies there-fore to have counted, recognized as such, and discarded a number of muons roughly 10^9 times larger than that of the - undetected - quarks.

Finally, in connection with the visual detection of quarks in cloud-chamber or similar techniques I only note here that a claim in 1969 to have seen tracks of particles with anomalously low ionization has not been supported by the evolution of this type of experiments; most probably the explanation of those events simulating quarks had to be found in the disappearence of some droplets of the track in the Wilson chamber in the time elapsed between the formation of the track and its photograph.

6. THE RELATIONSHIP BETWEEN THE UPPER LIMIT OF ϕ_q AND THAT OF σ_q

Take the upper limit for the quark flux (equation (2)) established by the cosmic-rays experiments. Which is the upper limit of the quark-production cross-section σ_q by the cosmic-rays primaries corresponding to the above flux?

We can write approximately:

$$\Phi_q^{(\text{top atmosphere})} \cong \Phi(> E_T) \cdot P_q(E_T) \quad . \tag{3}$$
$$\text{(cosmic primaries)}$$

In words: the flux of quarks produced at the top of the atmosphere is equal to the flux of cosmic-rays primaries with energy larger than the threshold E_T for producing a quark times the average probability $P_q(E_T)$ of production of a quark at energy larger than E_T.

Because

$$P_q(E_T) = \frac{\sigma_q(E_T)}{\sigma_{tot}(E_T)} \tag{4}$$

we get from the equations (3) and (4):

$$\sigma_q(E_T) = \frac{\Phi_q^{(\text{top})}}{\Phi_{prim}(>E_T)} \sigma_{Tot}(E_T) \cong 40 \frac{\Phi_q^{(\text{top})}}{\Phi_{prim}(>E_T)} \text{ mb} \quad . \tag{5}$$

The integral flux of cosmic-ray primaries with energy larger than E_T is given by the expression:

$$\Phi_{prim}(>E_T) = 2.08 \ (E_T(\text{GeV}))^{-1.67} (\text{cm}^2 \cdot \text{sr} \cdot \text{sec.})^{-1} \tag{6}$$

and because

$$E_T = \frac{2m_q^2}{m_P} \cong 2m_q^2 \ (\text{GeV}) \tag{7}$$

we deduce, on inserting (7) into (6):

$$\Phi_{prim}(>E_T) = 0.65 \, m_q^{-3.3} \, (GeV).$$

Thus:

$$\sigma_q(E_T) = \Phi_q^{(top)} \cdot 0.6 \cdot 10^{-27} \, (m_q(GeV))^{3.3} \, .$$

If we assume that the absorption of quarks in the atmosphere is small-in line with the standard rule $\sigma_{q-Nucleon} = \frac{1}{3}\sigma_{NN}$ and, in fact we neglect this absorption altogether, writing:

$$\sigma_q^{(top)} = \Phi_q^{(sea \, level)} \; \gtrsim \; \Phi_q < 10^{-11} \, (cm^2 \cdot sr \cdot sec)^{-1} \quad (8)$$

we obtain:

$$\sigma_q(E_T) < 0.6 \cdot 10^{-36} \, m_q^{3.3} \, cm^2 \, (m_q \text{ in GeV})$$

and, for $m_q = 10$ GeV,

$$\sigma_q(E_T) < 1.2 \cdot 10^{-33} \, cm^2 \quad\quad\quad (9)$$

the value already indicated in the dotted line at the right-up corner of the Fig. 1. The effect on (9) of modifications to the drastic assumption (8) is easy to calculate.

7. SEARCH FOR QUARKS IN MATTER

As already stated the main assumption underlying
the search for quarks in matter is the stability of at
least one quark; this is ensured, in particular, if the
quarks have fractional charge and if charge conservation
holds.

There are two different possible origins for the
quarks in the matter around us: a) they can be relics
of primordial quarks, a few quarks that remained un-
burnt in the original big fire combining the quarks
into hadrons. b) They can have been produced by the
cosmic rays in the course of the centuries.

Two remarks are appropriate concerning the search
for quarks in matter: 1) as I pointed out long time
ago [10] and as it has been recently reemphasized by
De Rujula [11], attempts to find a quark Klondike have
been in vain! The concentration of quarks in the matter
around us is totally unknown. 2) This total ignorance
is in fact the basic weakness of the searches for
quarks in matter.

In a sense the experiments trying to produce
quarks in accelerators and those dealing with the
search for quarks in matter are complementary. The
success of the accelerator experiments will depend
entirely on a sufficiently low value of the unknown
quark mass; whereas the outcome of the searches of
quarks in matter does not depend on the mass of the
quark but depends on the unknown concentration. In
the accelerator experiments one keeps going on increasing the collision energy. In the search for quarks

in matter one tries to explore larger and larger
quantities of matter.

   Can one set an upper limit on the number of
quarks produced in the course of the centuries by the
cosmic rays on the Earth? Yes, but only if we make
several assumptions. Assume that: a) since the formation
of the Earth ($\sim 3.10^9$ years ago) the flux of the cosmic
rays has always remained the same as it is now. b) That
the quarks produced during this long period are all to
be found in a layer having a depth of 3 Km ; we assume
in other words that, on the average, the turbulence of
the Earth's crust during the life of the Earth has
affected a layer of $\sim$ 3 Km. Then the upper limit to
the number of quarks created in a column of height 3 Km
and cross section 1 cm^2 by the cosmic rays during the
life of the Earth is:

$$N < \pi \cdot (31 \cdot 10^6) \cdot (3 \cdot 10^9) \cdot 10^{-11} \simeq 3 \cdot 10^6$$

where the first factor π comes from the integration over
half-solid angle, the second and third factors give the
number of seconds in 3.10^9 years and the last factor
10^{-11} is the upper limit to the quark flux from the
cosmic rays discussed previously (equation (2)). We
thus obtain (dividing N by 3.10^5 cm) for the upper
limit to the concentration c:

   c < 10 quarks per cubic cm.

Experiments to detect such small concentrations would
be very difficult; so it has to be hoped that, if
isolated quarks do exist, some relics from the
primordial quarks have remained with us.

Should not one exploit the presumed chemical
properties of the quarks and search them in places
where they are likely to be more abundant than elsewhere? Of course the answer is positive, in principle;
the difficulty is, however, that it is not easy
predict the history of a quark in the Earth. One
should note that, as far as I know, there exists so
far no generally accepted explanation of the formation
of the mines in some particular spots of the Earth.
I will therefore omit entirely this interesting but
controversial chapter and refer to the survey of
Kim [2] for a short discussion and the appropriate
literature.

8. A SUMMARY OF THE TECHNIQUES USED IN THE SEARCH
FOR TERRESTRIAL QUARKS

Several ideas have been used in the search for
terrestrial quarks; as stated at the beginning I will
discuss in the next lecture the experiments by the
magnetic levitation electrometer and consider here
briefly the following types of experiments:
1) Optical spectroscopy.
2) Emission of negative ions from filaments.
3) Energy spectroscopy.
4) Cyclotron mass-spectroscopy.
5) Gas efflux.

All the above experiments (except nr. 4) have
one aspect in common; namely the material that is
explored is first subjected to an enrichment procedure
trying to concentrate all the quarks that it might
contain into a very small filament or foil. In principle

this procedure is a sensible one and can certainly increase the chance of finding a quark. But if quarks
are not found, this procedure raises some questions
when we try to express the negative results in terms
of an upper limit to the quark concentration; indeed
to know really the enrichment factor that has been
used - often assumed to be very large (10^6 or more) -
implies a knowledge of the quark chemistry larger than
we do really have.

Let us now briefly describe some of the above
experiments. The status of the situation, as far as
the <u>optical spectroscopy</u> is concerned, is exemplified
by an experiment of Rank [12] searching for lines in
the ultraviolet from atoms of the type $(q^{2/3}e^-)$. The
spectrometer is sensitive up to 7000 Å, and this fact
limits the search to the Lyman lines of the $(q^{2/3}e^-)$
($\sim$ 2500 Å) excluding the Balmer lines or the lines
from $(\bar{q}^{+1/3}e^-)$. The samples examined were taken from
sea water, lake water, oysters, and plankton and the
quarks contained in a few kilos of material were
collected on foils of Au or Pt by vaporizing the sample
in an electric field. Because the spectroscopic
sensitivity is low (a signal can be seen, only if there
is more than a quarkic atom per 10^{10} normal atoms), a
positive result would correspond - for an assumed
enrichment factor of 10^7 - to have more than one quark
per 10^{17} atoms in the original material. The result of
the search is negative and is expressed by the author
as follows: "At a wave length corresponding to a
possible Lyman line a signal from some sample was
seen, but there was no evidence for other (expected)
Lyman lines".

<u>The negative ion emission by filaments</u>
<u>enriched in quarks</u> has been the basis of a long
series of experiments conducted by the Argonne group
[13,14]. In one of the methods used a filament (of Pt)
is "enriched" in quarks trying to transfer on it all
the quarks contained in large amounts of several substances: sea water, dust from air conditioners, Lunar
soil, manganese nodules from the ocean, etc. The idea
of the method is that the behaviour of a quark emitted
from a filament, on heating, should be similar to that
of a negative ion. If this is so the enriched filaments,
when heated, should - according to the authors - emit
more negative ions than the non-enriched filaments. Because in a typical experiment less than 10^5 counts of
excess are found, it is concluded that quarks, if present,
have an abundance of less than 10^{-21} quarks/nucleon in
the enriched filaments. Mass spectroscopy is also used
to explore in some cases the masses of the emitted
negative ions, but with inconclusive results.

<u>Energy spectroscopy.</u> Under this heading I refer
to an experiment performed by Cook et al. [15], the idea
of which is very simple, although in practice it is of
difficult realization and interpretation. The idea is
the following: "enrich" in quarks a Ta foil and
evaporate (by heating) the quarks it contains. If their
charge is e/3 (e is the electron charge) they should
acquire, when accelerated through a potential V, a
kinetic energy precisely equal to one third of the
kinetic energy of a charge e accelerated by the same
potential difference. The question is, how to measure
this kinetic energy; this is not easy because the ions
are presumably massive - either because the quark it-

self is massive, or because the atom to which it is
attached is such. Due to this high mass the velocity
of the ions arriving on a Si detector when accelerated
by a potential difference, say 50 kV (the value used
in the experiment [15]) is too low to be in the region
where the detector gives a signal proportional to the
kinetic energy. This difficulty appears clearly in the
Fig. 3. The full curve gives the signal from an <u>electron</u>
current accelerated by V = 50 kV. Due to the high
velocity of the electrons a very clean peak appears
at T = 50 kV as expected. If the electrons are then
swept away by an appropriate magnetic field, the remaining negative ion signal is given - in the scale
of the right - by the dashed line. One can see that this
signal has a peak at 40 kV rather than 50 kV; the small
bump around 15 kV might be, in principle, a quark peak
masked by a bad resolution; however, on comparing the
signal from foils enriched differently no clear conclusion can be reached and only an upper limit to the
quark abundance in the enriched samples can be given:
$< 10^{-24}$ quarks per nucleon.

Finally we list the use of a <u>cyclotron as mass
spectrometer.</u> It was in this way that, in 1939, Alvarez
et al.[16] discovered the He^3. So far this technique
has been mainly employed to search for integrally
positive charged quarks [17] ("anomalous Hydrogen");
in the case of quarked tungsten - the reason for this
particular choice will appear from the next lecture -
one has also looked by a similar technique [18] with
negative results for fractional quarks.

It would lead us too far to expand on this important chapter of the cyclotron mass-spectroscopy that

has already given, as a by-product, an extremely sensitive
new procedure to perform radiocarbon-dating experiments.
We refer to the original literature remarking only that
in the search for integrally charged quarks the
sensitivity of this method - at least in some range of
assumed masses - is considerable ($\sim 2.10^{-19}$ quarks per
nucleon in the mass range from 2 to 8 atomic mass units).

Finally we come to the <u>gas-efflux method</u> to be
described below. This method was invented long time ago
[19] to look for a possible tiny difference in charge
between the electron and the proton. Sometimes one finds
in the literature the statement that the results obtained
by this method [20] do establish, as a by-product a strong
upper limit for the non-existence of fractionally charged
quarks. Although experiments of search for quarks by this
method might perhaps be possible, it seems to me that the
experiments performed so far have little relevance for
the search for quarks.

To clarify this point I briefly describe the idea
of the method (Fig. 4). An insulated container A inside
a Faraday cage B is filled with a gas; this gas is
prefiltered in order to avoid the presence of electrified
dust. The electrometer E measures the static potential
difference between A and B (B is earthed) when the gas
is inside A. The gas is then allowed to go out from the
container A. If each atom or molecule of gas has a tiny
charge (due to a tiny difference in the absolute values
of the charges of the electron and the proton) the total
charge contained in A will change after the efflux and
the electrometer E will note this change. Of course to
increase the sensitivity of the method it is necessary

that possible electrified grains of dust still present, or ions, are not allowed to leave the container A because they might produce a large signal not due to the effect that one intends to measure. This is accomplished by inserting at the exit of the vessel an electrostatic filter F; an internal battery (bat. in Fig. 4) creates a potential difference of some 50 Volt between the two electrodes of the filter. The reason why I stated above that these experiments - which establish the neutrality of a molecule of N_2 to better than 6 ± 6.10^{-20} e - are possibly not relevant for the search for quarks has to do precisely with this filter. If this filter keeps the quarks inside A, no effect from the possible presence of quarks can be found.

To conclude I state again that I tried to cover in this schematical survey the various different methods used so far in the search for quarks, except for the experiments based on the use of the magnetic levitation electrometer to which I shall dedicate the next lecture.

REFERENCES FOR LECTURE II

[1] L.Jones, Rev. Mod. Phys. 49 (1977) 717.

[2] J. Kim, Contemp. Physics 14 (1973) 289.

[3] An old and much shorter survey by the author can also be consulted: G. Morpurgo, in Subnuclear Phenomena (B) edited by A. Zichini (Academic 1970) p.640.

[4] G. Morpurgo, Physics 2 (1965) 95.

[5] G. Morpurgo, Phys. Lett. 20 (1966) 684.

[6] They explore, however, different kinematic regions; compare, e.g.: D.Antreasyan et al.,Phys. Rev.Lett. 39 (1977) 513.

[7] R. Adair and N. Price, Phys. Rev. 142 (1966) 844.

[8] F. Chilton et al., Phys. Lett. 22 (1966) 91.

[9] T. Gaisser and F. Halzen, Phys.Rev.D11 (1975) 3157.

[10] G. Morpurgo, Annual Rev. Nuclear Science 20 (1970) 105.

[11] A. De Rujula et al., Phys. Rev. D17 (1978) 285.

[12] D.M. Rank, Phys.Rev. 176 (1968) 1635.

[13] W.A. Chupka et al., Phys. Rev. Lett. 17 (1966) 60.

[14] C.M. Stevens et al., Phys. Rev. D14 (1976) 716 and references listed there.

[15] D. Cook et al., Phys. Rev. 188 (1969) 2092.

[16] R. Muller, L. Alvarez, W. Molley, E. Stephenson, Science 196 (1977) 521.

[17] L. Alvarez and R. Cornog, Phys. Rev. 56 (1939) 379.

[18] R.N. Boyd, D. Elmore, A.C. Melissinos, and E. Sugarbaker, Phys. Rev. Lett. 40 (1978) 216.

[19] A. Piccard and E. Kessler, Arch. Sci. Phys. et nat.7 (1925) 340.

[20] A. Hillas and T. Cranshaw, Nat. 184 (1959) 892; J. King, Phys. Rev. Lett. 5 (1960) 562.

FIGURE CAPTIONS FOR LECTURE II

Fig. 1. The upper limits to the quark-production cross-
sections in accelerator experiments (from L.Jones,
ref.[1]) as a function of the assumed mass. As
stated in the text the various curves a,b,c,d,e
refer to different experiments (the references
are listed below). The two curves labelled c
give the upper limit to σ_q from the same experiment using a uniform phase space in the
upper curve and a constrained phase space in
the lower one.
a) J. Allaby et al., Nuovo Cim. A64 (1969) 75.
b) A. Antipov et al., Phys.Lett. B29 (1969) 245.
c) T. Nash et al., Phys.Rev.Lett. 32 (1974) 858.
d) L. Leipuner et al., Phys. Rev.Lett. 31 (1973)
1226.
e) C. Fabjan et al., Nucl.Phys. B101 (1975) 349.

Fig. 2. A schematic drawing of a typical cosmic-rays
apparatus with electronic techniques (from
E.P. Krider et al., Phys. Rev. D1 (1970) 835).
The spark chambers are set in between the six
scintillation counters to ensure that each event
of interest is caused by a single particle. The
shower detector was initially inserted to reduce
the number of event triggers due to soft showers
but due to its small efficiency was not used for
most of the dataruns. The typical size of each
scintillation counter is 72 x 72 x 6 (inch)^3.

Fig. 3. Schematic plots (taken from ref.[2]) of the experiment of Cook et al.[15]. The full curve
refers to the spectrum from an unquarked foil

at 600°C. The counts are almost entirely
electrons. The scale for this curve is on the
left. The dashed curve refers to the measured
spectrum of a quarked foil at 530°C. The
electrons have been swept away by a magnetic
field; the scale to this figure is given on
the right. The bump at B is compatible with
O^- or OH^- ions, that of A to objects as H_2^-.

Fig. 4. The apparatus used in the gas-efflux experiment
by Hillas and Cranshaw, ref.[20]. The drawing -
a schematical reproduction of the original from
ref.[20] - is taken from ref.[2] - A and B are
aluminium cylinders insulated from each other;
F is the ion trap with its battery b; E is a
galvanometer; the dark areas are polyethilene
insulators.

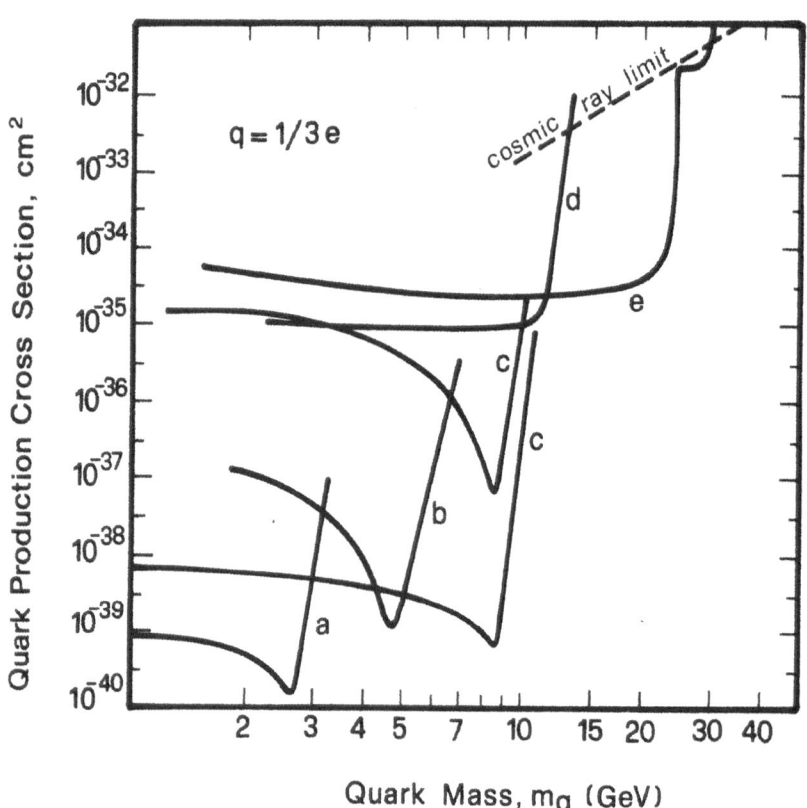

Fig. II-1

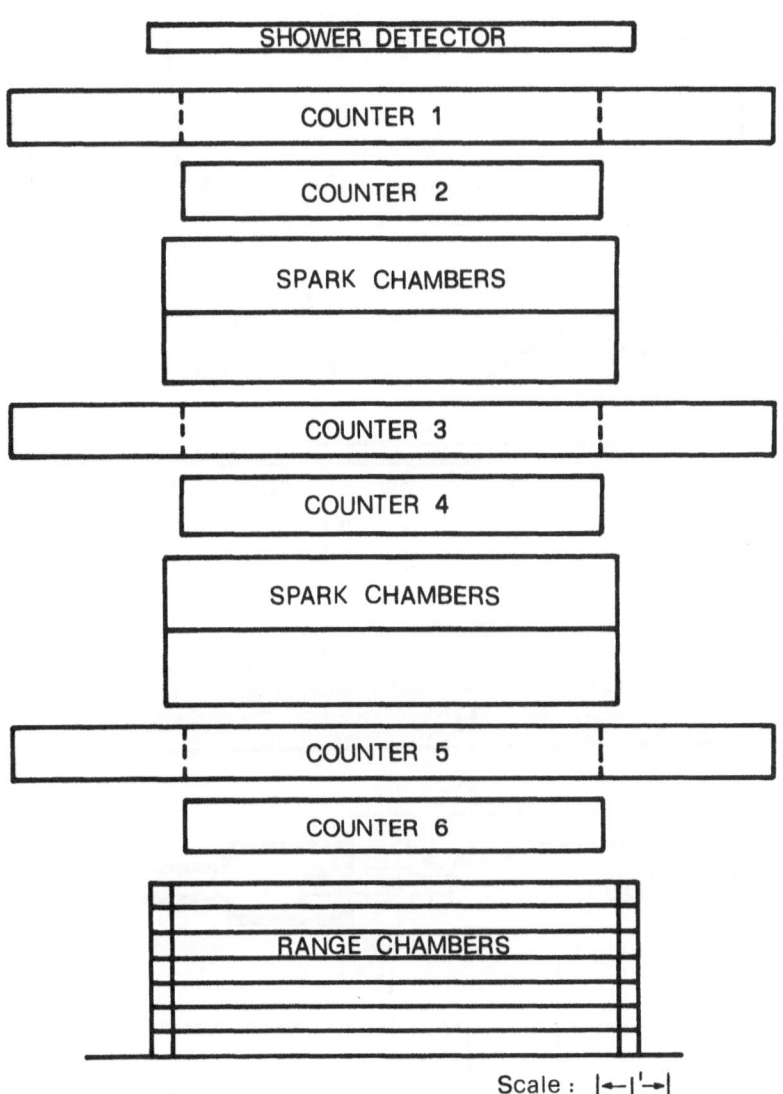

Fig. II-2

48

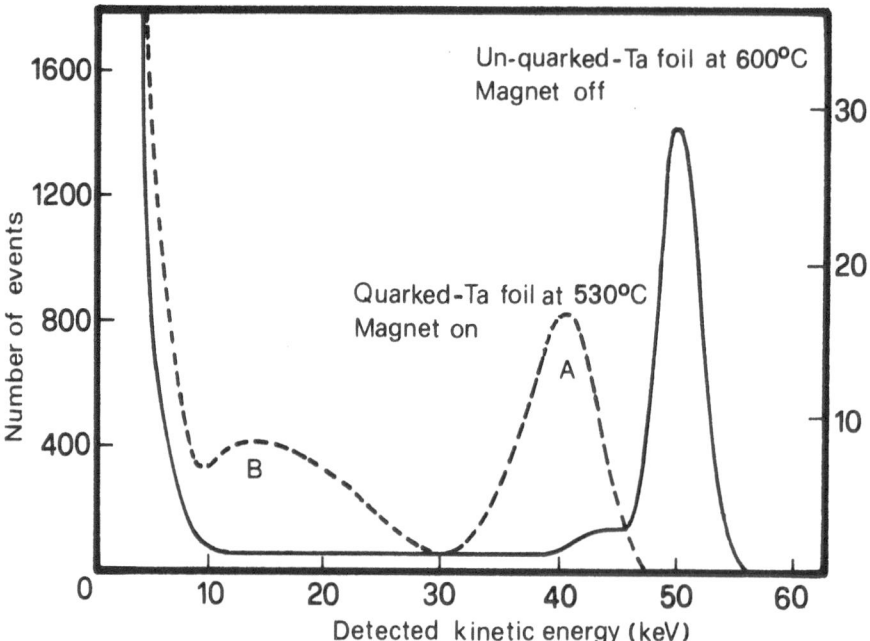

Fig. II-3

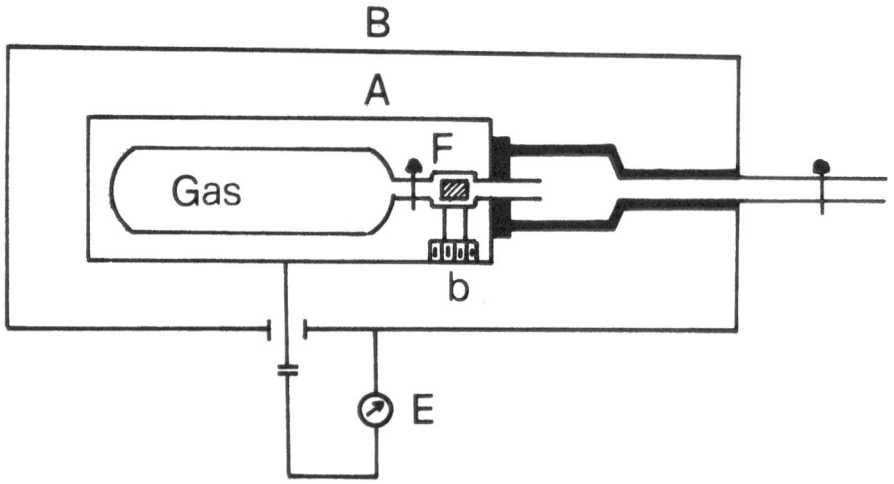

Fig. II-4

LECTURE III: SEARCH FOR QUARKS.
THE MAGNETIC LEVITATION ELECTROMETER[+]

1. THE GENERAL IDEA OF THE EXPERIMENTS

Since 1965 a number of searches for quarks have
been performed with a new technique - the magnetic
levitation electrometer - first developed by the
Genoa group [1,2] and subsequently used in other
laboratories [3]. As far as we know these experiments
continue, at the time of writing, in Genoa [4] and
Stanford [5].

The idea underlying this type of experiments is
very simple. A grain of matter containing a fractionally
charged quark can never be brought (by addition or
expulsion of electrons) into a state of charge zero.
We cannot neutralize it by adding or subtracting
electrons. Its residual charge Q_R will remain non-zero
and always equal (if quarks have charges $\pm$ e/3 or
$\pm$ 2e/3) to $\pm$ e/3. By measuring this residual charge
one can know if the grain does or does not contain
a quark.

2. THE MILLIKAN EXPERIMENT

To determine the residual charge Q_R one can, in
principle, perform a Millikan experiment [6]. Note
that the original Millikan experiments had two purposes:
a) to show the discreteness of the charge and b) to
measure the absolute value of the electron charge. By
now this second purpose is uninteresting because there

[+]This lecture was prepared and written together with
Dr. M. Marinelli.

are better ways to measure e. To prove the discreteness
of the charge Millikan's experiment proceeds as follows:
a potential difference is applied between the plates of
a parallel-plate capacitor (schematically drawn below in
Fig. 1). The time t_i taken by a given oil droplet with
charge Q_i to go up from the height z_1 to the height z_2
under the action of the applied electric field is inversely proportional to the charge Q_i

$$t_i = \frac{K}{Q_i} \tag{1}$$

where K is a proportionality constant depending on the
applied electric field and on the mass of the droplet.
One can measure, for the same droplet, the times t_i
corresponding to different values of the charge Q_i,
changing the charge by photo ionization. If the
charges Q_i are multiple of an elementary charge e,
as follows:

$$Q_i = n_i\, e \tag{2}$$

the sequence of times that one measures must then have
the form:

$$t_i = \frac{K}{e} \frac{1}{n_i} \tag{3}$$

that is:

$$t_i = \frac{const.}{n_i} \quad . \tag{4}$$

If the relationship between charge and n_i is instead
of the form:

$$Q_i = (n_i \pm \tfrac{1}{3})\, e$$

(that is if the droplet contains a quark) we have, instead of (4) the equation:

$$t_i = \frac{\text{const.}}{n_i \pm \tfrac{1}{3}} . \tag{5}$$

It is possible to recognize whether the equation (4) or (5) holds and thus determine the presence or absence of a quark inside the droplet. The Millikan experiment is, however, not really convenient to this purpose because the size of the droplet to be used is very small and the hope of finding a quark inside one is correspondingly small. The reason why the droplets have to be small is that the electric field between the plates is used both to counteract the gravity and to measure the charge of the droplet. If the droplets are too heavy, huge potential differences have to be applied to the plates of the capacitor; this is unfeasible for more than one reason. The typical droplets used by Millikan had a mass $\sim 10^{-11}$ g (and a radius of $\sim 1\mu$).

Subsequent experiments [7] were performed with somewhat heavier droplets, up to 4.10^{-9} g (radius 7μ). Indeed a search for quarks by Rank [8] was performed by the Millikan method in oil droplets of that mass; but the result has been somewhat disappointing in the sense that both for "enriched" and not enriched oil droplets a plot of the frequency of occurrence of a

given residual charge shows the same wide distribution
of residual charges from $\sim$ -0.25 e to $\sim$ +0.3 e. Another
search for quarks by the Millikan method was performed
by the Argonne group [9] on small polyethilene spheres.
No quark was found. Finally a Millikan experiment has
been performed [10] on small fragments of tungsten -
for a reason to be explained later - again with
negative results.

The idea of the magnetic levitation electrometer
has been that, to perform a measurement of charge on
objects substantially heavier than the Millikan
droplets, it seems necessary to use, for suspending
the object, a force different from the electric force
used to measure the charge. In other words we should
be able to "levitate" an object possibly in vacuum
so that its charge does not change due to the presence
of ions and, once the object is levitated, we should
measure its charge.

3. METHODS OF MAGNETIC LEVITATION

Several different methods of magnetic levitation
can be used, each with its advantages and disadvantages.
We list them below.

1) Diamagnetic levitation of graphite at room temperature
   (Braunbeck [11]). This method was used in the first
   apparatus built by the Genoa group [1]. It was also
   used subsequently by a group in Moscow.

2) Diamagnetic levitation of a superconductor (Niobium)
   at low temperatures. This method, conceptually

identical to the diamagnetic levitation of graphite, is in use at Stanford [5].

3) Ferromagnetic levitation with feedback (Beams [12]). This method was first set up by Moran et al. [13] and then used by Braginsky et al. [3], and by Garris and Ziock [3]; it is used in the present version of the Genoa experiment [4].

I will not enter here into the details of the methods referring for this to the literature. I only state that a diamagnet can have a stable equilibrium position under the action of the gravity and of an appropriate magnetic field. On the other hand a ferromagnet cannot have such a stable equilibrium position (a general theorem that can be traced back to Earnshaw forbids it); a feedback system is necessary to levitate a ferromagnet.

4. THE PRINCIPLE OF THE EXPERIMENT

No matter which method of magnetic levitation is used the levitating object can be thought as a pendulum moving freely in vacuum without any suspending string. The object is levitated between two parallel plates kept at different (opposite) potentials that create an electric field E at the position of the object. Ignoring for the moment any non-uniformity of the electric field the electric force acting on the suspended object is then:

$$F_x = Q\ E_x \tag{6}$$

where the x axis has been chosen along the direction of
the normal to the plates, that is along the electric
field; and Q is the charge of the levitating object.

The displacement Δ_i of the object from its
equilibrium position on inserting the electric field
E_x is therefore, when the charge is Q_i,

$$\Delta_i = \frac{Q_i \; E_x}{4\pi^2 m \nu_x^2} \tag{7}$$

where ν_x is the frequency of the pendulum along the
x axis and m is the mass of the levitating object.

The principle of the method lies in the
proportionality of Δ_i to Q_i. A levitating object is
prepared with an excess of a few electrons; ejecting
these one by one, e.g. by photoionization, one
records a sequence Δ_i of displacements. If no
fractional charge is contained in the object, this
sequence goes through the value $\Delta = 0$ (for instance
in arbitrary untis, a sequence might be 24, 12, 0,
-12, -24....); if instead the object contains a quark
of charge 1/3, the minimum Δ cannot vanish: it equals
1/3 of the displacement corresponding to one electron
charge (the above sequence would then be: 28, 16, 4,
-8, -20,...). We call Δ_R the minimum displacement and
Q_R the corresponding minimum charge (R stays for
residual). If, for some object, Q_R is found to be non-
zero one has to be sure, before concluding that a
fractional charge is contained in the object, that the
only force acting on the object is the force (6). In

other words one has to be sure that the residual charge
is really a charge and not some spurious effect due to
forces additional to (6).

Before discussing in detail the forces acting
on the object   in addition to (6)   a remark is
appropriate: to increase the sensitivity of the method
one can use an oscillating electric field   instead of
a static one; if the frequency of oscillation of the
electric field is chosen equal to the natural
oscillation frequency of the pendulum, the resonance
phenomenon produces an amplification by a factor:

$$f = \frac{4N_{1/2}}{\ln 2} \qquad (8)$$

where $N_{1/2}$ characterizes the damping of the pendulum
(it is the number of free oscillations after which
the oscillation amplitude reduces to a half. The
equation (8) refers to an applied square wave). The
amplitude of oscillation is then given by the
equation (7) for the displacement multiplied by f.
Continuing to call Δ the amplitude of oscillation -
rather than the static displacement - we then have,
instead of (7), the equation

$$\Delta_i = \frac{Q_i E_x N_{1/2}}{(\ln 2) \pi^2 m v_x^2} \qquad (9)$$

where E_x is now the amplitude of the square-wave
electric field.

5. THE RESULTS OF THE ORIGINAL EXPERIMENTS
WITH GRAPHITE

The first experiment using the diamagnetic
levitation of graphite was done in the static version,
that is with a static electric field (not in the
resonance version just described); irregular grains
of pyrolitic graphite of mass typically a few times
10^{-9} g were used. The first grains measured (masses
from $8.5 \cdot 10^{-10}$ to $4.8 \cdot 10^{-9}$ g) did show a sequence
of charges beautifully passing through zero. However,
we found soon grains having a non-zero apparent
residual charge. As already stated, before attributing
these to the presence of quarks we had to examine
whether these non zero Q_R were due to forces acting
on the grain in addition to the force (6); forces
produced, for instance, by the non-uniformity of the
electric field. Indeed this was precisely the case.
We shall discuss at length in the next Section 6
these forces and the corresponding spurious effects
because the whole success of the experiment depends
on their elimination or control.

From the static version we then proceeded to
the resonance version of the graphite experiment. The
sensitivity increased by a factor ~ 200 (the typical
value of the factor f in the equation (8)) and we
could then explore grains of mass a few times 10^{-7} g.
The measurement of the oscillation amplitude was
visual by looking at the amplitude of oscillation on
a television screen. In spite of having increased the
mass of the grain by some ten thousand times with
respect to the Millikan technique, the signal to noise
ratio was still so good that a visual measurement of

the oscillation amplitude was quite acceptable. Note
(this remark will be of interest later) that the
graphite grains used in the experiments described
in this section have an irregular shape: on inserting
the electric field they aligned themselves with their
longer axis in the direction of the electric field
and, once aligned, there was no rotation and, therefore, no transfer of excitation from rotational to
translational motion.

For a detailed description of these graphite
experiments - that led to an upper limit for the
abundance of quarks in pyrolitic graphite of less
than one quark in 2.10^{18} nucleons - the interested
reader should consult the references [1,2]; in
particular [2] contains a detailed description of
the experimental apparatus and a complete discussion.

6. FORCES ADDITIONAL TO QE_x GIVING RISE
TO POSSIBLE SPURIOUS EFFECTS

If the electric field at the position of the
object is not uniform, the charge force QE_x is not
the only force acting on the object. This force has
to be completed by at least two terms - as discussed
in ref.[2] - the polarization force $\underset{\sim}{F}^{(pol)}$ and the
permanent dipole force $\underset{\sim}{F}^{(dip)}$. Let us begin with the
polarization force due to the fact that an object in
an electric field $\underset{\sim}{E}$ gets an induced dipole moment $\alpha\underset{\sim}{E}$
(α is the polarizability). The x component of $\underset{\sim}{F}^{(pol)}$
is:

$$F_x^{(pol)} = \alpha \underset{\sim}{E} \cdot \frac{\partial \underset{\sim}{E}}{\partial x} \qquad (10)$$

where $\underset{\sim}{E}$ is the total electric field acting on the object and, for a conducting sphere of radius r, it is:

$$\alpha = 4\pi \, \varepsilon_o \, r^3 \; . \qquad (11)$$

The first thing to be noted in (10) is the fact that $\underset{\sim}{F}^{(pol)}$ is quadratic in $\underset{\sim}{E}$, that is invariant with respect to the inversion $\underset{\sim}{E} \rightarrow - \underset{\sim}{E}$; therefore, if the total electric field appearing in (10) could be identified with the applied electric field $\underset{\sim}{E}^{(a)}$, the force would give absolutely no trouble, because, on inverting $\underset{\sim}{E}$, would remain invariant whereas the charge force, linear in $\underset{\sim}{E}$, changes sign; or equivalently in the resonance version one can say that, if $\underset{\sim}{E}$ in (10) could be identified with $\underset{\sim}{E}^{(a)}$ - the applied electric field - the right-hand side of (10), being quadratic in $\underset{\sim}{E}^{(a)}$, would contain the frequencies zero and $2\nu_x$ (the latter is, however, absent if $\underset{\sim}{E}^{(a)}$ is a strict square wave); because these frequencies are completely rejected being out of resonance - no contribution from (10) would arise. However, $\underset{\sim}{E}$ in the equation (10) cannot be identified with $\underset{\sim}{E}^{(a)}$: in fact a tiny static electric field $\underset{\sim}{E}^{(s)}$ produced by Volta contact-potentials between the plates or by a static electrification due to foreign substances on the plates can be present.

Therefore we have to insert in (10) for E:

$$\underset{\sim}{E} = \underset{\sim}{E}^{(a)} + \underset{\sim}{E}^{(s)} \qquad (12)$$

and two cross terms arise:

$$F_x^{(pol.s.)} = 4\pi\epsilon_0 r^3 (E^{(a)} \cdot \frac{\partial E^{(s)}}{\partial x} + E^{(s)} \cdot \frac{\partial E^{(a)}}{\partial x}). \qquad (13)$$

Both these terms, linear in $E^{(a)}$, oscillate at v_x and, therefore, produce the same effects of a non-zero residual charge. Unless we eliminate this force or its effects on the residual charge, we will not be able to decide, if an object contains a quark, at least if this force is of the same order of magnitude of the charge force corresponding to a residual charge $e/3$.

In our geometry - in which $\frac{\partial E^{(a)}}{\partial x}$ is very small - the first term on the right-hand side of the equation (12) prevails, and we can write the force $F_x^{(pol.s.)}$ as

$$F_x^{(pol.s.)} = Q_R^{(s)} E_x^{(a)} \qquad (14)$$

where we have introduced the spurious residual charge due to the Volta force:

$$Q_R^{(s)} = 4\pi\epsilon_0 r^3 \frac{\partial E_x^{(s)}}{\partial x} . \qquad (15)$$

$Q_R^{(s)}$ increases with the volume of the sphere and, for a given volume, can only be reduced by decreasing the static field $E^{(s)}$.

In the experiments with graphite discussed in the section 5 the geometry was such that $E^{(s)}$ and

therefore $Q_R^{(s)}$ could be reduced increasing the distance
of the plates from the object. Indeed many objects did
show at a small distance between the plates a non-zero
residual charge; however, for all of them this residual
charge decreased to zero when, on the same grain, the
measurements were repeated at a progressively larger
distance between the platelets. It should be mentioned
that the graphite grains in the resonance version of
these experiments had a radius around 30 microns and
that the spurious charge (15) is proportional to r^3.

In the present version of the experiment with
iron spheres - to be described in a moment - the
volume of the spheres is $\sim$ 100 times larger than the
volume of the graphite grains; this fact - and others
to be mentioned later - makes the reduction of the
electrostatic spurious effects more difficult in the
present version of the experiment.

Consider now the permanent dipole force $F^{(dip)}$;
this is present if the levitating object has a
permanent electric dipole moment d. It is given, of
course, by:

$$F_x^{(dip)} = d \cdot \frac{\partial E^{(a)}}{\partial x} \quad . \tag{16}$$

At the geometrical center between the two plates -
assumed to be parallel - $\frac{\partial E^{(a)}}{\partial x}$ is zero. In the graphite
experiment the geometrical center could be established
to a precision of a few microns by the aid of a microscope.
Either due to this fact or due to the circumstance that
the graphite grain has a very small permanent electric

dipole-moment, there was no indication - in the graphite experiments of the force (16). In the ferromagnetic experiment we eliminate by spinning - as we shall explain - two components of $\underset{\sim}{d}$ and check that the third (d_z) does not give problems, due to the smallness of $\frac{\partial E_x{}^{(a)}}{\partial z}$.

In addition to the two forces $\underset{\sim}{F}^{(pol)}$ and $\underset{\sim}{F}^{(dip)}$ discussed above there are other possible effects that must be considered but can be eliminated altogether. Here we shall mention only two. Take the term $\alpha\underset{\sim}{E}^{(a)}$. $\frac{\partial\underset{\sim}{E}^{(a)}}{\partial x}$ obtained on inserting the equation (12) into (10). If the high-voltage power-supplies giving rise to $\underset{\sim}{E}^{(a)}$ are not well balanced, the time average of the potential of one or both the plates with respect to the walls of the levitation chamber (grounded) can be non-zero. This produces a force of the type

$$F_x^{(bal)} = \alpha\underset{\sim}{E}^{(a)} \cdot \frac{\partial\underset{\sim}{\varepsilon}^{(bal)}}{\partial x} \tag{17}$$

where $\underset{\sim}{\varepsilon}^{(bal)}$ is the electric field at the position of the object arising from this non-zero value of the average potential mentioned above; the force (17) can be dangerous for large value of the spacing between the plates and for large unbalances. Another force $F_x^{(2nd\ harm.)}$ arises in a similar way due to the presence of the second harmonic in $\underset{\sim}{E}^{(a)}$ because of the slightly different switching times of the reed relays producing the square wave.

Both $\underset{\sim}{F}^{(bal)}$ and $\underset{\sim}{F}^{(2nd\ harm.)}$ can however be measured and eliminated rather easily by an appropriate balancing of the power supplies. We will not expand on

them here, because their control and elimination is
straightforward.

This concludes our summary of the graphite
experiments, the results of which - in terms of an
upper limit to the ratio number of quarks to number
of nucleons - were already stated in Sect. 5.

7. THE FERROMAGNETIC LEVITATION EXPERIMENT

The next step in the experiments of our group
consisted in trying to increase further the sensitivity
by a factor $\sim$ 500 by increasing correspondingly the
mass of each object being explored; we also thought it
advisable to change the material under exploration because, after all, the quarks might be rare in the
pyrolitic graphite and more abundant in other materials.
So we decided to switch to steel spheres of mass
$\sim 10^{-4}$ g and to use for their levitation the feed-
back technique. The increase by a factor $\sim$ 100 in the
volume of the object (the density of graphite is $\sim$ 1/4
that of iron) produces difficulties in eliminating the
spurious Volta forces because such effects are proportional to the volume of the object (compare the
equation (15)) as repeatedly stated. Also the use of
perfect steel balls instead of irregular graphite
grains introduces some problems due to the absence
of a strong alignment (compare the Sect. 9).

I shall describe in what follows this ferromagnetic experiment in the following order: 1) a
general scheme of the experiment, 2) the elimination
of the effects from the permanent dipole force and

from the lack of a strong alignment, 3) a discussion of
the Volta effects, 4) a presentation of the results obtained so far, 5) future programs; this final section
will contain some remarks in connection with the experiment going on at Stanford.

8. A GENERAL SCHEME OF THE EXPERIMENT

This is presented without the constructive
details in the Fig. 2. The Figure Caption explains
the situation. A basic feature is the use of the lock-
in to improve the signal-to-noise ratio. The square-
wave electric field produces the oscillation of the
levitating object and - at the same time - creates
the reference signal for the lock-in. The frequency
of the lock-in is thus centered - and locked in phase -
to the oscillation frequency of the electric field
with a width determined by the integration time of
the lock-in usually set at 300 sec.

Before going on it is appropriate to give the
order of magnitude of the amplitude of oscillation Δ
and the corresponding signal from the lock-in for a
typical object with unit charge. For m = 100 µg,
v_x = 1 Hz, $N_{1/2}$ = 30, $E_x^{(a)}$ = 1.4 kV/cm we have Δ = 1µ.
In typical conditions of illumination the signal from
the horizontal photodetector HD (compare the Fig. 2)
corresponding to the above value of Δ = 1µ produces a
lock-in signal $\approx$ 15 mV. The signal to noise ratio in
good seismic conditions for an object of mass 10^{-4} g
with an integration time 300 sec is very good as can
be seen in the Fig. 3 (compare the Sect. 10).

9. THE SPINNING PROCEDURE

The lack of a strongly fixed orientation in space
of the levitating object could produce a signal simulating
a non-zero residual charge. The reason is the following:
under the action of the alternating electric field the
object may have periodic rotations, in phase with the
electric field, by a small angle around some internal
axis. If the object, due - for instance - to the
presence of some dust on it, is not perfectly spherical,
these rotations produce a periodic change of the shadow
on the horizontal photodetector simulating a residual
oscillation, that is a residual charge. Or it may happen
that due to some coupling between rotation and translation
the above small-angle periodic rotation produces a
translational oscillation again simulating a residual
charge. For these reasons - as well as for eliminating
the **permanent** dipole force - we decided to spin the
levitating object at a high frequency (a few hundred
revolutions per second). This was achieved in the
initial stage of the experiment by a synchronous
procedure exploiting the feedback system; later we
constructed using four coils an induction motor
having as rotor the levitating sphere. The objects -
initially cylinders, now spheres - spin at $\sim$ 500 Hz.
Due to this spinning the effects described above are
eliminated; in particular the gyroscopic stability
prevents any rotation of the type previously mentioned
under the action of the electric field. It should also
be added that by reducing the spinning from 500 to
25 Hz we could check that the amplitude of oscillation
of the object was totally independent from the spinning
frequency in this wide range.

In addition to eliminating the possibility of the spurious effects described above the spinning also eliminates any effect of those components (d_x and d_y) of the permanent electric dipole moment $\underset{\sim}{d}$ of the object, normal to the rotation axis z. Clearly the part of the force (16) due to these components is averaged away. The only remaining part of the dipole force is:

$$F_x^{(dip)} = d_z \frac{\partial E_z^{(a)}}{\partial x} = d_z \frac{\partial E_x^{(a)}}{\partial z} \quad . \tag{18}$$

To be sure that the effects of this force are un-important we have inserted in our apparatus the possibility of increasing artificially $\frac{\partial E_x}{\partial z}$ by a factor 100 with respect to the value - due to a small possible lack of parallelism of the plates - that this quantity has ordinarily. For all the objects for which this test was made the force (18), <u>after</u> this amplification by a factor 100, remains below $(1/2) e\ E_x^{(a)}$; in other words, for the above objects the residual apparent charge due to the force (18) was always less than $0.5 \cdot 10^{-2} e$.

10. A PRESENTATION OF THE RESULTS OBTAINED SO FAR IN THE FERROMAGNETIC VERSION OF THE EXPERIMENT

i) So far the results of the ferromagnetic version of the experiment are inconclusive. We describe them below. The first five spinning cylinders reported in the ref. (2b), totalling a mass of about 10^{-3} g , did all show a residual charge Q_R less than 0.1 e. A sub-sequent cylinder - the sixth-did show, however, in two

repeated measurements (separated in time by about 15 days)
substantial drifts (although, if this has any meaning,
the long-time average of Ω_R gave $\langle \Omega_R \rangle = 0$). All these
measurements were done with graphite plates.

ii) After completion of the new spinning system -
by the induction rather than the synchronous "motor"
(compare the Sect. 9) - we used steel spheres instead
of cylinders, mainly because spheres of the appropriate
sizes (0.2, 0.3 or 0.4 mm diameter) were more readily
available; initially the same graphite plates were used.
For a while we got "erratic" results as we shall now
explain.

iii) Take one sphere (negatively charged) and
measure its Ω_R by the usual sequence of ionizations.
Then open the box, recharge the same sphere negatively,
and proceed to a new determination of Ω_R in exactly the
same conditions as before, in particular with the same
distance between P_0 and P_1. Note that, depending on the
seismic noise, each determination of Ω_R takes from 12
to 24 h. The "quality" of the measurements is excellent
in the sense that the Ω_R determination is often good to
1 or 2%. The Fig. 3 gives a few examples of measurements
(compare the figure caption for more details). A number
of spheres were subjected to several (two or three) such
repeated determinations of Ω_R. By "erratic" we mean that
the values of Ω_R obtained in such determinations (all
referring to the same sphere) differ considerably among
them (and often differ substantially from zero). We insist
that the only operation performed between two such subsequent measurements was to open the box and touch the
sphere. The Fig. 4 gives a summary of all the measurements
performed on the spheres; values on the same horizontal

line refer to the same sphere. The distance between the plates used most frequently was 3 cm. In some cases variations in the distance between the plates were performed (without opening the box) and these are noted by an appropriate symbol (see the figure caption).

iv) For quite some time we could not understand the reason for the erratic behaviour; the contrast with the first five cylinders was disturbing. Initially we were inclined to ascribe the erratic results to the use of the spheres instead of the cylinders or to the change in the spinning system although we could hardly imagine a specific reason; a sequence of tests (including a few measurements on cylinders) excluded this kind of "explanation". At some point, as we are going to explain below, a regularity emerged; we presume that this regularity indicates the correct explanation of the erratic behaviour but stress that we are still at the level of a conjecture.

v) Each time we did a measurement either using new plates (brass or gold-sputtered brass) or after cleaning the old ones, Ω_R always was less than e/10. Altogether this behaviour appeared in eight different spheres. However, if the clean plates were left in the box for some days - the box being connected to a continuously operating diffusion (silicon) oil pump maintaining a vacuum of about 10^{-5} Torr - then the plates appeared to deteriorate. We ascribe this "deterioration" to a change in the surface-electrical properties of the plates taking place especially when the box is opened after a few days of connection to the diffusion pump. We conjecture that a substantial $\underline{E}^{(s)}$ can arise in this way leading, by equation (5), to

a non-zero and erratic Q_R. The paper by Palevsky et al.
[15] giving an accurate description of the preparation
of the capacitor plates in a vibrating reed electrometer
is exceedingly instructive in showing how strongly $E^{(s)}$
can depend on the preparation procedure.

vi) If the above conjecture is correct and if,
therefore, only the subset of measurements with the
clean plates is meaningful, the data so obtained
improve by a factor 1.75 the upper limit to the ratio
N(quarks)/N(nucleons) given in ref. [2b] for the first
five cylinders leading to a new upper limit:

$$N(quarks)/N(nucleons) < 10^{-21} \, .$$

However, as we repeat, this is only a conjecture.
Our main task for the future will be to check it, that
is to be able to guarantee clean conditions of
operations; by the use, in particular, of a clean
vacuum and of a standard preparation procedure - and,
possibly, non-contact electrostatic control - of the
plates. Also we shall slightly modify the levitation
chamber so as to test several spheres with the same
radius in succession without opening the box (compare
Sect. 12).

11. THE PATCH EFFECTS

At this point one can reasonably ask why the
procedure of increasing the distance of the plates from
the object that was used in the graphite experiment to
eliminate the residual charge due to the static Volta
fields is not so effective in the present situation.

The reason is the following: the plates have now a
diameter 10.5 cm much larger than that of the platelets
used in the graphite experiment; this is necessary to
ensure small electric gradients which is now more important than before due to the increased size of the
levitating object.

Due to the large diameter of the plates what
happens now on increasing the distance of each plate
from the object (passing from, say, 0.5 to 2 cm -
distance between the plates = 4 cm -) is that we do
in fact change the effect of the static Volta fields
on the object, but we cannot be sure to have eliminated
them; indeed when the plates are near to the object,
this feels especially the effect of the patches near
to the center, and due to the subtended angle the
peripheral patches are less important; on increasing
the distance of the plates the effect of the central
patches decreases but that of the peripheral ones increases. One should note again that - for a given
distance, patch potential etc. - the patch effect is
now larger than it was in the graphite experiment by a
factor from 50 to 150 due to the increased size of the
object.

We can get an idea, using a specific model, of
the effects of possible patches on the plates. The model
is not necessarily realistic so that the result of the
calculation should be taken only as an indication. The
model is this: we assume that on each plate a circular
patch of radius R exists having its center coincident
with the center of the plate; we also assume that all
the points in the circular patch are at a potential that
differs by V_o from the rest of the plate. For each

assumed distance between the plates (call this distance 2a) we select R so as to give the largest possible value of $\frac{\partial E_x^{(s)}}{\partial x}$ at the position of the object; and therefore the largest Q_R (compare the eq. (15)). If both plates have equal patches of the kind just assumed, it can be shown that the expression of

$$\frac{\partial E_x^{(s)}}{\partial x}\bigg|_o$$

at the position of the object (indicated by the subscript o) is given by:

$$\frac{\partial E_x^{(s)}}{\partial x}\bigg|_o = 6V_o aR^2 \sum_{k}^{\infty}{}_o \; (-1)^k \; \frac{2k+1}{[R^2+(2k+1)^2 a^2]^{5/2}} \; . \qquad (19)$$

If $\frac{\pi R}{2a} < 1$, that is:

$$R < \frac{2}{3} a \qquad (20)$$

this expression simplifies into:

$$\frac{\partial E_x^{(s)}}{\partial x}\bigg|_o \approx \frac{6 \, V_o \, a \, R^2}{(R^2 + a^2)^{5/2}} \; . \qquad (21)$$

For given a the value of R maximizing (21) is $R = \sqrt{\frac{2}{3}}a$; the condition (20) is almost satisfied. Using, therefore, the expression (21) we get:

$$\text{Max} \; \frac{\partial E_x^{(s)}}{\partial x}\bigg|_o \approx \frac{4 \, V_o}{3.5a^2} \approx \frac{V_o}{a^2} \; . \qquad (22)$$

Thus from the equation (15)

$$\text{Max } \Omega_R^{(s)} \cong 4\pi \varepsilon_o r^3 \frac{V_o}{a^2} \; . \tag{23}$$

For $r = 0.15$ mm, $V_o = 0.1$ Volt, $a = 1.5$ cm, we get from (23):

$$\text{Max } \Omega_R^{(s)} \cong 0.9 \text{ e} \; . \tag{24}$$

This is a very large residual spurious charge and, if values of $V_o = 0.1$ Volt were typical, the erratic behaviour of the Fig. 4 might be explained. There is, however, one aspect of the situation that is not entirely clear to us, especially after having measured - using a Monroe Model 162 electrostatic Voltmeter - the values of the patch effects on plates exposed to the air. This is the following: although patch potentials V_o are always present, we now know that a magnitude of 0.1 Volt is rather uncommon; a more typical value for a plate exposed to the air is, say, 30 mV. Therefore, if we consider the measurements performed on spheres of radius 0.1 mm done precisely with the purpose of decreasing the patch effects (by a factor $(1.5)^3 = 3.37$) we should have typically, instead of $\Omega_R^{(s)} = 0.9$ e, a value of $\Omega_R^{(s)} = 0.1$ e; if we then take into account that many measurements were done at a distance a between the plates of 4 cm - rather than 3 cm appearing in the above estimate - we should reduce $\Omega_R^{(s)}$, on the basis of (22), by a further factor $2.25/4$; and if we note in addition that the calculation just presented should give us the worst possible effect (it was done so as to maximize $\Omega_R^{(s)}$), values larger than

O.1 e at 2a = 4 cm for spheres with radius r = O.1 mm
appear to be somewhat unexpected. A possible explanation
is that the surfaces of the plates in the box evacuated
by a diffusion oil pump may have patches with values of
V_o particularly large, that however disappear, or
strongly decrease, exposing the plates to the air. However, if this is the explanation, it is strange that
large drifts of Q_P with time are unfrequent; indeed the
process of formation of these patches should be
accompanied - and signalled - by a drift in Q_R and the
vast majority of the measurements do not show any drift.
(Although occasionally we find measurements with drifts,
large drifts are unfrequent; of course it is possible
that the formation of patches takes place before the
measurement, immediately after the box is closed and
the vacuum is prepared.)

12. FUTURE PROGRAMS

Clearly if there are no drifts during the
measurements the patch effects are the same for two
different spheres of the same radius measured in
succession without opening the box. So far we had to
open the chamber to change the sphere, but we are
now constructing a sphere injector to do this operation
without opening the chamber and without modifying the
vacuum or anything else. For spheres with exactly the
same radius - as our steel spheres are - all the spurious
effects (except those from the permanent dipole moment,
that can be measured and - compare Sect. 9 - are
typically less than one per cent) should remain the
same; therefore on comparing two subsequent spheres

we hope to determine, if both have the same residual
charge or if the residual charges differ by ± e/3,
or - let us hope not - by some different value. This
procedure, in which the box is not opened, has been
used in the Stanford experiment where differences of
± e/3 have been in fact found. (The residual charge
on a sphere in the Stanford experiment appears to
change frequently - by e/3 - if the sphere is washed
in acetone or subjected to electric discharges). In
a few months, the time for the sphere injector to be
ready, we hope to be able to see what happens with
our steel spheres.

Concerning the Stanford experiments I only say
here that the procedure of ref.[5b] improves in
several aspects the one used in the first version of the
experiment [5a]; in particular the use of spheres of
practically the same radius, thus avoiding extra-
polations like that of the Fig. 4c of ref. [5a], is
a definite improvement. I don't know, if the presumed
correlation between the presence of quarks and heating
on tungsten has now disappeared (in agreement with some
experiments [10,14] that did not confirm this peculiar
property of W) or is still present.

Certainly the chemical properties of quarks that
would emerge from ref. [5] (disappearance or appearance
of quarks with discharges or washing in acetone) appear
very strange, but they cannot be excluded a priori.
I feel that only more measurements both in the Stanford
and in our version of the experiment (once the injection
system has been set up) are necessary to confirm or
disprove the presence of quarks.

Although a standard question is why, if quarks do exist with the abundance and properties implied by the Stanford results, all the previous terrestrial searches by the techniques discussed in the previous lecture have failed to detect them, explanations for these things are often found "a posteriori". Therefore I do feel that a continuation of this kind of experiments is very important.

REFERENCES FOR LECTURE III

[1] a) C.Becchi, G.Gallinaro,and G.Morpurgo, Nuovo Cimento 39 (1965) 409.

b) G.Gallinaro and G.Morpurgo, Phys. Lett. 23 (1966) 609.

[2] G. Morpurgo, G.Gallinaro,and G. Palmieri, Nucl. Instr. and Meth. 79 (1970) 95 (this paper contains an account of the work performed up to 1970).

[3] E. Garris and K. Ziock, Nucl. Instr. and Meth.117 (1974) 467.

V. Braginsky et al., Phys. Lett. 33B (1970) 613.

A.F. Hebard and W. Fairbank, Proc. of the 12th Int. Conf. on Low-Temperature Physics, Kyoto 1970 (Keigaku Publ. Co., Tokyo 1971) p. 855.

[4] a) G. Gallinaro, M. Marinelli,and G. Morpurgo, INFN/AE 76/1 Report (Feb. 1976).

b) G. Gallinaro, M. Marinelli,and G. Morpurgo, Phys. Rev. Lett. 38 (1977) 1255.

c) Same authors, Proc. of the 4th General Conf. of EPS at York (September 1978 - In course of publication).

[5] a) G. La Rue, W. Fairbank,and A. Hebard, Phys. Rev. Lett. 38 (1977) 1011.

b) G. La Rue, W. Fairbank, and J. Phillips, Phys. Rev. Lett. 42 (1979) 142.

[6] R. Millikan, The Electron (The Univ. of Chicago Press 1963). An extemely interesting hystorical presentation of the Millikan experiments is givne by G. Holton in: Historical Studies in the Physical Sciences - ninth vol. (1968) 161.

[7] V.D. Hopper, Proc. Roy. Soc. (London) 178 (1941) 243.

[8] D.M. Rank, Phys. Rev. 176 (1968) 1635.

[9] W.A. Chupka et al., Phys. Rev. Lett. 17 (1966) 60.

[10] R. Bland et al., Phys. Rev. Lett. 39 (1977) 369.

[11] W. Braunbeck, Z. Physik 112 (1939) 75, 764.

[12] J. Beams and F. Linke, Rev. Sci. Instr. 8 (1937) 160.

[13] A. Moran et al., Phys. Rev. 164 (1967) 1599.

[14] R.N. Boyd et al., Phys. Rev. Lett. 40 (1978) 216.

[15] H. Palevsky et al., Rev. Sci. Instr. 18 (1947) 298.

FIGURE CAPTIONS FOR LECTURE III

Fig. 1. Schematic idea of the Millikan experiment.

Fig. 2. A schematic view of the apparatus. The coils A and B provide the feed-back levitation of the object O illuminated by the lamp L. The vertical photodiode VD detecting the shadow of O (through the half-transparent mirror M) controls the feed-back levitation system and, in particular, the current in the coil B (as well as, to some extent, A). The horizontal photodetector HD measures the oscillation of the object and

transfers the signal to the lock-in amplifier.
In addition HD governs to the coils C, D that
create the damping. The four coils used for
the spinning (and many other components) are
omitted in the figure for simplicity. For a
somewhat more complete explanation compare
the ref. [4a].

Fig. 3. Measurements on three objects. The sequence
of oscillation amplitudes Δ for three different
objects: those indicated as 16, 14 black and 3
(dotted point on the right) in the ensuing
Fig. 4. In the upper (a) and middle (b) graphs,
both referring to spheres of diameter 0.3 mm
(mass 110 µg), the sequence of Δ's is: 62.7, 42.2,
0, 21 and, respectively 61, 20, 0, 20 corresponding obviously to $\Omega_R = 0$. The sphere of $\phi = 0.2$ mm
in the lower graph (c) - number 3 in Fig. 4 -
has a sequence of Δ's: +12.5, -19.5, -52 and
therefore a residual charge $\Omega_R = 0.384$ (compare,
however, the other measurements on the same sphere
indicated in Fig. 3). In (a), (b), (c) the signal-
to-noise ratio is very good. This is not always
so. When the traffic or seismic conditions are
bad, the signal-to-noise ratio is worse. However,
the statistical error on Ω_R is always very small
(usually equal or smaller than the circles or
squares in Fig. 4) because, if an object keeps
its charge for, say, 6h, this is equivalent to
twelve independent measurements of Δ (the lock-
in is normally used with an integration time of
300 s at 12 db corresponding to a 98% cancellation
of correlations in 30 min). The graphs (a), (b)

and (c) above are totally exempt for drifts:
the Δ levels are horizontal. Some cases of
measurements showing drifts have been excluded
from consideration: they are not included in
Fig. 3. (Note: the Δ's in (a), (b), (c) are
given in <u>arbitrary</u> units).

Fig. 4. A summary of the measurements. Each horizontal
line of the figure gives the values of Q_R found
for the same object in different measurements;
the figure contains 22 different objects, as
indicated on the left. Squares refer to spheres
of mass 1.1×10^{-4} g ($\Phi = 0.3$ mm), circles to
spheres of mass 3.3×10^{-5} g ($\Phi = 0.2$ mm). Most
measurements were performed at a separation between the plates of 3 or of 4 cm. Measurements on
the same object performed after a change in the
separation between the plates - without opening
the box - are indicated with a dot below. All
the other measurements (without dots) on the
same line are separate measurements on the same
object performed having opened the box and
handled the object between one measurement and
the next. Black circles or squares refer to
measurements made with new or clean plates;
the measurements of this kind are all compatible
with $Q_R < e/10$. In the graph the sign of Q_R is
that for which $|Q_R| < 0.5$ e.

oil droplet
z_2
z_1
condenser plates

Fig. III-1

22
20
18
16
14
12
10
8
6
4
2
Q_R
-.5
-.4
-.3
-.2
-.1
0
.1
.2
.3
.4
.5

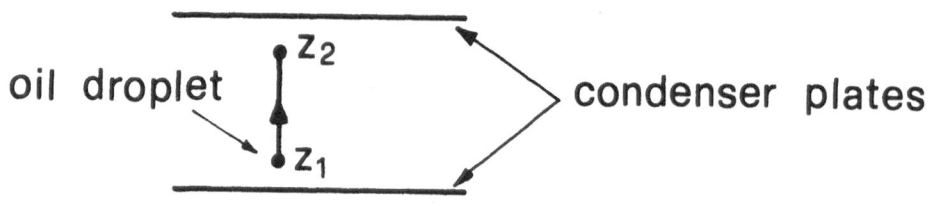

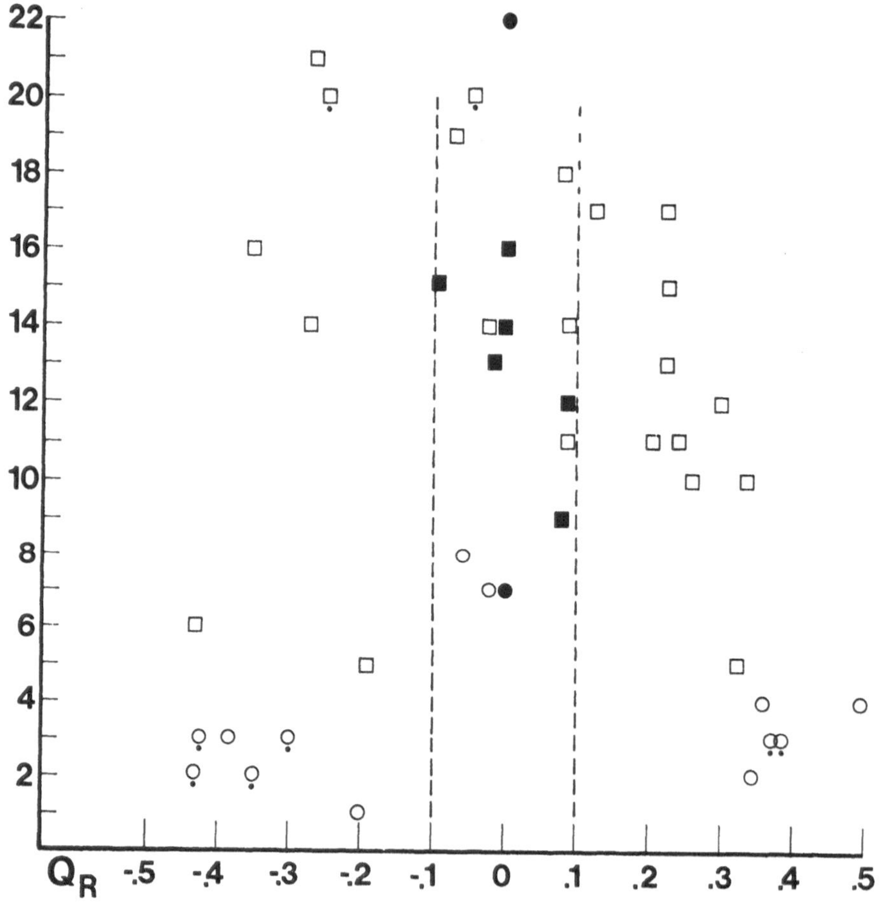

Fig. III-4

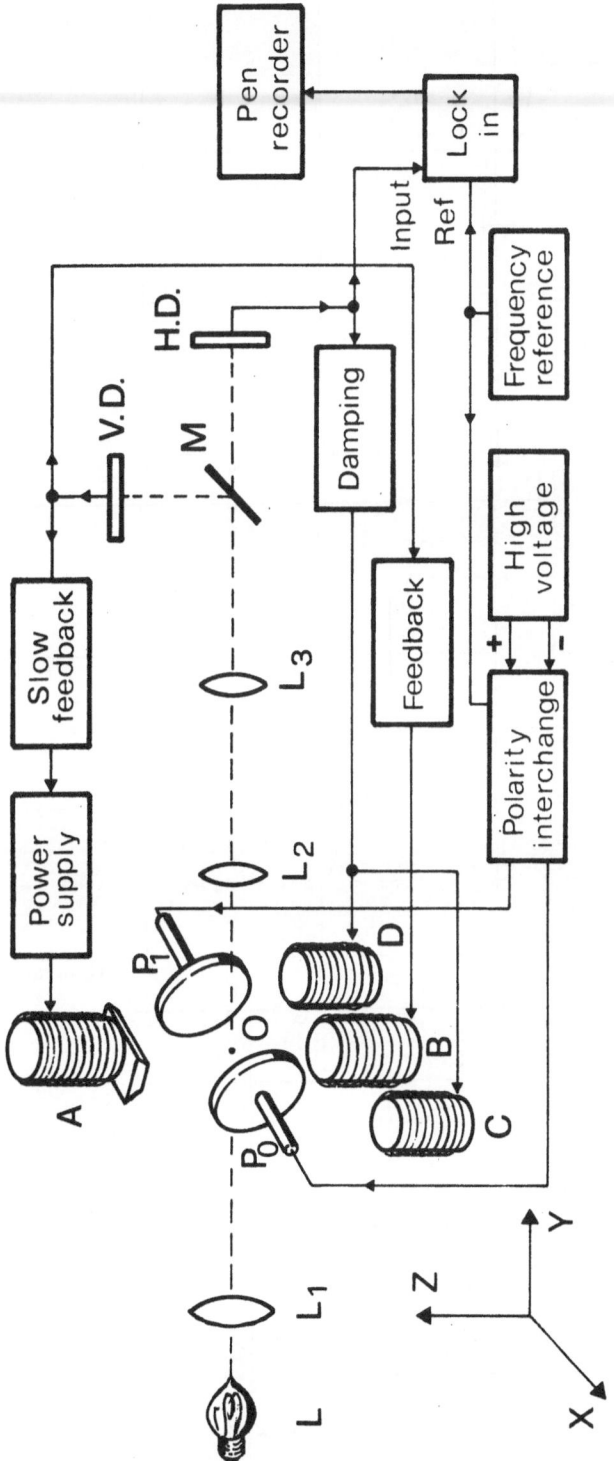

Fig. III-2

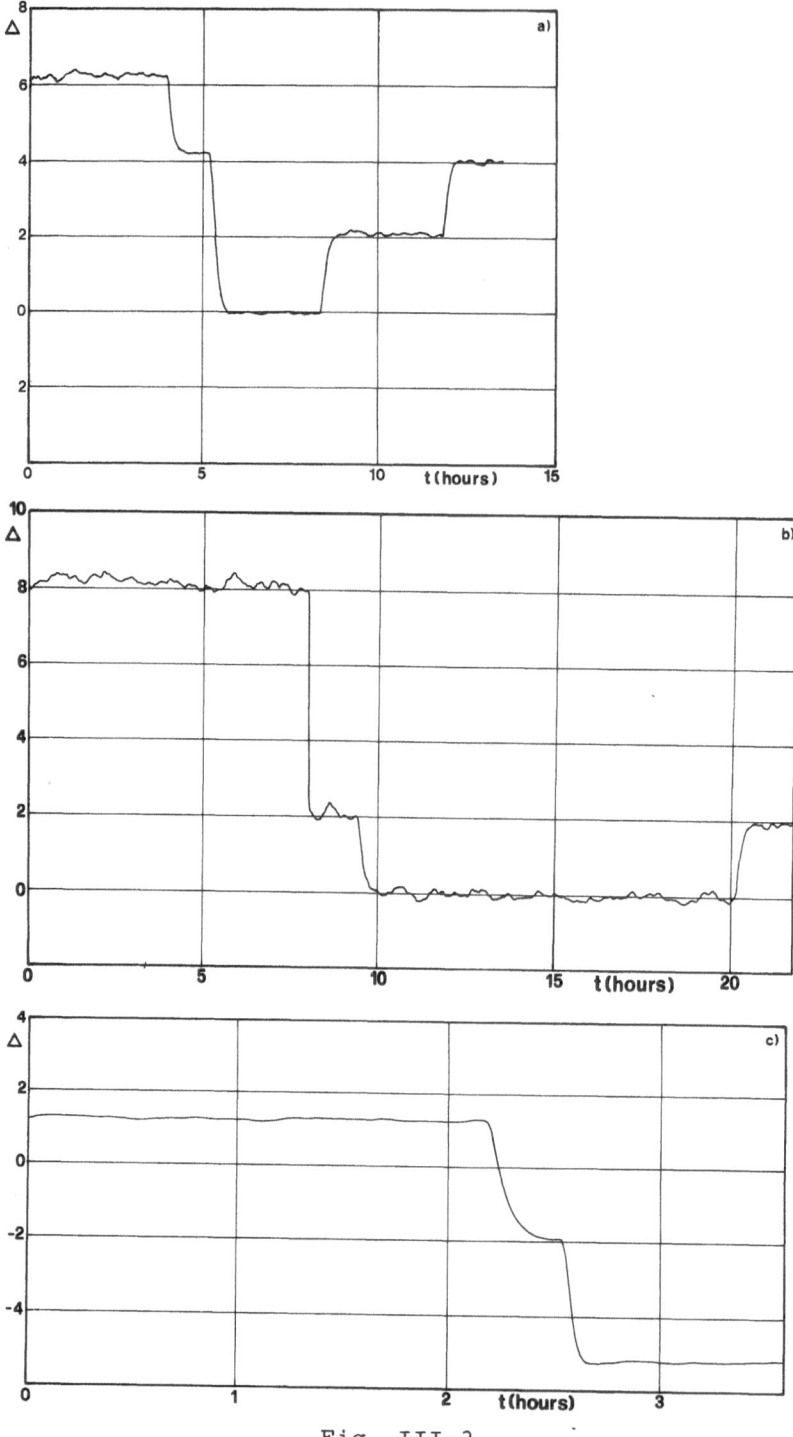

Fig. III-3

Acta Physica Austriaca, Suppl. XXI, 81—258 (1979)

RECENT EXPERIMENTS AT DESY[+]

by

G. FLÜGGE

Deutsches Elektronen-Synchrotron DESY

Notkestraße 85, D-2000 Hamburg 52

CONTENTS

INTRODUCTION

I. GENERAL REMARKS ON e^+e^- REACTIONS

    1. e^+e^- storage rings

    2. Cross sections

    3. Measurement of σ_{had} below 5 GeV

    4. Bumps and peaks below 5 GeV: charm

II. THIRD GENERATION OF QUARKS AND LEPTONS

    1. The heavy lepton τ

        a) τ mass

        b) τ decay

        c) Summary

[+]Lectures given at the
XVIII. Internationale Universitätswochen für Kernphysik
Schladming, Austria, February 28 - March 10, 1979.

2. The b quark
   a) The Ypsilon story
   b) Determination of parameters
   c) Ypsilon prime
   d) Quark charge and number of T states
   e) Summary

III. EVENT TOPOLOGY
   1. Jets below 9.4 GeV
      a) Quantities to measure jets
      b) SLAC-LBL results
      c) PLUTO results at DORIS
      d) Conclusion on jets
   2. T decay topology
      a) Experimental procedure
      b) PLUTO data
      c) NaJ detector data
      d) Tests of coplanarity (PLUTO)
      e) Further tests of the three-gluon model
      f) Conclusion

IV. FIRST PETRA RESULTS
   1. The machine
   2. First round detectors
   3. Results
      a) QED processes
      b) Total cross section
      c) Even topology
      d) Two-photon physics
   4. Physics plans at PETRA
   5. Summary
   REFERENCES

INTRODUCTION

The main task of an experimental talk at a theoreticians school should probably be a tempering one. In this respect, e^+e^- physics may have been a bad choice. The field has so rapidly developed and discoveries are chasing each other that much of the optimism of theory has passed over to e^+e^- experimentalists.

A vast amount of experimental material arose from the simple reaction of e^+e^- annihilation. I, therefore, have to limit myself to recent results - most of them less than one year old. The paper will be organized as follows:

In the first lecture (chapter I and II) I will give a short introduction to e^+e^- machines and cross sections. In particular I will discuss the total cross section and - after a short summary on charm - concentrate on the third generation of quarks and leptons: the heavy lepton τ and the T family. In my second lecture the various aspects of event topologies in the DORIS energy range will be discussed, including the T decay. In the third lecture I will then describe the new storage ring PETRA and present first results on QED checks, total cross section, jet structure, and two-photon processes.

I. GENERAL REMARKS ON e^+e^- REACTIONS

1. e^+e^- storage rings

The history of e^+e^- storage rings dates back to 1960 when B. Touschek in FRASCATI built the first machine of this kind [1]. The original motivation for e^+e^- storage rings was to study QED limits at large energies. Very soon, however, the prime interest turned to hadron production [2]. The annihilation of electrons and positrons into hadrons via the one-photon channel presents several advantages. Contrary to hadron collisions the system has well defined quantum numbers of the photon. In (symmetric) storage rings the full energy of both beams becomes available in the head-on collisions of the stored particles.

Storage Rings

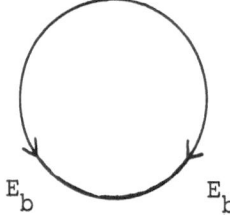

$E_{CM} = 2 E_b$ (for zero crossing angle)

E_b = beam energy

E_{CM} = center of mass energy

The laboratory frame is identical with the center-of-mass system (for zero crossing angle and equal energies). This highly facilitates the data analysis although it requires large angular acceptance of the apparatus.

Table 1 gives a survey of e^+e^- machines which have been built [2,3]. I will mainly concentrate on two machines belonging to the third and fourth generation of these e^+e^- storage rings: DORIS [4] and PETRA [5] which have been built in Hamburg at DESY.

Fig. 1 shows the storage ring DORIS which was originally built for beam energies ranging from 1.5 to 4.3 GeV. Electrons and positrons are produced and pre-accelerated in the two LINACs I and II and then injected

Name	Location	First Beam	Maximum Beam Energy (GeV)
AdA	Frascati	1961	0.25
Princeton-Stanford	Stanford	1962	0.55
ACO	Orsay	1966	0.55
VEPP-2	Novosibirsk	1966	0.55
ADONE	Frascati	1969	1.55
BYPASS	Cambridge (USA)	1970	3.5
VEPP-3	Novosibirsk	1970	3.5
SPEAR	Stanford	1972	3.9
DORIS	Hamburg	1974	5.0
VEPP-2M	Novosibirsk	1975	0.67
DCI	Orsay	1976	1.8
PETRA	Hamburg	1978	19
PEP	Stanford	(1979)	18
CESR	Cambridge (USA)	(1979)	8
VEP-4	Novosibirsk	(1979)	8

Table 1. History of electron storage rings.

into the DESY synchrotron. Here they are accelerated to their final storage energy and then transferred into the DORIS storage rings. Electrons and positrons are stacked separately in two rings on top of each other and eventually brought to collision in two intersection regions.

Next to the center-of-mass energy the rate of interaction is, of course, of largest importance in an

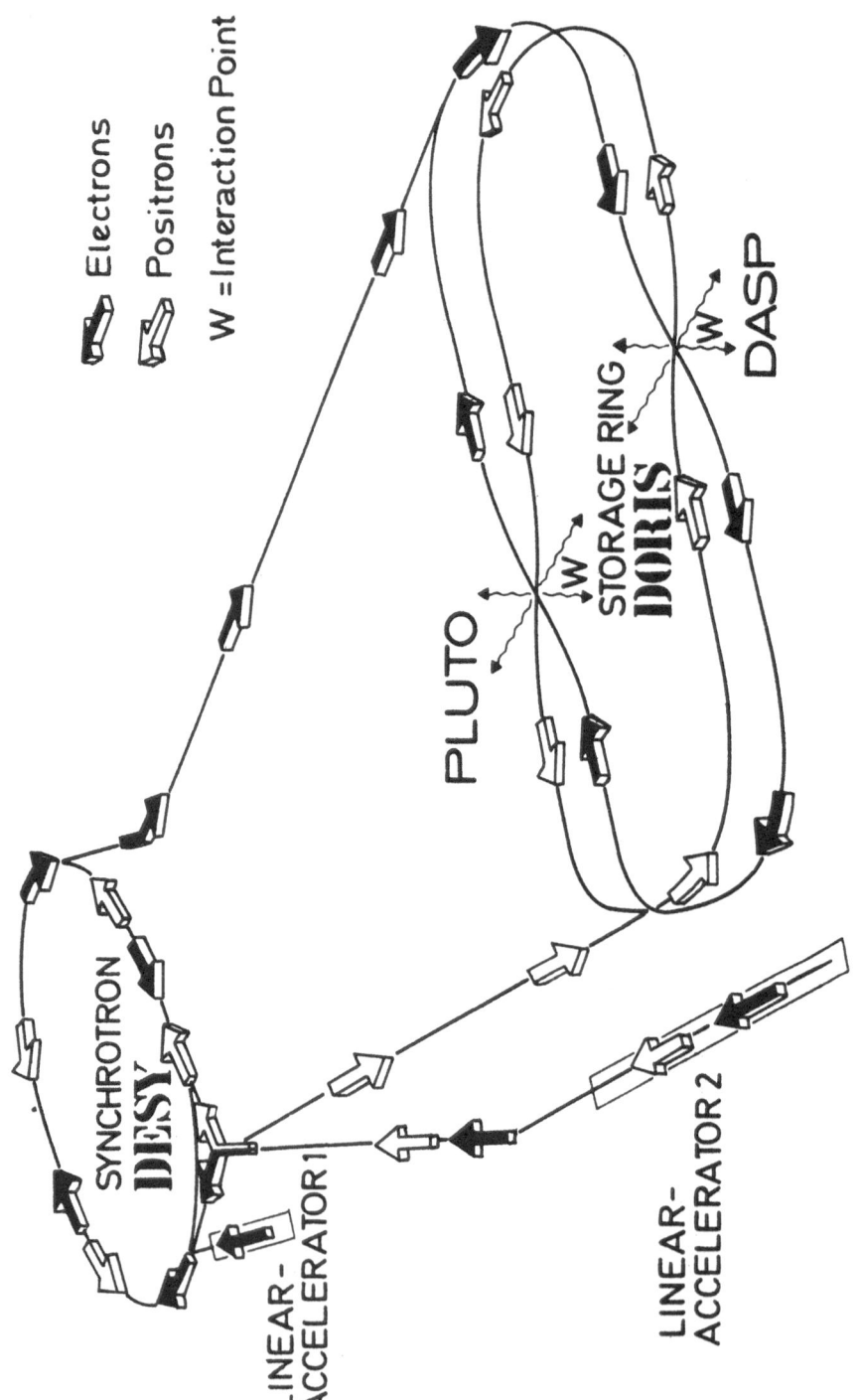

Electrons
Positrons
W = Interaction Point

SYNCHROTRON
DESY

LINEAR-
ACCELERATOR 1

LINEAR-
ACCELERATOR 2

PLUTO

W
STORAGE RING
DORIS

W
DASP

Fig. 1. The e^+e^- storage ring DORIS.

e^+e^- storage ring. This interaction rate is given by the formula

$$N = \frac{n_+ n_-}{t\,F}\,B.\sigma$$

where the proportionality constant between the cross section and the interaction rate is called the luminosity:

$$L = \frac{n_+ n_-}{t\,F}\,B\ .$$

With $\quad$ B $\quad\quad\quad\quad\quad$ number of bunches

$\quad\quad\quad$ $F = 4\pi\sigma_x\sigma_y$ $\quad$ interaction area

$\quad\quad\quad$ $I_\pm = n_\pm e/t$ $\quad$ total currents per beam

$\quad\quad\quad$ f $\quad\quad\quad\quad\quad$ revolution frequency

one finally obtains

$$L = \frac{1}{4\pi e^2}\,\frac{I_+ I_-}{fB\sigma_x\sigma_y}\ .$$

This formula shows that the luminosity is proportional to the product of the two currents I_+ and I_-. To see the energy dependence of the luminosity we have to insert machine parameters. Without further derivation let me just quote the result of a rather lengthy calculation which can for instance be found in ref.[6]:

$$L \sim \frac{(\Delta Q)^2}{\beta_y}\,B\,\varepsilon_x\,E_b^2$$

ΔQ is a measure of the frequency range which is available to the circulating beam without hitting a higher order resonance. It cannot easily be predicted. Experience shows, however, that ΔQ is usually $\lesssim$ .06. The vertical amplitude function β_y is proportional to the vertical beam dimension σ_y. Therefore the beam has to be focussed vertically to get a high luminosity. The emittance ε_x finally measures the available phase volume in the machine. For fixed beam optics this quantity is proportional to E_b^2. Thus we finally get

$$L \sim E_b^4 \; .$$

At most e^+e^- machines this theoretical variation of luminosity is in practice limited to a power behaviour $L \sim E_b^2$. At a certain energy the synchrotron-radiation losses in the machine take over and the luminosity drops rapidly roughly as E_b^{-8}. The energy dependence of L is sketched in fig. 2.

Typical parameters of the DORIS storage ring are

maximum energy:	8.6 GeV E_{CM}	(we will see later that this was up-graded to 10 GeV)
average luminosity:	$3.10^{29}/cm^2s$	
beam lifetime:		typically 5 to 10 hours for single beams.

Let me mention in this context a technical term which is of practical importance in e^+e^- physics, the 'integrated luminosity'

$$\int L \, dt \; .$$

The dimension commonly used for this quantity is

$$1 \ nb^{-1} \equiv 10^{33} \ cm^{-2} \ .$$

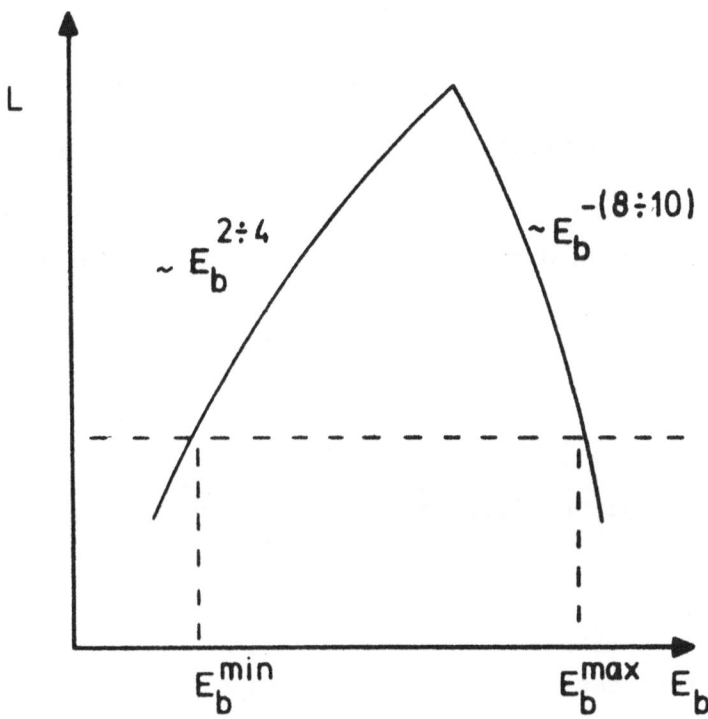

Fig. 2. Typical energy dependence of the luminosity in $e^{+}e^{-}$ storage rings.

As an example let us take a luminosity of 3.10^{29} integrated over one day.

$$\int L \ dt = 3.10^{29} \cdot 24 \cdot 3600 \approx 26.10^{33} \ cm^{-2} = 26 \ nb^{-1} \ .$$

The usefulness of this quantity is apparent from the equation

$$\sigma(nb) \cdot \int L \ dt \ (nb^{-1}) = \text{number of events.}$$

2. Cross sections

In this section I want to give a short introduction to the main processes encountered in e^+e^- physics. The expected cross sections will be estimated to provide a feeling for the rates we are dealing with.

μ-pair production

The μ-pair production

$$e^+e^- \to \mu^+\mu^-$$

is kind of a pilot reaction in e^+e^- physics since it is represented by the simple graph

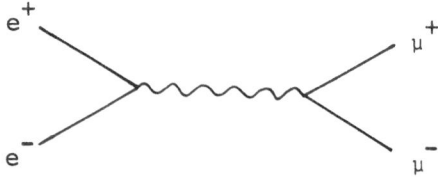

The total cross section for this reaction is given by the formula

$$\sigma_{\mu\mu} = \frac{4\pi\alpha^2}{3E_{cm}^2} \sim \frac{21.7 \text{ nb}}{E_b^2 (\text{GeV}^2)} (\text{spin } 1/2, \text{ pointlike, } E_{CM} \gg m_\mu).$$

Cross sections in e^+e^- reactions will mostly be given in terms of $\sigma_{\mu\mu}$. Let us, therefore, calculate the $\mu\mu$ rates to get a feeling for the number of events one expects in e^+e^- physics. Assume an average luminosity of 3.10^{29} at $E_{cm} = 5$ GeV varying like E_b^2. Then the energy dependence cancels out in the product $L.\sigma_{\mu\mu}$ and the expected average rate will be

$$N_{\mu\mu} = L.\sigma_{\mu\mu} \approx 10^{-3} \text{ s}^{-1} \approx 90 \text{ d}^{-1}.$$

To demonstrate the usefulness of the 'integrated luminosity' let me quickly redo the calculation in nb^{-1}: Within a day one integrates 26 nb^{-1}. Multiplying this by $\sigma_{\mu\mu} = 3.5$ nb we again get 90 events/day.

Bhabha scattering

Another important QED cross section is the Bhabha scattering

$$e^+ e^- \rightarrow e^+ e^-$$

which is represented by the two graphs

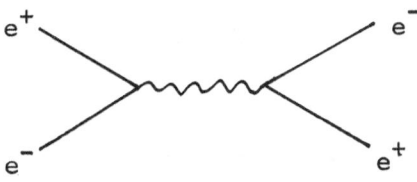

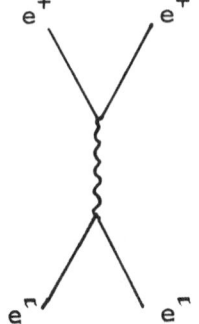

Bhabha scattering is used as a monitor reaction in e^+e^- collisions because it gives a large calculable cross section at small angles where the validity of QED is proven (low momentum transfer).

Hadron production

The most important cross section in e^+e^- physics is, however, hadron production via the one-photon channel.

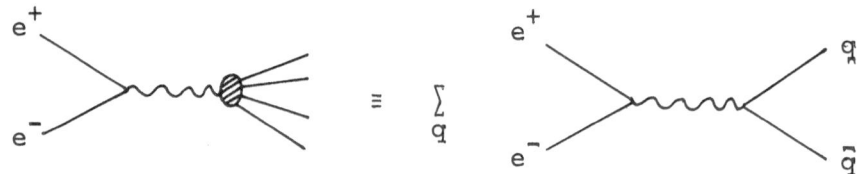

In the quark-parton model this process is simply described by the sum over all quark pair cross sections. It is therefore related to $\sigma_{\mu\mu}$ by the formula (assuming point-like spin 1/2 massless coloured quarks)

$$R = \frac{\sigma_{had}}{\sigma_{\mu\mu}} = 3 \sum_q Q_q^2 \qquad\qquad q = \text{quark flavours .}$$

Thus R is just the sum over all quark charges where the sum runs over all quark flavours and colours. The expected values for R are summarized in table 2. This table also contains the expectation for R if we include QCD correct-ions in first order [7]. The energy dependence of R is schematically demonstrated in fig. 3.

Near to a new flavour threshold bumps and peaks appear in the cross section. In addition also new leptons show up by their hadronic decay modes. Like in the case of $c\bar{c}$ and $\tau^+\tau^-$ production lepton and quark thresholds may (accidentally?) overlap. In the following I will first talk about the asymptotic behaviour of the total cross section and then come back to the threshold region.

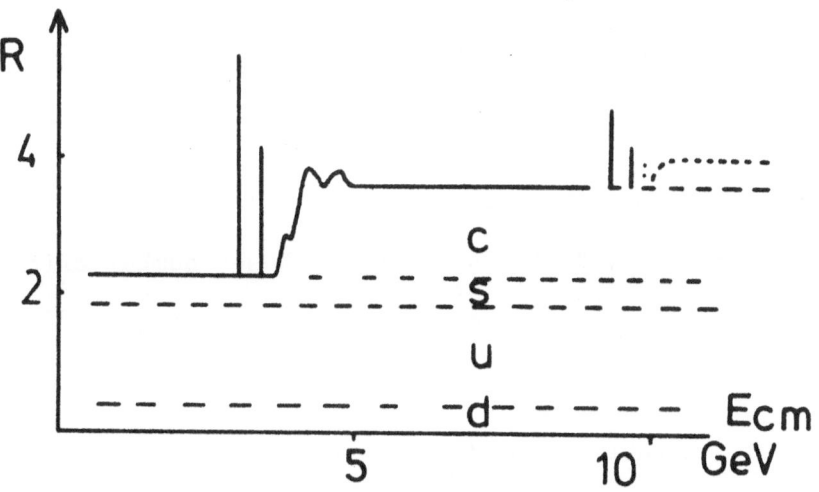

Fig. 3. Schematic behaviour of $R = \sigma_{had}/\sigma_{\mu\mu}$ as a function of energy.

Quark q	Charge Q_q	$R_{QPM} = 3\sum Q_q^2$	R_{QCD}^+
u	2/3		
d	- 1/3	2	~ 2.3 $(E_{cm} = 3.6$ GeV$)$
s	- 1/3		
c	2/3	3 1/3	~ 3.9 $(E_{cm} = 5.0$ GeV$)$

Table 2. Theoretical predictions for $R = \sigma_{had}/\sigma_{\mu\mu}$.

+Gluonic corrections in first order QCD:

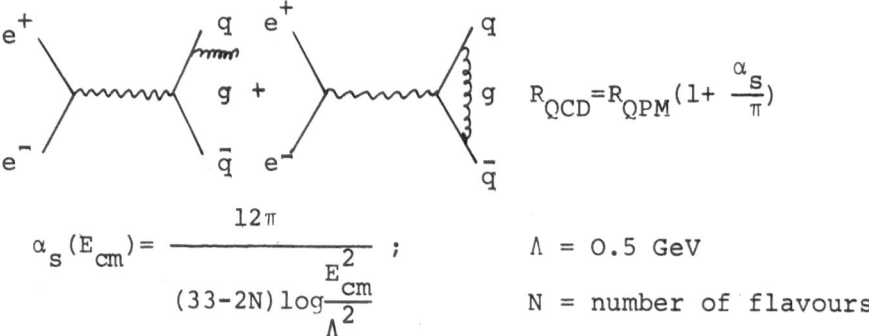

$$R_{QCD} = R_{QPM}\left(1 + \frac{\alpha_s}{\pi}\right)$$

$$\alpha_s(E_{cm}) = \frac{12\pi}{(33-2N)\log\dfrac{E_{cm}^2}{\Lambda^2}} \quad ;$$

$\Lambda = 0.5$ GeV

N = number of flavours

3. Measurement of σ_{had} below 5 GeV

The total hadronic cross section in e^+e^- reactions is measured according to

$$\sigma_{had} = \frac{N}{\varepsilon.\int L \, dt}$$

where N is the number of events seen in the detector. ε is the acceptance of the detector and $\int L \, dt$ is the integrated luminosity.

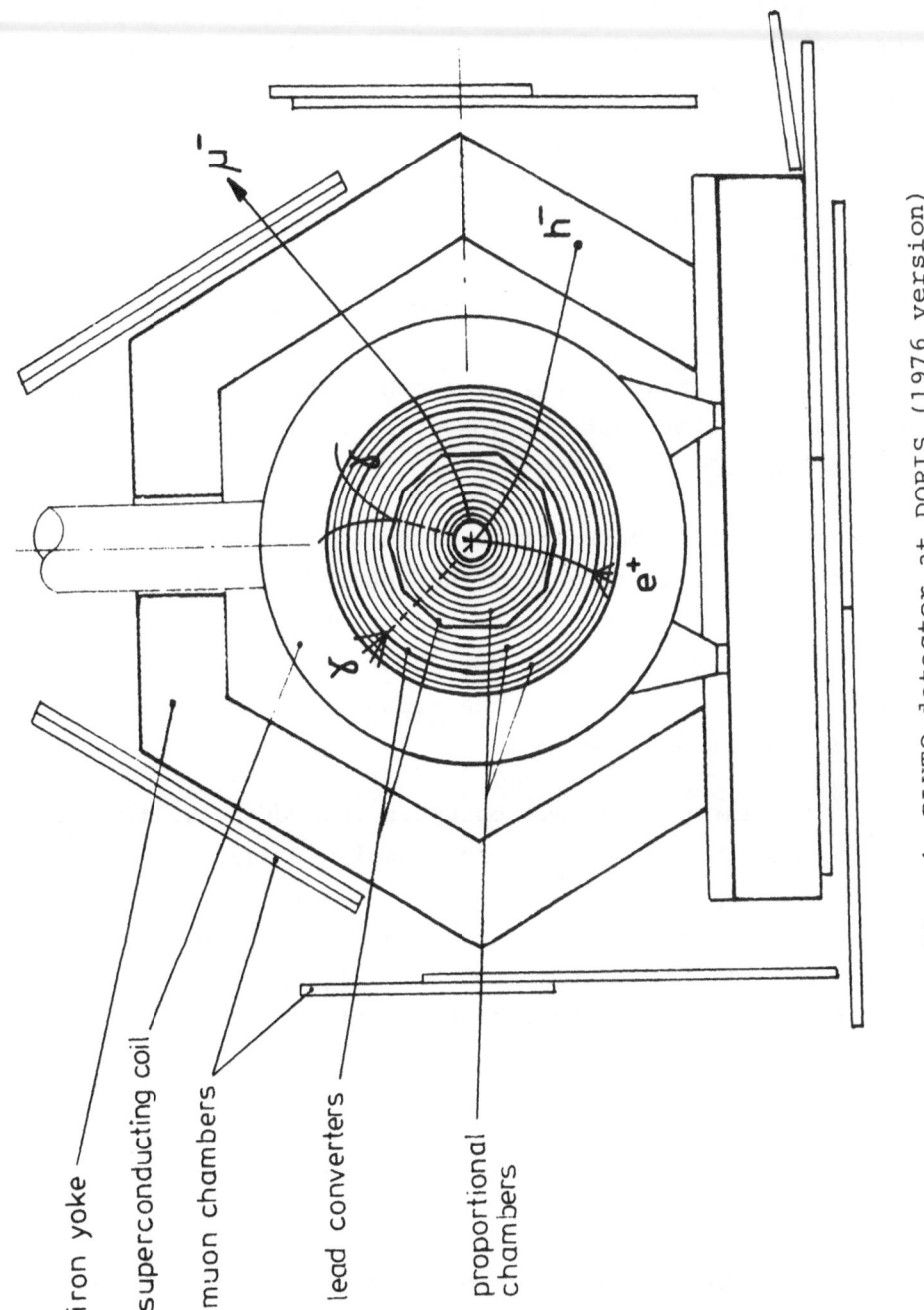

Fig. 4. PLUTO detector at DORIS (1976 version)

A typical detector for measuring total cross sections is the PLUTO spectrometer [8] shown in fig.4. It consists of a superconducting coil which produces a magnetic field of 1.7 T. The magnetic field volume is filled with a set of cylindrical proportional chambers to detect the tracks of charged particles. In addition two lead cylinders are inserted to detect photons and electrons. A set of proportional tube chambers outside the iron flux return yoke is used to separate hadrons and muons. The acceptance of such a detector for hadronic events is slightly energy dependent. In the energy range around 5 GeV for instance the event acceptance of PLUTO is of the order of 80%.

Hadronic events are accepted if they have at least two tracks originating from the interaction point. In addition non-collinearity of the tracks is required to remove QED contributions. Note that such a signature includes at least part of the cross section for the production of new leptons.

The integrated luminosity is recorded simultaneously to the measurement of σ_{had} by counting small angle Bhabhas in the forward direction. Fig. 5 shows the results of the total cross section measurements of the PLUTO collaboration [9] in terms of $R = \sigma_{had}/\sigma_{\mu\mu}$. The systematic error of ± 15% is indicated in the figure. For comparison [10] also the results of three other experiments are shown: SLAC-LBL [11], DASP [12] and DELCO [13]. Below charm threshold around 3.5 GeV all data agree remarkably well. Also far above threshold in the asymptotic region around 5 GeV there is good agreement between all four experiments.

How well do these 'asymptotic' values above and

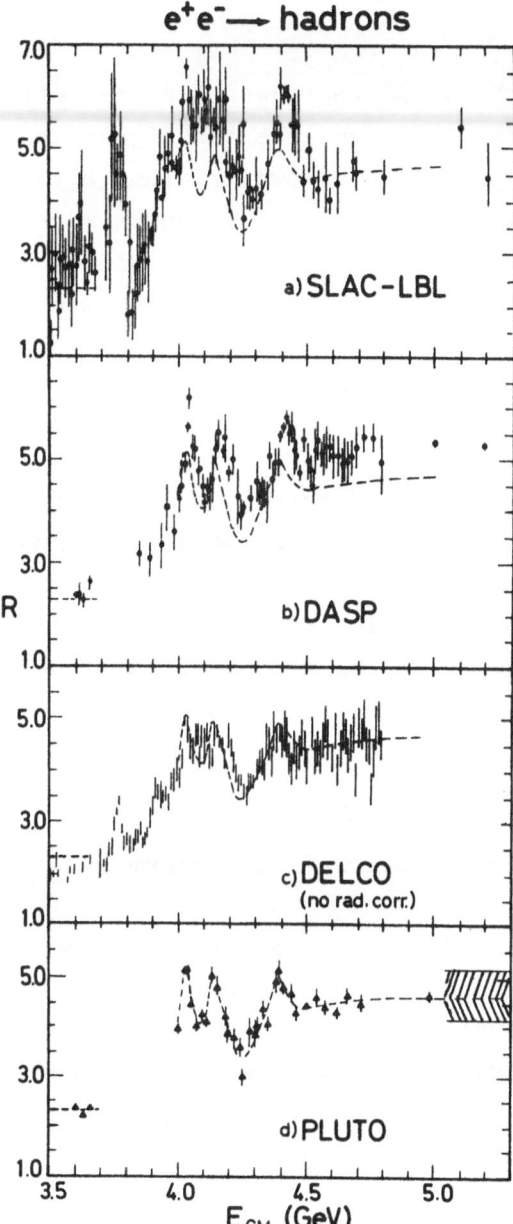

Fig. 5. Measurements of R as a function of energy
a) SLAC-LBL group [11]
b) DASP group [12]
c) DELCO group [13](no radiative corrections)
d) PLUTO group [19]
Adopted from G. Feldman [10].

below charm threshold reproduce the theoretical predictions? All measurements are higher than the simple quark-parton prediction. However, they agree well with the expectation of four quark flavours, heavy-lepton production and gluonic corrections. Since, however, both the gluonic corrections and the systematic errors are of the same level of 10 to 15% we cannot draw any definite conclusions about QCD contributions in the total cross section.

In the resonance region near charm threshold there are considerable differences between all four experiments, in width, height and also position of the resonances. All data agree about the dip in the cross section around 4.2 GeV which shows that charm production drops down to a very low level between the resonances even above threshold.

4. Bumps and peaks below 5 GeV: charm

Since the discovery of the J/ψ resonance [14] in 1974 enormous progress has been achieved in the study of charmed particles and charmonium [15]. During the past year, however, the interest has rather moved to the higher energy region. Therefore I will only give a very short summary of the situation of charm and charmonium in this lecture [16].

Our experimental knowledge on charm is schematically summarized in fig.6. The odd C-parity ^3S state J/ψ, its radial excitations ψ', and the ^3D state $\psi''(3.77)$ show up in the total e^+e^- cross section, the latter due to its mixing with the nearby ^3S state. The existence of the

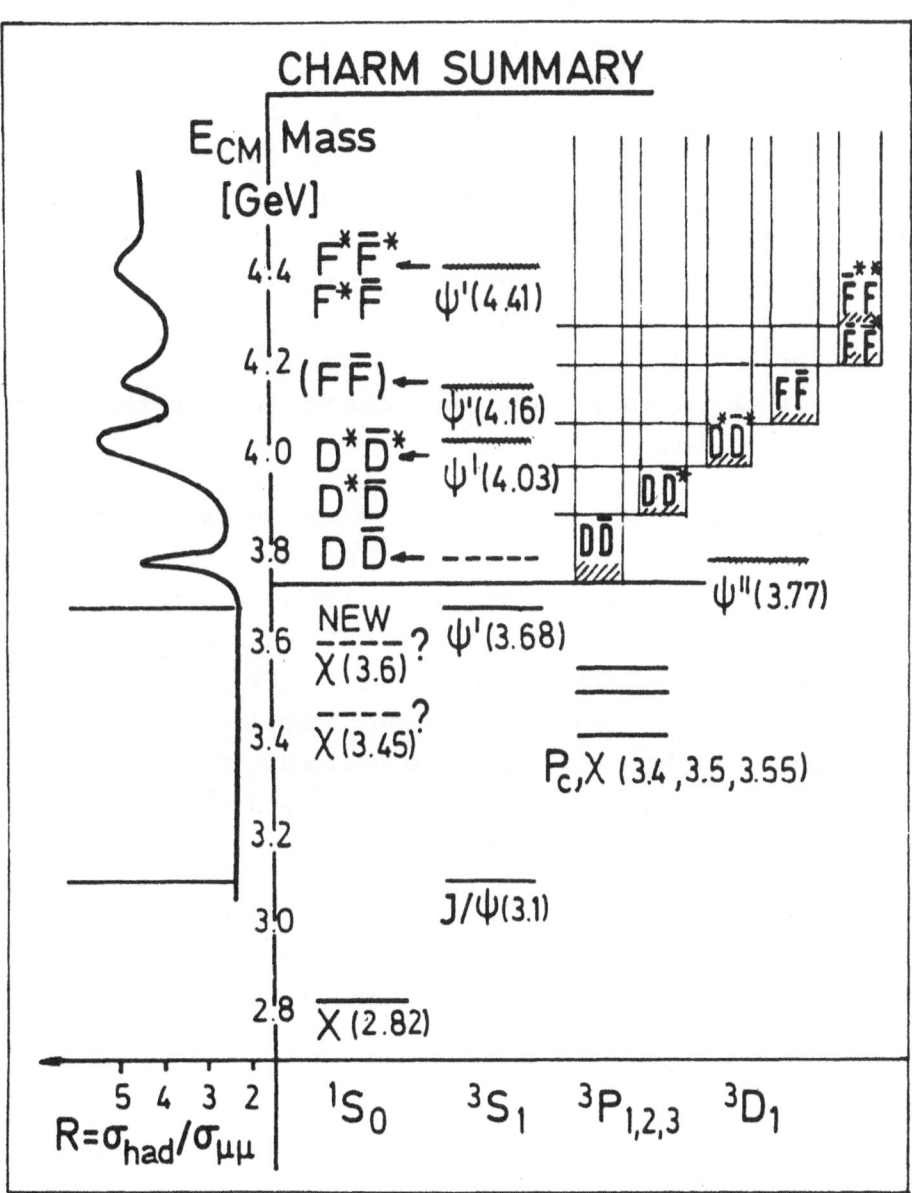

Fig.6. Schmatic summary of the experimental situation on
CHARM.

$\psi'(4.16)$ is somewhat controversial [10]. Quarkonium models would like it to be a ^3D state [17].

The ^3P states are established, although their quantum number assignment is not rigorously proven [15]. Evidence for a X(2.82) state based on a 5 standard deviation signal has been reported from the DASP group. However, preliminary results from the crystal ball experiment do not show any comparable signal [18]. The existence of the states $\chi(3.45)$ and $\chi(3.59$ or 3.18) is not established [19]. The quantum numbers of all three states (if they exist) are unknown, except for their even C-parity.

The upper part of fig. 6 indicates, how the production of D, D^*, F and F^* mesons comes in with increasing energy: $D\bar{D}$ at the $\psi''(3.77)$, $D^+\bar{D}$ and $D^*\bar{D}^*$ at $\psi'(4.03)$ [15,20], FF at $\psi'(4.15)$ and $F^*\bar{F}$ and/or $F^*\bar{F}^*$ at $\psi'(4.42)$ [21]. The evidence for $F\bar{F}$ production at the $\psi'(4.16)$ is suggestive but not compelling, since it is only based on the inclusive η signal of the DASP group. No clear distinction between $F^*\bar{F}^*$ and $F^*\bar{F}$ production at the $\psi'(4.42)$ can be made.

In summary, although many aspects of the charmonium system and the charmed mesons D and F are well understood, the pseudoscalar components η_c and $\eta_{c'}$ are still not established.

II. THIRD GENERATION OF QUARKS AND LEPTONS

Since the discovery of a new lepton τ in 1975 [22] and of a new quark b in 1977 [23] our interest has rather

turned to the study of a third generation of quarks
and leptons.

Generation	1	2	3
Quarks	u d	c s	t b
Leptons	ν_e e^-	ν_μ μ^-	ν_τ τ^-

The rest of this lecture I will, therefore, concentrate
on the achievements which have been obtained in the
study of heavy leptons and Ypsilon particles during
the last year.

1. The heavy lepton τ

The new heavy lepton τ was first seen in the
celebrated eμ events observed at SLAC in 1975. Fig. 7
shows an example of the eμ events seen at PLUTO in 1976
which confirmed [24] the existence of the heavy lepton τ.
The figure clearly shows the apparent nonconservation
of energy and momentum in this event. Both groups, SLAC-
LBL and PLUTO, were able to demonstrate that only
neutrinos could be responsible for the missing energy.
In 1977 it could then be shown that also inclusive μ
two prong events could only be explained by the
production of a new type of heavy lepton in e^+e^-
collision [24,25,26].

a) τ mass

Until 1977 the mass of the τ was rather unprecisely

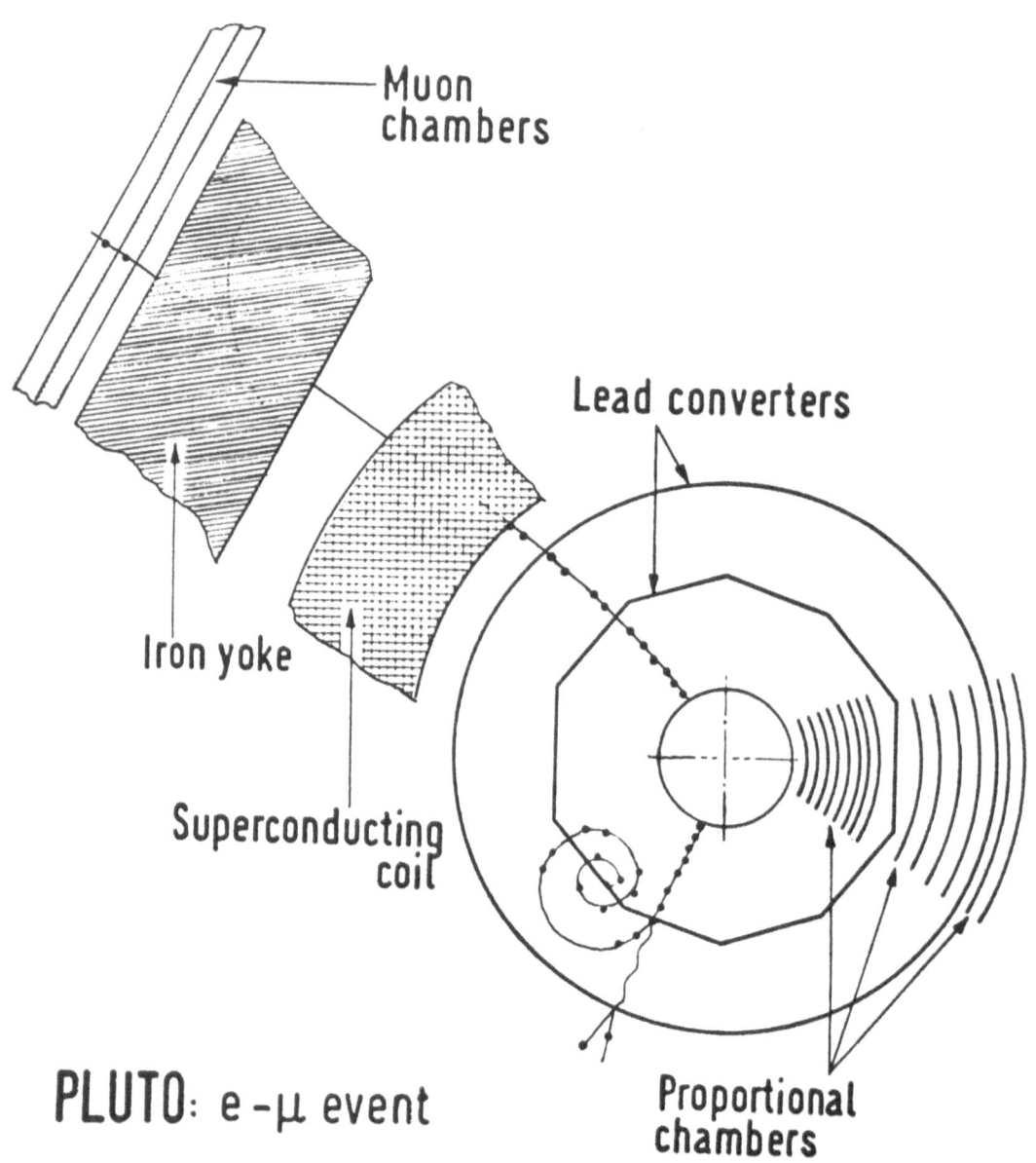

Muon
chambers

Lead converters

Iron yoke

Superconducting
coil

PLUTO: e -μ event

Proportional
chambers

Fig. 7. eμ event measured at PLUTO.

determined [27]. Mainly for this reason a certain
scepticism remained that the τ might be confused with a
charm particle. A major break-through in this issue came
with the discovery of the DASP group that τ production
was already present at the ψ' resonance [28] (fig. 8a).
From the inclusive electron production of fig. 8a a mass
of $M_\tau = 1.807 \pm .02$ GeV could be determined by the DASP
group. The DESY-Heidelberg group (fig. 8b) followed very
quickly with an even better determination of the mass
[29] $M_\tau = 1.787 \, ^{+.010}_{-.018}$ GeV. Both values were finally
topped by the excellent measurement of the DELCO group
[30] at SPEAR which is shown in fig. 8c. This measurement
of the inclusive electron production in two-prong events
sets a mass value of $1.782 \, ^{+.003}_{-.004}$ GeV to be compared with
the $D^\pm$ meson mass of $M_D = 1.868 \pm .001$ GeV.

The very precise determination of the τ mass
definitely excludes charm as a source for these
particles. In addition all three measurements of fig. 8
show very clearly that only a spin 1/2 hypothesis
describes the energy dependence of the measured cross
sections.

b) τ decays

According to the heavy sequential lepton hypothesis
this new particle τ has a conventional weak coupling
to its own massless neutrino. In this model we can
readily predict its decay branching ratios (fig.9).
In 1977 the experimental situation was such that the
leptonic decay branching ratios were in fair agreement
with the theoretical predictions. The same was true for
most of the semihadronic decays of the τ. The only

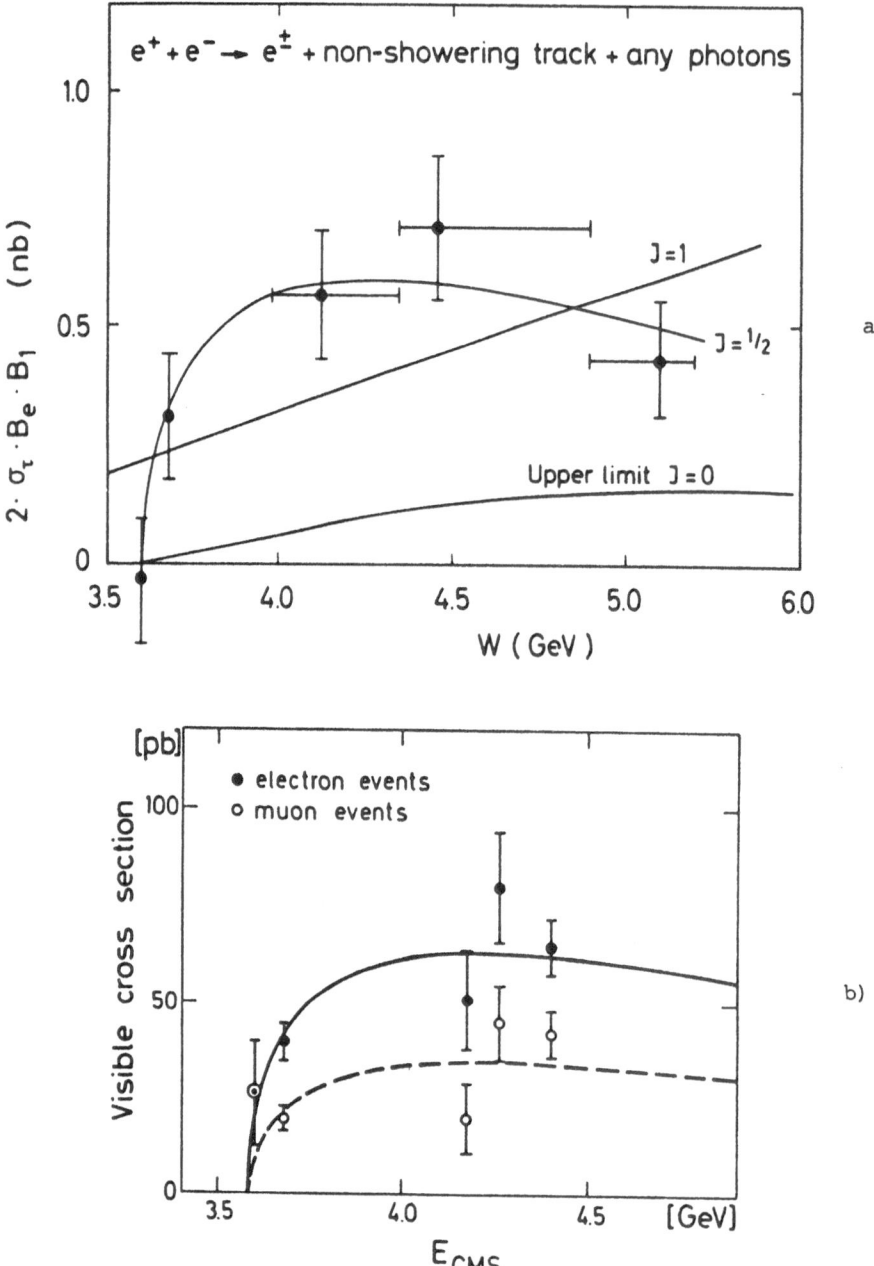

Fig.8. Inclusive lepton production near the τ threshold
a) DASP results on $e^+e^- \rightarrow e^{\pm}$ + non-showering
  track + any photons
b) DESY-Heidelberg data on $e^+e^- \rightarrow e^{\pm}(\mu^{\pm})$ + non-
  showering track + no photons.

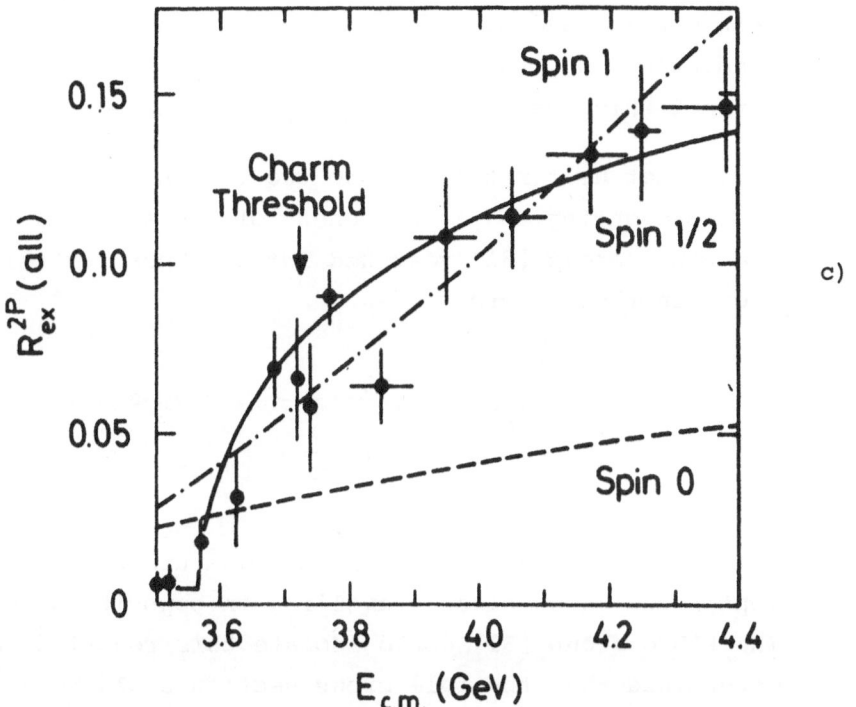

Fig. 8. Inclusive lepton production near the τ threshold
c) DELCO result on $e^+e^- \to e^\pm$ + non-showering
track.

exception was the π decay which was found to be too low (S. Yamada in ref.[19]).

This situation was extremely puzzling since the "A_1" decay which is due to the axial component of the weak current had been found in good agreement with the prediction. Why should now the divergence of the axial current which causes the π decay be absent?

Let us first look into the $\tau \rightarrow$ "A_1" ν decay which is representing the 1^+ component of the axial current. The PLUTO group [31] searched for evidence for this decay in the process

$$e^+ e^- \rightarrow \tau \tau \begin{array}{l} \longrightarrow \mu\,\nu\,\nu,\ e\,\nu\,\nu \\ \\ \longrightarrow \pi^+\ \pi^-\ \pi^{\pm}\ \nu \end{array}$$

This reaction gives a very nice signature with one lepton and three pions travelling in opposite directions. The PLUTO group [31] could isolate this reaction and determined that the full cross section could be accounted for by the decay $\tau \rightarrow \rho^o \pi \nu$. The branching ratio for this decay mode was found to be

$$BR(\tau \rightarrow \rho\pi\nu) = (10.4 \pm 2.4)\% \qquad (I(\rho\pi) \overset{!}{=} 1)\ .$$

The existence of a $\rho\pi$ final state with negative G-parity in itself proves that an axial piece is present in the hadronic weak current in τ decays. (Provided only first class currents are present; by definition of first class currents.) To get a statement independent of the latter assumption, the spin parity of the $\rho\pi$ system was studied. The density distribution in a

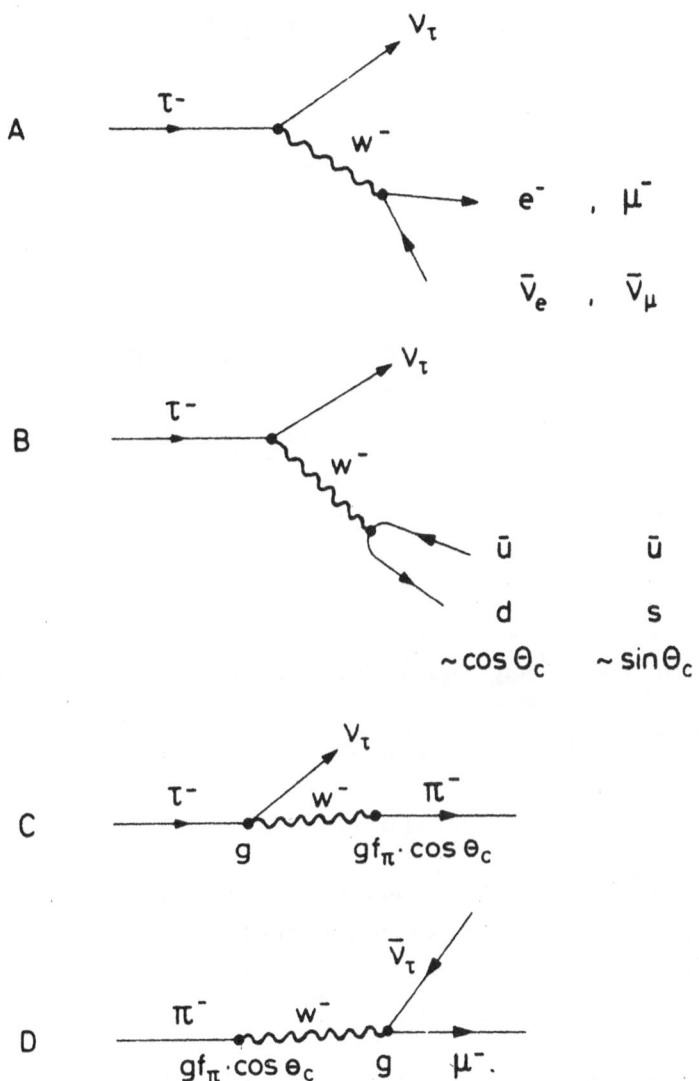

Fig.9. Decays of the heavy sequential lepton τ.

3-dimensional Dalitz plot of the masses of the two $\pi^+\pi^-$
combinations and the $\rho\pi$ system was investigated. Only
the $J^P = 1^+$ s-wave and the $J^P = 2^-$ p-wave gave an
acceptable description of the data. Fig. 10a shows the
mass distribution of the $\rho\pi$ system together with the
expectation from a Monte-Carlo calculation for different
partial waves. The p- and d-waves give a very bad account
of the data. Only the

$$J^P = 1^+ \text{ s-wave}$$

is acceptable. This proves again the existence of an
axial part in the hadronic current. In particular, there
are no indications for a 1^- s-wave from second class
axial currents.

The $\rho\pi$ mass distribution is much better described
if one assumes a resonance with M = 1 GeV and $\Gamma = 0.475$
GeV in the 1^+ s-wave (Fig. 10b). This indicates that the
observed decay may indeed be due to

$$\tau \rightarrow A_1\nu \rightarrow \rho\pi\nu.$$

The evidence is not compelling, however.

Since the axial vector current is not conserved,
its divergence can also contribute to the hadronic
current. Therefore also $J^P = 0^-$ final states are allowed
and the τ will decay into $\nu\pi$.

This decay plays a central role in the discussion
of the weak current involved in τ decay since it
constitutes the "inversion" of the μ decay. It can,
therefore, unambiguously be predicted from the pion
coupling constant f_π (Fig. 9c,d).

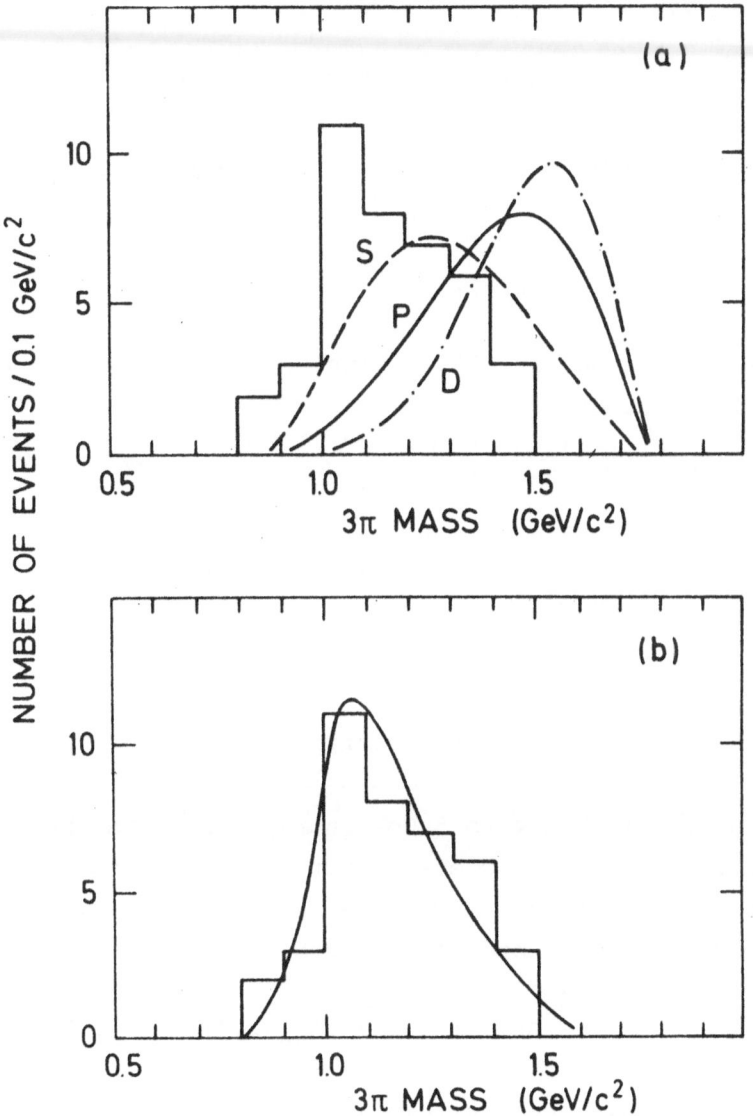

Fig.10. PLUTO: 3π mass distribution of the decay τ→ρπν.
a) Mass distribution of the 3π system in the
   ρ band (0.68≤M_ππ≤0.86 GeV). The curves are
   phase-space calculations for different partial
   waves of the ρπ system.
b) The same distribution. The curve is a resonant
   s-wave with M_{A_1}=1.0 GeV and Γ_{A_1}=475 MeV.

The relative width is given by [32]

$$\Gamma(\tau \to \pi\nu)/\Gamma(\tau \to e\nu\nu) = 12 \ \pi^2 \ f_\pi^2 \ \cos^2\theta/M_\tau^2 \ .$$

With $BR(\tau \to e\nu\nu) = 16.8\%$, $f_\pi = 0.129$ GeV/c^2 and $M_\tau = 1.8$ GeV/c^2 this yields

$$BR(\tau \to \pi\nu) = 9.5 \ \% \ .$$

The PLUTO group studied inclusive pion production [33] from the reaction:

$$\to \pi\nu$$
$$e^+e^- \to \tau \ \tau$$
$$\to 1 \ \text{prong} + \text{no photons}$$

32 events of the signal class

$$e^+e^- \to \text{hadron} + 1 \ \text{charged track} + \text{no photon}$$

were seen in the 4 to 5 GeV energy range. On the other hand, only 8.9 ± 1.0 events were expected from hadron misidentification, $\tau \to \rho\nu$ decay, and hadronic sources. The PLUTO group obtains a branching ratio of

$$BR(\tau \to \pi\nu) = (9.0 \pm 2.9) \ \%$$

with an additional systematic uncertainty of 2.5 %. The DELCO experiment at SLAC achieved [35]

$$BR(\tau \to \pi\nu) = (8.0 \pm 3.2)\% \ (1.3 \ \% \ \text{additional systematic error})$$

from a study of the reaction $e^+e^- \to e^\pm + 1$ hadron.

Two other groups at SLAC - SLAC-LBL and MARK-II - obtained very similar results [10]. The present world average of [10]

$$BR(\tau \rightarrow \pi \nu) = (8.3 \pm 1.4) \text{ \%}$$

is in good agreement with the theoretical expectation.

c) Summary

Table 3 gives a summary of the experimental knowledge about τ, which is now clearly established as a new heavy lepton with the mass $M_\tau = 1.782 \, ^{+.03}_{-.04}$ GeV. All properties of this new particle are as expected for a sequential left-handed lepton with conventional weak coupling to its own massless neutrino. It should be noted, however, that the orthoelectron hypothesis (the neutrino being of the ν_e type) as well as pure V or pure A coupling can not firmly be excluded (table 3).

Till now the new lepton τ has remained a domaine of e^+e^- physics. Within three years, most of its properties have been established. It was the particle that destroyed the four lepton - four quark symmetry (which had just been established) and gave a new impetus to the old puzzle of μ-e universality. Today it is the corner stone of a third generation of quarks and leptons.

Parameter	Units	Prediction	Exp. Value	Experiments
Mass	GeV/c^2	—	$1.782 ^{+.003}_{-.004}$	PLUTO,SLAC-LBL, DASP,DESY-Heidelberg, DELCO
Neutrino mass	MeV/c^2	0	< 250 (95% C.L.)	SLAC-LBL, PLUTO, DELCO
Spin		1/2	1/2	PLUTO,DASP,DELCO, DESY-Heidelberg
Lifetime	10^{-13} s	2.8	< 23 (95% C.L.)+	PLUTO,SLAC-LBL, DELCO
Michel parameter ρ		0.75++	$0.72 \pm .15$	DELCO
Leptonic branching ratios				
$B_e: \tau^- \to \nu_\tau\, e^-\, \bar{\nu}_e$	%	16.8 $\Big\}$	16.7 ± 1.0	SLAC-LBL,PLUTO, Lead-Glass-Wall, Ironball,MPPS, DASP,DELCO
$B_\mu: \tau^- \to \nu_\tau\, \mu^-\, \bar{\nu}_\mu$	%	16.4		
B_μ/B_e		.98	$.99 \pm .20$	SLAC-LBL, PLUTO, DASP
Semihadronic BR				
$\tau^- \to \nu_\tau\, \pi^-$	%	9.5	8.3 ± 1.4	PLUTO,SLAC-LBL, DELCO
$\tau^- \to \nu_\tau\, \rho^-$	%	25.3	24 ± 9	DASP
$\tau^- \to \nu_\tau\, A_1^-$	%	8.1	10.4 ± 2.4	PLUTO,SLAC-LBL
$\tau^- \to \nu_\tau + \geq 3$ prongs	%	~26	32 ± 4	PLUTO,DASP,DELCO
$\tau^- \to K^- ...\,/\tau^- \to \pi^- ...$		.05	$.07 \pm .06$	DASP

+In conjunction with upper limits on $\nu_\mu \to \tau$ production this value excludes $\nu_\tau \equiv \nu_\mu$. Similar limits on $\nu_e \to \tau$ do not exist to exclude $\nu_\tau \equiv \nu_e$ (ortho-electron hypothesis) [34].
++V-A prediction. $\rho(V+A)=0$ is excluded, $\rho(V$ or $A)=0.375$ disfavoured by the data[10,35].

Table 3. Summary of τ parameters. World averages or best values are given.

2. The b quark

a) The Ypsilon story

Since the discovery of the Ypsilon meson by the
Columbia-Fermilab-Stony Brook collaboration at FNAL in
1977 [23] (Fig.11) the new particle has been produced in
various hadron experiments [36]. The discoverers themselves improved both the statistics and the resolution
of their experiment [37]. From these data we have firm
evidence for the existence of at least two T states and
some indications of even a third one. The challenge for
e^+e^- physics was, of course, to search for these new
states as narrow resonances in e^+e^- collisions and thereby reveal their potential nature as bound states of new
quarks. Therefore after the announcement of the discovery
in June 1977 the PLUTO collaboration proposed in July 1977
to upgrade DORIS to reach the 10 GeV region.

In the original layout of DORIS the beam energy was
limited to 4.6 GeV from the magnet design. However, the
available high-frequency power would have done only for
4.3 GeV. So first of all the power limitations had to be
overcome.

Due to synchrotron radiation losses

$$U_{syn} = 88 \; \frac{E^4 \; (GeV^4)}{\rho \; (m)}$$

the power consumption

$$W = U_{syn} \cdot e \cdot f \cdot B \cdot (I_+^b + I_-^b) \sim I^b \cdot B$$

114

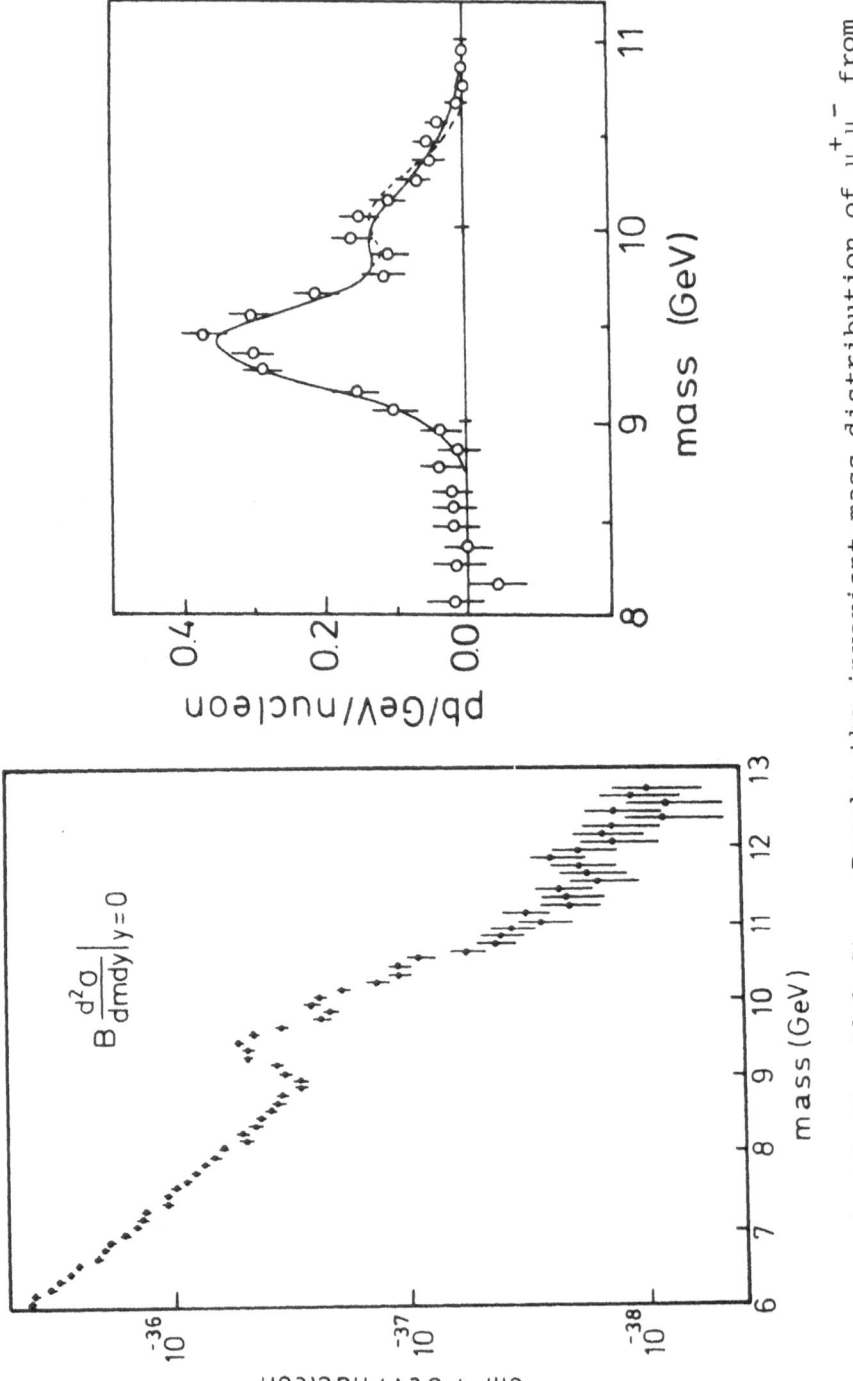

Fig. 11. Columbia-Fermilab-Stony Brook: the invariant mass distribution of $\mu^+\mu^-$ from $p + Be \rightarrow \mu^+\mu^- + X$: evidence for Υ and Υ'.

rises linearly with the bunch current I^b and the number
of bunches B. On the other hand the luminosity

$$L \sim I_+^b \cdot I_-^b \cdot B \sim (I^b)^2 \cdot B$$

increases quadratically with I^b. Therefore, the most
economic way of reaching high energies with sufficient
luminosity is a single bunch machine with large I^b. Once
a single bunch mode is envisaged, separate rings for
electrons and positrons are not needed any more. In
addition, changing to a single ring operation one
profits from a zero crossing angle. At DORIS one wins
a factor of 10 in luminosity as compared to the finite
crossing angle of the two-ring mode. Consequently it
was decided to transform DORIS into a single-ring single-
bunch machine. Apart from the technical effort of this
reconstruction various other problems remained: The
magnets were designed for 4.6 GeV and **had** to be driven
into saturation ($\sim$ 5% at 5 GeV). PETRA cavities had to
be introduced to deliver sufficient high-frequency power.

In parallel, the PLUTO detector was upgraded. Shower
counters were introduced to replace the lead converters.
Fig. 12 shows the new detector. The barrel and endcap
shower counters consist of lead scintillator sandwiches
and cover 94% of 4π. This adds to the charged-track re-
cognition (87% of 4π) in the proportional chambers.

On April 12, 1978, the preparations were finished
to start the search. Already on May 2, 1978, thanks also
to the precise determination of the mass by the Columbia-
Fermilab-Stony Brook collaboration, the T was found at
DORIS by the PLUTO [38] and DASP [39] collaborations

116

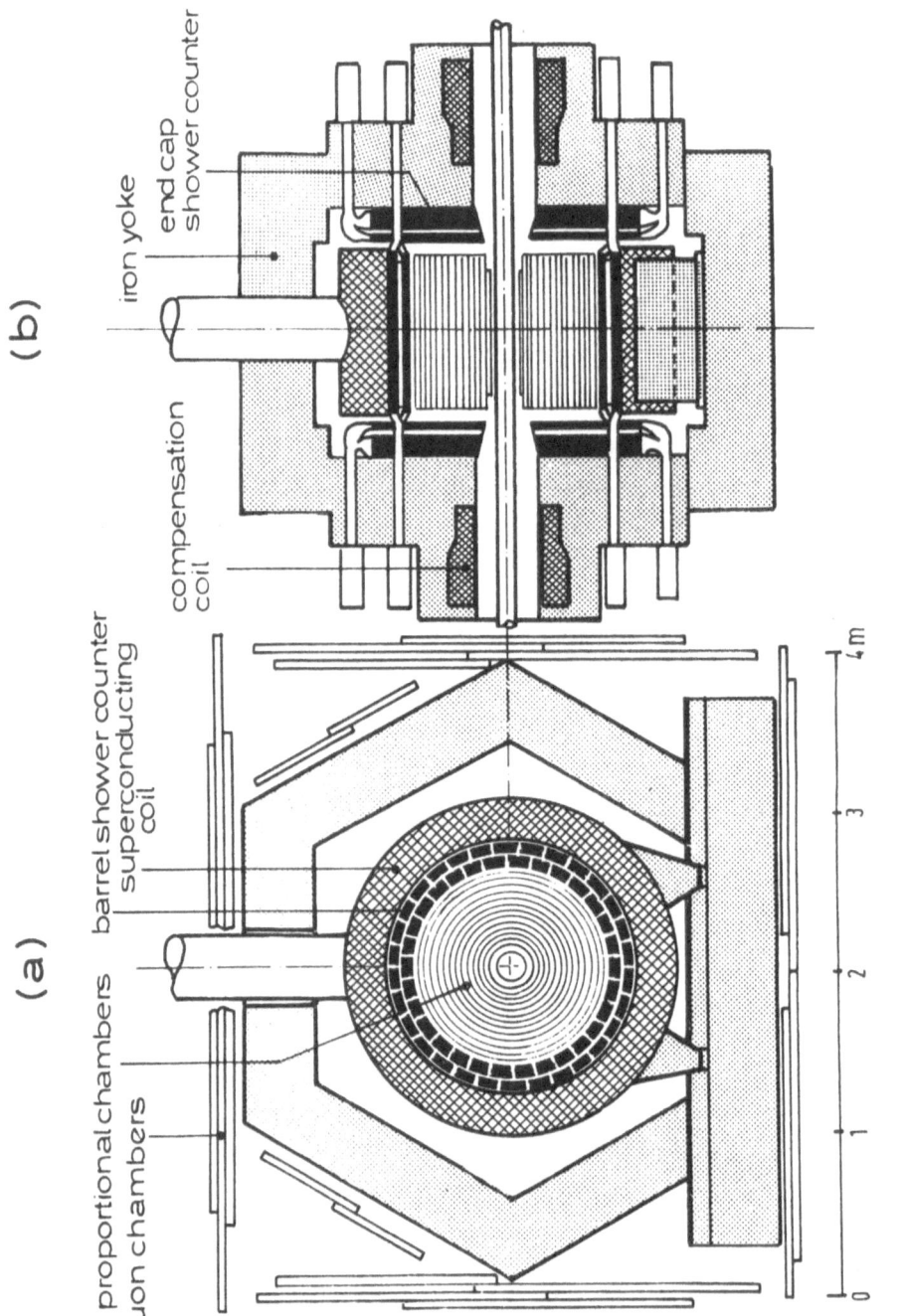

(b)

iron yoke

end cap
shower counter

compensation
coil

(a)

barrel shower counter
superconducting
coil

proportional chambers
muon chambers

0   1   2   3   4 m

Fig. 12. Upgraded PLUTO detector.

simultaneously. The original data of this search are
shown in Fig. 13 which displays the visible cross
section in both detectors as a function of energy.
A clear signal at 9.46 GeV is seen in both experiments.

From these original data both groups agreed on a
mass value of M_T = 9.46 $\pm$ .01 GeV, an electronic width
of Γ_{ee} = 1.3 $\pm$ .4 keV and a total width of the resonance
Γ_{tot} < 18 MeV. Note that the error on the mass is due to
the DORIS calibration uncertainty and the width corresponds
to the DORIS energy spread. These values already strongly
favoured an interpretation of the T being a bound state
of a new quark-antiquark pair with a charge of 1/3 [38].

b) Determination of parameters

The immediate issue of e^+e^- physics of the T is,
of course, a determination of the leptonic and the total
width of the resonance. The leptonic width Γ_{ee} can be
inferred directly by integrating the hadronic cross
section of the resonance according to the formula

$$\frac{M^2}{6\pi^2} \int \sigma_{had} \, dE = \frac{\Gamma_{ee} \Gamma_{had}}{\Gamma_{tot}} \approx \Gamma_{ee}$$

assuming Γ_{ee}/Γ_{had} << 1. The integral extends to infinitely high energies which in practice means that
radiative corrections have to be applied properly.
The absolutely normalized results of the PLUTO group
[40] are shown in Fig. 14. The results of two other
experiments, the DASP2 group [42] and the NaJ-lead glass
detector [41], which replaced the PLUTO detector after

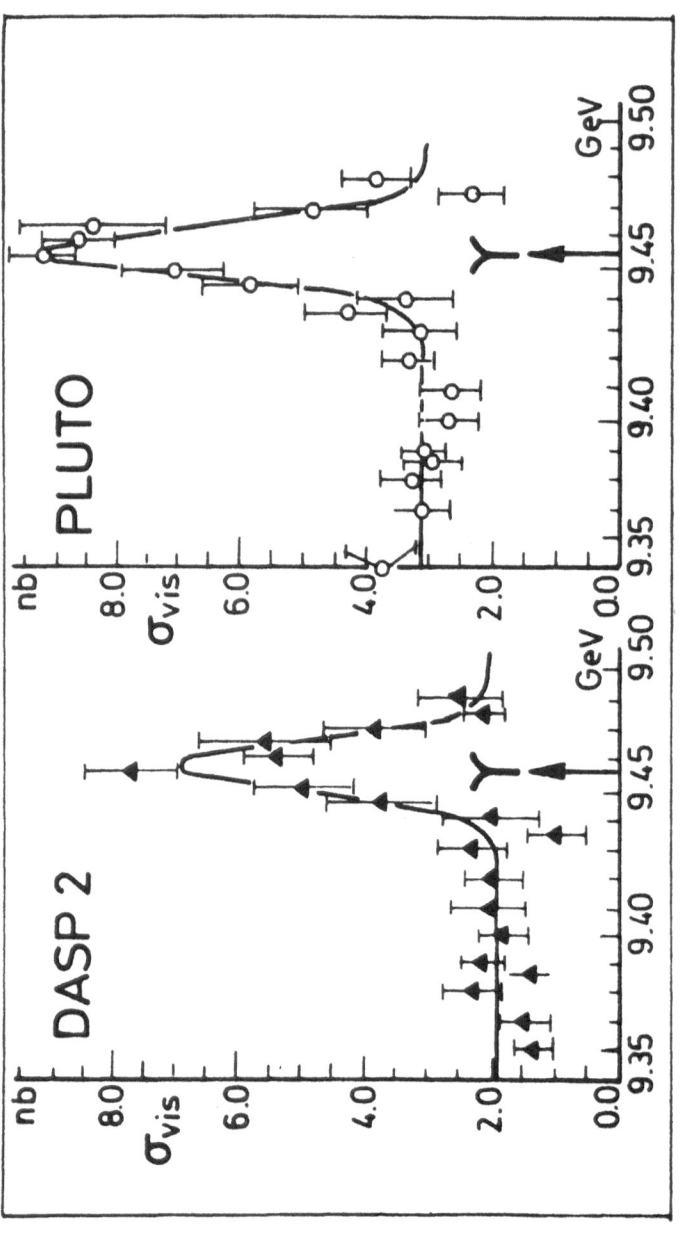

Fig. 13. PLUTO and DASP 2: the original evidence for τ production in e^+e^- annihilation.

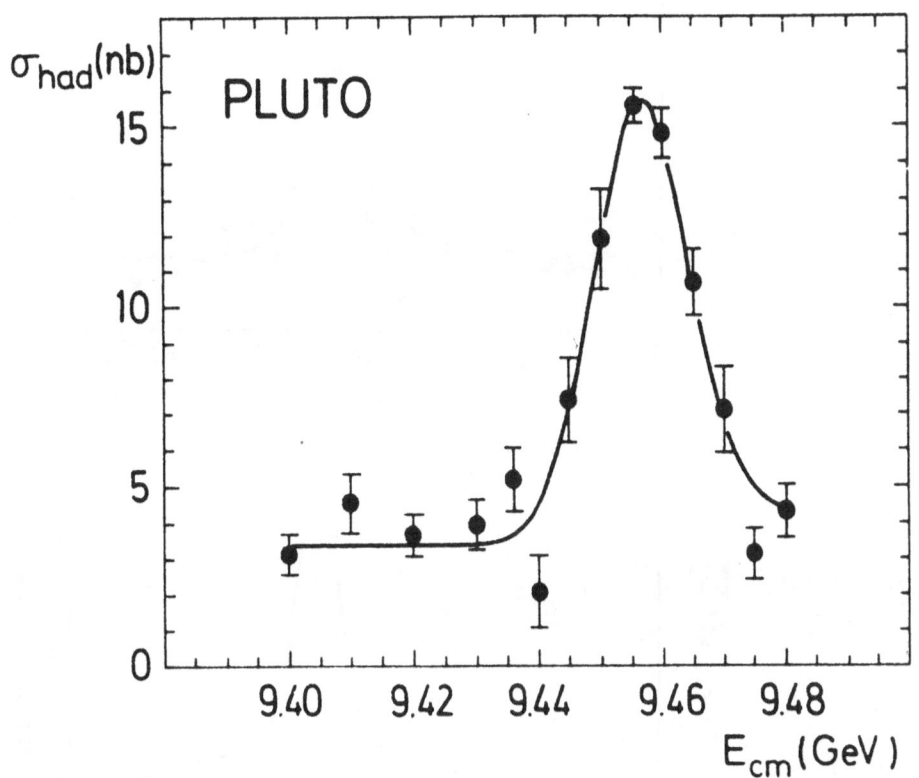

Fig. 14. PLUTO: Absolutely normalized hadronic cross section in the Ypsilon region. The curve is a Breit-Wigner fit including radiative corrections. Γ_{ee}, Γ_{tot} and the machine resolution are free parameters.

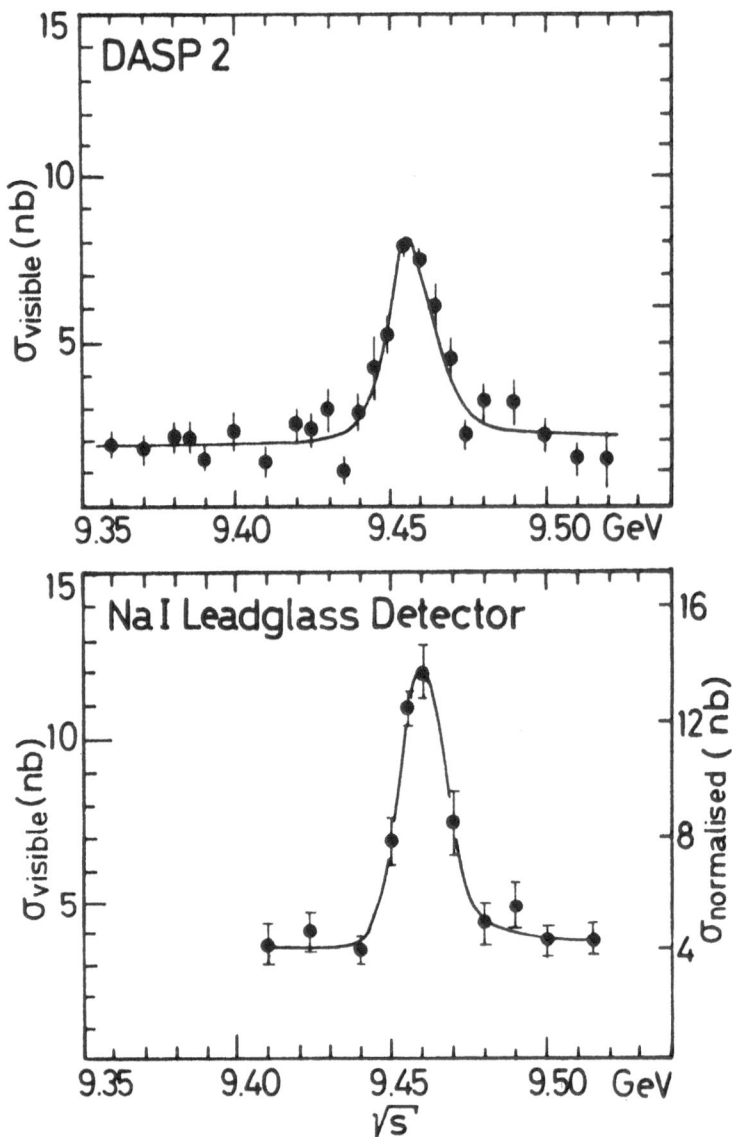

Fig. 15. DASP 2 and DESY-Hamburg-Heidelberg-München
collaboration: visible cross section for
$e^+e^- \rightarrow$ hadrons in the Ypsilon region.

its move to PETRA, are shown in Fig. 15.

(The latter detector was operated by a DESY-Hamburg-Heidelberg-München collaboration.) Their values are not absolutely normalized. For a determination of the leptonic width Γ_{ee} both detectors used the PLUTO value of R at 5 GeV. The results of the three experiments are summarized in table 4.

	M(T) (GeV)	Exp.Width (MeV)	Γ_{ee}(T) (keV)	$B_{\mu\mu}$ (%)	Γ_{tot} (keV)
PLUTO	9.46±0.01	17.2±0.2	1.33±0.14	2.2±2.0	>23 (2s.d.)
DASP2	9.46±0.01	18 ±2	1.5 ±0.4	2.5±2.1	>20 (2s.d.)
D-H II	9.46±0.01	17 ±2	1.04±0.28	$1.0^{+3.4}_{-1.0}$	>15 (2s.d.)

Mean Values:
$$\Gamma = (1.32 \pm 0.09)\ \text{keV}$$
$$B_{\mu\mu} = (2.3 \pm 1.4)\ \%$$
$$\Gamma_{tot} > 25\ \text{keV}\ (95\%\ \text{c.l.}).$$

Table 4. Results on T (9.46).

An attempt was made by the three groups [40-42] to determine the total width of the resonance. To this end one has to measure the μ-pair branching ratio $B_{\mu\mu}$ on the resonance. Assuming μe universality, the total width can then be obtained from $\Gamma_{tot} = \Gamma_{ee}/B_{\mu\mu}$. In all three experiments the determination of $B_{\mu\mu}$ suffers from very low statistics. The angular distribution of the PLUTO events [40] is shown in Fig.16. The data are in good agreement with the expectation of $1 + \cos^2\theta$. The values

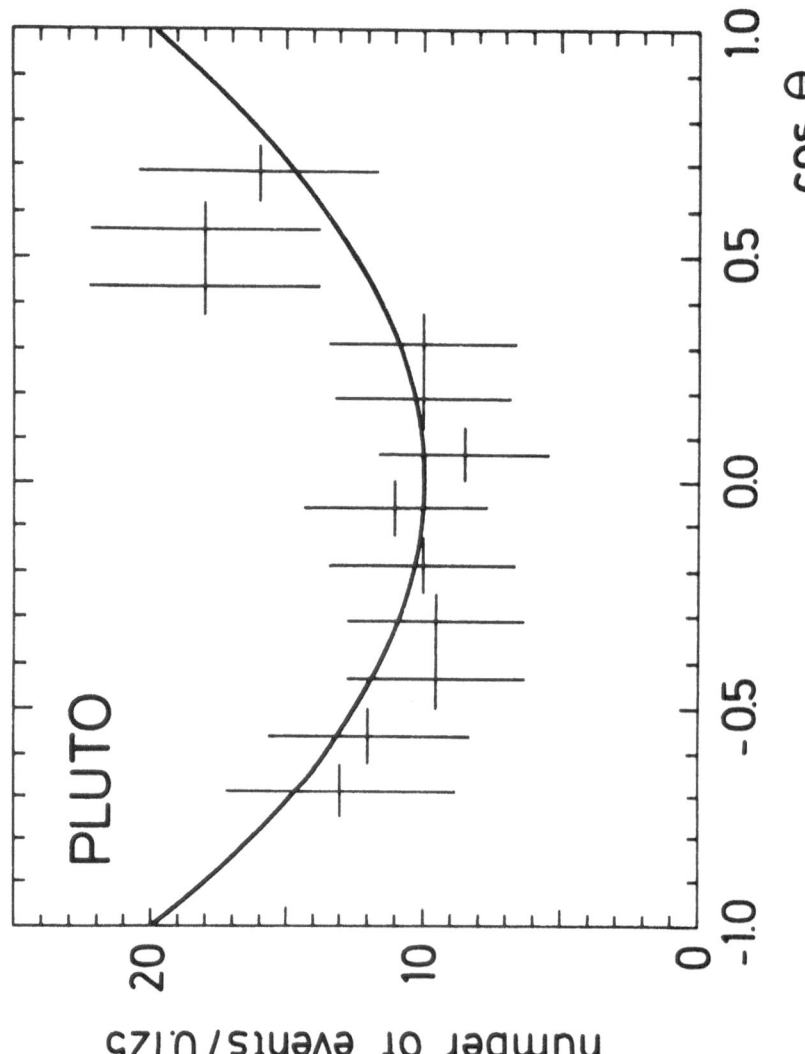

Fig. 16. PLUTO: angular distribution for muon pairs produced in the Ypsilon region. Data on and off resonance are combined. The curve is proportional to $1+\cos^2\theta$.

of $B_{\mu\mu}$ obtained from the three experiments are summarized
again in Table 4.

Due to the large error on $B_{\mu\mu}$ only lower limits can
be given for the total width of the resonance. Even if
all values are combined the error is still too large to
obtain a two standard deviation upper limit on the total
width. Again one can only obtain a lower limit of 25 keV
on a 95% confidence level. (Note that this limit justifies
our previous assumption $\Gamma_{ee}/\Gamma_{had} \ll 1$.) If we take
$B_{\mu\mu} = 2.3\%$ at face value we find the 'best' total width
of

$$\Gamma_{tot} = 57 \text{ keV} .$$

c) Ypsilon prime .

In August 1978 the DASP2 [42] and the NaJ-lead
glass groups [41] advanced into the region of 10 GeV to
search for the first excitation in the T family (T'').
Fig. 17 shows their results. There is a resonance structure
around 10.02 GeV with a width compatible with the resolution
of the e^+e^- machine DORIS. In table 5 the parameters of the
T' as found by the two groups are compiled together with
the mean values. A surprising feature of these data is
the relatively low mass difference between T and T'. In
particular $\Delta M(T) = 558 \pm 10$ MeV is smaller than $\Delta M(\psi) =$
$= 589 \pm 1$ MeV.

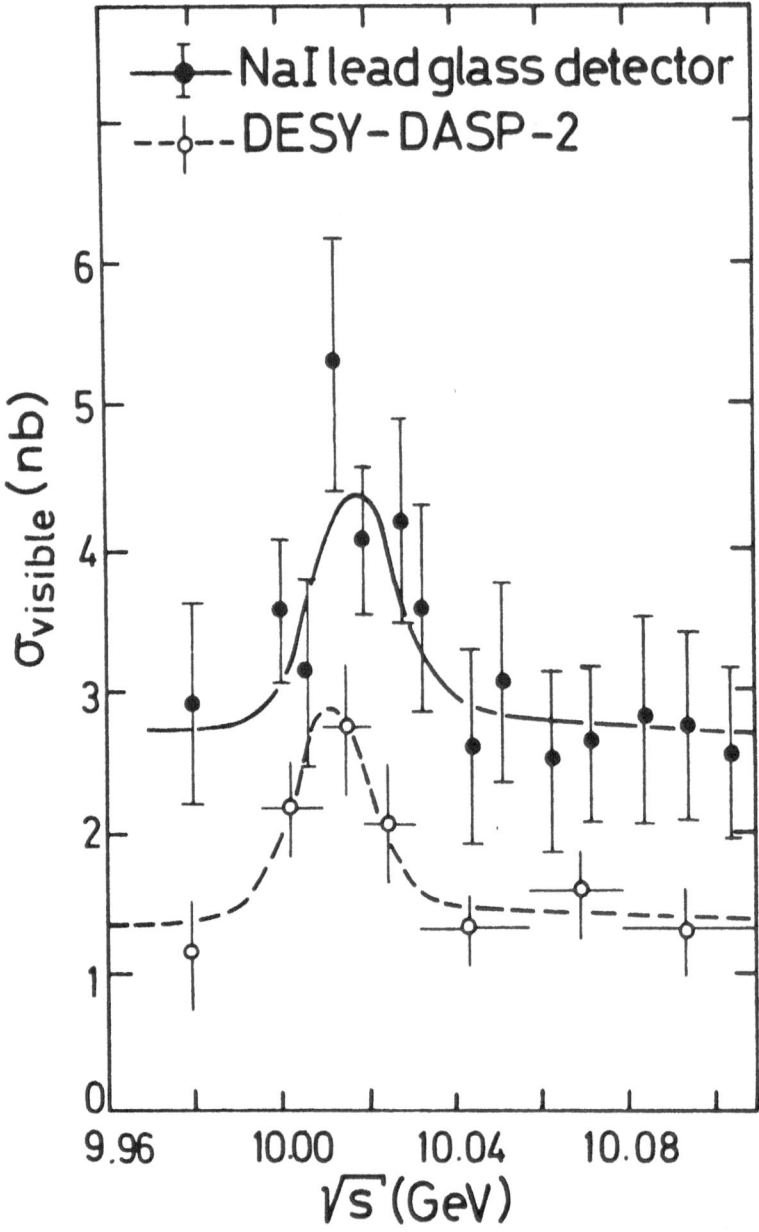

Fig. 17. DASP 2 and DESY-Hamburg-Heidelberg-München
        collaboration. Evidence for the Ypsilon
        Prime in e⁺e⁻ annihilation.

	M(T') (GeV)	M(T')-M(T) (MeV)	Γ_{ee}(T') (keV)	$\dfrac{\Gamma_{ee}(T)}{\Gamma_{ee}(T')}$
DASP2	10.012±0.020	555±11	0.35±0.14	4.3±1.5
D-H II	10.02 ±0.02	560±10	0.32±0.13	3.3±0.9
Mean	10.016±0.020	558±10	0.33±0.10	3.6±0.8

Table 5. Results on T'(10.02).

d) Quark charge and number of T states

What are the conclusions that can be drawn from these results? The leptonic widths of the states are Γ_{ee}(T) = 1.32 ± 0.09 keV and Γ_{ee}(T') = 0.33 ± 0.10 keV. Since $\Gamma_{ee} \sim Q_q^2$ is sensitive to the quark charge, a comparison with model predictions may decide on the type of constituent quark [17]. Fig.18 shows lower bounds on Γ_{ee} for 1/3 and 2/3 charge deduced from rather general assumptions [43]. The measurements indicated in the figure exclude charge 2/3. They coincide nicely with the shaded region indicating the area allowed for Q_b = = -1/3 in a large class of potential models [43].

Duality arguments [44] predict that the reduced leptonic width Γ_{ee}/Q_q^2 should be mass independent. Fig. 19 shows that this quantity is in fact surprisingly constant up to the T mass if we assume Q_b = -1/3. Note the reduction of error bars since 1978.

Fig. 20 shows a comparison of the FNAL and DESY data shown at the Tokyo conference in August 1978. The

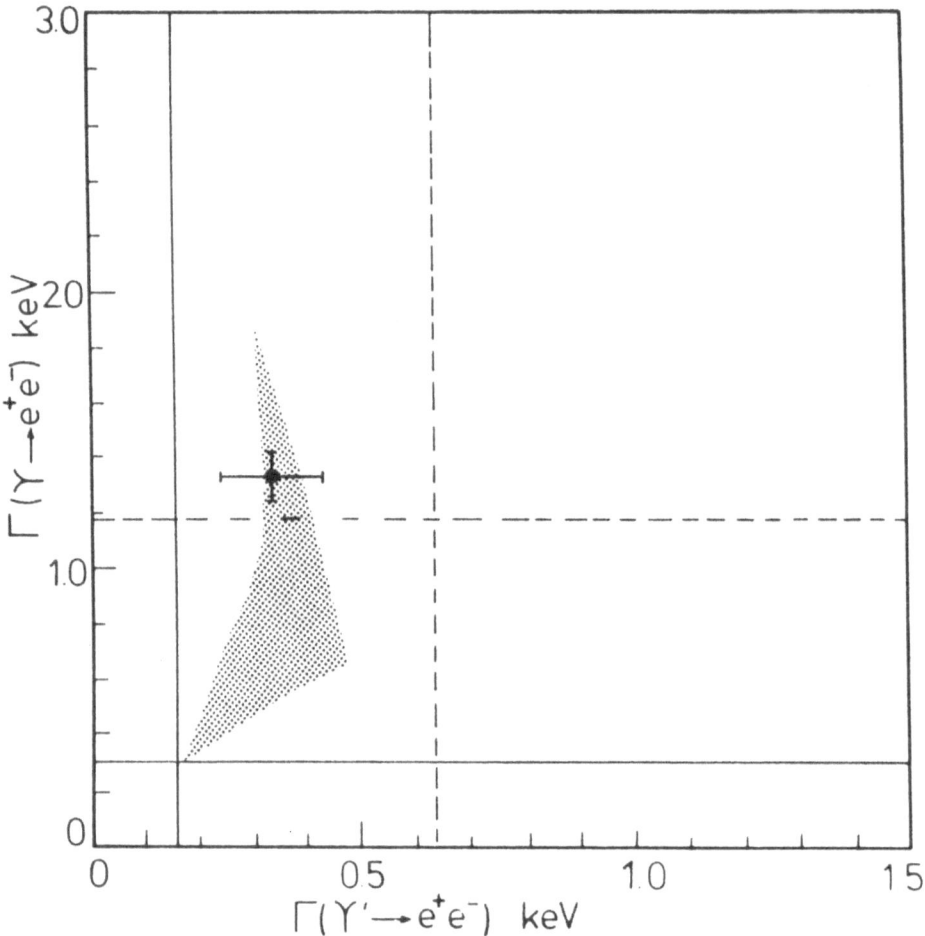

Fig.18. Lower bounds on Γ_{ee} for T and T' for different
quark charges. The shaded area indicates the
charge 1/3 predictions of 20 different potential
models. A comparison with the DORIS data shows
good agreement with charge 1/3 and excludes
charge 2/3.

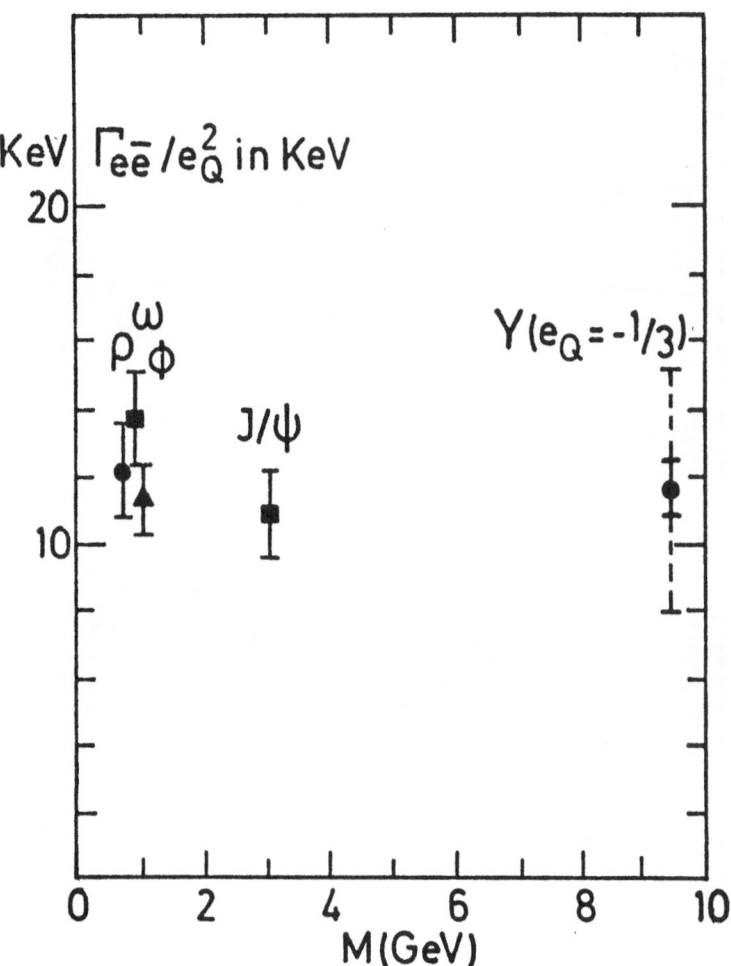

Fig. 19. Reduced leptonic width Γ_{ee}/Q_q^2 for vector meson ground states as a function of mass (dashed error bars: August 1978).

128

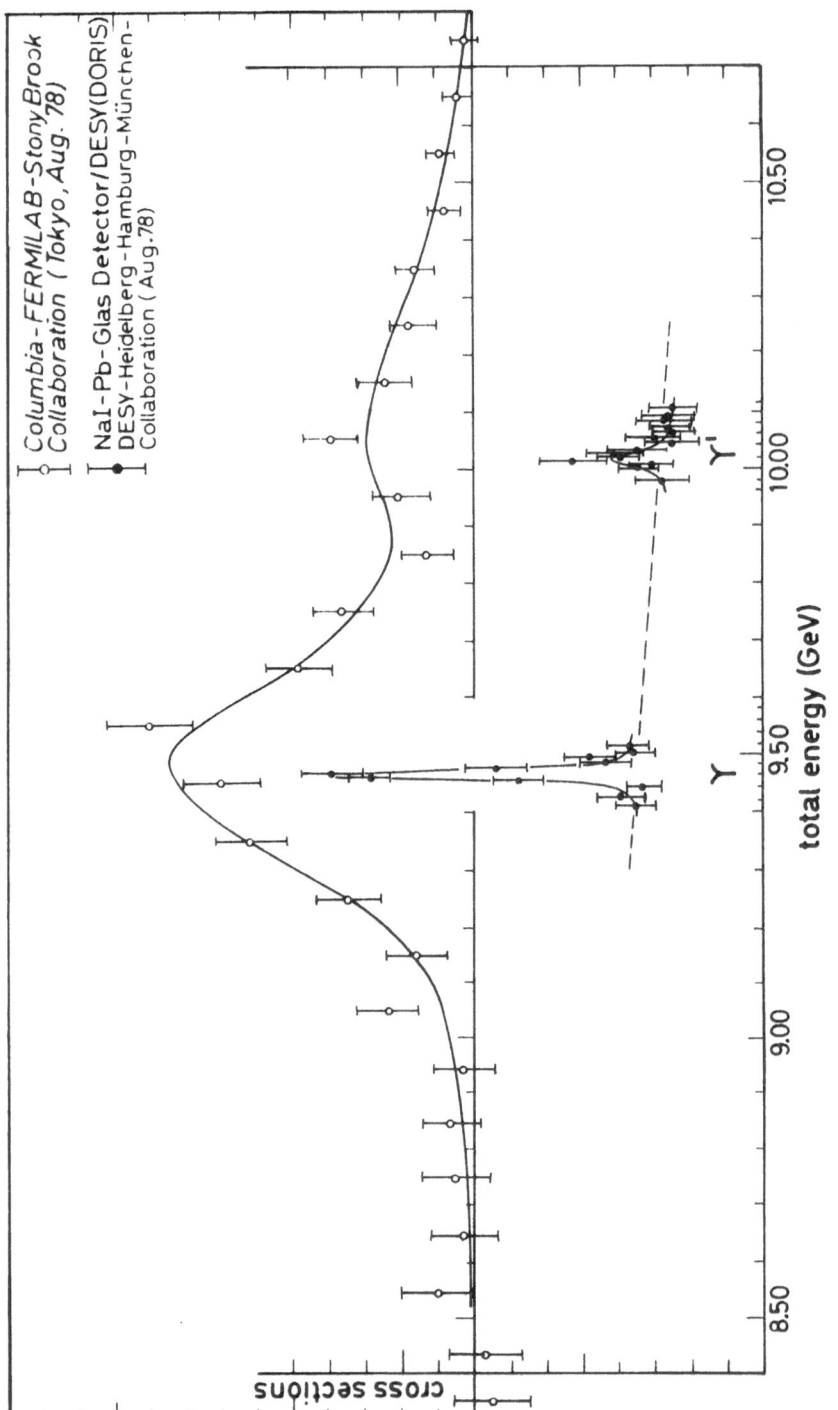

Fig. 20. Columbia-Fermilab-Stony Brook and DESY-Hamburg-Heidelberg-München Collaborations: The Υ family in hadronic and e⁺e⁻ reactions.

mass difference of M = 558 ± 10 MeV is more precise
and lower than the one suggested by the Columbia-Fermilab-Stony Brook collaboration. This group, therefore,
imposed the DORIS value on a fit of their data [37].
Thus they increased the evidence for a third state in
the T family from 4 to 11 standard deviations. They
obtained a mass value of

$$M(T'') = 10.41 \pm 0.05 \text{ GeV}.$$

e) Summary

In summary we have seen that Ypsilon and Ypsilon
Prime are produced as narrow resonances in e^+e^-
collisions. Their masses are $M(T) = 9.46 \pm 0.01$ GeV
and $M(T') = 10.02 \pm 0.01$ GeV. The leptonic widths are
in agreement with a charge 1/3 of the constituent quark
and exclude a charge 2/3. The low mass difference improves
the evidence for a third Ypsilon resonance in the FNAL
data.

Thus there is strong evidence for a b quark, the
counterpart of the τ in the proposed third generation
of quarks and leptons. However, there is still a long
way to go to reveal any detailed properties of this new
quark.

What about the sixth components? The τ neutrino
is not rigorously proven and t is still waiting for
discovery (if it exists).

III. EVENT TOPOLOGY

1. Jets below 9.4 GeV

We have seen in chapter I that the asymptotic behaviour of R is in good agreement with the simple description of the quark-parton model. Let us, therefore, assume that quark pair production really governs the process

$$e^+e^- \to \text{hadrons.}$$

In this picture the two quarks should fragment to form two back-to-back jets of particles (Fig.21).

What are these jets like? In the quark-parton model jets are described in a phenomenological way by a fragmentation of quarks with limited transverse momentum with respect to the original quark axis. On the other hand QCD tells us that jets are produced automatically by gluon bremsstrahlung in the framework of perturbative QCD [45]. Since in this process the transverse momentum increases with energy it will eventually win over the quark-parton process once the energy is high enough [46].

For the poor experimentalist finally a jet is a bunch of particles, which he has to isolate within the limited acceptance and the measurement errors of his detector.

Fig. 22 shows a very clean jetlike event from the PLUTO detector. Two distinct back-to-back bunches of particles are clearly visible. Also the neutral energy of the two jets is clustered and follows the charged

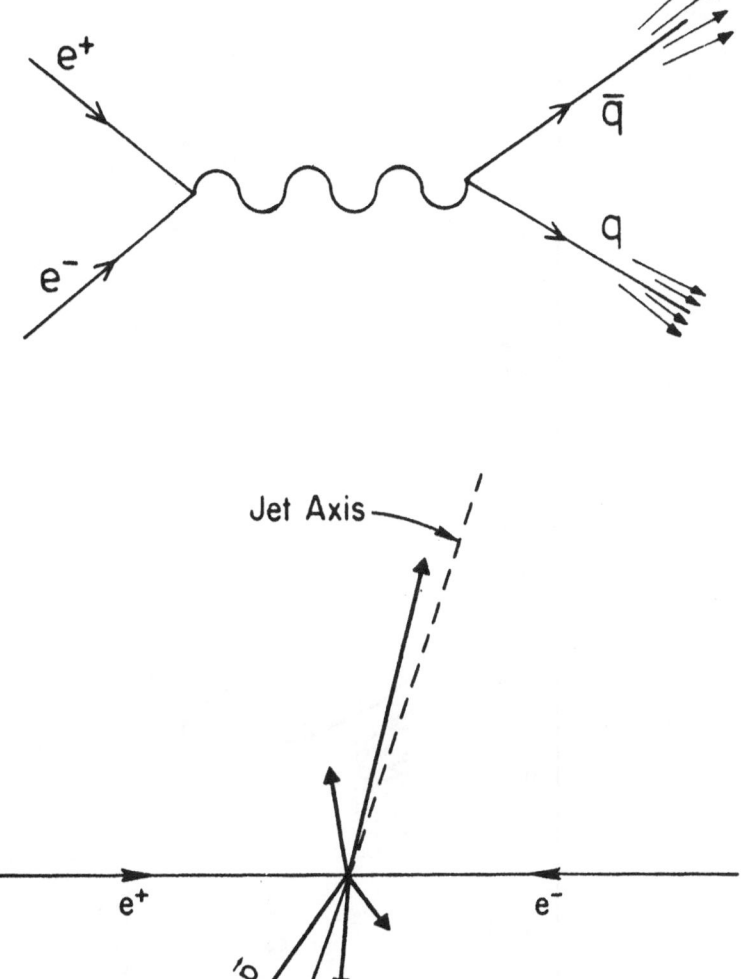

Fig. 21. Definition of quantities used in the jet
analysis.

132

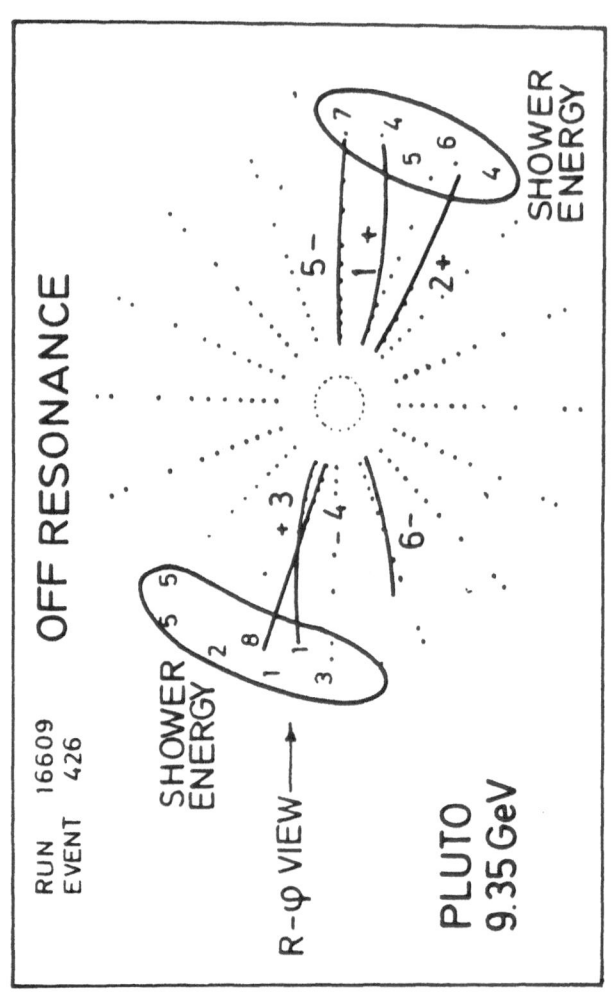

Fig. 22. PLUTO: a non-typical nice jetlike event of E_{CM} = 9.35 GeV.

energy. Unfortunately, even at these high energies of
about 9.4 GeV only few events show a jet structure in
such a nice way. Therefore a more realistic random
selection of events around 9.4 GeV is given in Fig.23.
To reveal any topological structure in these events we
certainly need quantitative measures of event topologies.

a) Quantities to measure jets

Several quantities have been proposed to measure jets.
I will only use two of them here, namely sphericity [47]

$$S = \frac{3}{2} \min \frac{\sum p_{Ti}^2}{\sum_i p_i^2}$$

and thrust [48]

$$T = \max \frac{\sum_i |p_{Li}|}{\sum |p_i|} \quad .$$

(This definition is slightly different from the
original one, where the sum for p_L runs over one
hemisphere only.)

Both quantities simultaneously define the jet axis
and give a measure for the topological structure of the
event. The axis is found in a variational method by
either minimizing the sum of the transverse momentum
squared (p_T^2) or maximizing the sum of the absolute
longitudinal momentum component ($|p_L|$) with respect
to a given axis (Fig.21). Extreme values of the two
quantities for isotropic or ideally jetlike events are
summarized in table 6.

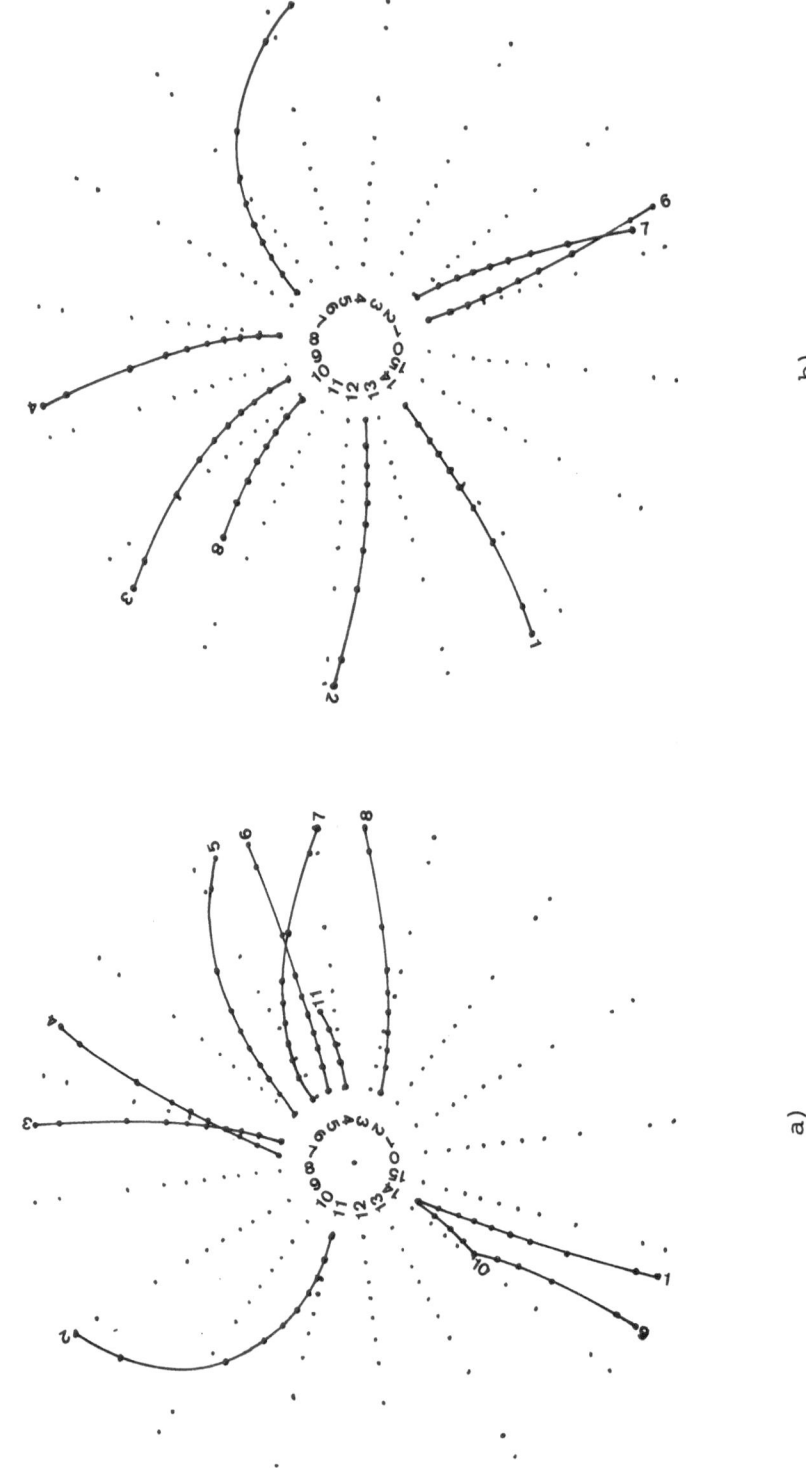

a)

b)

Fig.23. PLUTO: a random selection of events around E_{CM} = 9.4 GeV.

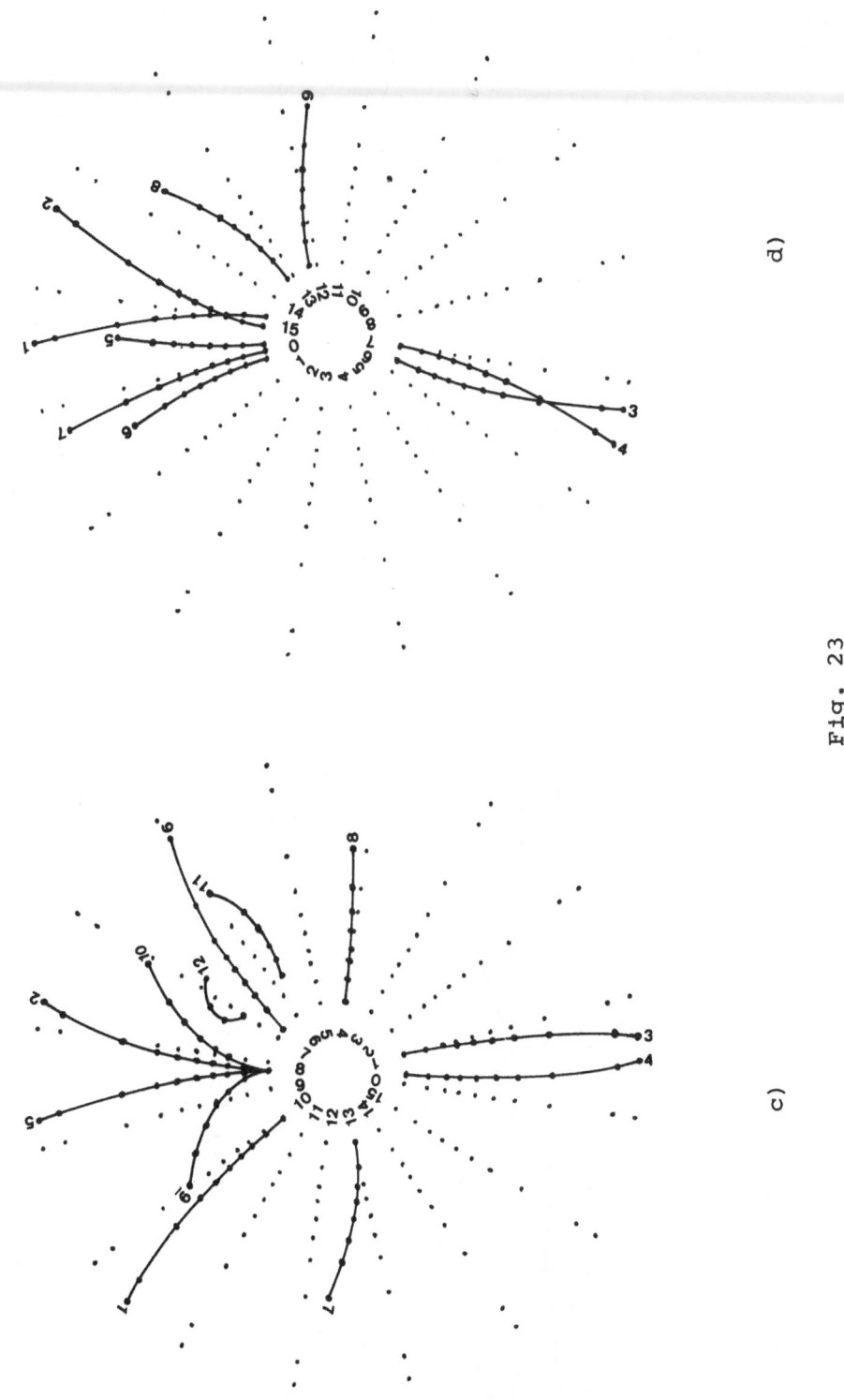

d)

c)

Fig. 23

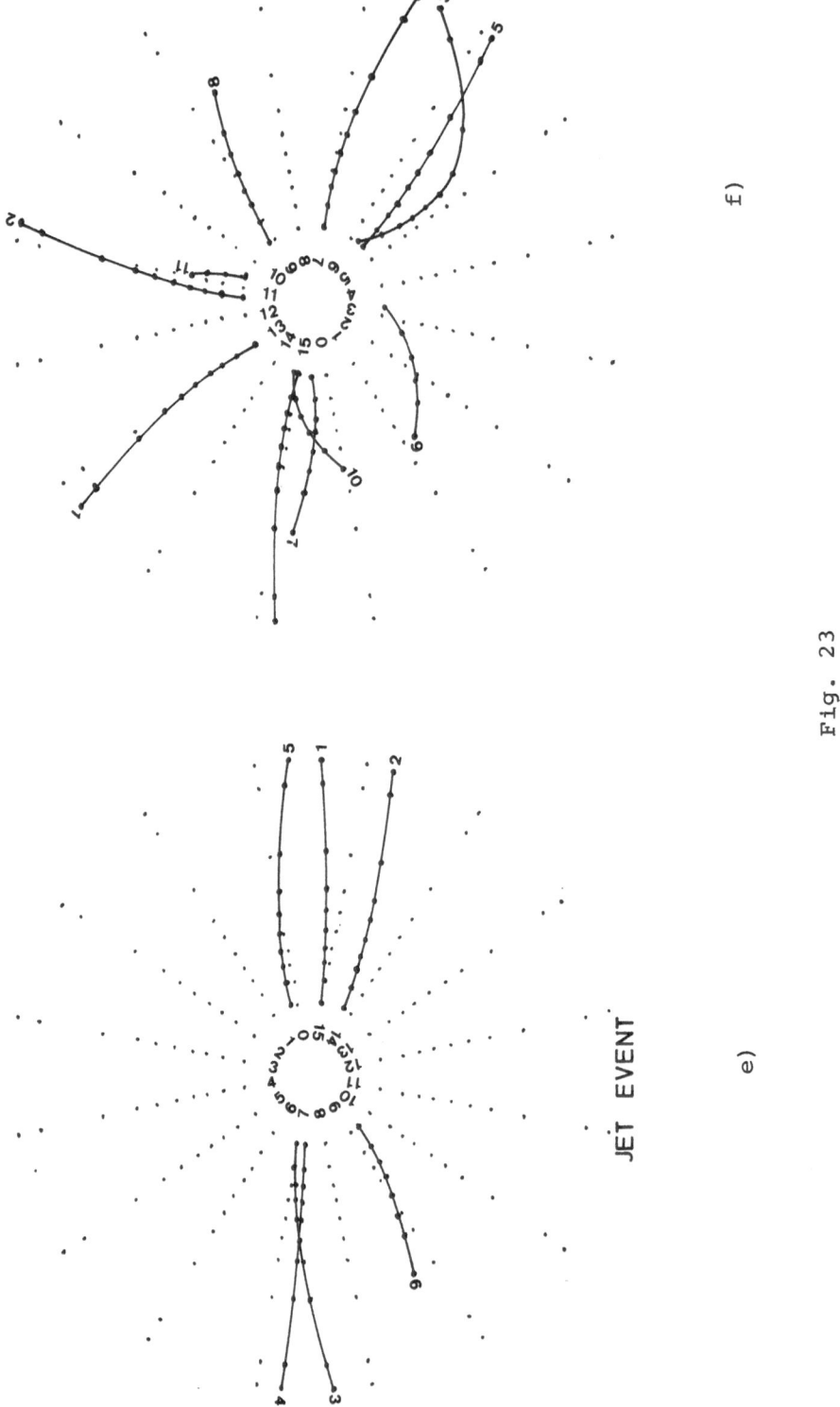

JET EVENT

e)

f)

Fig. 23

Events		S	T
isotropic		→ 1	→ 1/2
ideally jetlike		→ 0	→ 1

Table 6. Values for S and T in extreme topologies.

If we assume that p_T is about constant the quantities
$<S>$ or $<1-T>$ will both fall with increasing energy
(assuming that the multiplicity is only slowly varying).

Sphericity has the virtue that it can be easily
calculated by standard diagonalization methods. It will,
therefore, be used in most of the following analysis.
However, since the momenta enter quadratically, high
momentum particles get a stronger weight in $<S>$. Also
it is not invariant against clustering (multiplicity)
of particles. Therefore, theorists dislike it, because
it depends on details of the fragmentation. They rather
introduced a third jet measure, spherocity [49], which
is linear in the momenta. However, this quantity suffers
from discontinuities and, therefore, cannot easily be
applied to data analysis (Brandt and Dahmen in ref.[48]).

b) SLAC-LBL results

In 1975 the topological structure of charged particles
emerging from e^+e^- reactions was studied at SLAC [50].

This analysis covered a c.m. energy range up to 7.8 GeV.
The results in terms of sphericity are shown in Figs. 24
and 25. We realize a drastic decrease of <S> with increasing c.m. energy between 2.4 and 7.8 GeV. These
features of the data are in good agreement with a quark-
parton model and cannot be described by a simple phase
space distribution (Figs. 24 and 25). The results were
generally taken as first evidence for jets in e^+e^-
reactions. In addition it was shown that the jets are
governed by limited transverse momentum as expected in
a simple quark-parton model. The angular distribution
of the jets was found to follow a $1 + \cos^2\theta$ distribution
as expected for the production of spin 1/2 particles.

c) PLUTO results at DORIS

New results on the subject have become available recently
from the PLUTO collaboration [51]. These data cover an
even larger energy range from 3.6 to 9.4 GeV c.m. energy.

Like in the SLAC-LBL data results are so far mainly
based on an analysis of the charged tracks. Since the
momentum resolution is degrading rapidly in the very
forward direction only tracks with $\cos\theta$ less than 0.85
have been used. Hadronic events were selected according
to their total energy and their charged mutliplicity:

$$E_{tot} = E_{neut} + E_{ch} > 0.6\ E_{CM}$$

$$n_{ch} \geq 4\ .$$

The beam gas background is low. It is estimated from

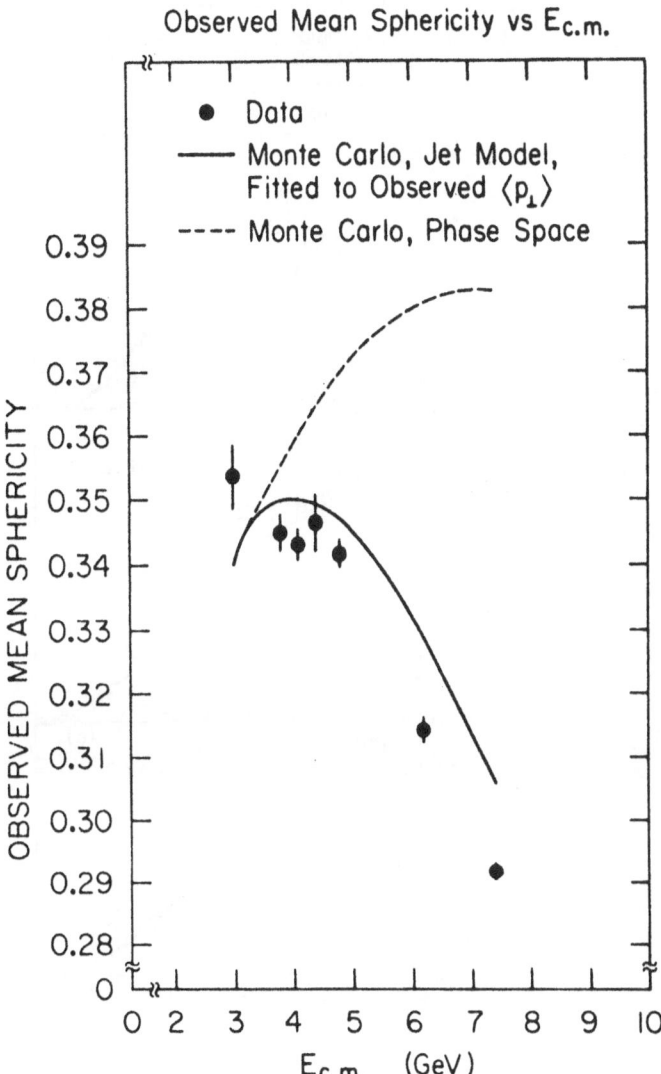

Fig. 24. SLAC–LBL Collaboration: First evidence for jets in e⁺e⁻ annihilation. Observed mean sphericity as a function of energy. The full and dashed line show Monte-Carlo simulations of a jet and phase space model, respectively.

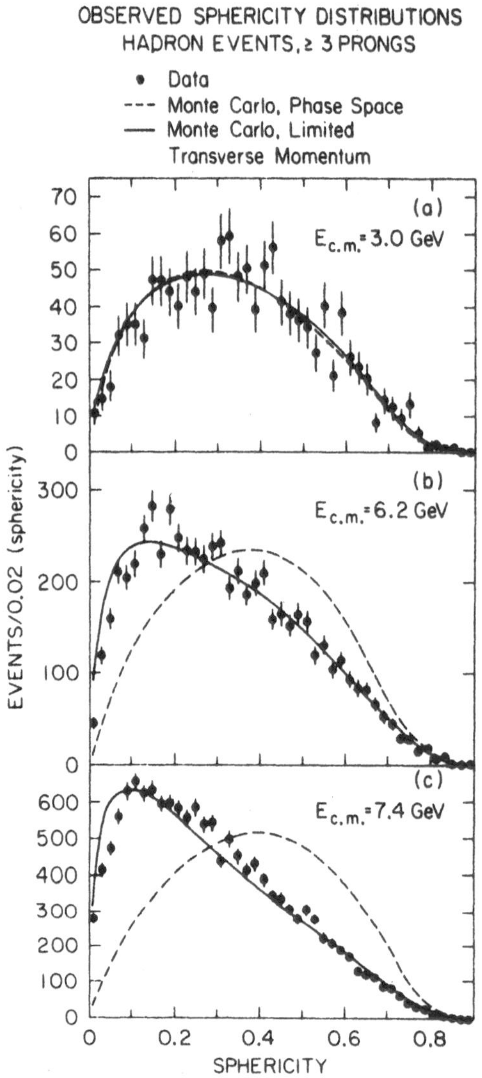
OBSERVED SPHERICITY DISTRIBUTIONS
HADRON EVENTS, ≥ 3 PRONGS
Data
Monte Carlo, Phase Space
Monte Carlo, Limited Transverse Momentum
(a)
E c.m.= 3.0 GeV
(b)
E c.m.= 6.2 GeV
(c)
E c.m.= 7.4 GeV
EVENTS/0.02 (sphericity)
SPHERICITY

Fig. 25. SLAC-LBL: sphericity distributions observed at E_{CM} = 3.0 GeV (a), 6.2 GeV (b), and 7.4 GeV (c). Data are compared with a jet model (solid curve) and a phase-space model (dashed curve).

events outside the interaction point and subtracted. The remaining background due to various QED processes (mainly e^+e^-, $\gamma\gamma$ processes and τ pair production) is estimated to be of the order of less than 10%.

Fig. 26 shows the observed mean sphericity as a function of c.m. energy in the range from 3.6 to 9.4 GeV. <S> decreases drastically with increasing energy. Fig. 27 shows the same trend for <1-T>. Again mean observed values are given. Qualitatively both quantities behave as expected for a jetlike structure of the hadronic final state. To get a quantitative comparison with models the following procedure was applied throughout this section.

- We define a model.
- We generate events according to this model and pass them through a realistic Monte-Carlo simulation of our detector.
- We compare the results of these Monte-Carlo calculations with our observed quantities.

The data were compared with two models.

(1) The phase-space model

The particles were produced according to phase space assuming π production only. The mean charged multiplicity and the neutral energy were adjusted to the experimentally observed values.

(2) Two jet Monte Carlo (Feynman-Field parametrization)

The Feynman-Field model [52] is a parametrization of

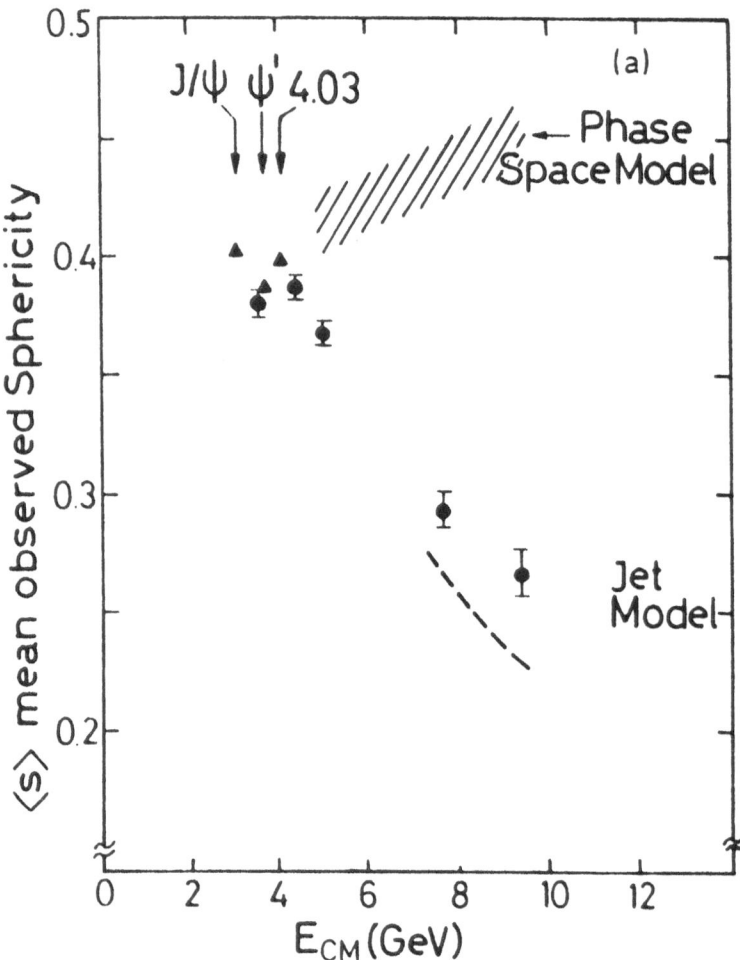

Fig. 26. PLUTO: mean observed sphericity of charged
particles (>4 prongs) as a function of E_{CM}.
The shaded area represents the phase-space
prediction, the dashed curve the two-jet model
without radiative corrections.

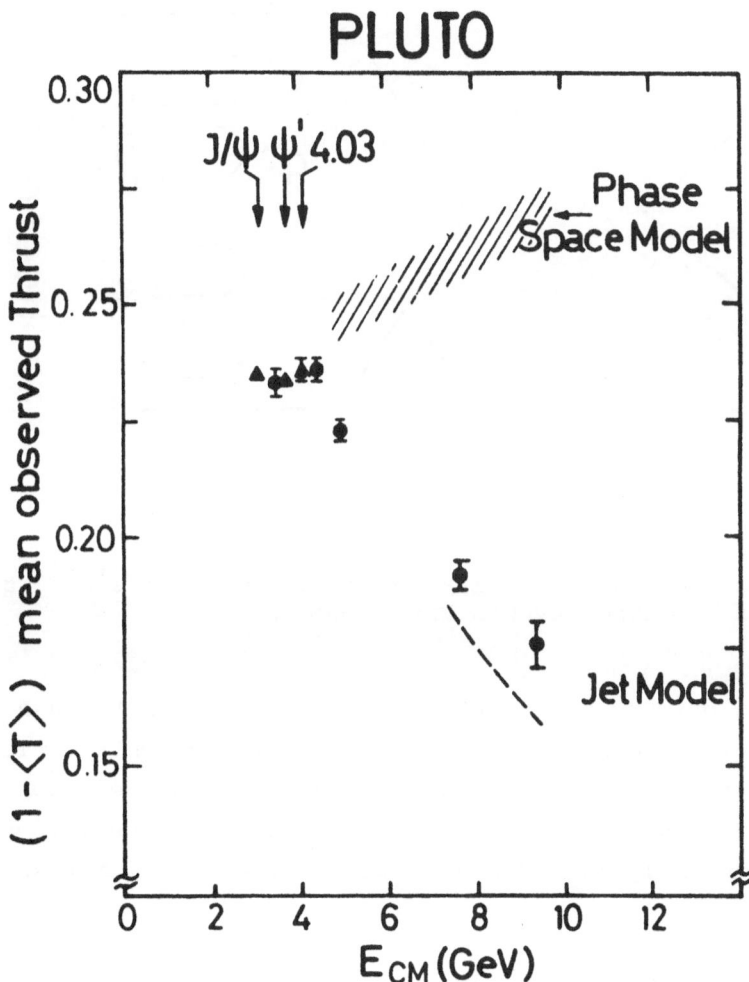

Fig. 27. PLUTO: mean observed thrust as a function of
E_{CM}. <1-T> is plotted for easier comparison
with <S>. Model predictions like in Fig. 26.

quark parton jets. The application of this model to
e^+e^- reactions is sketched in Fig. 28. The dressing of

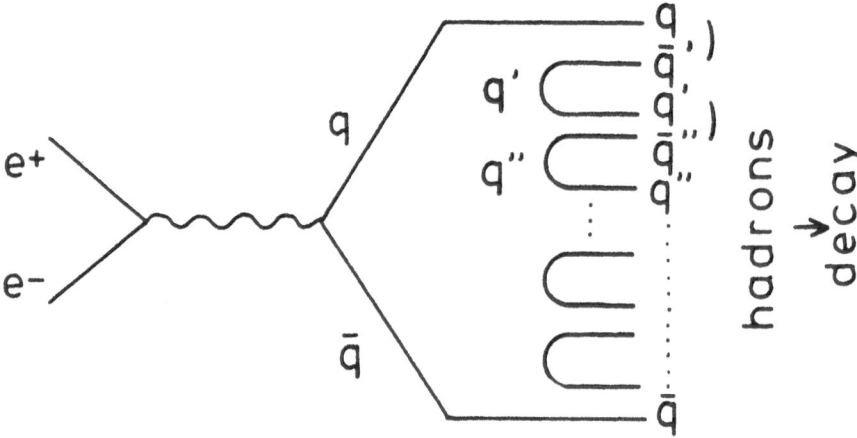

Fig. 28. The fragmentation of quarks from $e^+e^- \rightarrow q\bar{q}$ in
the Feynman-Field model.

the primary quark antiquark pair proceeds through a
chain of other quark-antiquark pairs which are picked
up from the vacuum. Thus mesons are produced down the
chain until all the energy is used up. This process is
completely determined by the following assumptions:

Fragmentation function

In each step a probability $f(\eta)$ is calculated that a
hadron takes $(1-\eta)$ of the available momentum with it.

The production of s$\bar{s}$ pairs is assumed to be half as likely as u$\bar{u}$ production.

The production of vector mesons and pseudoscalar mesons is assumed to be equally probable.

Finally the decay of all primary hadrons is included in the model. The function $f(\eta)$ is chosen such that the fragmentation function $D(z)$ fits deep inelastic lepton and e^+e^- data. ($D(z)$ describes the momentum distribution of particles along the original quark direction.)

Limited transverse momentum

Particles in the chain of Fig. 28 are produced with limited transverse momentum. It is chosen such that the final mean transverse momentum of the decay products is about 330 MeV. This is again adjusted to lepton and e^+e^- data.

With these assumptions all parameters are fixed. The model has been applied without further adjustments. Note that charm production is not included in the present version of the Feynman-Field model.

The results of these two model calculations for sphericity and thrust are given in Figs. 26 and 27, respectively. We observe good agreement between the data and the Feynman-Field calculations whereas a simple phase-space description is completely ruled out. The agreement with Feynman-Field may look surprising since charm is not included in the model. However, adding charm has only little effect on <S>. This can be

qualitatively explained by the fact that far above
threshold all quark jets behave very similarly. I will
come back to this point in the last chapter.

How well is the axis defined?

We have seen that the gross features of our events expressed in terms of sphericity or thrust agree very
well with the features of a simple jet model. Before
we go any further into the description of detailed
features of these data we should get an impression how
well the jet axis is defined. Fig. 29a shows distributions of the relative angles between the axis as defined
by sphericity, thrust or the fastest particle in the jet.
The mean width of these distributions is about 15°. This
gives a rough estimate of the uncertainty with which the
jet axis is defined. Monte-Carlo studies indicate that
the relative angle between the original quark axis and
the axis defined by sphericity or thrust is even smaller
(Fig. 29b).

Energy distribution inside a jet

Our next question is whether the neutral and the charged
energy flow inside a jet are correlated. Fig. 30 shows
the two energy flow distributions with respect to the
jet axis (defined by thrust). The data are divided up in
different thrust bins. We note that qualitatively the
neutral energy distribution follows closely the charged
one. In detail there are, however, differences which are
most pronounced in the high thrust bin. The excess of

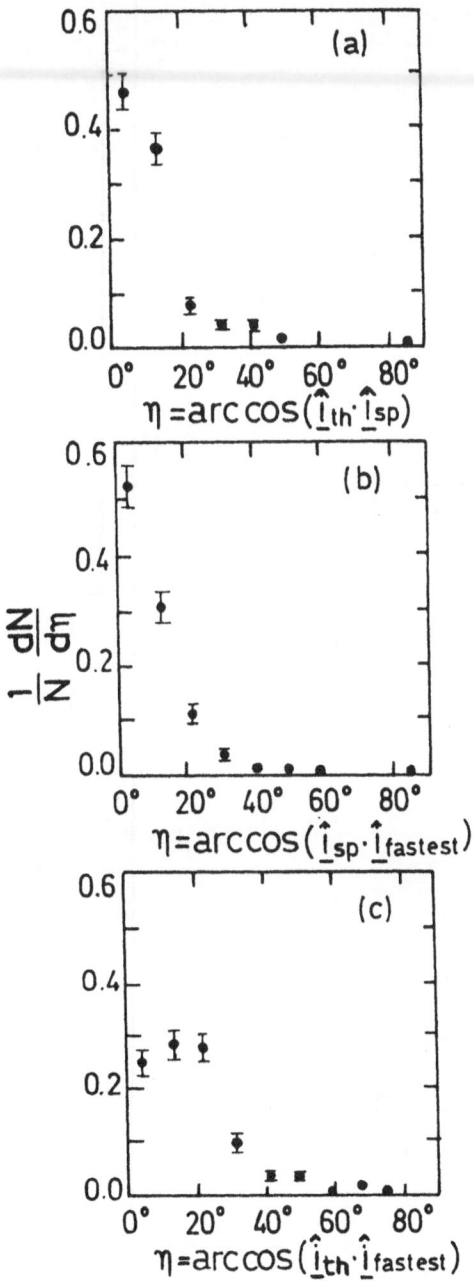

Fig. 29a. PLUTO: the relative angle between the jet axes as defined by thrust or sphericity or the fastest particle in the jet.

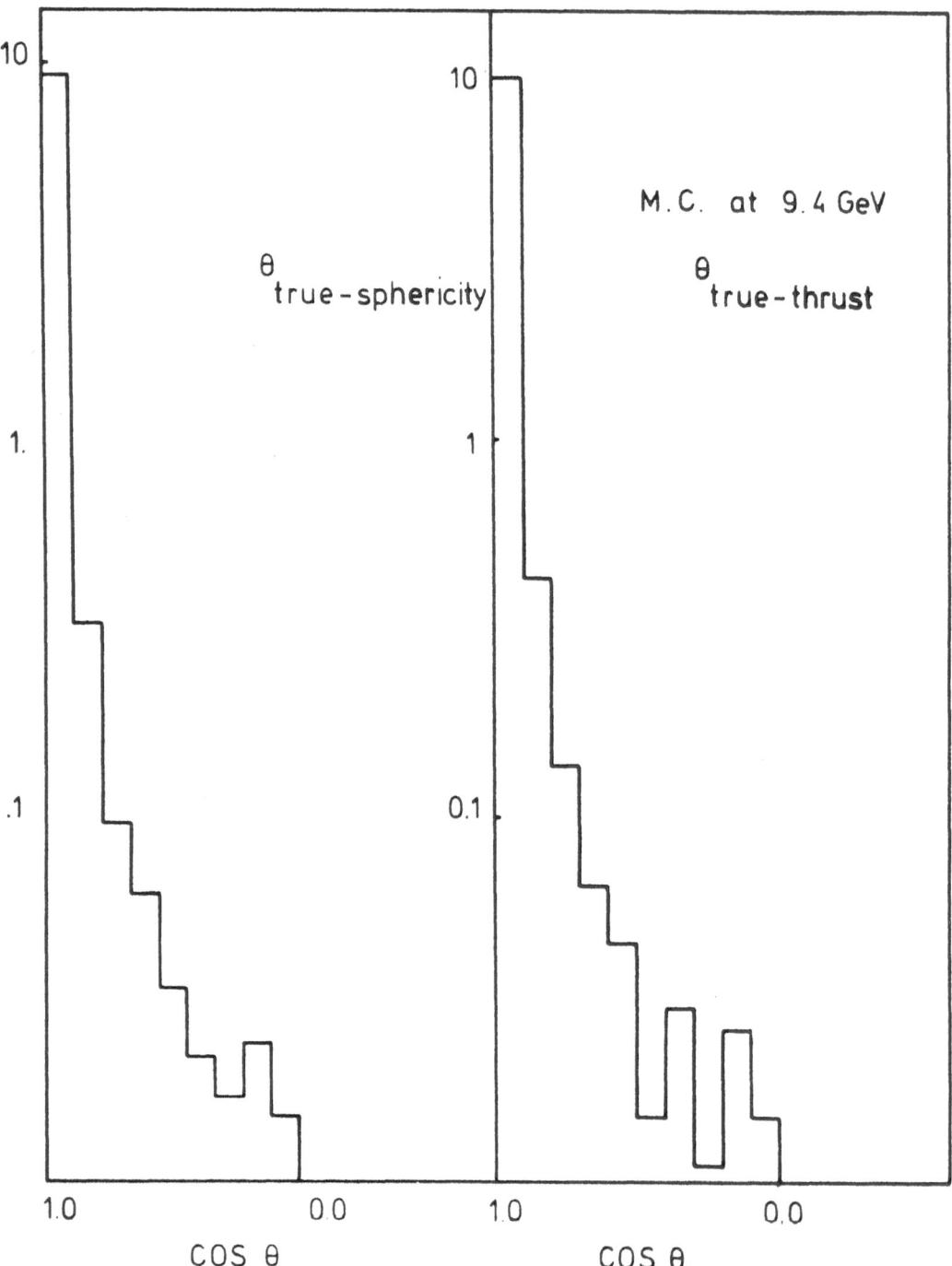

Fig. 29b. Two-jet model calculations: relative angle between the generated quark axis and those defined by thrust and sphericity.

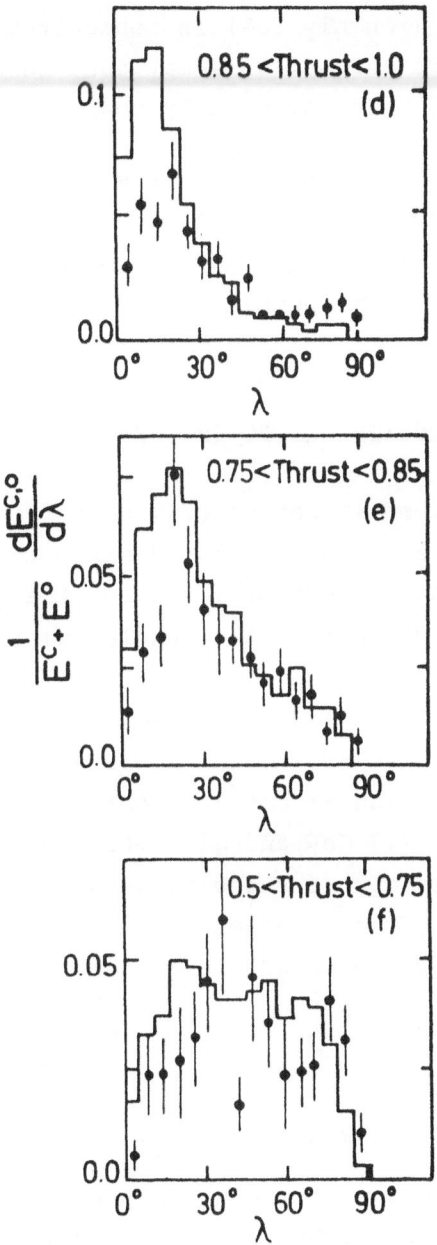

Fig. 30. PLUTO: angular distribution of the energy flow with respect to the thrust axis for three different thrust bins. The charged energy (histogram) and neutral energy (points) are treated separately.

charged energy (roughly 20%) is concentrated at small
angles and large thrust. This may just be a reflection
of the jet axis being defined by charged particles.

The opening angle of the jets is still rather
large. In the two higher thrust intervals the mean
half opening angle is 25° and 33° for charged and
neutral particles, respectively.

Angular distribution of the jet axis

If jets in e^{+}e^{-} annihilation originate from the production
of a pair of quarks the angular distributions of these
jets should contain information about the quark spin.
Neglecting the quark mass the angular distribution
should be 1+cos$^{2}\theta$ for spin 1/2 particles. Fig.31 shows
the angular distribution of the jet axis as defined by
thrust for energies of 7.7 and 9.4 GeV. If we fit these
data with a function 1+αcos$^{2}\theta$ we obtain values of α =
= 0.76 $\pm$ 0.3 at 7.7 GeV and α = 1.63 $\pm$ 0.6 at 9.4 GeV.
Within the large errors these values are consistent with
the theoretical prediction of 0.8 and 0.88 at the two
energies [53]. These predictions include finite mass
effects of the quarks.

p_T and p_L distributions

The basic assumption of the quark-parton model is that
the average p_T approaches a limiting value with increasing energy. Fig. 32 shows the mean p_T and mean p_L
distributions with respect to the thrust axis as observed
in the PLUTO detector.

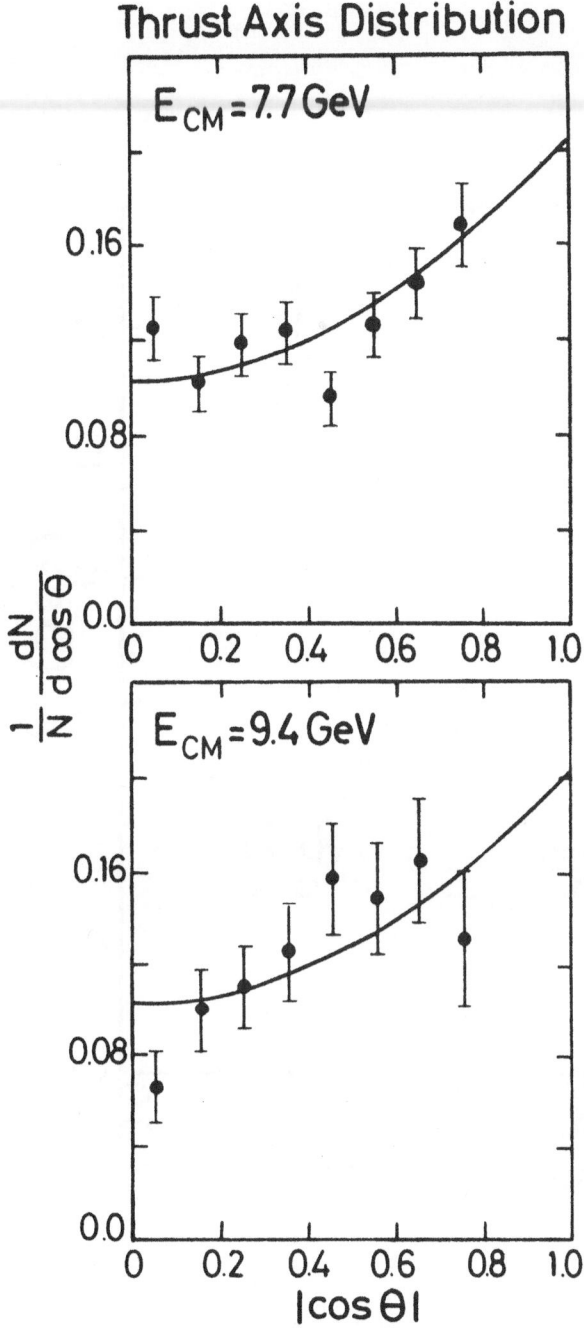

Fig.31. PLUTO: angular distribution of the jet axis as defined by thrust at 7.7 GeV and 9.4 GeV. The curves are $1+\cos^2\theta$ distributions.

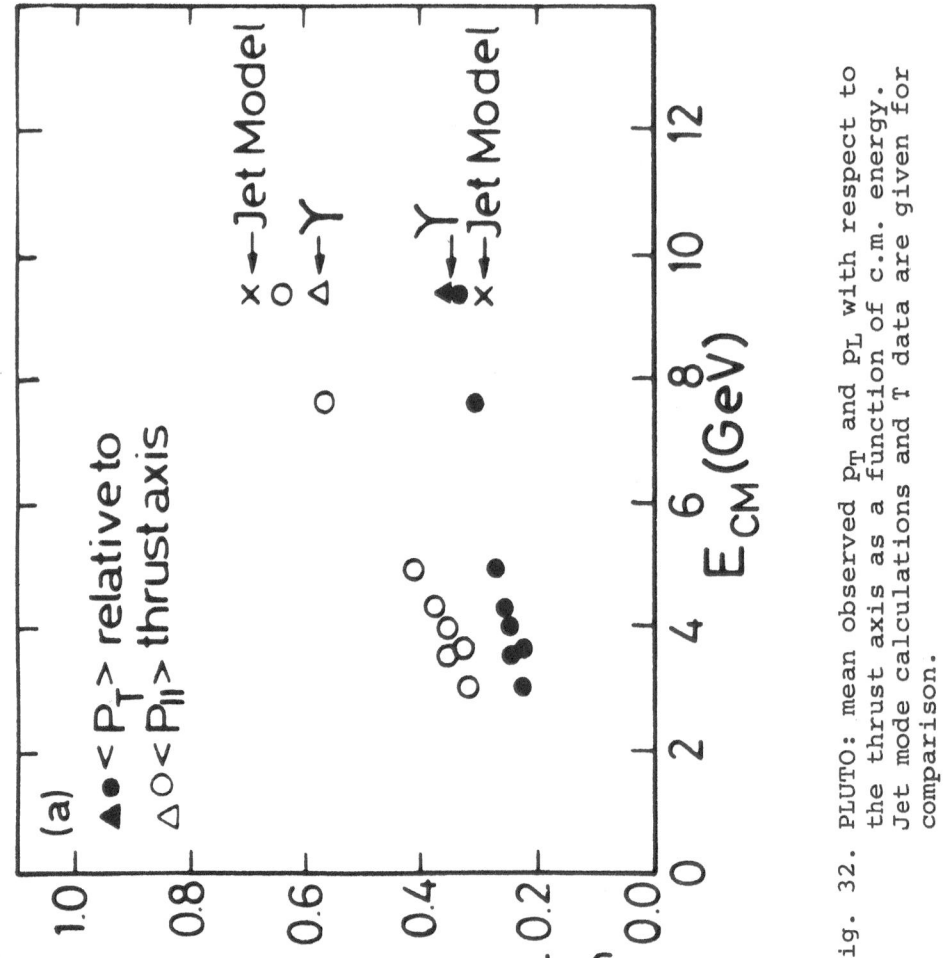

Fig. 32. PLUTO: mean observed p_T and p_L with respect to the thrust axis as a function of c.m. energy. Jet mode calculations and Υ data are given for comparison.

The figure shows indeed that the increase in p_T
is slowing down at the highest energies whereas p_L keeps
growing nearly linearly. The figure nicely demonstrates
again how the jet structure evolves with increasing energy.
The half opening angle defined by the transverse and the
parallel momentum is 28° at 9.4 GeV in good agreement
with the value obtained from the energy-flow distributions.

Differential distributions and experimental cuts

So far we have only compared gross features of our data
like mean observed sphericity and mean observed transverse
momentum with different models. We proceed now to
differential distributions. At the same time we will try
to get a feeling of the effects caused by limited
acceptance and measurement errors. We therefore compare
our measured values with two kinds of Monte-Carlo predictions:

- Values generated in the Feynman-Field model without
  passing the events through the detector.
- Distributions after simulating the detector and
  applying initial state radiative corrections.

Figs. 33 and 34 show the distributions of p_T and S at
9.4 GeV. The comparison with model predictions indicates
reasonable agreement with the data once the detector
simulation and radiative corrections are applied. However, there is still some discrepancy between model
and data at low sphericity values. The comparison between the generated and the final Monte-Carlo distributions
shows how far the real distributions are distorted by
the detector. The effect is particularly large in the
sphericity distribution. Table 7 shows how the mean values

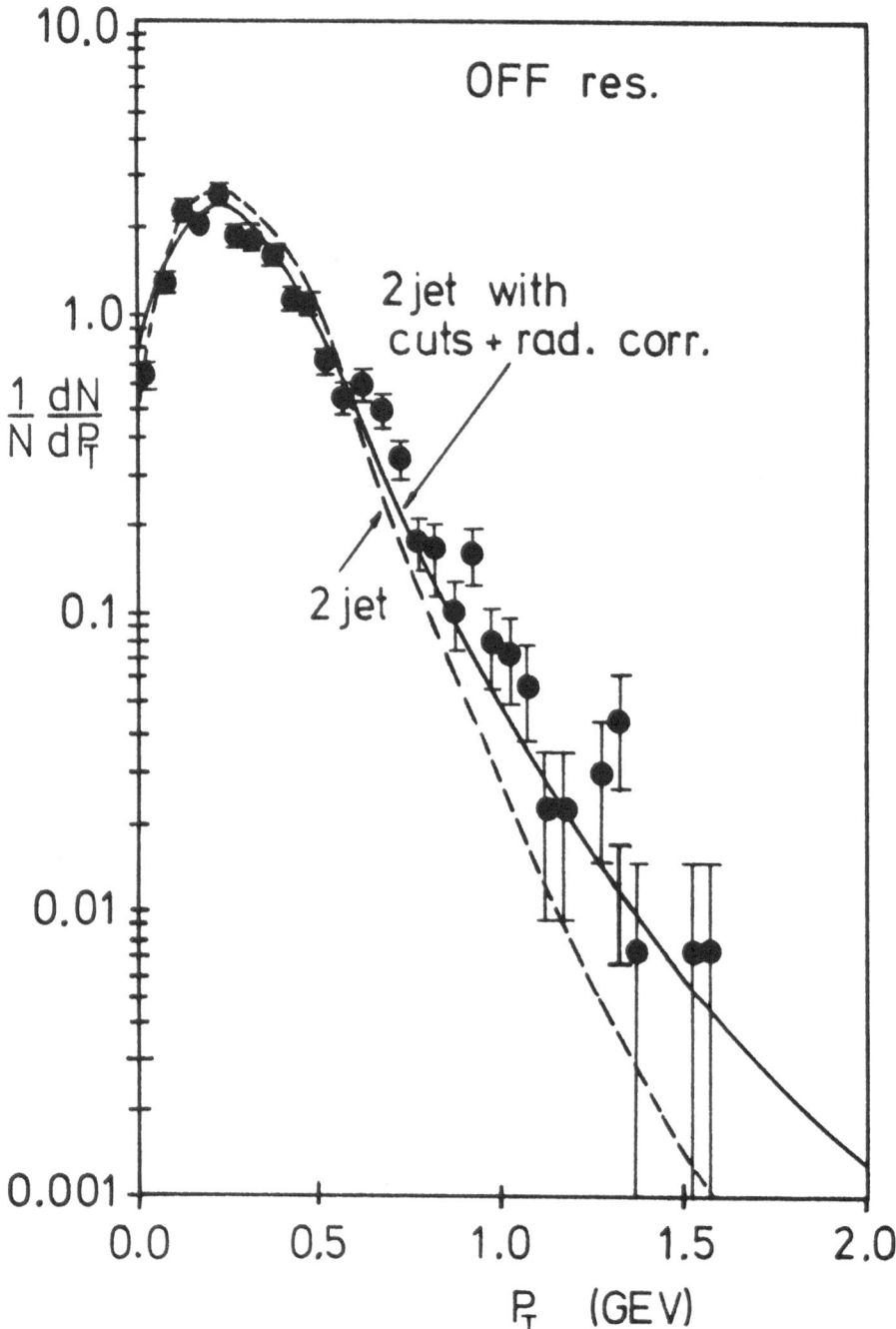

Fig.33a. PLUTO: (preliminary) observed p_T distributions at 9.4 GeV. OFF resonance: the curves are two-jet model predictions including (——) and not including (– – –) detector cuts and radiative corrections.

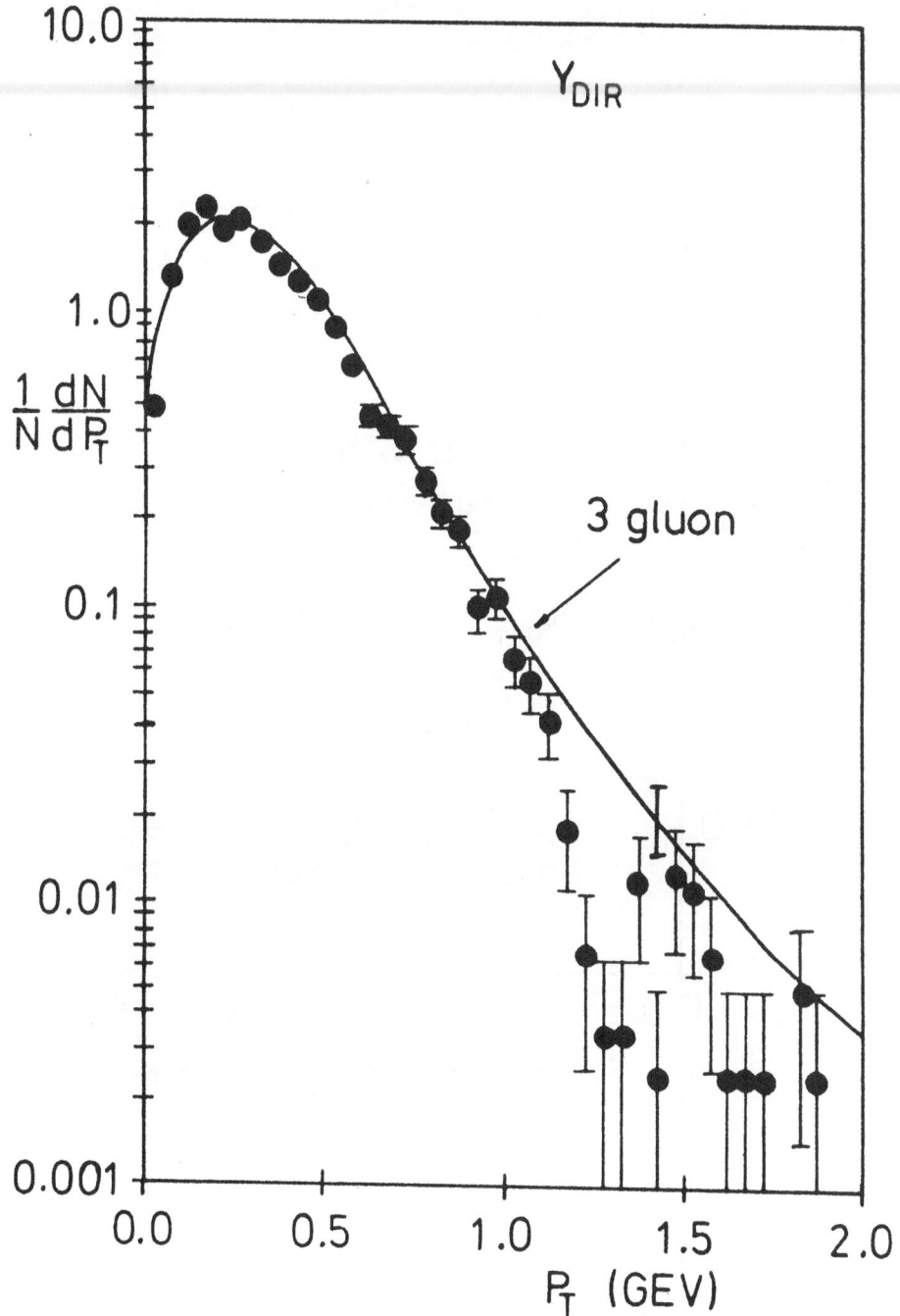

Fig.33b. ON resonance: the curve is a three-jet model prediction including all corrections.

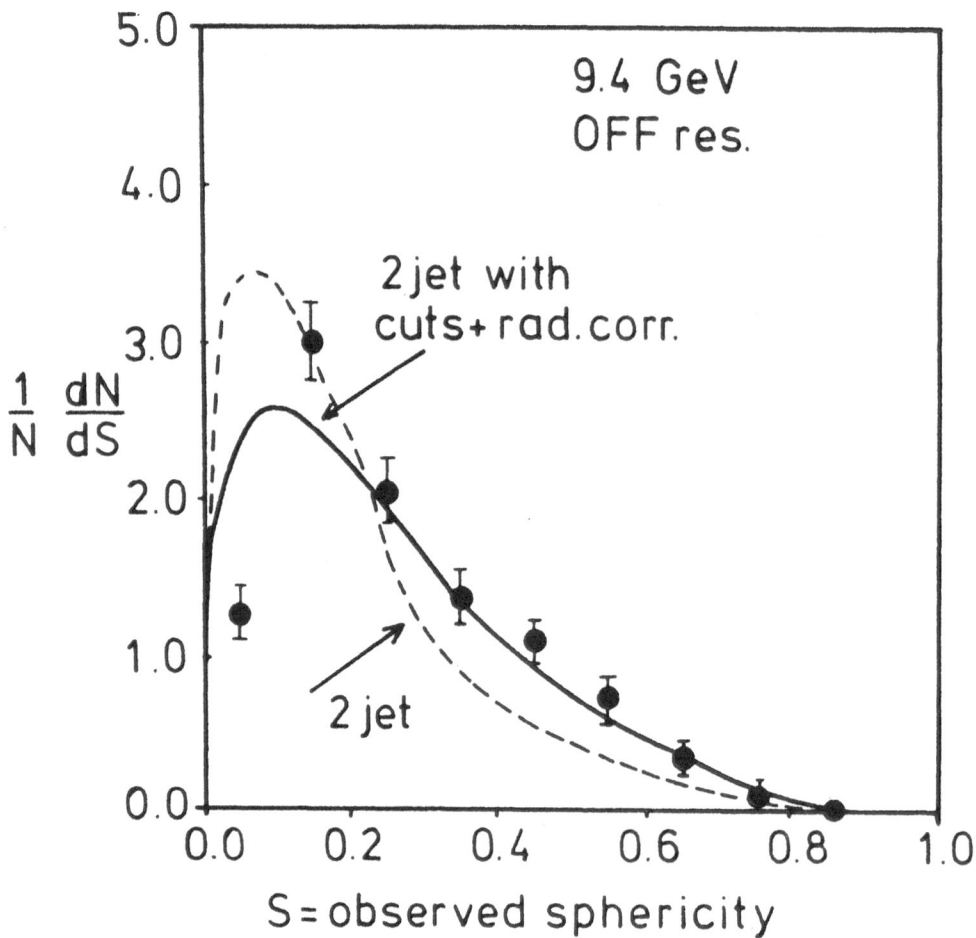

Fig.34a. PLUTO: (preliminary) observed S distributions
at 9.4 GeV. OFF resonance: curves as in Fig.33.

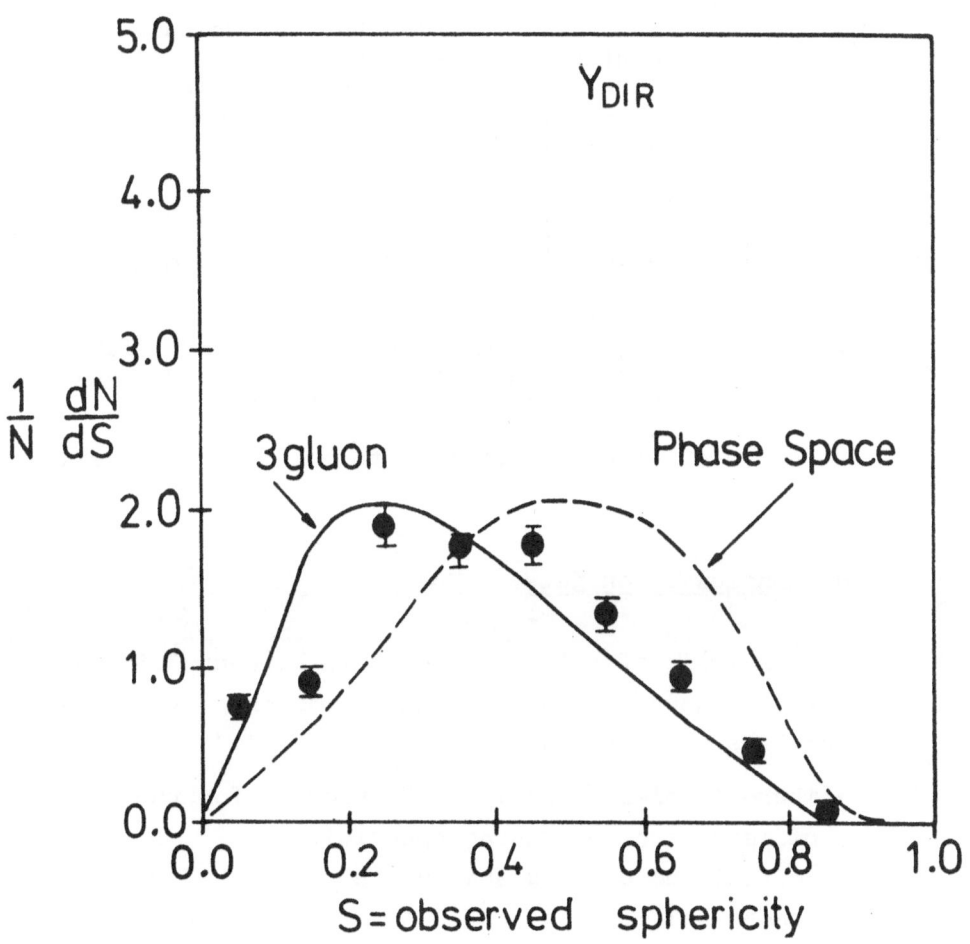

Fig.34b. PLUTO: (preliminary) observed S distributions
at 9.4 GeV. ON resonance: curves as in Fig.33.

of p_T, p_L and S evolve from the generated values to the final ones including corrections. The mean transverse momentum increases only slightly because the distortions in the high momentum part of the distribution get very low statistical weight.

	generated value	including resolution and efficiency	including resolution, efficiency and radiative corrections
$<p_T>$	302	317	317
$<p_L>$	722	715	655
$<S>$	0.20	0.22	0.25

Table 7. The effect of detector resolution, efficiency and radiative corrections on the mean values of p_T, p_L and S. Two jet MC events were generated at 9.4 GeV. All values are calculated with respect to the sphericity axis.

d) Conclusion on jets

The analysis of off-resonance data between 3.6 and 9.4 GeV has shown evidence for quark-parton jets in e^+e^- annihilation into hadrons.

- There is clear (confirming) evidence for the existence of jets with decreasing sphericity.
- Neutral and charged energy are correlated with a common half-opening angle of about 30° at $E_{CM} = $ = 9.4 GeV.
- The mean transverse momentum levels off at high energies as expected in the simple quark-parton model.
- The angular distribution of the jet axis is compatible with the production of spin 1/2 particles.

So far we have not discussed any test of QCD gluon bremsstrahlung. I will come back to this point in the last
chapter on PETRA data.

2. T decay topology

In the frame-work of QCD a $q\bar{q}$ bound state couples
to three gluons. Once the energy of the $q\bar{q}$ state is high
enough a fragmentation of these three gluons into jets
will become the preferred decay mode (Fig. 35a) [7].

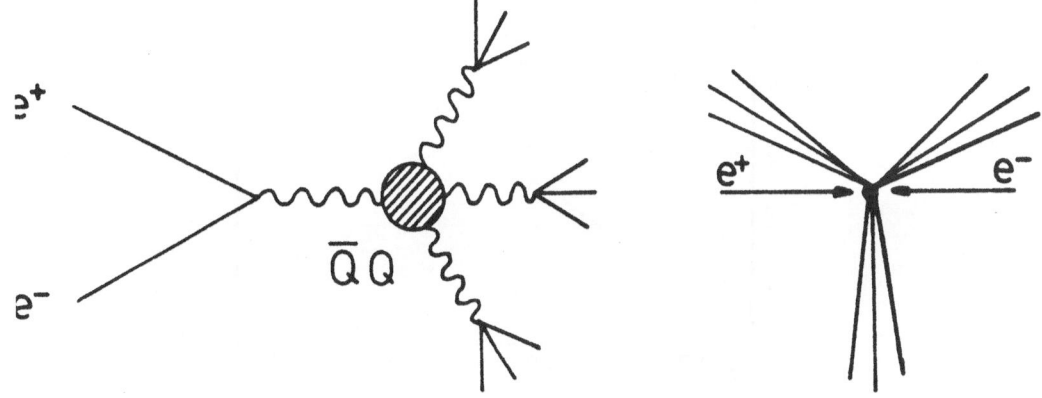

Fig.35a. Proposed Ypsilon decay into three
gluons.

The observation of a three jet structure in the T decay
would therefore be a decisive test on the existence of
gluons and the validity of QCD [54-56]. This conjecture
leads to the following predictions.
(1) Topological quantities like sphericity and thrust
change drastically as one passes through the resonance.

(2) A three-jet structure would of course lead to a planar configuration of the events.

(3) Eventually three separated jets may be visible.

Although a possible observation of (1) and (2) may be indicative only (3) could be really decisive. Unfortunately it turns out that an asymmetric partition of energy among the three jets is preferred [54] which leads to a nearly back-to-back structure of the events instead of a symmetric three-star structure (Fig.35b).

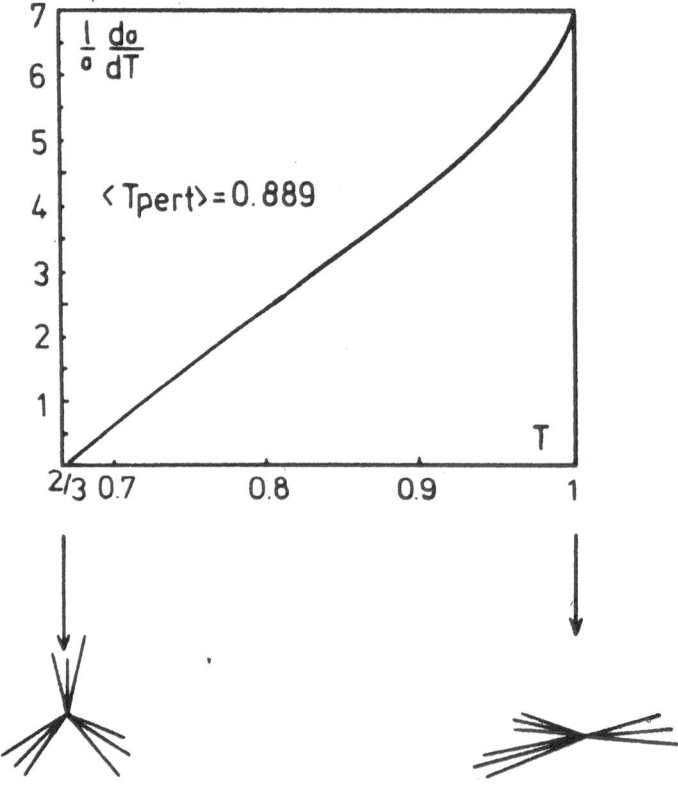

Fig.35b. Relative cross section for symmetric and asymmetric three-gluon events.

In addition at the present stage of theory and analysis
any interpretation of the data suffers from the following
problems.

- The fragmentation of gluons is not known and theoretical
  predictions are rather vague and controversial [57].
- At the T resonance the energy of the proposed gluon jets
  is still very low (about 3 GeV/jet).
- Resonance events can only be separated statistically
  from the continuum.

a) Experimental procedure

This last point complicates the study of direct Ypsilon
decays. The analysis proceeds in three steps.

- Isolate the direct decay mode.
- Define models.
- Compare the uncorrected data with these models.

In this section I will mainly concentrate on experimental
results from two groups - the PLUTO collaboration [58]
and the NaJ-lead glass experiment [59]. These two experiments are complementary to each other since the
first one concentrates on charged particles whereas the
second one measures neutral energy. There will not be
enough time to discuss the analysis of a third group -
the DASP2 collaboration [42] - which measured only
particle directions.

Data were compared with three models,
(1) the phase-space model,
(2) the two-jet Feynman-Field model,
both defined in section 1, and
(3) a three-gluon jet model,

as described in the following:

The basic assumption of this model is that gluons fragment like quarks [54]. The two ingredients of the model are then:

- A three-gluon matrix element [54,60] for the production of three massless gluons through an intermediate virtual photon.
- A fragmentation of the gluons with limited p_T. The mean p_T is adjusted to fit the two-jet data at 9.4 GeV below the resonance (at comparable jet energies). The charged multiplicity and neutral energy is adjusted to the τ data.

b) PLUTO data

In the resonance region three terms contribute to the hadronic cross section σ_{on} (Fig. 36): the continuum contribution denoted as σ_{off}, the one-photon or vacuum polarization term denoted as $\sigma(\gamma)$ and finally the direct decay mode of the τ into hadrons (σ_{dir}). The terms σ_{off} and $\sigma(\gamma)$ both show the characteristics of two-jet production discussed in the previous section. Only the direct decay term σ_{dir} is expected to show the features we are looking for.

Although it will turn out that there are large differences between these contributions there is no clean way to separate them from each other event by event. The extraction of σ_{dir} has therefore to be done on a statistical basis. From all differential distributions in the resonance region appropriately normalized distributions of the corresponding off-resonance data

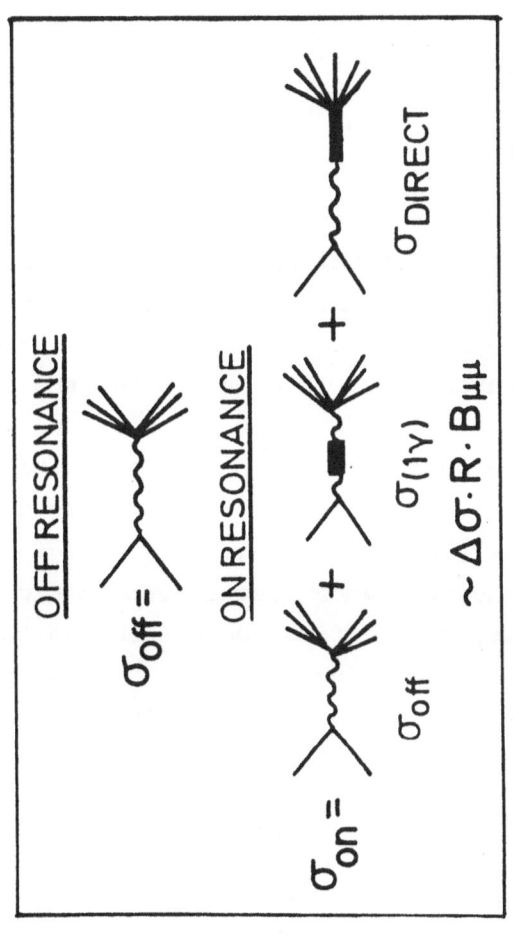

Fig. 36. Off and on resonance contributions to the annihilation cross section.

are subtracted:

$$\sigma_{dir} = \sigma_{on} - \sigma_{off} - \sigma(\gamma) = \sigma_{on} - \sigma_{off}(1 + \frac{\sigma_{\mu\mu}^{on} - \sigma_{\mu\mu}^{off}}{\sigma_{\mu\mu}^{off}}) \; .$$

The normalization of the off-resonance data is calculated from the relative luminosities on and off resonance and from the PLUTO measurements on the muon branching ratio $B_{\mu\mu}$ (see chapter II).

1872 events on resonance and 470 events off re-sonance enter into the final analysis. This corresponds to a total integrated luminosity of 396 nb^{-1} in the whole 9.4 GeV energy region [58]. The subtraction method de-scribed above leaves about 1250 events for the direct-decay contribution. Fig.37a shows the mean observed thrust of these events. For comparison also the un-subtracted Ypsilon data and the off-resonance data are given. We observe a drastic change of topology as one passes through the resonance (see insert in Fig.37a). We can, therefore, draw a first conclusion on the two-jet mechanism which describes the data outside the re-sonance: This process is certainly excluded as a major contribution to the direct T decay.

What about the three-gluon decay? Fig. 37b shows a detailed comparison with our three models (phase-space, three-gluon, and two-jet decay). The complete data (Fig. 37b) and the break down into multiplicity classes (Fig. 37c) are both in good agreement with the three-gluon Monte Carlo [61]. The data are in disagreement with the two-jet description and the phase-space model although the latter is only about 2 to 3 standard

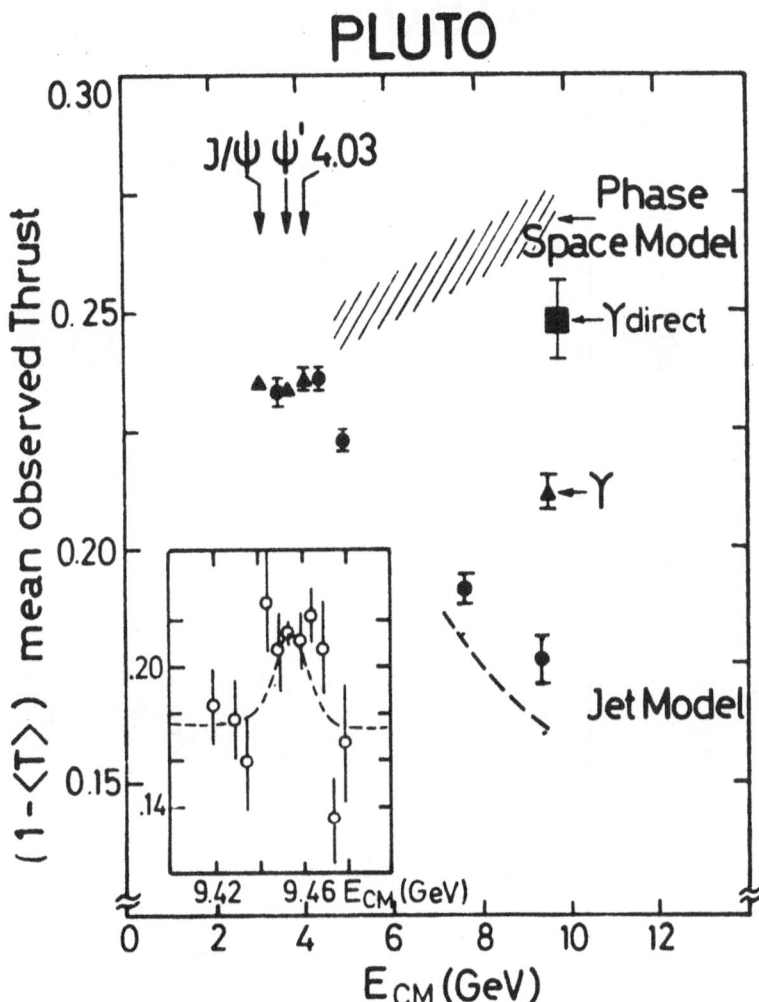

Fig. 37a. PLUTO: observed mean thrust and sphericity
for charged particles including the Ypsilon.
Observed mean thrust.

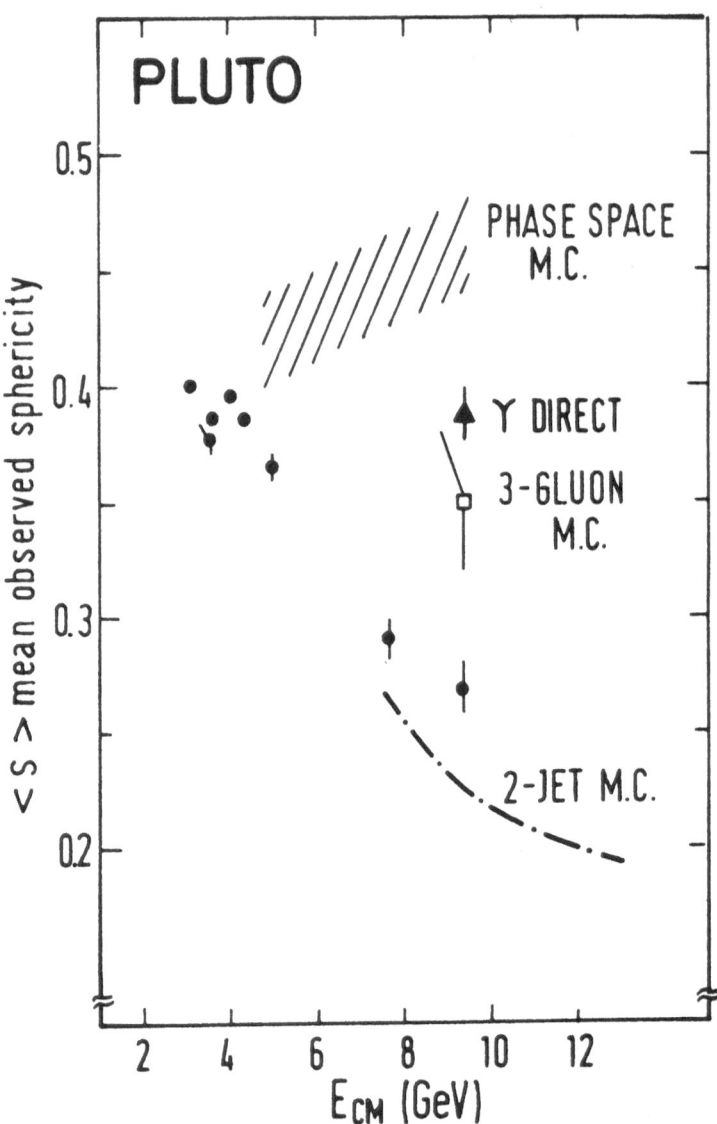

Fig. 37b. Observed mean sphericity. The predictions for phase-space, two-jet and three-gluon models are indicated.

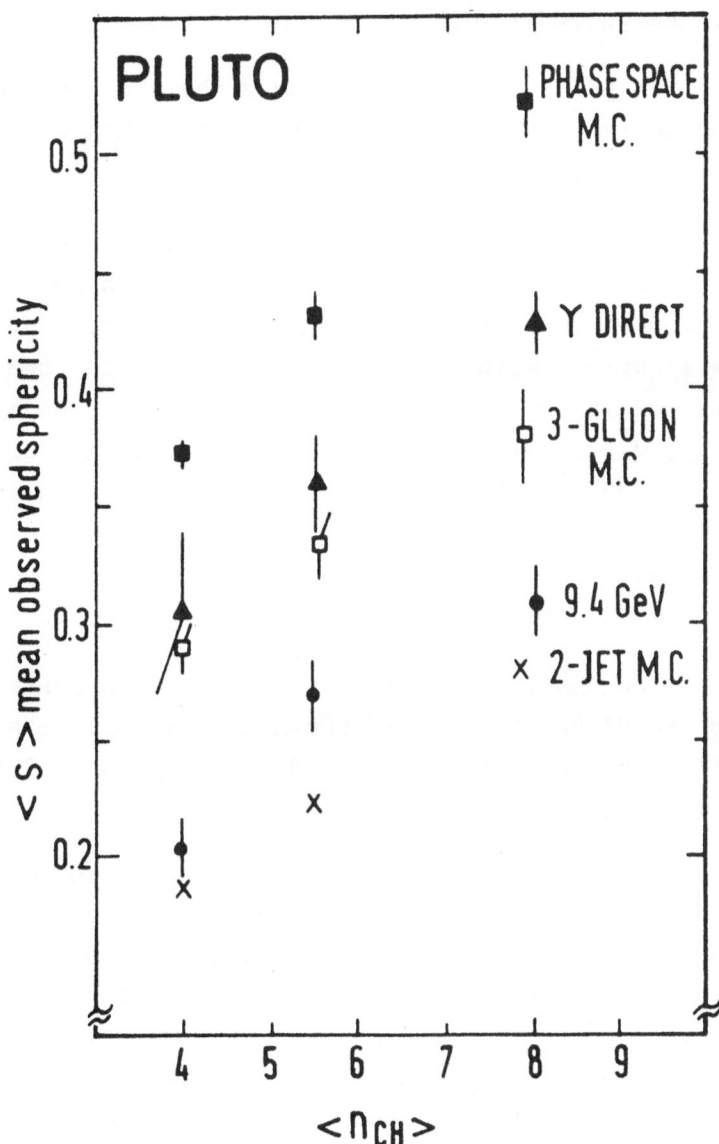

Fig. 37c. Data on and off resonance compared to model
prediction: breakdown into multiplicity
classes.

deviations away. Fig. 37c shows that the relative
values of data and models are nearly independent of
the charge dmultiplicity.

In particular this excludes that the change in
sphericity is merely a reflection of a change in
multiplicity.

Being encouraged by the reasonable agreement of
the mean sphericity with a three-gluon description let
us proceed to a detailed comparison of differential
distributions with the models. Fig.34b shows the
differential spherocity distribution on resonance.
The p_T distributions on and off resonance are compared
in Fig. 33a and b. Again we realize good agreement
between the resonance data and a three-gluon jet model.
A comparison of Fig. 33a and b shows that the p_T
distribution changes little as one crosses the resonance.

The angular distribution of the sphericity axis
for the direct Y decay is shown in Fig. 39. It agrees
well with the shape of $1 + 0.39 \cos^2\theta$ predicted from
the Ypsilon decay into massless spin-1 gluons [54].

Fig. 38 shows the mean p_T as a function of x_L on
and off resonance. x_L is the scaled longitudinal momentum
along the jet axis. Both distributions show a clear indication of the 'seagull' effect. The trend of the data
is again correctly described by the two models although
the increase of mean p_T at intermediate x_L is again
overemphasized by the three-gluon model.

In conclusion we see that the overall features of
the on-resonance data are reasonably well described by
a three-gluon model where the gluons are assumed to fragment

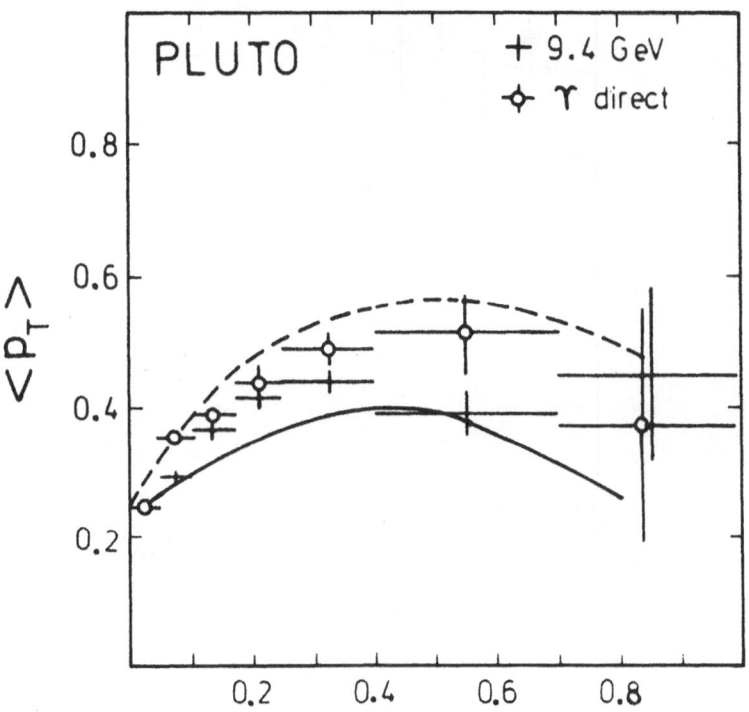

Fig. 38. PLUTO: (preliminary) "Seagull" effect - mean p_T
as a function of x_L (scaled momentum parallel to
the jet axis: $x_L = p_L/E_b$). The model predictions
for two jets (——) and three gluons (---) are
displayed for comparison.

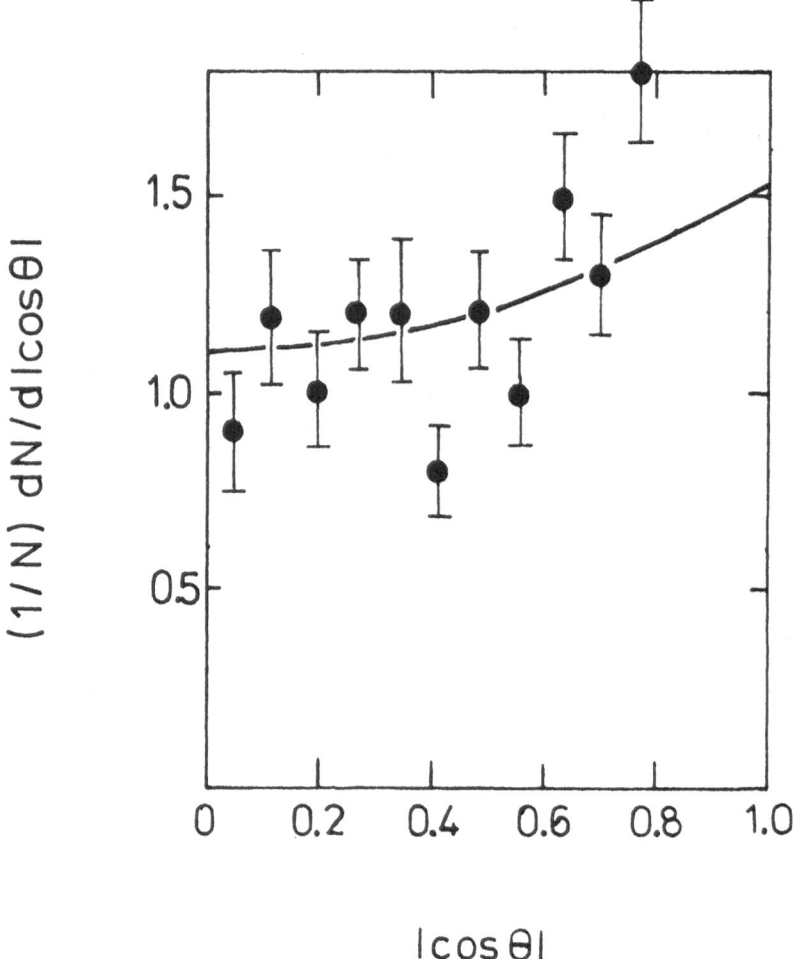

Fig. 39. PLUTO: angular distribution of the sphericity
axis on resonance. The curve is proportional
to $1 + 0.39 \cos^2\theta$ [54].

like quarks. In particular all observed quantities show
strong differences in comparing on- and off-resonance
data. These differences are correctly described by a
two-jet Monte Carlo off resonance and a three-gluon
Monte Carlo on resonance. On the other hand the three-
gluon model does not yield a quantitative agreement
with all features of the data. In particular details
of the p_T distributions are not correctly described
by the model. Taking the rough natur of this model this
is of course not surprising. For instance, since the
model was adjusted to the data outside the resonance it
automatically contains important charm contributions.
In QCD these contributions are not expected on the T
(compare section 2e). But even if this effect might
explain some of the differences between the model and
the data we are certainly not yet in a position to
really test the details of the gluon-fragmentation
functions.

c) NaJ-lead glass detector data

As a second example I want to discuss the results from
the DESY-Hamburg-Heidelberg-Munich collaboration [59].
This group took data at the upgraded DORIS storage ring
using the NaJ-lead glass detector of the DESY-Heidelberg
collaboration. The device consists of a nonmagnetic inner
detector with four layers of drift chambers surrounded by
NaJ and lead glass shower counters which cover 94% of 4π.
The topolical analysis in this experiment is based on the
energy deposited in the shower counters. Thus a large
fraction of the neutral energy enters into the analysis,
whereas only about one third of the charged energy is

counted by minimum ionizing losses. Like for the PLUTO
data observed quantities are shown and compared with
different models.

Figs. 40 and 41 show the observed sphericity
distributions on and off resonance for the T and for
T', respectively. The off-resonance data and the direct
terms of the resonance data are compared with the three
models mentioned in the previous section. Again there is
a clear distinction on and off resonance in both cases.
The resonance data are in agreement with a three-gluon
model. On the other hand the data are not sensitive to
the small differences between the three-gluon model
(full curve) and the phase-space model (dashed curve).

d) Tests of coplanarity

We have seen in the previous sections that the event
topology does in fact change as one passes from the
continuum to the T resonance. It was also shown that
differential distributions of particle momenta and
topological measures like sphericity and thrust change
in the way expected from QEC. Being encouraged in
particular by the good agreement of the data with our
three-gluon Monte Carlo let us, therefore, proceed to
a more detailed test of the proposed coplanarity of
the events.

To this end we introduce a generalized three-
dimensional sphericity in the following way [50]: let
us look at the expression

$$T_{\alpha\beta} = \sum_i (\delta_{\alpha\beta} (p^i)^2 - p_\alpha^i p_\beta^i)$$

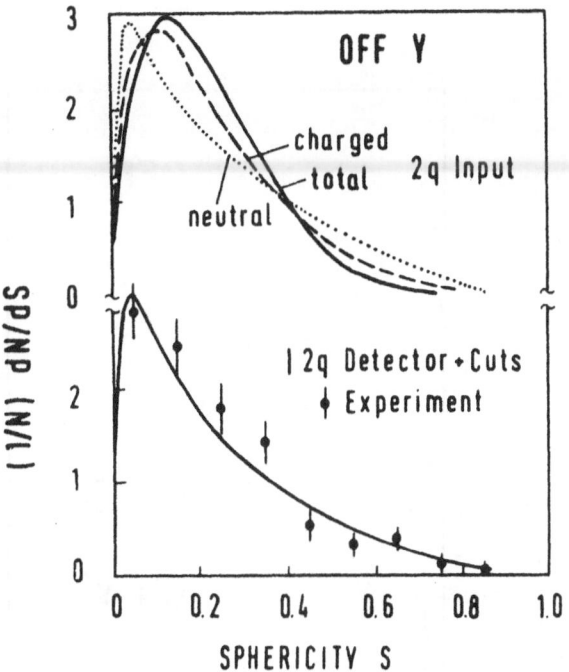

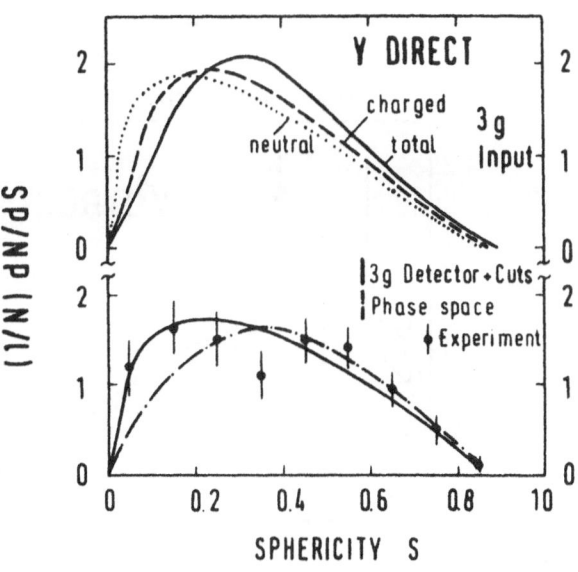

Fig. 40. NaJ-lead glass detector: observed sphericity
distributions in the T region. OFF resonance
data are compared to a two-jet model, T DIRECT
data to three-jet (——) and phase-space (---).
The upper curves show the composition of the
models (generated values).

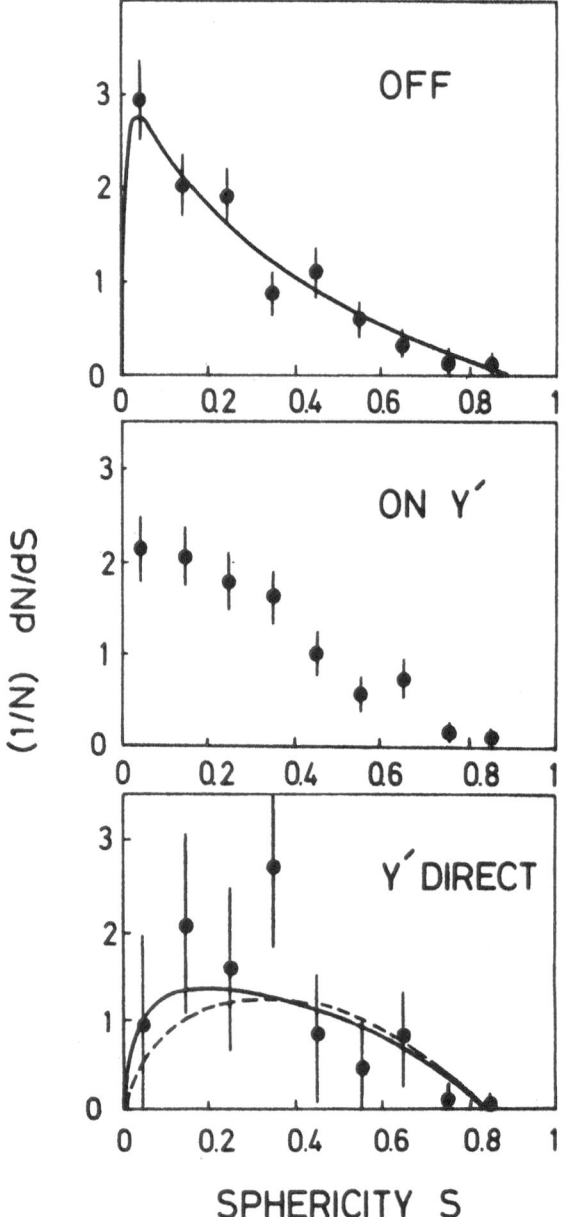

Fig. 41. NaJ-lead glass detector: observed sphericity distributions in the Υ' region. Curves defined as in Fig. 40.

defined in analogy to the inertia tensor. If we diagonalize this expression we obtain the eigenvalues λ_k which correspond to the three main axes of the event in momentum space (Fig.42). If we order these eigenvalues such that

$$\lambda_1 \geq \lambda_2 \geq \lambda_3$$

then λ_3 closely resembles our well known sphericity definition

$$\lambda_3 = \frac{\sum\limits_i (p_1^i)^2 + (p_2^i)^2}{\sum (p^i)^2} = \frac{\sum (p_T^i)^2}{\sum (p^i)^2} \; ; \quad S = 3\lambda_3 (\lambda_1 + \lambda_2 + \lambda_3) \; .$$

The physical meaning of λ_3 is again best understood from an analogy with the inertia tensor. λ_3 points into the direction of the smallest inertia moment in momentum space. To measure the flatness of events we have to study the other two eigenvalues in particular λ_1 which points into the direction of the smallest extent of the event in momentum space.

It is convenient to define the following quantities [58,62] (Fig. 42):

$$Q_k = 1 - \frac{2\lambda_k}{\lambda_1 + \lambda_2 + \lambda_3} = \frac{1}{2} \frac{\sum\limits_i (p_k^i)^2}{\sum\limits_i (p^i)^2} \; .$$

Q_k points into the same direction as λ_k, however, it measures the sum of the momentum components parallel to the axis λ_k. Consequently the Q_k are ordered in

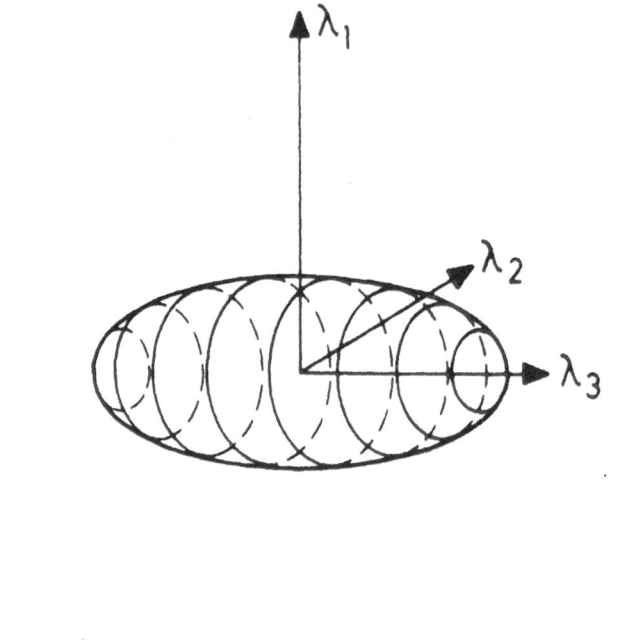

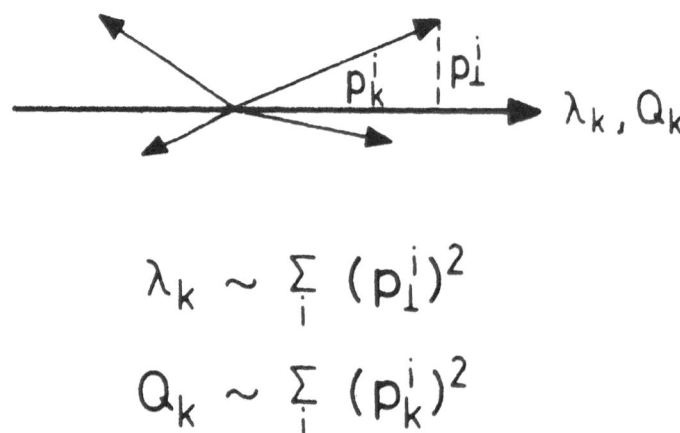

$$\lambda_k \sim \sum_i (p_\perp^i)^2$$

$$Q_k \sim \sum_i (p_k^i)^2$$

Fig. 42. The sphericity tensor in analogy to the inertia
tensor. Definition of λ_k, Q_k.

a rising sequence

$$Q_1 \le Q_2 \le Q_3$$

for a falling sequence of λ_k. We define now a triangle
plot in analogy to a Dalitz analysis where the two
perpendicular axes measure Q_1 and $(Q_3 - Q_2)/\sqrt{3}$ (Fig.43).
Since Q_1 measures the shortest extension of the event
in momentum space, planar events will be characterized
by small Q_1 values. Such events will then congregate
near to the $(Q_3 - Q_2)$-axis. The extrem topologies of a
sphere, a disk and a jet will fall onto the vertices A,
B and C in the Q_k triangle, respectively.

The Q_k values are calculated diagonalizing the
sphericity tensor for each event. All events are then
entered into their position in the Q_k triangle plot.
Continuum and vacuum polarization terms are subtracted
statistically from these three-dimensional distributions
in the same way as described in section 2b. The result
is shown in Fig. 43 together with Monte Carlo distributions.

The off-resonance data are indeed concentrated
in the corner C which corresponds to a jet topology.
The two-jet Monte Carlo shows a very similar distribution.

The data on resonance also show a peak in the jet
corner. However, there is a broad distribution along the
$(Q_3 - Q_2)$-axis as expected for a planar configuration.
There are very few events only in corner B. This shows
that a very symmetrical planar event structure is in fact
very rare as expected in a three-gluon decay mechanism.
The observed increase towards a jetlike configuration is
qualitatively expected from the abundant asymmetric three-
gluon jets. These events tend to exhibit a two-jet structure

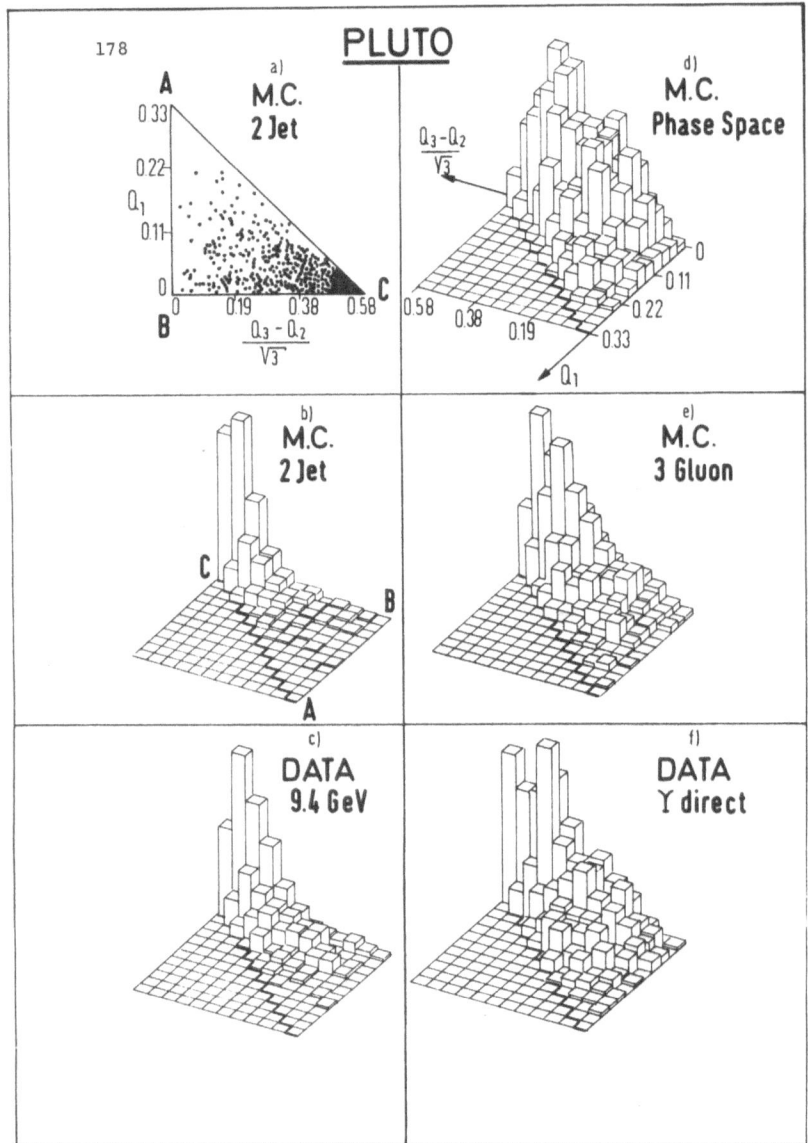

Fig. 43. PLUTO: triangle plots of the momentum space configuration. Q_1 is plotted versus $(Q_3-Q_2)/\sqrt{3}$. Data
are compared with the three models of two jets,
three jets, and phase space. The vertices A,B,C
correspond to perfect sphere, disc, and two jet
structures.

in the direction of the two most energetic gluons. A
quantitative comparison with the three-gluon model
(Fig.43e) shows remarkable agreement with the data.
On the other hand, a phase space Monte Carlo distribution as shown in Fig.43d has a broader nonplanar extension than the data.

For a more quantiative comparison we study the
projection onto the Q_1 axis. Fig.44 shows the mean
Q_1 as a function of energy for the off-resonance and
the on-resonance data. A comparison with the three
models shows again that the off-resonance data agree
well with the two-jet Monte Carlo description whereas
the on-resonance data are in good agreement with the
three-gluon Monte Carlo.

In the framework of a sphericity study Q_1
measures the planarity of an event through a sum over
the squared momentum component outside the event plane.
The corresponding measure in the thrust or spherocity
frame work has been baptized acoplanarity [46]

$$A = 4 \min \{ (\Sigma_i |p_{out}^i|)/\sum_i |p_i|) \}^2 .$$

Like in the thrust definition the acoplanarity is based
on a measurement of the absolute value of the momentum
component perpendicular to the event plane. A direct
measurement of the flatness of an event is given by the
mean-momentum component which points outside the event
plane. We denote this quantity p_{out} and define it with
respect to the plane given by the Q_1 direction. The mean
value of p_{out} is plotted in Fig. 45. Again the comparison
with our models shows good agreement of the off-resonance

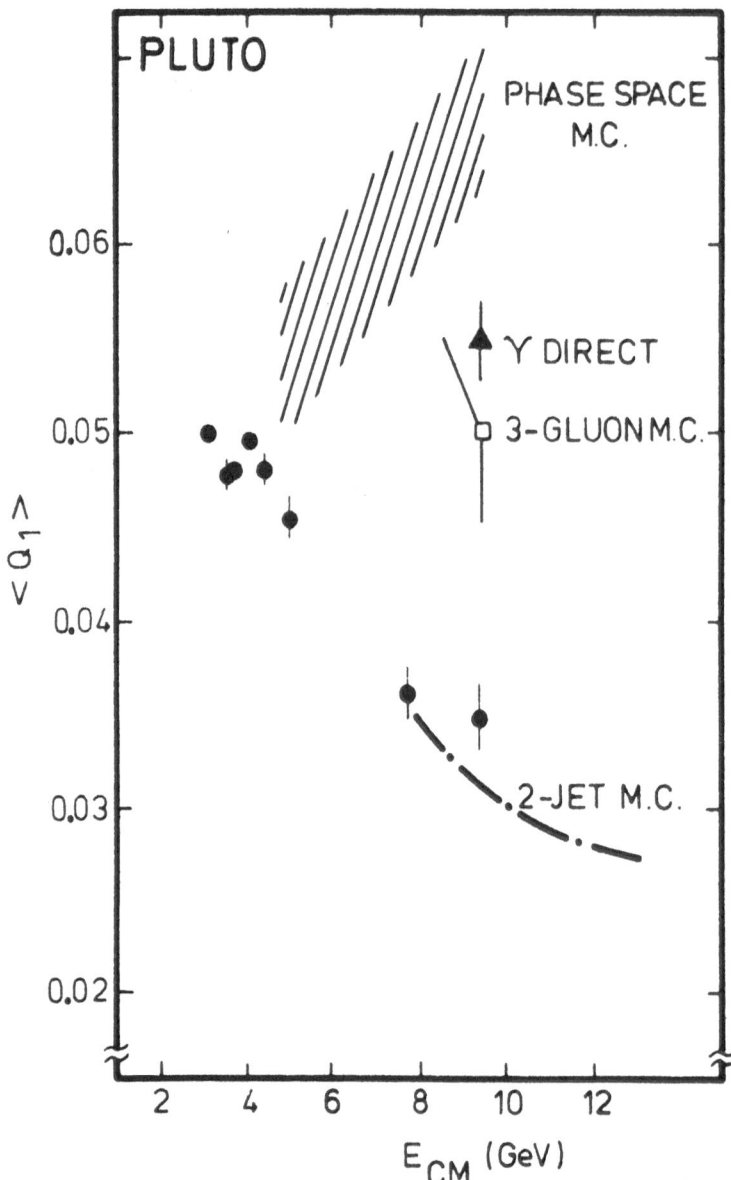

Fig. 44. PLUTO: mean observed Q_1 as a function of energy.

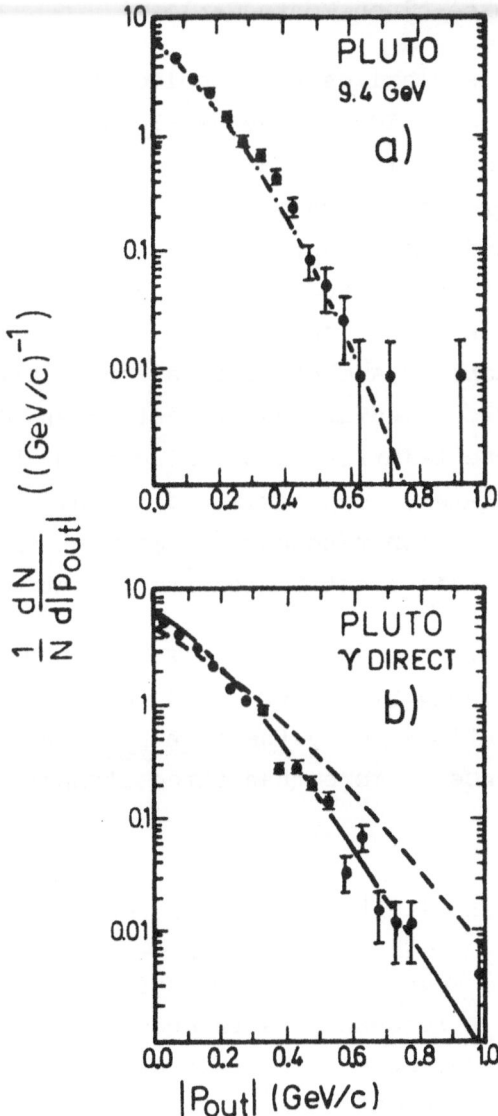

Fig. 45. PLUTO: p_{out} distributions with respect to the Q_1
plane in the 9.4 GeV region.
a) OFF resonance, compared with the two-jet model.
b) ON resonance, compared with the three-jet (——)
and the phase-space (---) model.

data and a two-jet Monte Carlo and of the on-resonance
data and a three-gluon Monte Carlo.

Table 8 summarizes all topological quantities
measured at the PLUTO and the NaJ-lead glass detector.
It shows an impressive agreement in all variables between the off-resonance data and the two-jet Monte Carlo
on one side and the direct T data and the three-gluon
Monte Carlo on the other side. For a detailed comparison
we have to look at the error bars. The errors given for
the data in table 8 are statistical only. The errors on
the Monte-Carlo calculations are estimates of the
systematic uncertainties which arise from the adjustment
of model parameters to experimental data. Taking these
errors all values on resonance agree with the three-gluon
Monte Carlo within about 1 standard deviation.

We notice, however, that also the phase-space
Monte Carlo generally differs by no more than two
standard deviations. Only the mean p_{out} value shows a
clear difference of more than three standard deviations.

e) Further tests of the three-gluon model

So far we have seen that the toplogical structure of
events on the T resonance is in good agreement with a
three-gluon decay model. In particular we get satisfactory
agreement with the data assuming that gluons and quarks
fragment the same way. In this section I want to study
whether the mean multiplicity and the K_s^o production observed on and off resonance are in agreement with our
conceptions.

The data are compared with three different models. Errors indicated at the experimental values are statistical only, whereas those on the models include estimates of the systematic uncertainties in determining the model parameters. M.C. values are without radiative corrections.

	M.C. two jet	DATA 9.4 GeV	DATA T direct	M.C. three gluon	M.C. phase space	
$\langle S \rangle$	0.22	0.27 ±0.01	0.39 ±0.01	0.35 ±0.03	0.46 ±0.02	
$\langle \Omega_1 \rangle$	0.030	0.036±0.002	0.056±0.003	0.050±0.005	0.067±0.005	PLUTO group [58]
$\langle p_{out} \rangle$	0.115	0.118±0.003	0.129±0.003	0.140±0.006	0.177±0.006	
$\langle T \rangle$	0.84	0.82 ±0.01	0.76 ±0.01	0.76 ±0.01	0.73 ±0.01	
$\langle A \rangle$	0.084	0.099±0.005	0.15 ±0.01	0.14 ±0.01	0.16 ±0.01	
$\langle S \rangle$	0.26	0.23 ±0.01	T direct 0.37 ±0.01	0.36	0.40	NaJ–lead glass detector [59]
$\langle T \rangle$	0.79	0.82 ±0.01	0.74 ±0.01	0.74	0.72	
$\langle S \rangle$	0.25	0.25 ±0.01	T'direct 0.32 ±0.04	0.34		
$\langle T \rangle$	0.80	0.82 ±0.01	0.77 ±0.02	0.75		

Table 8. Observed mean values for different measures of event topology.

Rapidity and multiplicity

Fig. 46 shows the observed rapidity distributions with
respect to the sphericity axis for off- and on-resonance
data. Both distributions clearly exhibit a plateau in
the fragmentation region very similar to secondary
particles from hadron collisions [63-65]. A detailed
comparison of the on- and off-resonance data shows some
very remarkable features.

- The plateau region is wider in the off-resonance data
  than in the on-resonance data.
- We can readjust the data in such a way that they
  coincide in the fragmentation region. A plateau
  lengthening of the order of 0.7 in rapidity becomes
  particularly clear.
- The plateau is higher on than off resonance.
  Quantitatively we get 4.2 particles per unit
  rapidity on resonance and 3.0 particles per unit
  rapidity off resonance.

In our simple picture these observations are readily
explained at least qualitatively. Since the energy of
two jets off resonance is larger than the mean energy
of three jets on resonance the resulting plateau length
is larger. The observed lengthening of 0.7 is well explained by the expected logarithmic increase with energy.
Even the relative height of the rapidity plateau corresponds
rouhgly to the naive expectation of three to two. As we
remember from hadron physics a plateau in the rapidity
distribution results in a logarithmic increase of the
mean multiplicity. To estimate the mean multiplicity [66]
of the off- and on-resonance data we simply assume that
the mean multiplicity is the sum over all jet multiplicities.

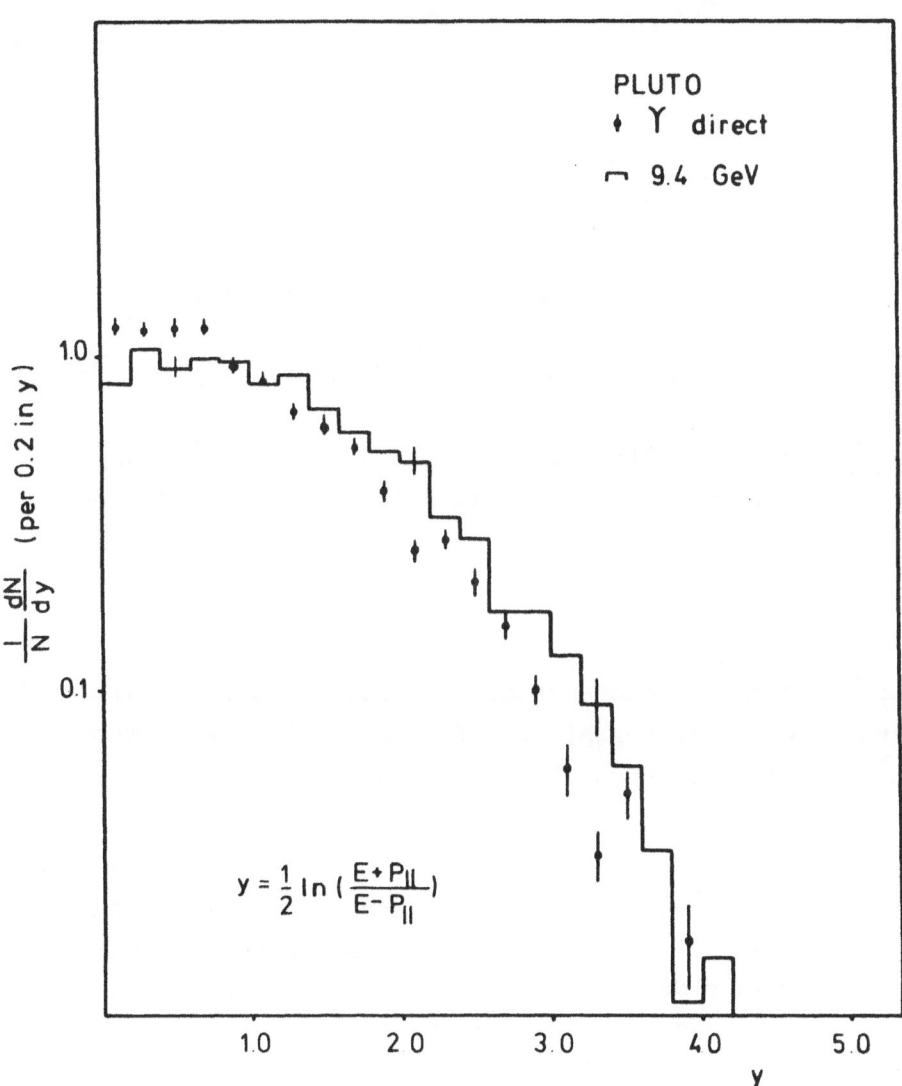

Fig. 46. PLUTO: (preliminary) observed rapidity
distribution with respect to the sphericity
axis on and off resonance.

If the jet multiplicities increase like

$$\langle n_{CH} \rangle_{jet} \simeq a \cdot \ln E_{jet} + b$$

we get for the off-resonance data

$$\langle n_{CH} \rangle_{off} \simeq 2 \cdot a \cdot \ln E_{CM}/2 + 2b \, .$$

With the additional simplifying assumption that all three jets have an energy of $E_{CM}/3$ we get on resonance

$$\langle n_{CH} \rangle_{on} \simeq 3 \cdot a \cdot E_{CM}/3 + 3b \, .$$

We compare this with the corrected values of the mean charged multiplicity as obtained in the PLUTO collaboration

$$\langle n_{CH} \rangle_{off} = 6.3 \pm 0.4 \qquad\qquad \langle n_{CH} \rangle_{on} = 8.0 \pm 0.3$$

and the rapidity plateaus of 3 (= 2a) and 4.2 (= 3a) ·particles per unit rapidity off and on resonance. We observe reasonable consistency within the errors for

$$a \simeq 1.4 \, , \qquad\qquad b \simeq 1.0 \, .$$

An increase of the mean (uncorrected) multiplicity from 6.44 ± 0.14 to 7.2 ± 0.18 has also been observed by the NaJ-lead glass detector [59]. The numbers correspond to observed charged particles and photon conversions with no acceptance corrections.

To conclude the observed multiplicities and rapidity distributions are in good agreement with our three-gluon model.

K_s^o production

Electron-positron annihilation shows a strong increase
in K production [67] as one crosses the charm threshold
at about 4 GeV. Above this energy most of the K's can
be attributed to charm decays. In the QCD picture Υ
decays into massless gluons. Charm production mediated
by gluons

$$g \rightarrow c\bar{c}$$

should be strongly suppressed because of the c quark
mass [68]. From these arguments one would therefore
not expect any additional K_s^o production as one crosses
the resonance.

The experimental result from the PLUTO group [69]
(Fig. 47) shows, however, a very strong increase. With
the appropriate scaling of Fig.47 one may conclude that
the inclusive K_s^o production is proportional to the total
cross section. At first sight this may suggest that the
same production mechanism for K_s^o is at work on resonance
as off resonance. This would, of course, contradict QCD.

On the other hand QCD does in fact predict another
source for K_s^o production on the Υ resonance. Since gluons
are flavour blind they should in first approximation
couple with equal strength to u, d and s quarks. Therefore, on the average one gluon per event produces an $s\bar{s}$
pair. Thus to a very simple approximation one would expect
half a K_s^o per event in the direct Υ decay. From Fig. 47
we conclude that indeed about 0.4 K_s^o per event are
produced on the Υ. The decisive test, however, whether
the observed K_s^o production is really due to this mechanism
and thus conformal with QCD predictions can only come from
a further check on the suppression of charm production.

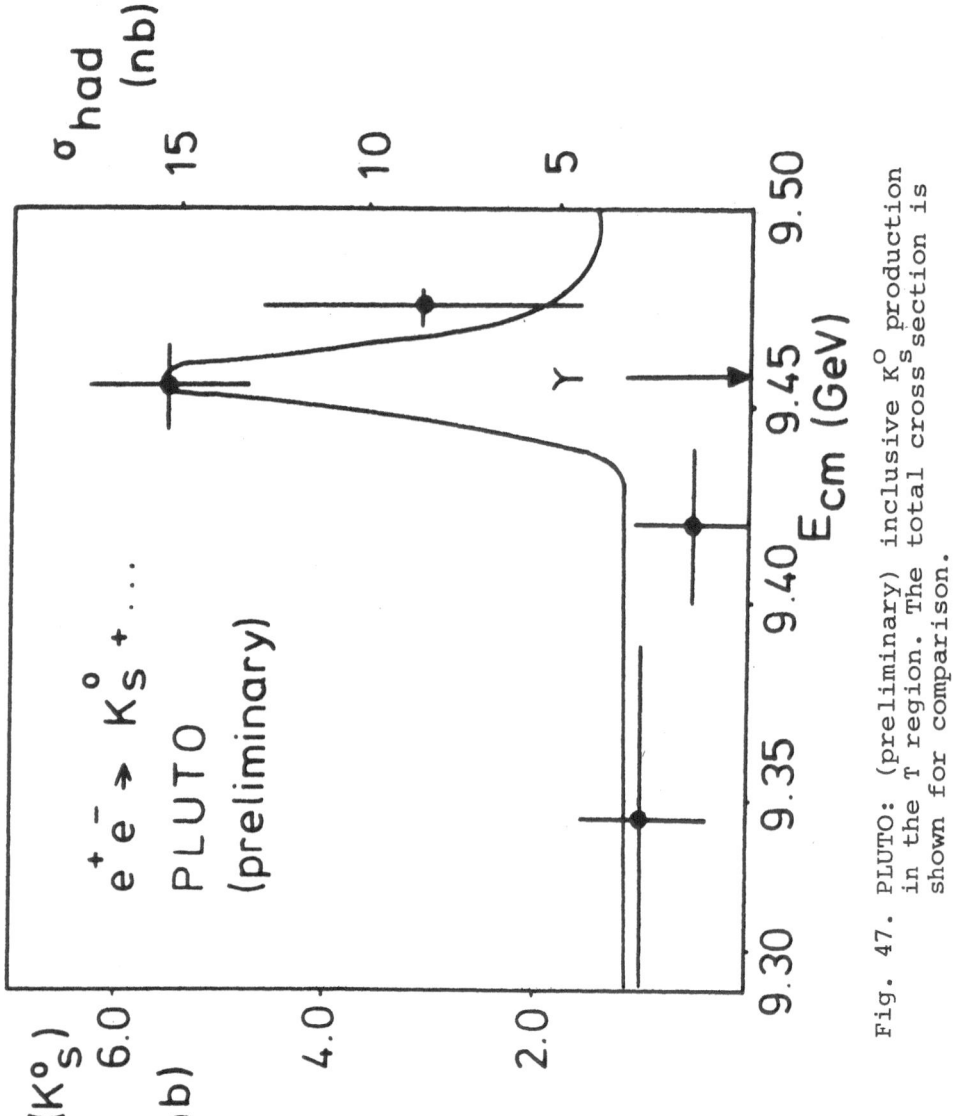

Fig. 47. PLUTO: (preliminary) inclusive K_s^0 production in the Υ region. The total cross section is shown for comparison.

f) Conclusion

Off-resonance data between 3.6 and 9.4 GeV

e^+e^- annihilation off resonance shows a clear two-jet
structure which is getting more and more pronounced
with higher energies. In first approximation there is
good agreement between the data and a naive quark-parton
model (Feynman-Field parametrization). In the next
lecture I will return to some necessary refinements of
the model.

Ypsilon data

All aspects of the data like topology, particle
distributions, and multiplicity are in fair agreement
with a naive three-gluon model prediction in which the
gluons fragment like quarks. The observed excess in K_s^o
production can be interpreted from the flavour blindness
of gluons. The data definitely rule out two jets as a
prominent source for T decays. A simple phase-space
distribution gives a bad description of the data. In
most variables which have been considered so far the
phase space is, however, not excluded by more than 2
standard deviations. Therefore, further work is
necessary and in progress to obtain definite statements
about QCD. In particular, the neutral energy will be
included into the analysis of the PLUTO data.

IV. FIRST PETRA RESULTS

In this chapter I am going to describe experimental
results which have been obtained very recently on the new
e^+e^- storage ring PETRA [5]. PETRA is the first of a new
generation of storage rings entering into the 30 GeV c.m.
energy region. Since it came into operation only a couple
of months ago I will spend some time on describing the
machine and the first generation of experiments.

1. The maschine

The history of PETRA is summarized in table 9. It
may be interesting to realize that the submission of the
proposal coincides historically with the discovery [14]
of J/ψ in November 1974. It took less than a year until
the machine was authorized on October 20, 1975. Again
one year later in Autumn 1976 decisions were taken on
the first round of experiments [70] : PLUTO [71],
MARK J [72], CELLO [73], JADE [74] and TASSO [75]. In
the following one and a half years the construction of
PETRA and of the five experiments went ahead. In July
1978 already - less than three years after authorization -
an electron beam was stored and accelerated in the machine.
In fall 1978 and beginning of 1979 first physics runs
could be scheduled and take data successfully. During
these first physics shifts three detectors were installed
in the machine: PLUTO, MARK J and TASSO.

Fig. 48 shows a bird's view of the DESY site with
the storage ring PETRA. PETRA - Positron-Electron-Tandem-
Ring-Accelerator - is an e^+e^- storage ring designed for a
maximum beam energy of 2 x 19 GeV. Its diameter is about

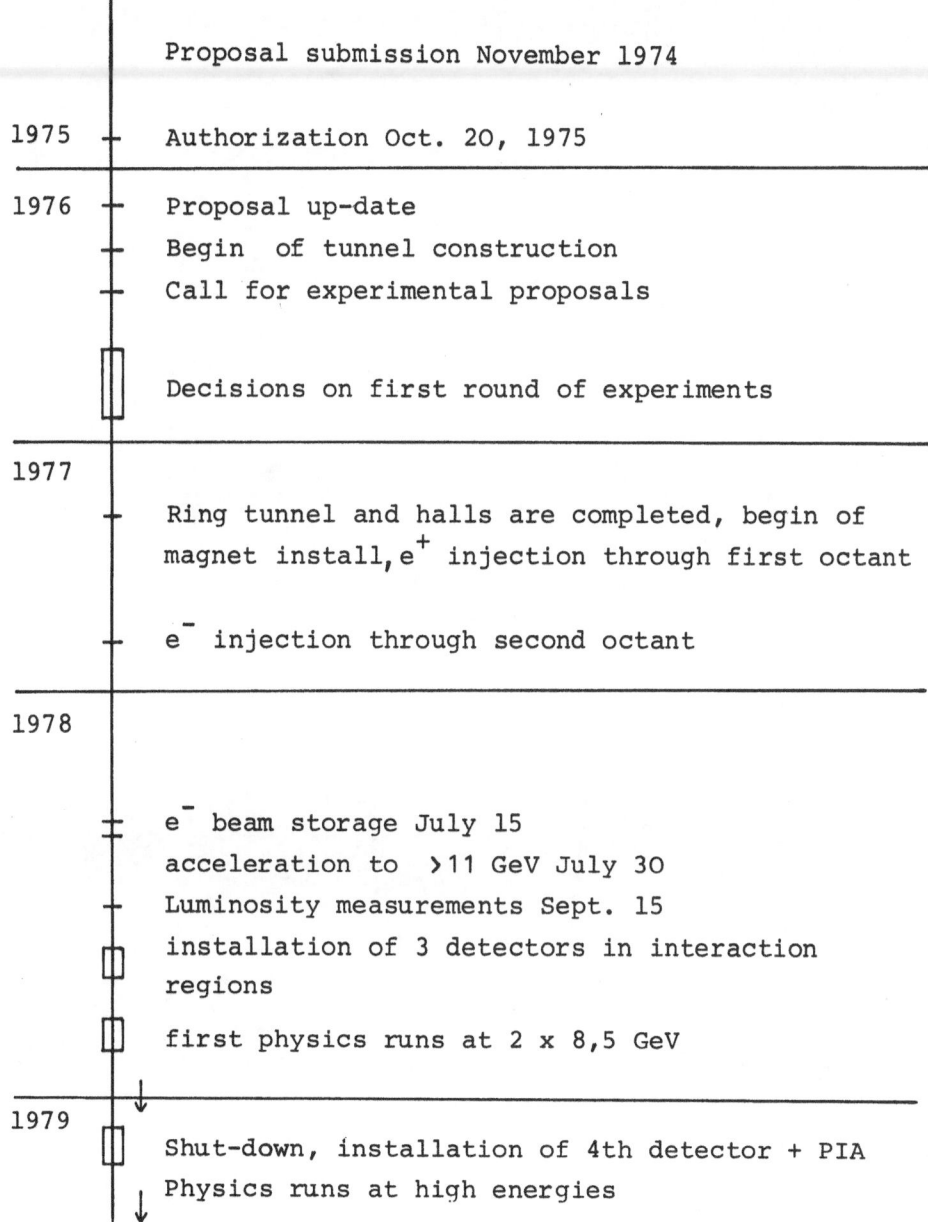

Proposal submission November 1974

1975 — Authorization Oct. 20, 1975

1976 — Proposal up-date
— Begin of tunnel construction
— Call for experimental proposals

Decisions on first round of experiments

1977

— Ring tunnel and halls are completed, begin of
magnet install, e^+ injection through first octant

— e^- injection through second octant

1978

— e^- beam storage July 15
acceleration to >11 GeV July 30
— Luminosity measurements Sept. 15
installation of 3 detectors in interaction
regions

first physics runs at 2 x 8,5 GeV

1979

Shut-down, installation of 4th detector + PIA
Physics runs at high energies

Table 9. History of PETRA

192

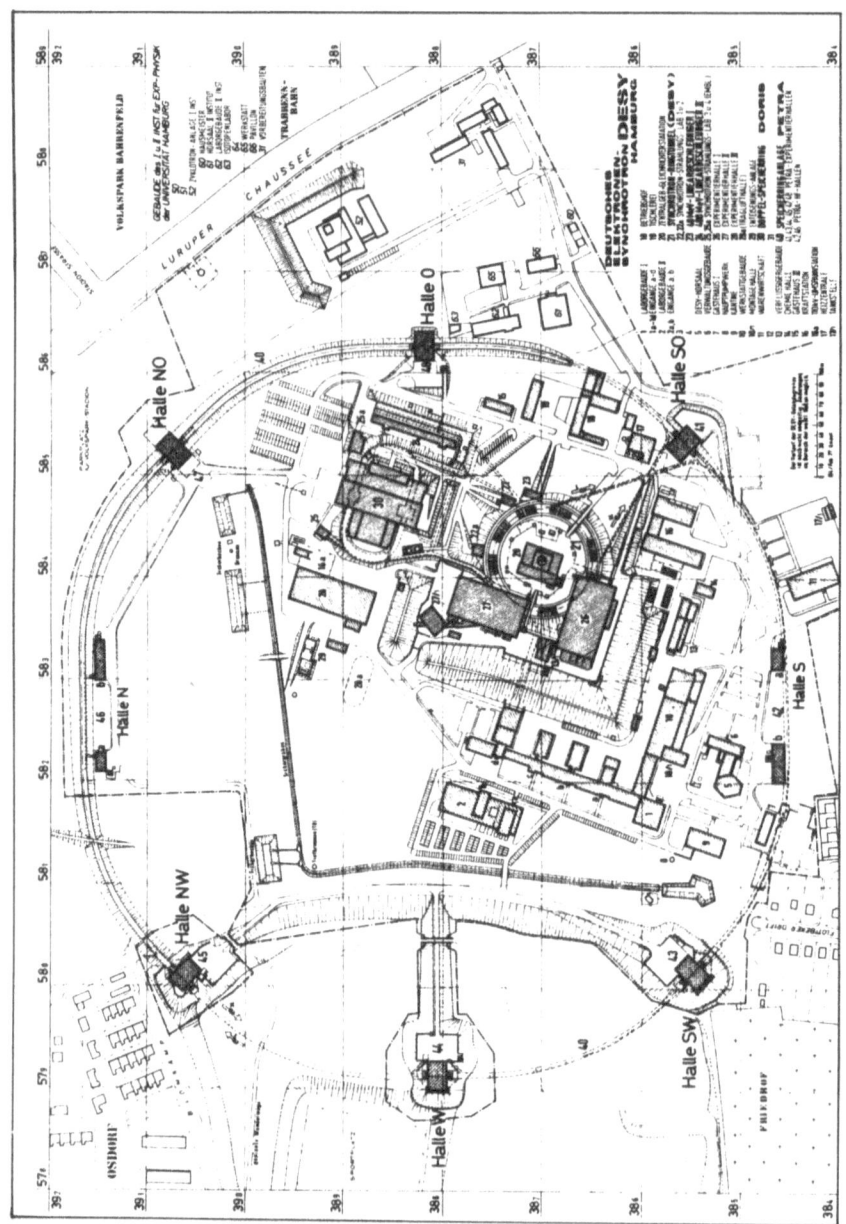

Fig. 48. The e^+e^- storage ring PETRA.

800 m. In the present configuration all other DESY
machines are used to fill the new PETRA storage ring.
Two linear accelerators produce electrons and positrons.
The positrons are preaccelerated to 2.2 GeV in the DESY
synchrotron and then stacked into the DORIS storage ring.
After accumulation they are reinjected into the DESY
synchrotron and, like the electrons, accelerated to their
final injection energy of 6.5 GeV. At this energy particles
are injected into the PETRA tunnel where they are stacked.
Typical currents of several mA are then circulating in
the PETRA storage ring and can be accelerated to their
final energy. Contrary to the original design of DORIS,
PETRA is a single-ring few-bunch machine. Depending on
the number of experimental areas which have to be served,
the number of bunches per beam varies between one and
four. Some relevant figures on the PETRA performance [76]
in January 1979 are summarized in table 10.

The accomplished values for the single beam life-
time, the bunch length, and the single bunch current are
close to the design figures. A maximum beam energy of
11.1 GeV could be obtained. (Meanwhile after introducing
additional cavities more than 15 GeV were reached.) This
indicates that there are no principle obstacles to reach
30 GeV c.m. energy.

In the light of possible resonance searches it is
important to note that the design value for the momentum
spread of 6.5×10^{-5} GeV x p^2 (p in GeV) was observed.
This guarantees an energy resolution of $\sigma = 2.3$ MeV at
10 GeV and 21 MeV at 30 GeV c.m. energy. The best
luminosity which was obtained at 2 x 6.5 GeV was
1.2×10^{30} cm^{-2} s^{-1}. Compared to the designed luminosity
of 1.4×10^{31} cm^{-2} s^{-1} this is still a factor of 10 too

Parameter	Accomplished values	Comments
– Lifetime	5 ./. 8 h	design: 9 h; improving
– Bunch length	11.4 mm (r.m.s.)	no bunch lengthening
momentum spread	0.065 MeV p^2 (r.m.s.)	p = beam momentum in GeV
energy resolution	0.023 MeV E^2_{CM}(r.m.s.)	E_{CM} = c.m. energy in GeV
– Single bunch current	$\sim$ 18 mA	design: 20mA
– Energy per beam	$\leq$ 11.1 GeV	in March more than 15 GeV have been reached
– Luminosity $\begin{cases}\text{max}\\ \text{typical}\end{cases}$ at 2x8.5 GeV	1.2×10^{30} cm^{-2} s^{-1} 0.5×10^{30} cm^{-2} s^{-1}	$\left\{\begin{array}{l}\text{2 x 2 bunches; design value}\\ \text{(2 x 4 bunches)}\\ 1.4 \times 10^{31}\ \text{cm}^{-2}\ \text{s}^{-1}\end{array}\right.$
tune shift $\Delta\Omega$	0.015 ./. 0.025	design 0.06
number of bunches	2 x 2	design 2 x 4

Table 10. PETRA performance (by January 1979).

low. To understand this discrepancy in detail let us
look again at the following expression for the
luminosity:

$$L \sim (\Delta Q^2) \cdot B \cdot \varepsilon_x \sim E^4 \quad .$$

The number of bunches for the measured luminosity was
two whereas four were assumed for the design luminosity.
This explains a trivial factor of 2. The remaining
difference is due to the tune shift ΔQ. Whereas $\Delta Q = 0.06$
was assumed in the design, the accomplished values so far
were only 0.025 under optimum conditions and 0.015
typically.

To appreciate these numbers one should certainly
not forget that the mean luminosity of about 0.5×10^{30}
was reached after a few months of machine studies only.
Fig. 49 shows again the development of the luminosity
during November 1978. It was then when the first
hadronic event was observed in the PLUTO detector.
The figure shows the rapid improvement in luminosity
delivered to the experiments. Note also that the mean
luminosity at 17 GeV was higher than at 13 GeV, as
expected.

2. First round detectors

The first experiments at PETRA were primarily
motivated by the possible discovery of new degrees of
freedom, in particular, the proposed new quarks [77] b
and t and may-be even further quarks and leptons. As
already discussed in the first lecture an appropriate
handle on new flavours is the total cross section

196

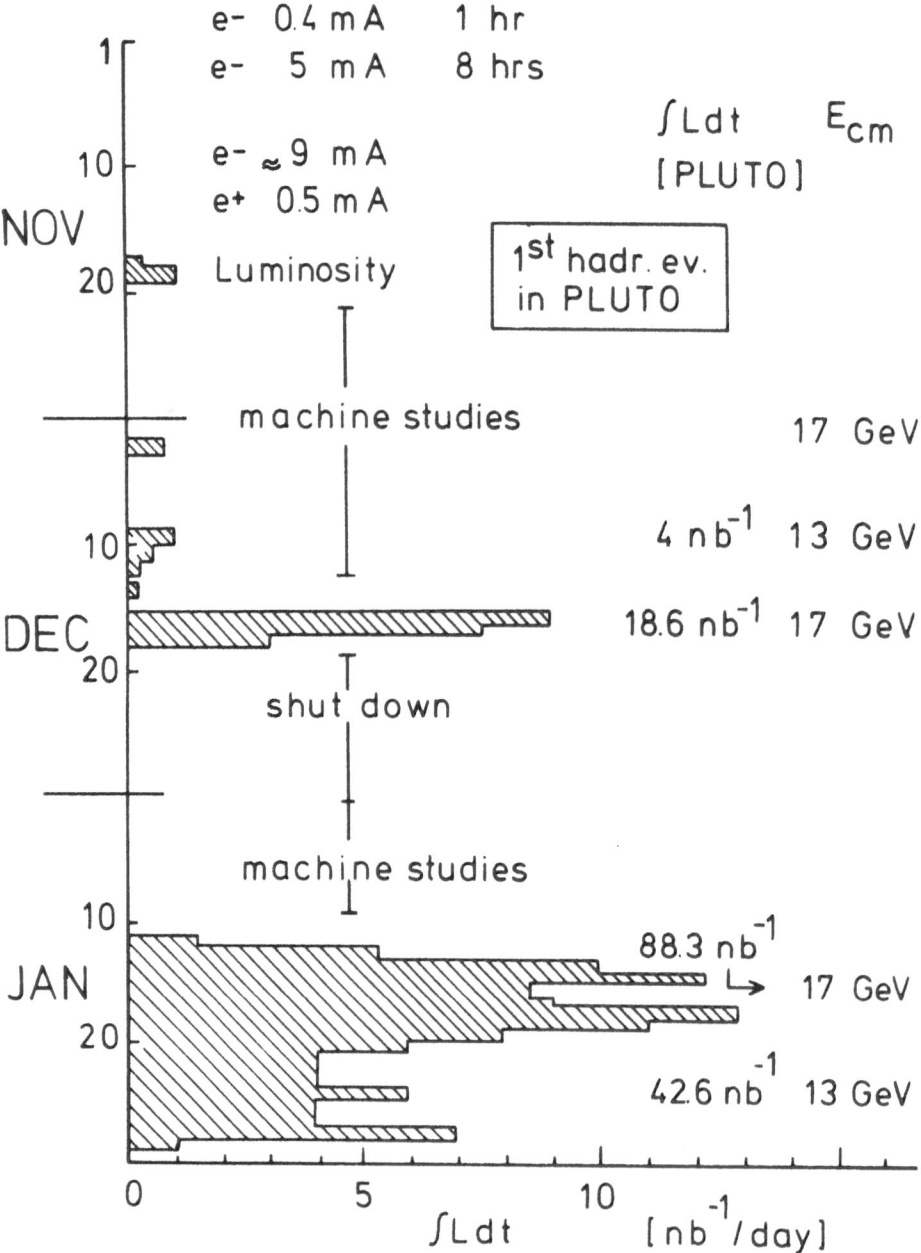

Fig. 49. PETRA development.

$$\frac{\sigma_{had}}{\sigma_{\mu\mu}} = R = 3 \times \sum_q Q_q^2 \; .$$

In addition topological quantities like sphericity or
thrust may be even more important looking for new
thresholds. Any first round experiment should be in a
position to measure these quantities. Therefore, a good
detector should have a large acceptance for charged and
neutral particles. For the topological studies good
energy resolution both for charged and neutral particles
is desirable. As we will see in the last section of this
chapter, two-photon processes become increasingly important
at larger energies. To discriminate against these processes
a good measurement of the total hadronic energy is in-
dispensable. Of course, this also ensures good discrimina-
tion against beam gas, beam wall, and synchrotron radiation
background.

I will briefly describe the three experiments which
have taken data during the first PETRA run. Fig. 50 shows
the detector PLUTO in its proposed final configuration at
PETRA. All components were ready and tested by November 78
when the first data taking started. In addition to the
DORIS configuration which has already been described in
chapter II mainly two new components have been added. The
magnet yoke has been surrounded by additional iron to
provide a total iron thickness of 1 m for muon filtering.
Large area drift chambers have been mounted outside the
new iron house. The complete setup will provide a muon
detection over 83% of 4π with a punchthrough and decay
probability of less than three percent up to a muon
momentum of 5 GeV.

Two forward spectrometers provide electron detection

in the angular region between 25 and 250 mrad. Each
spectrometer contains a small-angle tagger covering the
angular range up to 68 mrad. It consists of a fine
segments array of lead-glass blocks and two sets of
proportional chambers. The remaining angular range is
covered by the large-angle tagger which uses a lead-
scintillator sandwich preceeded by a layer of proportional
tubes. The energy resolution in the SAT and the LAT is
11%/$\sqrt{E}$ and 13%/$\sqrt{E}$, respectively. The forward spectrometers are important mainly for two reasons. They extend
the range for Bhabha scattering down to 25 mrad which is
particularly needed for monitor purposes and they serve
as tagging devices for two photon reactions. The importance
of this will become clear in the course of this chapter.

Fig. 51 shows the Mark J detector [72]. It was built
for the dedicated purpose of measuring weak-electromagnetic
interference through μ-pair production at high energies
[78,79]. The whole apparatus is therefore rotatable in
θ and ϕ. Essentially the setup consists of a central
electromagnetic shower detector (σ/E = 14%/$\sqrt{E}$) surrounded
by a hadron calorimeter. Several layers of track chambers
are inserted between these two calorimeters. For muon
detection the hadron calorimeter is surrounded by additional
iron which is covered by sets of multilayer drift chambers.
For momentum analysis of the muons the iron can be magnetized.
The January data were taken without magnetic field. Also the
endcaps of the detector were still missing during that
period.

Fig. 52 shows a sideview of the TASSO detector [75].
Until January only the central detector consisting of a
drift chamber, the time-of-flight system, the magnet and
the top and bottom μ counters were installed. The data

PLUTO

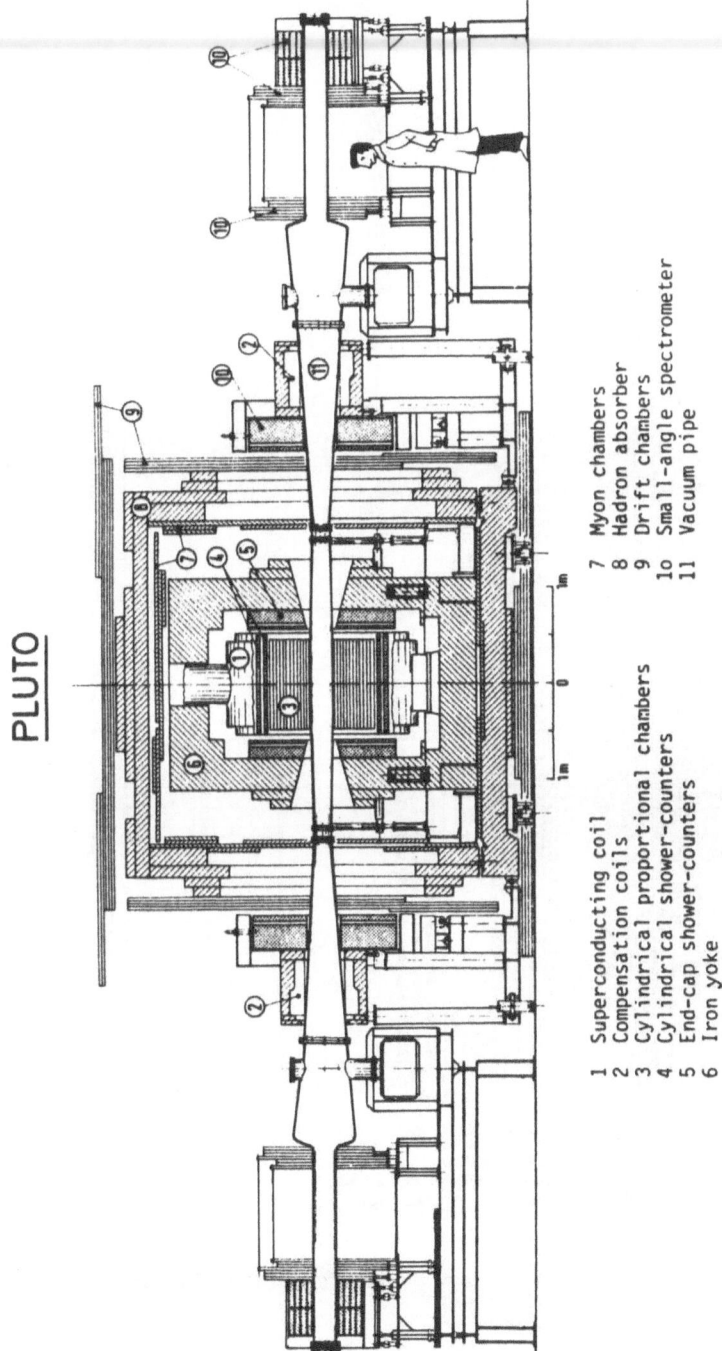

1 Superconducting coil
2 Compensation coils
3 Cylindrical proportional chambers
4 Cylindrical shower-counters
5 End-cap shower-counters
6 Iron yoke
7 Myon chambers
8 Hadron absorber
9 Drift chambers
10 Small-angle spectrometer
11 Vacuum pipe

Fig. 50. The PLUTO detector in its PETRA configuration (Aachen-Bergen-DESY-Hamburg-Maryland-Siegen-Wuppertal Collaboration).

200

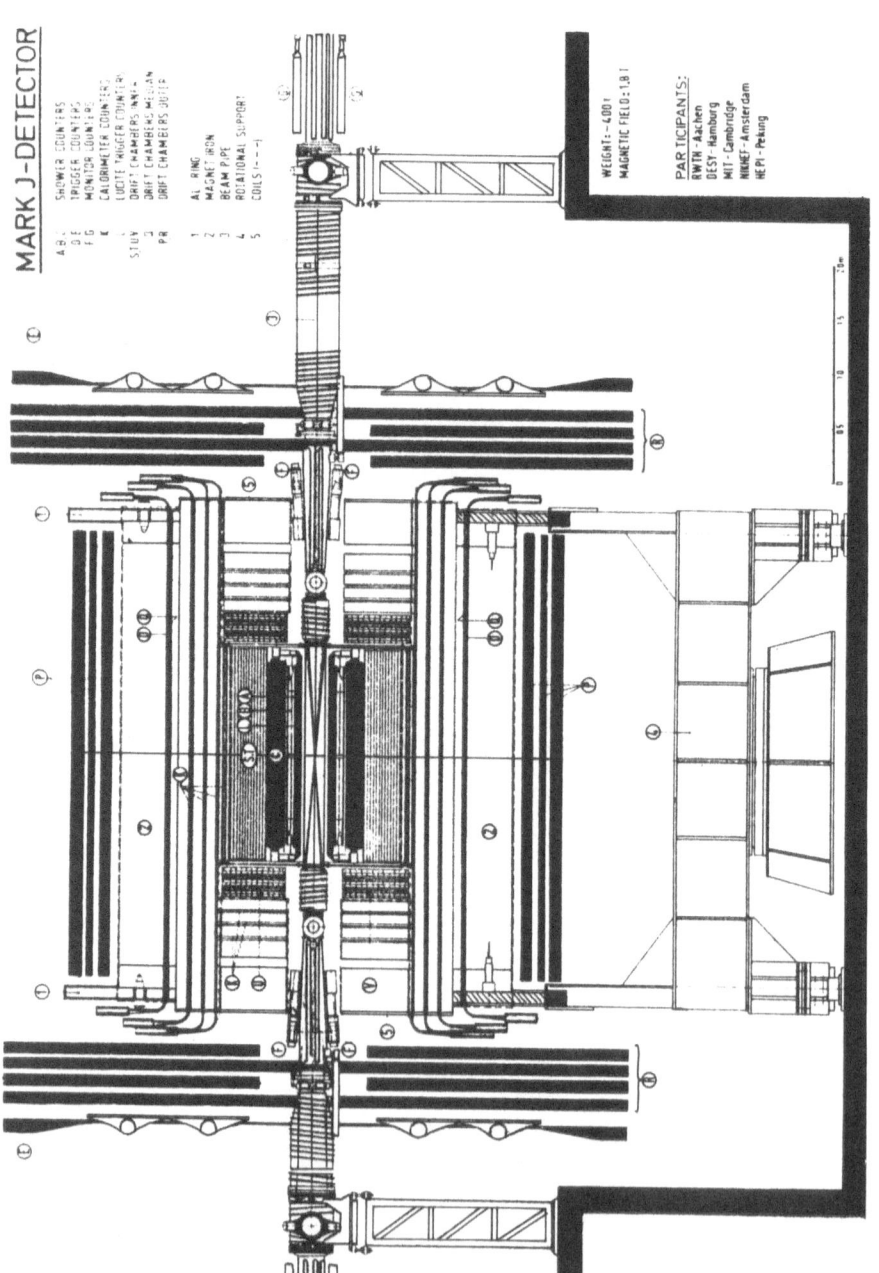

Fig. 51. The MARK J detector (Aachen-DESY-LAPP-MIT-NIKHEF-Peking Collaboration). The endcaps were still missing in the January run.

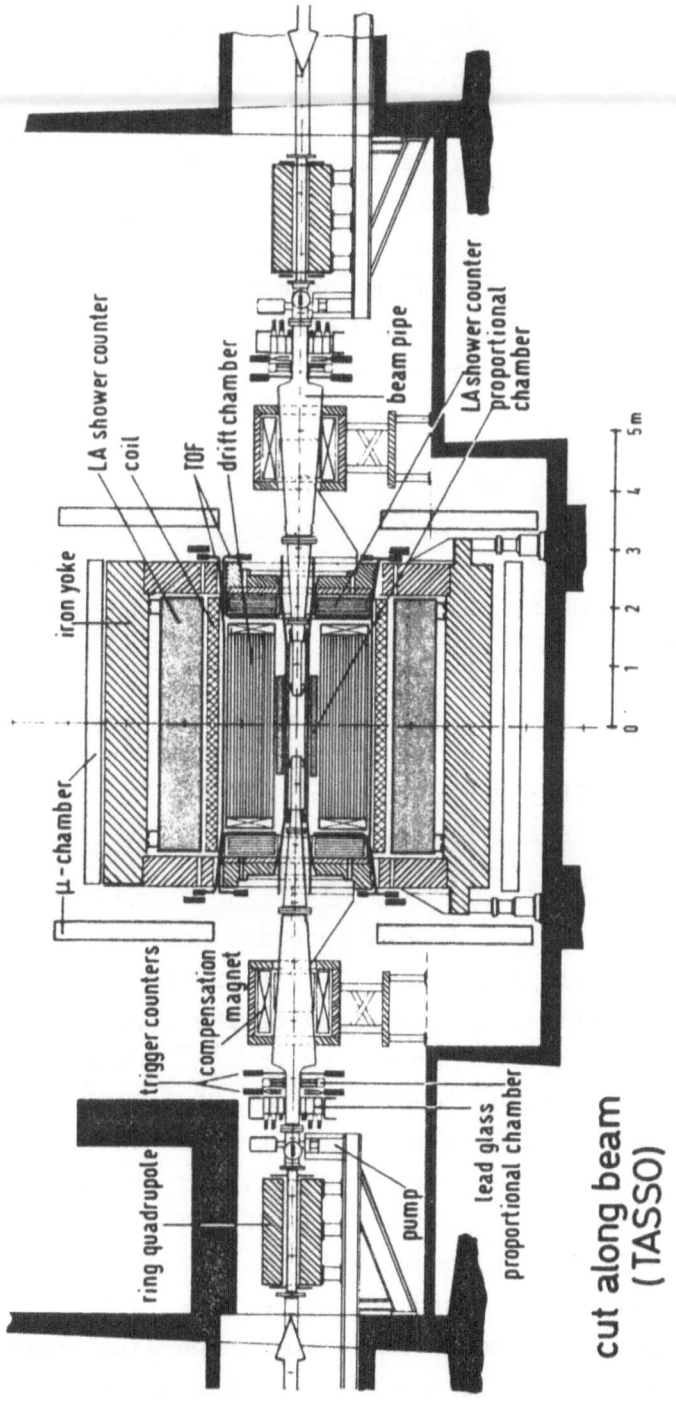

cut along beam
(TASSO)

Fig. 52. The TASSO detector (Aachen-Bonn-DESY-Hamburg-IC London-Oxford-Rutherford-Weizman-Wisconsin-Collaboration). Only the magnet with the central detector was installed for the January run.

I will report here were essentially obtained from the
magnetic detector. A warm coil provides a solenoide
field of .5 Tesla. The field volume of 4.5 m length
and 2.7 m diameter is filled with a large cylindrical
drift chamber with 15 sense wire plances, 9 radial and
6 with a stereo angle of $\pm 4^\circ$ to determine the z direction.
A single wire resolution of 200 microns (rms) is projected
which will yield a momentum resolution of .7% x p (rms).
So far $\lesssim$ 230 microns have been reached.

Background and rates

For a mean luminosity of 3 x 10^{29} we expect a hadronic
rate of

$$\dot{N}(\text{hadronic events}) = L \cdot R \cdot \sigma_{\mu\mu} \approx 0.5 \times 10^{-3}/s.$$

This is to be compared with a trigger rate of

$$\dot{N}(\text{fast trigger}) \approx 10^3/s$$

which is dominated by beam-gas interactions and off-
momentum particles. A comparison of these two numbers
may give the most vivid impression of the experimental
difficulties which are encountered in e^+e^- experiments.
The enormous data-reduction factor of 10^6 to 10^7 is
usually accomplished in a multistep trigger and data-
analysis chain. In the PLUTO detector, for example, the
fast trigger rate is typically some kHz. This is to be
reduced to a few Hz of events which are written on magnetic
tape. The reduction is done in three steps, a fast trigger,
a slow trigger [80] and a filter in the on-line computer.
These raw data are still dominated by background with a

considerable portion of cosmic events. 3 to 4 orders of
magnitude in data reduction are then still left to the
off-line analysis.

3. Results

a) QED processes

In this section I will concentrate on the QED process

$$e^+e^- \rightarrow e^+e^- \quad .$$

(The statistics on muon-pair production at PETRA is
still meager, within the limited statistics the MARK J
data are consistent with QED expectations.) The cross
section for Bhabha scattering is given by

$$\frac{d\sigma}{d\Omega} = \frac{\alpha^2}{2s} \{ \frac{q'^4 + s^2}{q^4} + \frac{2q'^4}{q^2 s} + \frac{q'^4 + q^4}{s^2} \} \qquad q = -s\cos^2\frac{\theta}{2}$$
$$q' = -s\sin^2\frac{\theta}{2}$$

where the first term corresponds to the spacelike and
the last term to the timelike contribution. The intermediate term describes the interference. I want to
discuss two important aspects of Bhabha scattering: The
high-rate forward scattering is a convenient monitor
reaction since it is governed by small q^2 where QED is
known to hold. The large angle Bhabha scatters can be
used to check the validity of QED at high q^2.

Bhabha scattering as monitor reaction

Bhabha scattering is experimentally defined by a time
coincidence between two colinear pairs of showering
particles having the energy of the circulating beam.
Fig. 53 shows the angular distribution in the small-
angle tagger for a subsample of the PLUTO data [81].
The solid line shows a fit with the QED expectation.
The relative normalization of the QED curve and the data
determines the integrated luminosity of this data sample.
Although this looks very simple and safe at first sight
there are several sources for systematic errors.

Since the SAT is mounted close to the beam pipe it
suffers from accidental coincidences of off-momentum
particles under bad beam conditions. Because the QED
function is very steep in forward direction ($\sin^{-4} \theta/2$)
the normalization is extremely sensitive to a precise
knowledge of the position of the SAT. Additional uncertainties come from an interplay of collinearity,
energy cuts, and radiative corrections.

To get a handle on these possible systematic
errors the SAT data were compared with e^+e^- large-angle
scattering in the central detector. Fig. 54 shows the
result of the central-detector Bhabha scattering. The
QED expectation from the SAT measurements is indicated
by the full curve. It agrees with the measurement to
within 6%. Therefore a 10% systematic error on the
luminosity measurement is a conservative estimate.

A vertex distribution of Bhabha events in the
central detector is shown in Fig. 55. The background
is apparently low (less than 1%). This figure provides

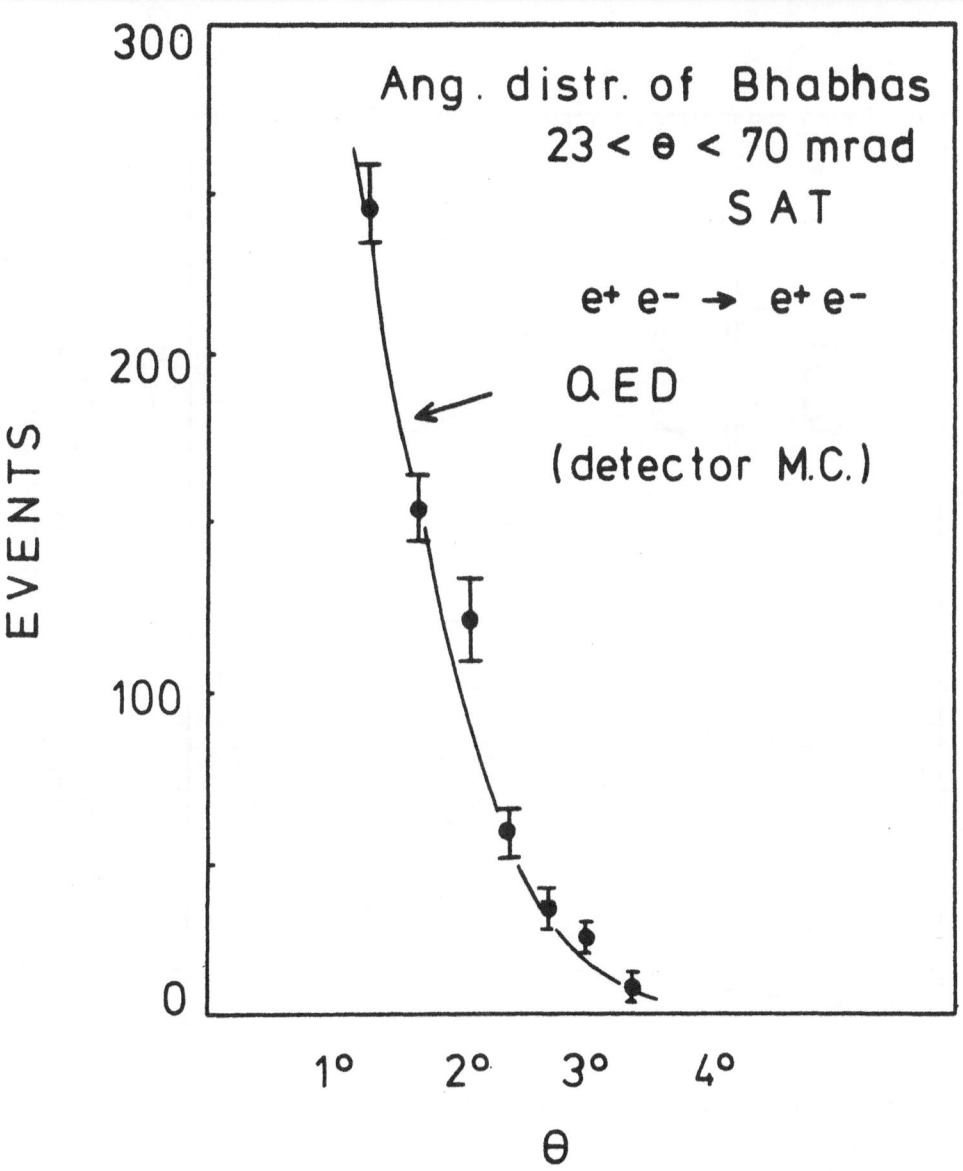

Fig. 53. PLUTO: (preliminary) angular distribution of small-angle Bhabha events in the small-angle tagger (SAT).

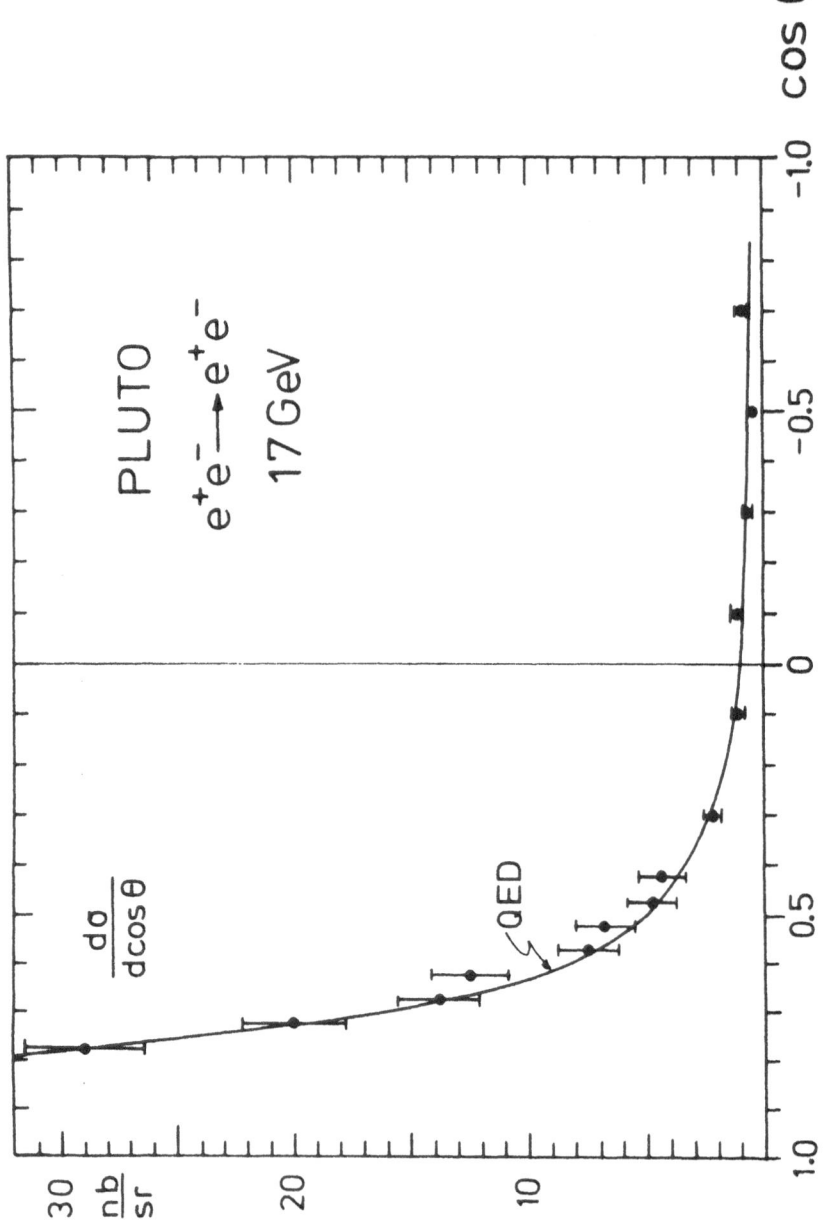

Fig. 54. PLUTO: (preliminary) Bhabha scattering angular distribution in the central detector (barrel + endcap). The curve indicates the QED extrapolation from the SAT including all radiative corrections.

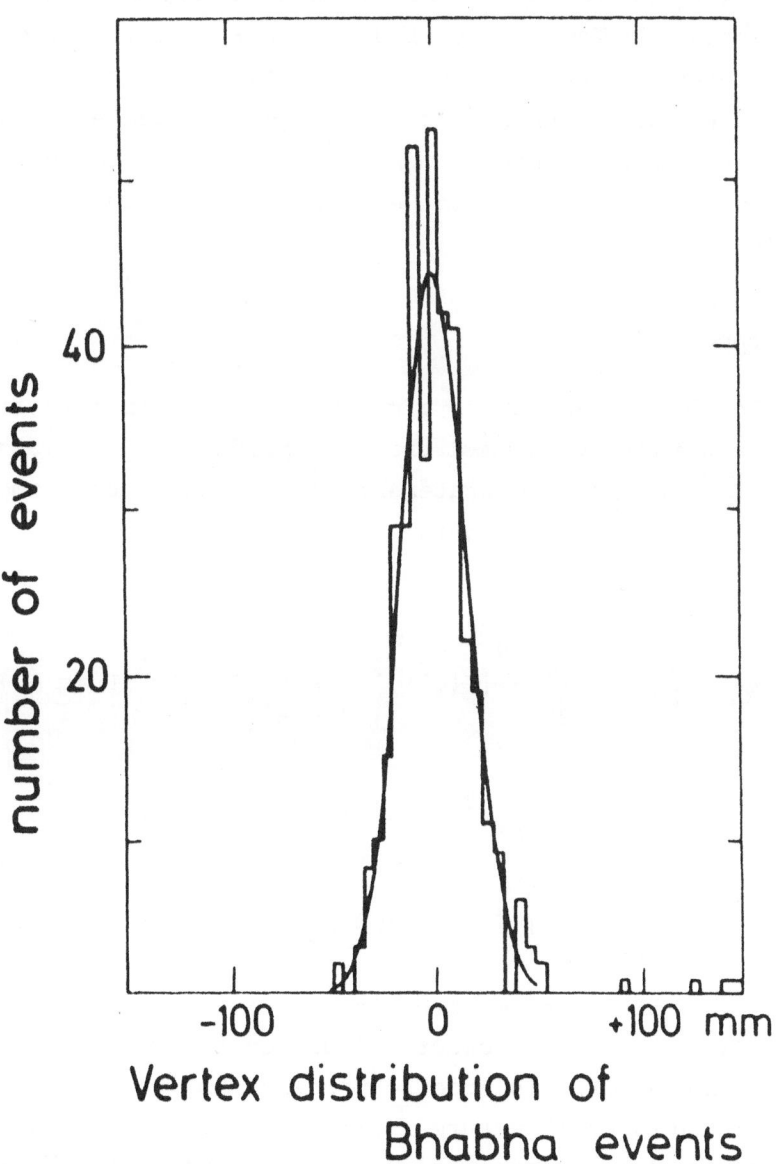

Fig. 55. PLUTO: Vertex distribution of Bhabha scatters
in the central detector.

us with a valuable extra information on machine parameters. From the width of the distribution one can
determine the length of the circulating bunches experimentally. The result is σ = 11.8 ± .5 mm. This is
in agreement with results from machine studies (σ =
= 11.4 mm) and corresponds to the theoretical expectation.
From this bunch length a momentum resolution per beam of
$6.5 \cdot 10^{-5}$ GeV·p (rms; p in GeV) can be inferred (compare
table 10).

QED check

In the above argument on a check of the luminosity it
has been tacitly assumed that QED holds even for large
q^2. To quantify this statement let us introduce form
factors in the differential cross section for e^+e^-
scattering [82].

$$\frac{d\sigma}{d\Omega} = \frac{\alpha^2}{2s} \{ \frac{q'^4 + s^2}{q^4} |F(q^2)|^2 + \frac{s^4}{q^2 q'^2} \mathrm{Re}\,(F(q^2) F^+(q'^2))$$

$$+ \frac{q^4 + s^2}{q'^4} |F(q'^2)|^2 \}$$

$$F(q^2) = 1 \mp q^2/(q^2 - \Lambda_\pm^2) \quad .$$

There are different ways of introducing deviations from
QED. Accordingly the exact definition and physical interpretation of Λ is model dependent. In the context of
this experimental lecture Λ with the above definition is
merely used as a parametrization to quantify possible
limits on QED.

All three experiments [83-85] at PETRA have made
an attempt to determine the cut-off parameters from their
Bhabha scattering data. Table 11 summarizes the results.
For comparison also the best values known so far from
SLAC [86] have been given. All values were obtained
fitting the experimental data with the above parametrization of the cross section and taking into account
radiative corrections [87]. The PLUTO result [84] was
deduced from the data shown in Fig. 54 the angular
distribution of the MARK J data [83] from which the cut-
off parameters were calculated is shwon in Fig. 56.

In conclusion we can say that QED holds even at
these high energies. Thus the cut-off parameter can be
pushed as high as 40 GeV.

Exp.		Λ_+ (GeV)	Λ_- (GeV)	Ref.
SLAC	1974	14.4	22.8	[86]
	1975[*]	15	19	[86]
	1976	33.8	38.0	[86]
MARK J		26	38	[83]
PLUTO[*] } (pre-		39	42	[84]
TASSO } liminary)		29	34	[85]

Table 11. QED cut-off parameters assuming equal timelike
and spacelike form factors

$$F(q^2) = 1 \mp q^2/q^2 - \Lambda_\pm^2 . \text{[*]}$$

95% C.L. lower limits for Λ_+ and Λ_- are given.

[*]In the limit $\Lambda^2 \gg q^2$ $F(q^2)$ can be written

$$F(q^2) = 1 \pm q^2/\Lambda_\pm^2 .$$

This parametrization has been applied to the data of
the MARK J and PLUTO detector.

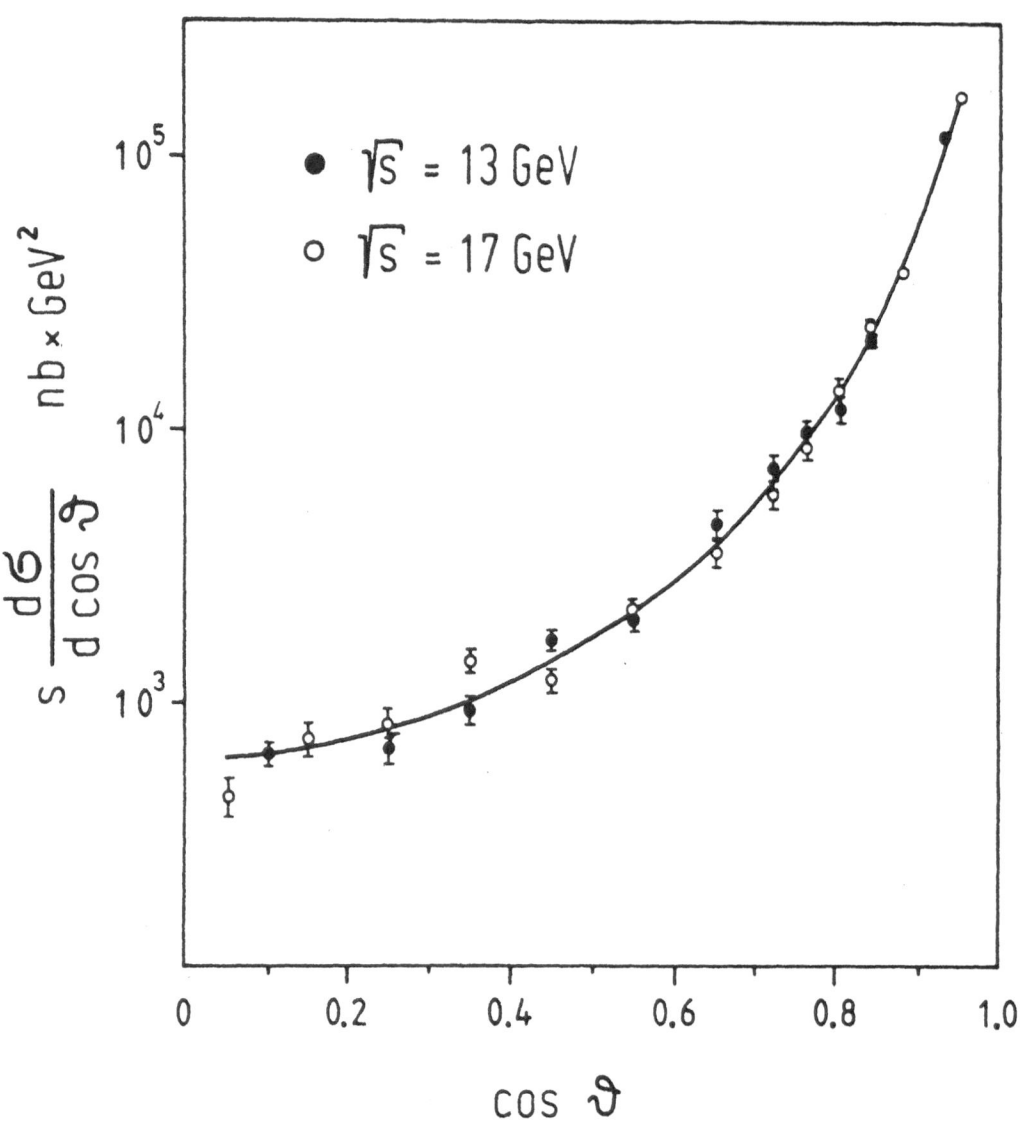

Fig. 56. MARK J: angular distribution of Bhabha scattering **at**
13 and 17 GeV. The curve is a common QED fit to
the data including all radiative corrections.

b) Total cross section

Entering a new region of energy the PLUTO, MARK J and
TASSO groups made an attempt to measure the total cross
section [81,88,89]. It has been emphasized already in
the first chapter why this quantity is of particular
interest in e^+e^- reactions. Any increase in $R = \sigma_{had}/\sigma_{\mu\mu}$
would indicate new flavours. To determine R the theoretical
value of $\sigma_{\mu\mu}$ is taken while σ_{had} is measured experimentally:

$$\sigma_{had} = \frac{N}{\varepsilon \cdot \int L \; dt} \; .$$

The total number of observed hadronic events N has been
normalized to the integrated luminosity as described in
the previous section.

The event acceptance ε is obtained by Monte Carlo
studies. Events are generated according to the Feynman-
Field parametrization and passed through a realistic
model of the detector. Of course, the larger the
acceptance the less does ε depend on details of the
model. The typical acceptance of the three detectors
is of the order of 70 to 80%. The systematic errors
of the total cross section measurements are mainly due
to acceptance and luminosity uncertainties.

The number of hadronic events N has to be separated
from a background of cosmic rays, QED events of the e and
μ type, beam gas, beam losses, synchrotron radiation, and
so on. Let me remind you again that this background is
about 6 to 7 orders of magnitude higher than the events
rate of about 1 event per hour under the January running
conditions. The event selection criteria are based on a
combination of energy and track requirements. In the

PLUTO detector [81] for example at least 2 charged
particles and a neutral energy deposition of more than
$0.3 \times E_{CM}$ was required. Very similar criteria were
applied in the MARK J detector [88]: at least 2 hadron
showers with at least 1 track pointing to the vertex
and a total energy of more than 10 GeV. Since no photon
measurement was available in the TASSO detector [89]
the track requirements had to be sharpened: at least 3
tracks originating from the vertex.

Fig. 57 demonstrates the quality of event selection
in the PLUTO detector. The neutral energy is plotted
against the event vertex on the beam axis. It shows that
after the neutral energy cut only very little beam-gas
background remains in the sample. This background can be
estimated from the tails outside the interaction region.
Fig. 58 shows an example of a hadronic event in the
detector PLUTO - by the way the first hadronic event
observed at PETRA.

The results of the three detectors are summarized
in table 12. The differences in integrated luminosity
are due to the fact that not all detectors were operational
during the full time of data taking. The radiative
corrections quoted are for the hadronic cross section
only. Radiative corrections to the monitor are included
in the quoted luminosities. Estimated beam-gas background
is very low in all three experiments. Residual $\gamma\gamma$-con-
tributions to the observed number of these calculations
have been cross checked with the measured two-photon
events which I will discuss at the end of this chapter.
The final number of hadronic events and the values of R
are given in the table. The estimated systematic errors
are of the order of 20%. Fig. 59 shows a plot of the

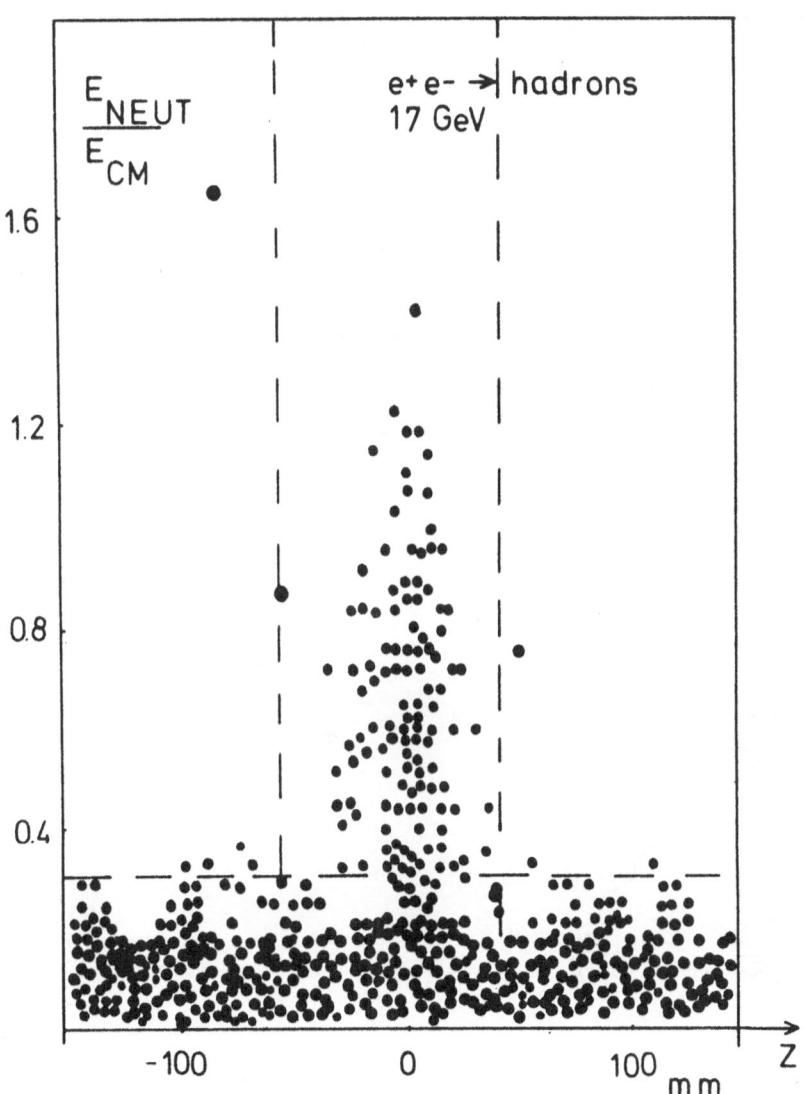

Fig. 57. PLUTO: event selection demonstrated in a two dimensional plot of neutral energy vs. event vertex along the beam axis.

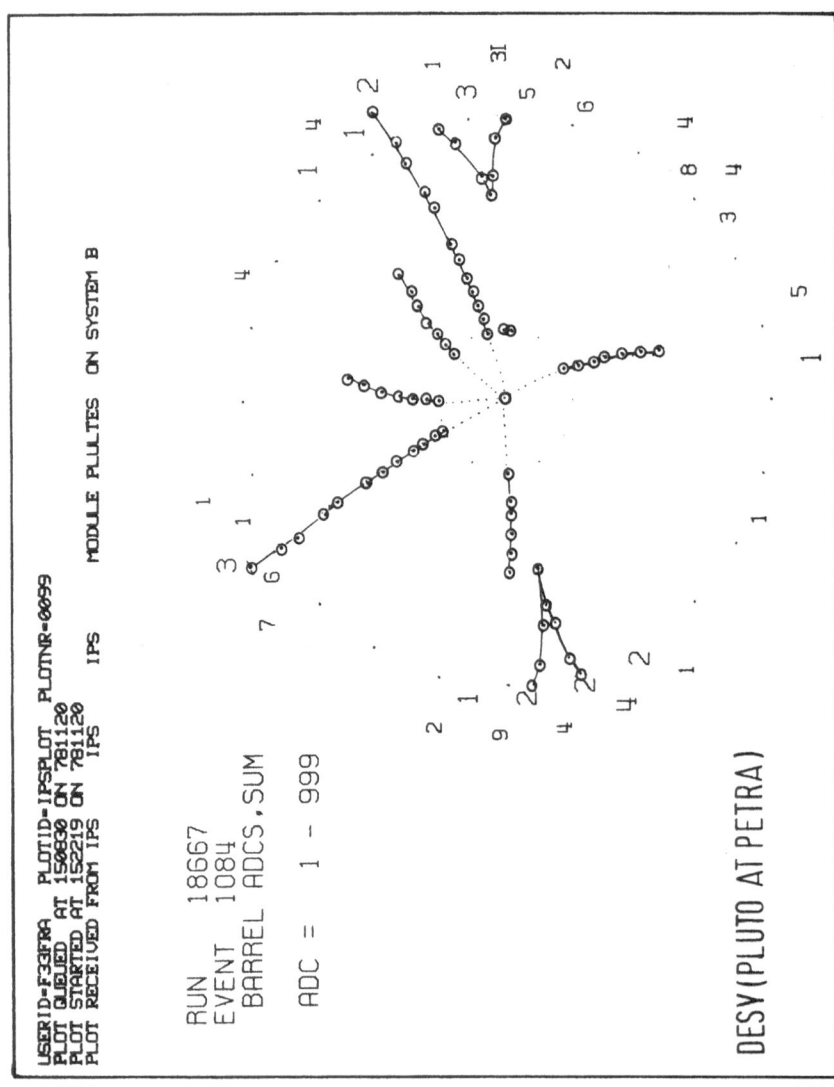

Fig. 58. PLUTO: first hadronic event seen at PETRA.

	PLUTO [81]		MARK J [88]		TASSO [89]	
E_{CM} (GeV)	13	17	13	17	13	17
Acceptance ε (%)	72	72	79	79	77	78
$\int L\,dt$ (nb^{-1})	42.6	88.3	45	60	29.6±3.0	39.2±3.5
Radiative corrections (%)	-10	-10			-8	-8
Estimated bg (ev.)	<1	<1	<1	<1	3	1.5
Estimated $\gamma\gamma$ (ev.)	6	13	a few %	a few %		
hadronic events (ev.)	96	108	83	68	75	40.5
$R = \dfrac{\sigma(\text{hadr.})}{\sigma_{\mu\mu}}$	5.0±0.5 (τ excluded)	4.3±0.5	4.6±0.5	4.9±0.6	5.6±0.7	4.0±0.7 (τ excluded)
Estimated systematic errors	20%	20%	±0.7	±0.7	20%	20%

Table 12. Total cross sections at 13 and 17 GeV.

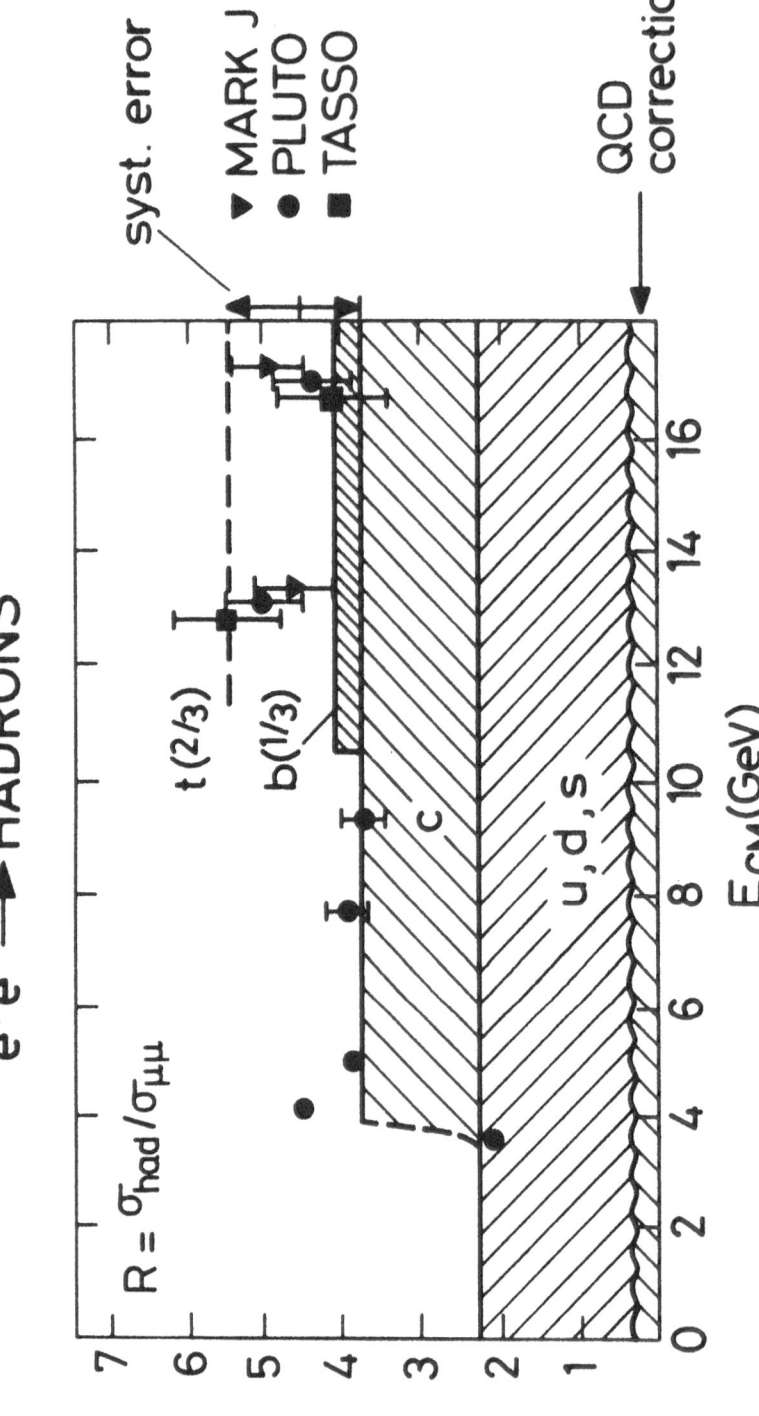

Fig. 59. PLUTO, TASSO, MARK J: measurement of $R = \sigma_{had}/\sigma_{\mu\mu}$ at 13 and 17 GeV. The PLUTO data below 10 GeV are given for comparison. The expectations for different flavours are indicated.

measured values of R together with the measurements of PLUTO below 9.5 GeV. The different levels of R expected for the old quarks and possible new contributions from 1/3 and 2/3 charged quarks are indicated.

We can draw the following conclusions:

- Above 9.5 GeV there is certainly room for an increase of R corresponding to a $b\bar{b}$ threshold. It is unlikely that this new quark b has a charge of 2/3 although this cannot be excluded due to the large systematic errors. (Compare also the evidence for $Q_b = -1/3$ given in chapter II.)

- Assuming that the systematic errors cancel in the relative measurement between 13 and 17 GeV we can exclude any new charged 2/3 thresholds coming up between these two energies.

c) Event topology

Events in e^+e^- annihilation exhibit a jet structure outside the resonances. As shown in chapter III this two jet topology becomes more and more pronounced if one increases the energy. In first approximation the event topology below 10 GeV is fairly well understood in the simple quark-fragmentation model. As the energy increases two additional effects are expected.

Thresholds

As one crosses a threshold of new flavour production the decay of new heavy quark pairs will contribute to the event topology [46]. The sphericity of these weakly

decaying slow quarks will be comparatively large [90].
Consequently the mean sphericity will increase. Under
favourable conditions (near threshold) the new
contributions might even be separable.

QCD jets

As Sterman and Weinberg [45] pointed out the effect
of gluon emission in the quark-pair production leads
to a natural broadening of the energy flow in the
final state [91-93]. The effect which is analogous
to QED radiative corrections can be calculated in
the first order QCD (Fig. 60).

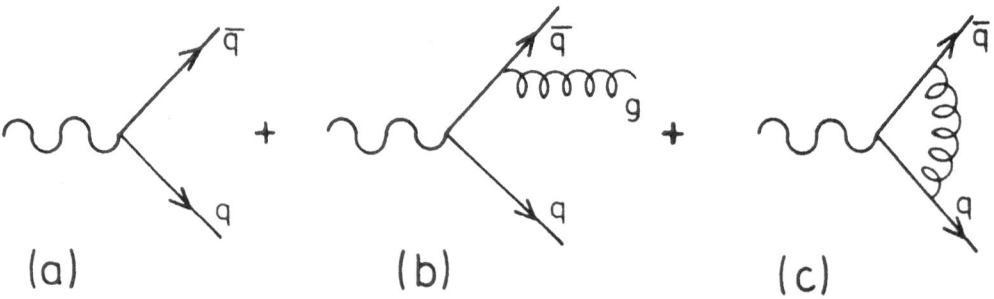

Fig. 60. Contributions to the annihilation cross section
in first order QCD.

At large energies this effect should lead to a broadening
of the jet fragmentation. One prediction is therefore
that the mean transverse momentum of the jets should
increase with energy.

Fig. 61 which is taken from ref.[46] shows the
expected behaviour of the different components which
contribute to the event topology.

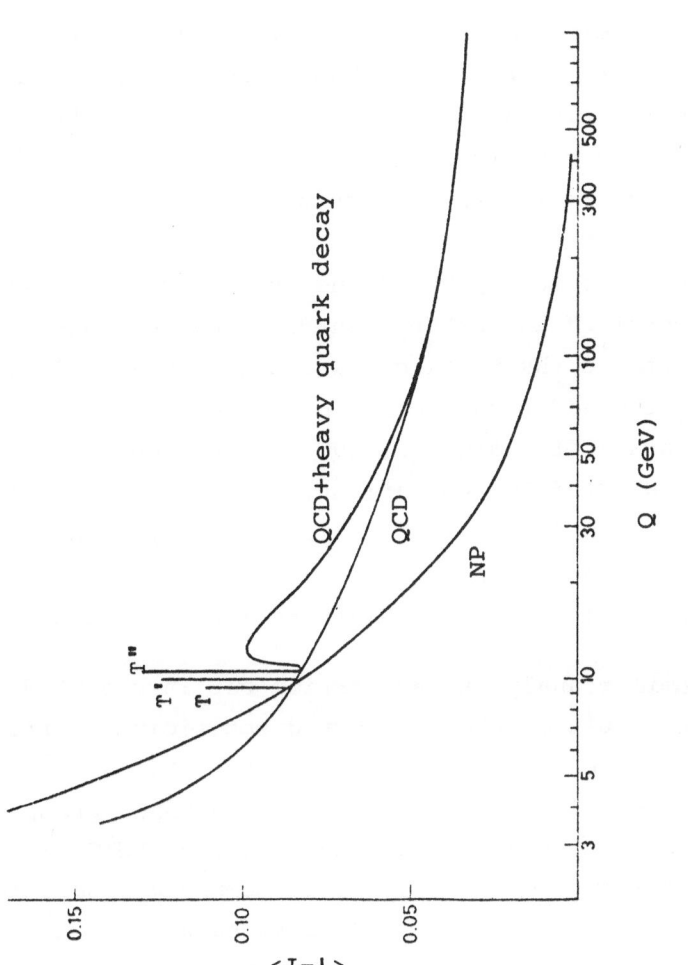

Fig. 61. Theoretical expectation for three contributions
to the event topology: first order QCD,
fragmentation (NP), and heavy quarks (from ref.[45]).

The PLUTO group has analysed the high-energy data
at 13 and 17 GeV in the same way described for the data
below 10 GeV. Fig. 62 shows the mean observed sphericity
as a function of c.m. energy now ranging from 3.6 to
17 GeV. The data of SLAC-LBL, TASSO and the NaJ-lead
glass detector are given for comparison. The dashed
curve shows the expectation of the Feynman-Field Monte
Carlo (without radiative corrections). The dash-dotted
line indicates the additional contribution from charm
production. Note that part of the charm production is
already contained in the Feynman-Field curve since it
is normalized to the total cross section. The full line
then shows the additional contribution expected from a
new heavy quark of charge 1/3 [90]. A charged 2/3 new
quark is indicated for comparison. It clearly shows that
the data are consistent with $b\bar{b}$ (charge 1/3) production
and that a charge of 2/3 can be excluded. Fig. 63 shows
the differential sphericity distribution at 13 and 17 GeV.

A similar analysis was carried out in the TASSO
group. The result of their observed sphericity distribution of charged particles is given in Fig. 64. We realize
that the sphericity distributions of the TASSO group are
narrower and that <S> is lower than in the PLUTO experiment. This is due to the better momentum resolution
in the TASSO detector. The distributions exhibit a tail
towards large values. For comparison the expectation for
b quark production is indicated in all figures [90]. We
see that there is room for this contribution in the data.
Note that the mean sphericity of the bottom events is
considerably higher than that of the old quarks. This
difference is particularly pronounced at 17 GeV. Again
a contribution of a new quark with charge 2/3 would be
four times as high in rate as the indicated b-quark
contribution. This is certainly excluded from the data.

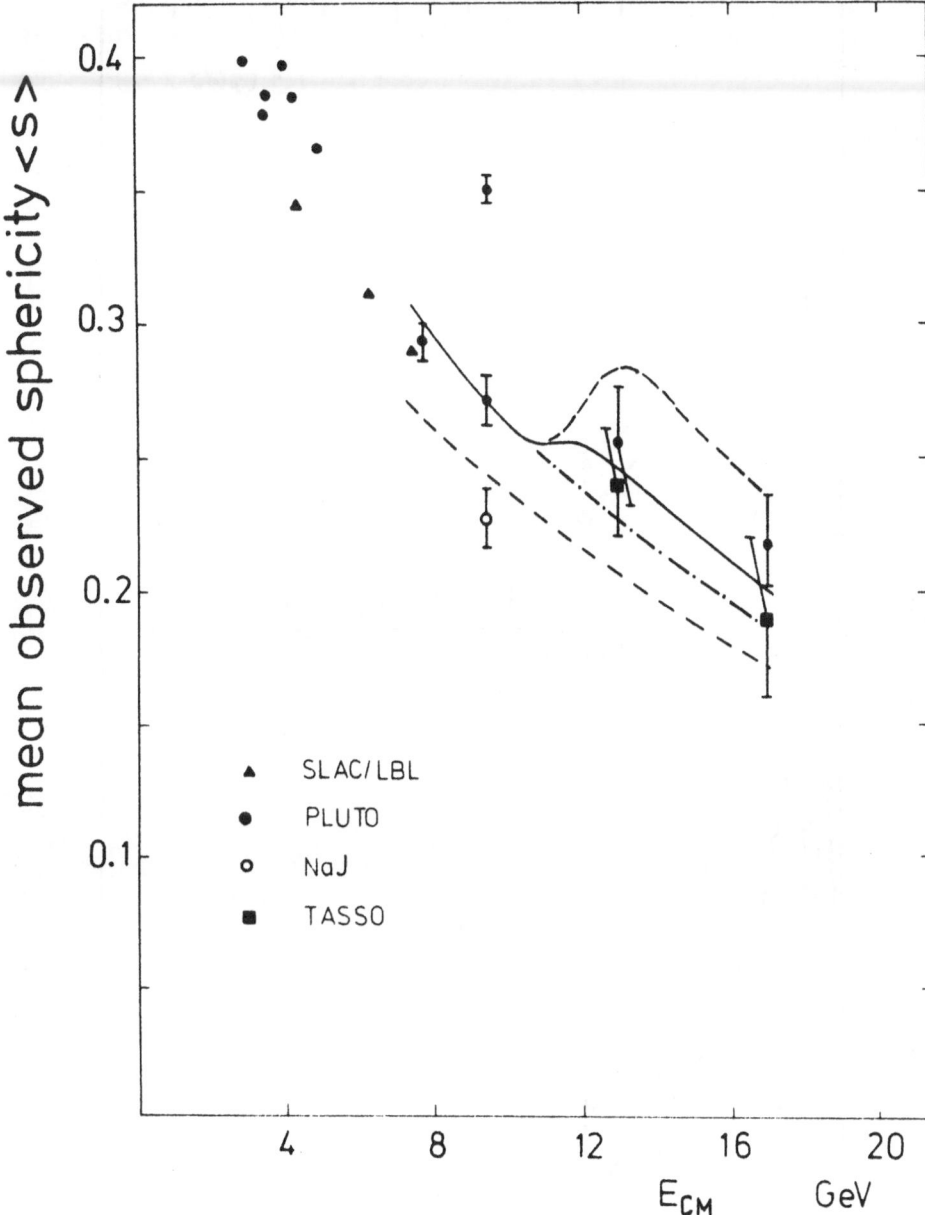

Fig. 62. SLAC-LBL, PLUTO, TASSO and NaJ-lead glass: mean
observed sphericity as a function of energy. The
dashed line indicates the simple two-jet model
(Feynman-Field). Inclusion of charm and bottom
leads to the dash-dotted and full curves [90].
A charge 2/3 contribution is indicated.

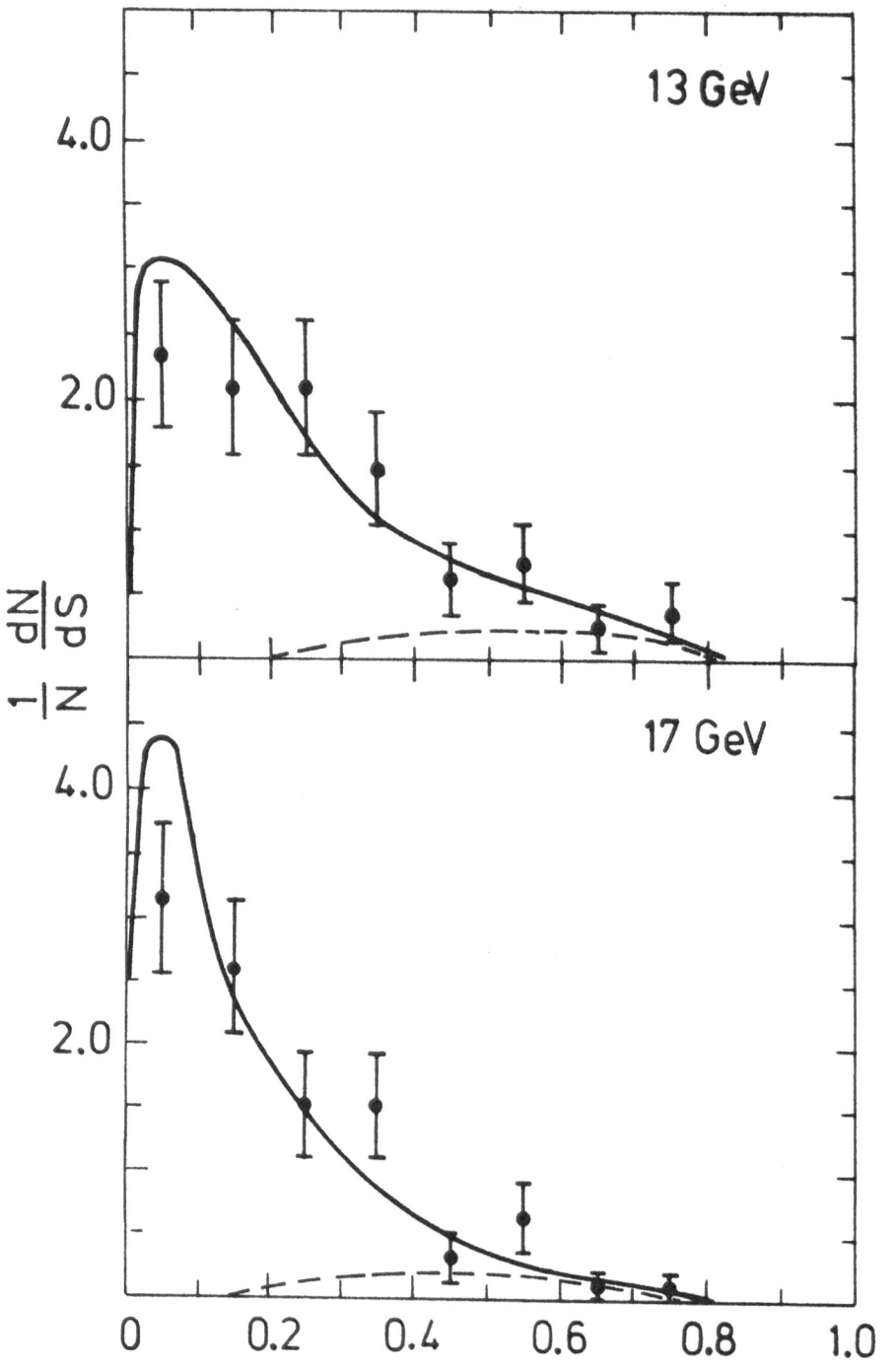

Fig. 63. PLUTO: observed sphericity distributions at 13 and 17 GeV. The expectation for b-quark production is indicated by a dashed line [90]. The full line is the two-jet model including all corrections.

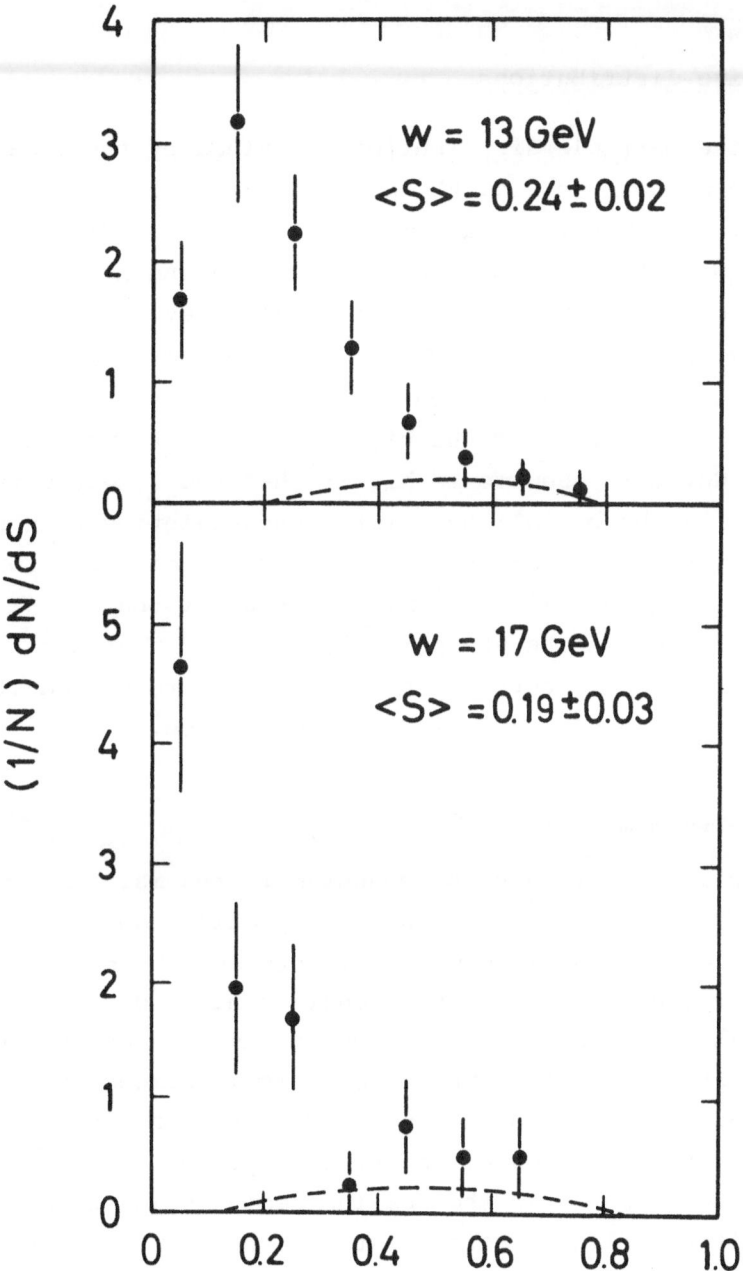

Fig. 64. TASSO: observed sphericity distributions at 13 and 17 GeV. The expectation for b-quark production is indicated by a dashed line [90].

Inclusive distributions

Inclusive particle distributions obtained by the TASSO
group are shown in Fig. 65. Data at 13 and 17 GeV are
compared with previous measurements of the DASP group at
5 GeV. We notice a remarkable scaling beyond x = 0.2
over the full energy swing from 5 to 17 GeV. Below x = 0.2
the distributions are filled up at high energies. Two
interesting observations can be made in this region.
Instead of approaching the high-energy limit from below
the 13 GeV data seem to be higher than the 17 GeV data.
Neglecting the overall systematic uncertainties this is
a two-standard deviation effect. Also the distribution
does not simply follow an extrapolation of the high-x
data but seems to be steeper below x = 0.2. Further
study of these effects which might be due to b production
is certainly needed and of prime importance.

Transverse momentum

Whereas a fixed transverse momentum is the main ingredient
of a simple quark-parton model, QCD predicts an increase of
p_T with energy due to gluon bremsstrahlung. Fig. 66a shows
the energy dependence of the longitudinal and transverse
momentum with respect to the sphericity axis as observed
in the PLUTO detector. The figure nicely demonstrates
again how jets are developing with increasing energy
through their limited p_T and rising p_L. In particular p_L
is rising linearly in the scaling region x_L greater than
0.1. After a slight increase at low energies where the
jets are not fully developed the mean p_T levels off at
high energies. Comparing the values at 13 and 17 GeV as
measured by the PLUTO and the TASSO groups (Fig. 66b)
one observes in fact constance between the two energies

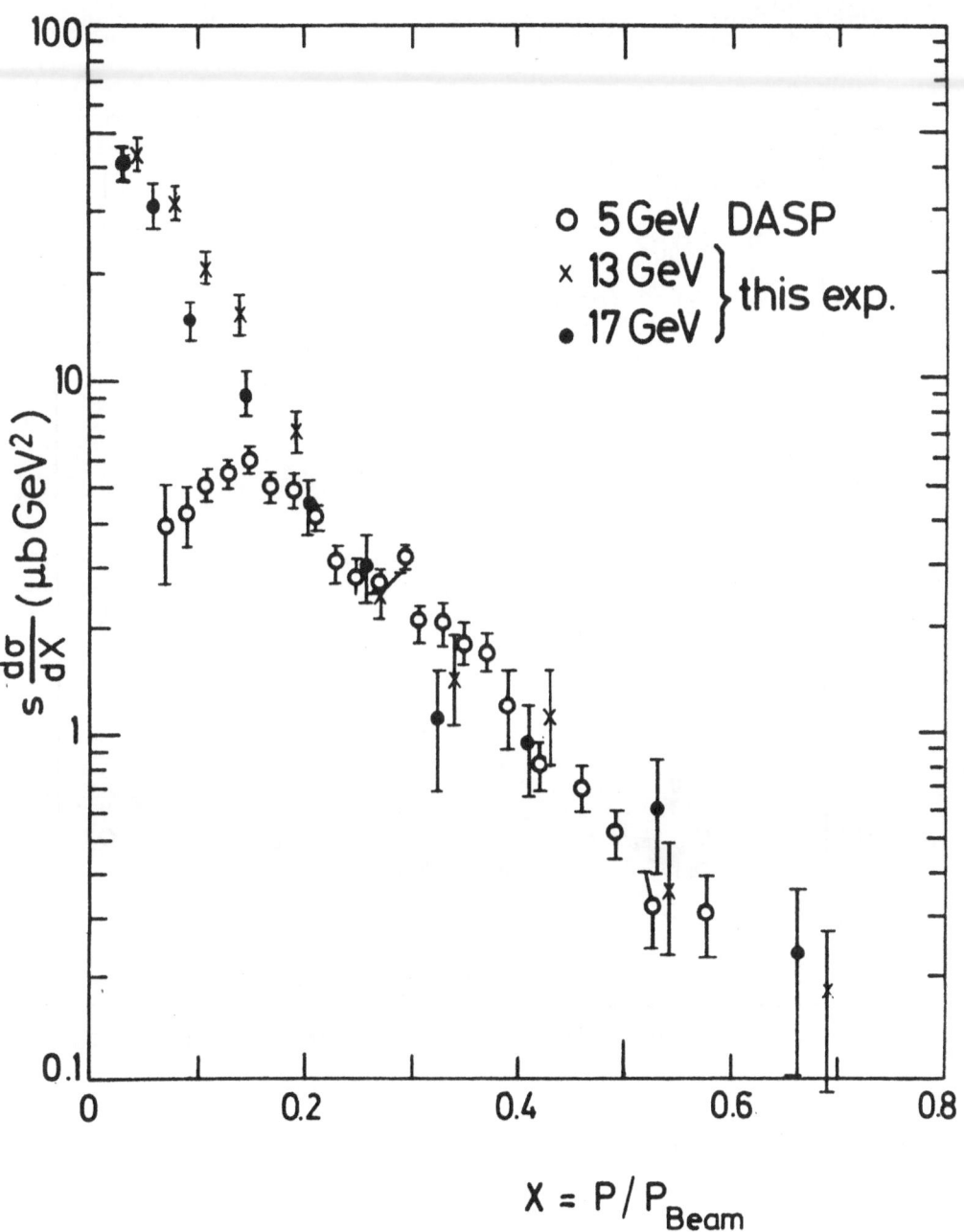

Fig. 65. TASSO: inclusive momentum distributions for charged particles.

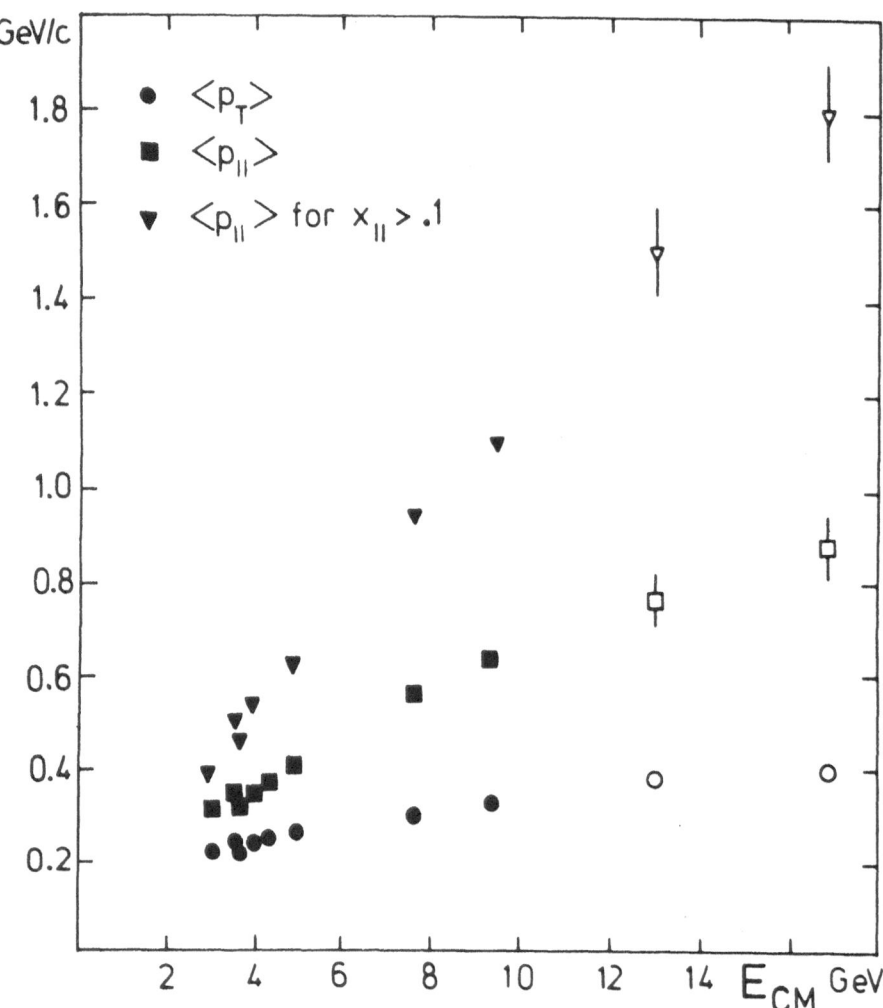

Fig. 66a. PLUTO: mean observed p_T and p_L as a function of energy. $\langle p_L \rangle$ is also shown for the scaling region $x_L > 0.1$.

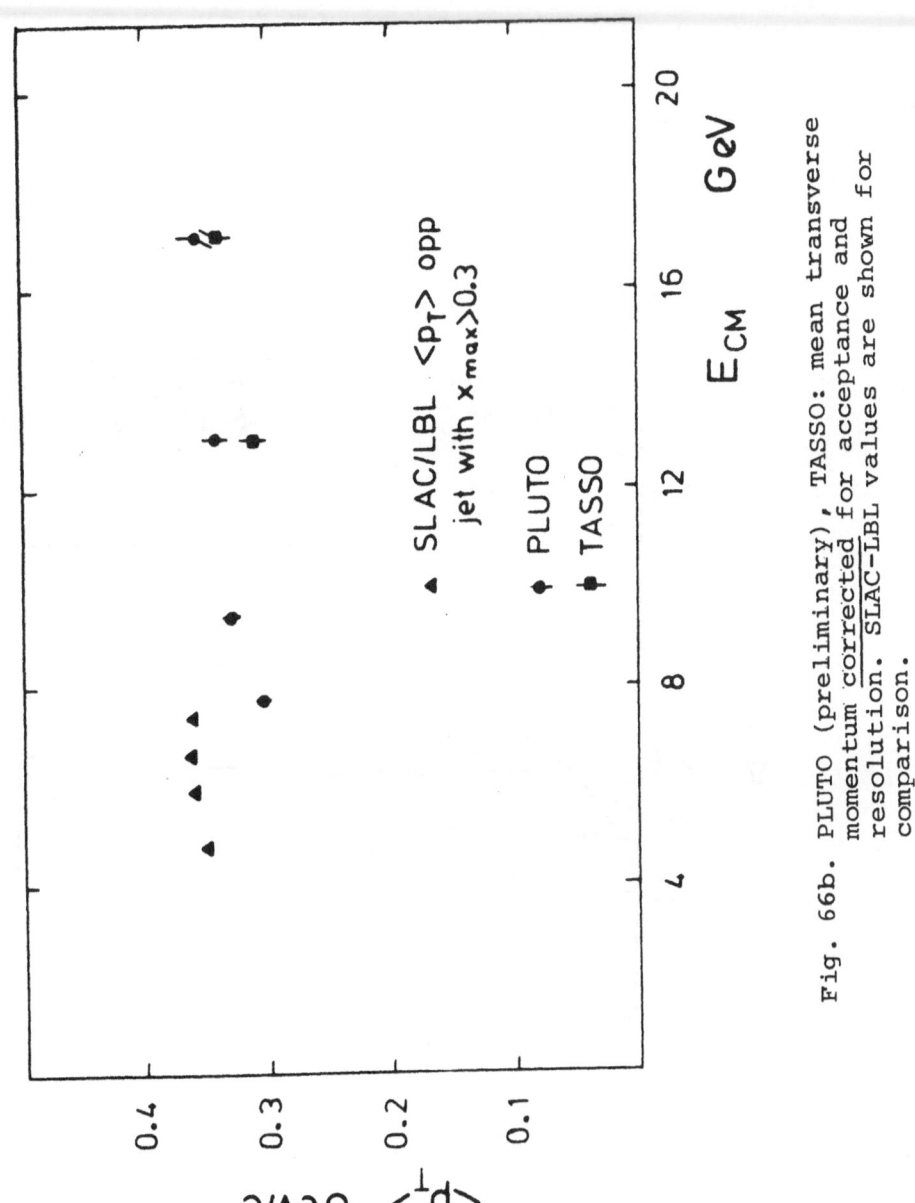

Fig. 66b. PLUTO (preliminary), TASSO: mean transverse momentum corrected for acceptance and resolution. SLAC-LBL values are shown for comparison.

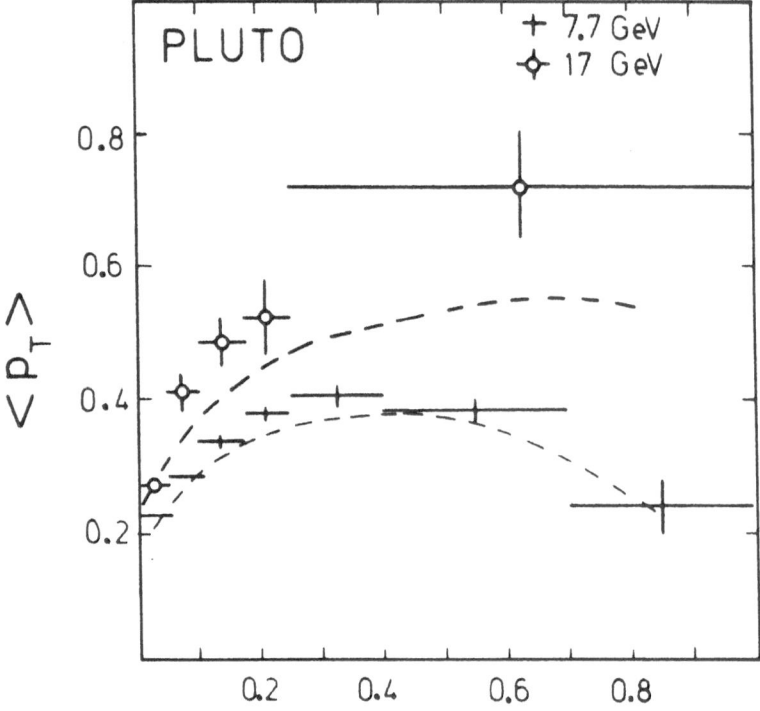

Fig. 67a. PLUTO: "Seagull" effect at 7.7 and 17 GeV. The mean observed p_T is shown as a function of $x_L = p_L/E_b$. The curves are two-jet model calculations.

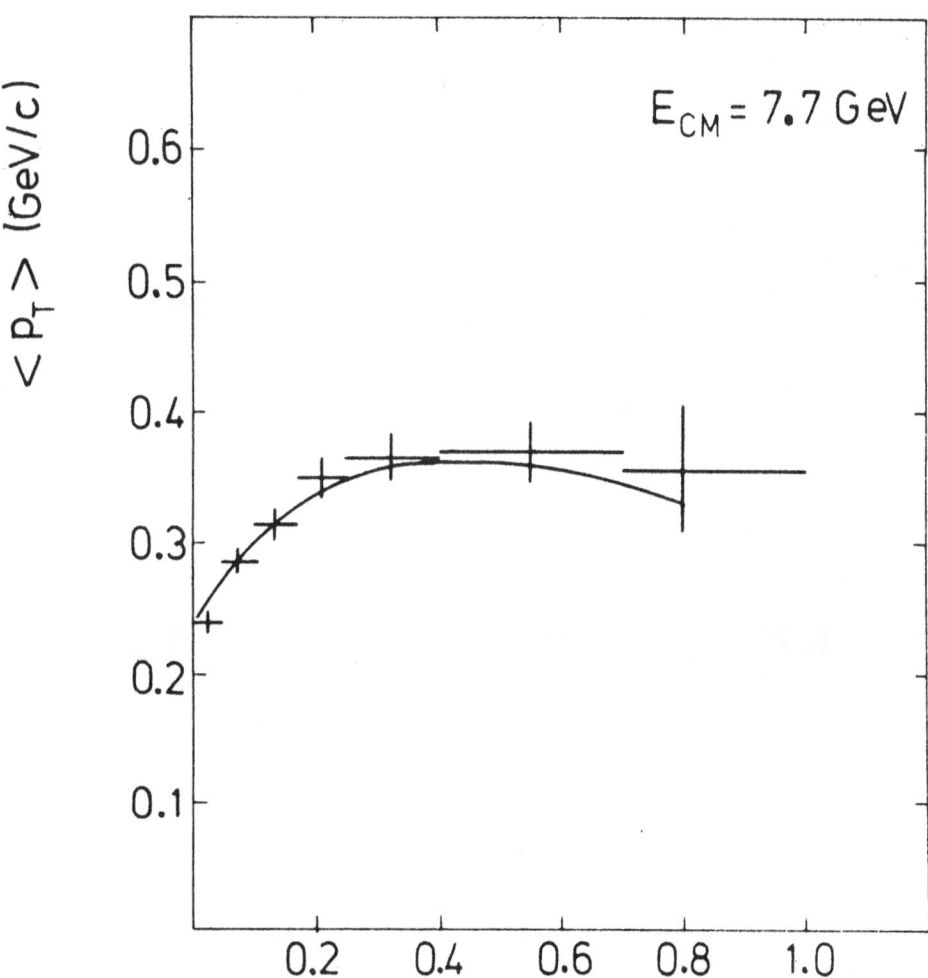

Fig. 67b. 7.7 GeV: $\langle p_T \rangle$ including all corrections. The two-jet model is shown for comparison.

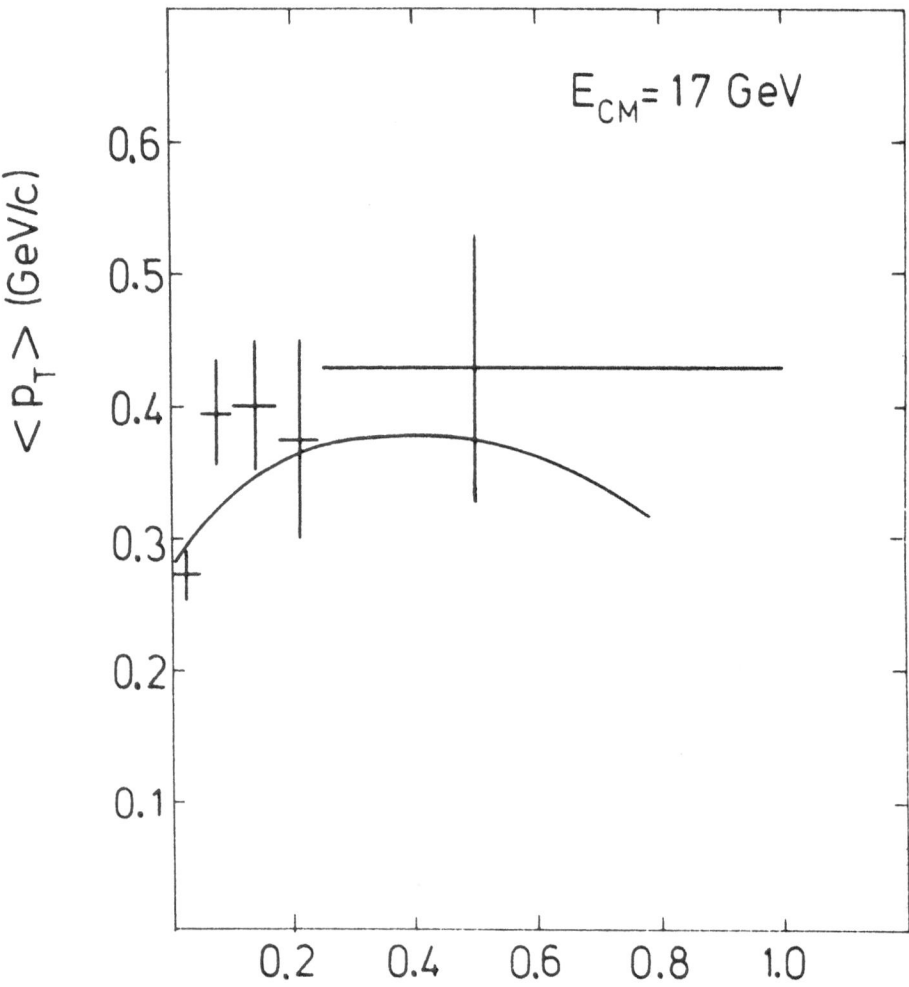

Fig. 67c. 17 GeV: $\langle p_T \rangle$ including all corrections. The two-jet model is shown for comparison.

within two standard deviations (Fig. 66b shows corrected values).

The value of mean p_T is, of course, governed by the plateau region. One might argue that this is the region of confinement effects and that, therefore, intermediate x_L might be a better place to look for QCD effects [94]. Indeed, if we compare the 'seagull' effect at 7.7 and 17 GeV (Fig. 67) we see a strong increase in mean p_T at intermediate x_L. However, the Feynman-Field Monte Carlo indicated in the figure shows a similar may-be somewhat weaker increase. Since, however, this effect is strongly dependent on a proper handling of secondary decays any conclusion on the energy dependence of the 'seagull' effect is premature.

d) Two-photon physics

The importance of two-photon physics was realized by many people in several publications [95,96] in 1971. The process is illustrated schematically in Fig. 68.

As the electrons approach each other they radiate photons which interact. The process can be visualized as the interaction of two photon beams which are produced by electron beams passing a radiator. The energy spectrum is given by:

$$N(k) \ dk \simeq \frac{2\alpha}{\pi} \ln (E_b/m_e) \ \frac{dk}{k}$$

k = photon energy.

Thus the photon rate is increasing logarithmically with

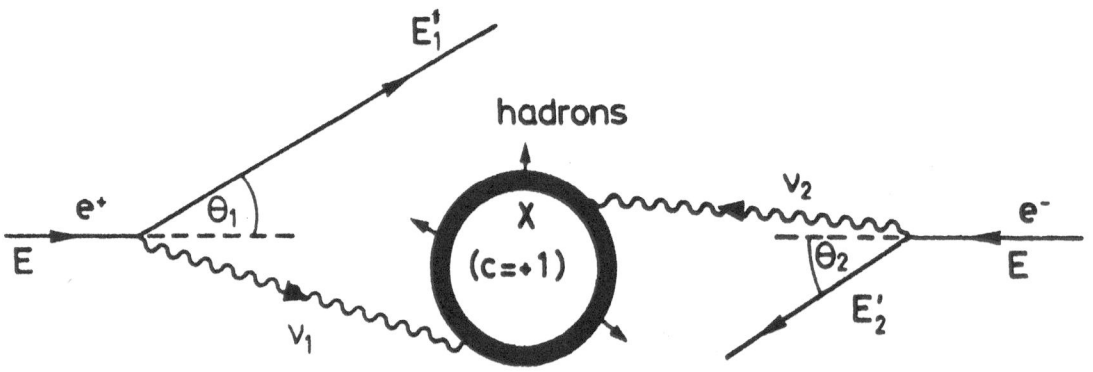

Fig. 68. The two-photon process.

beam energy. At E_b = 15 GeV the photon beam has reached
an intensity of

$$N(k)dk \simeq 0.05 \frac{dk}{k}$$

equivalent to a radiator of 5% radiation length. Experimentally the occurance of a two-photon reaction
in e^+e^- collisions can be detected by tagging devices
(Fig. 69).

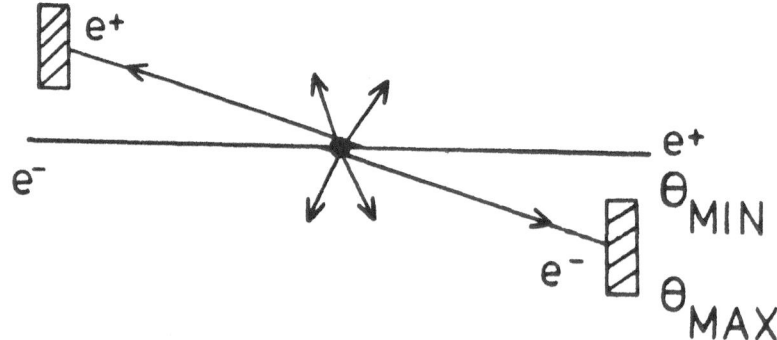

Fig.69. Tagging in two-photon physics.

They usually measure energy and direction of the outgoing electrons (positrons). Thus if both particles are tagged the kinematics of the γγ reaction can be determined. Since the tagging devices cover only a limited solid angle the tagged photon-beam intensity is reduced.

$$N(k)\,dk \simeq \frac{dk}{k} \cdot \frac{2\alpha}{\pi} \left(\ln \frac{E \sin\theta_{max}}{m_e} - \ln \frac{E \sin\theta_{min}}{m_e} \right) \qquad .$$

For a typical tagging device ranging from 25 to 250 mrad this corresponds to a reduction of the tagged photon flux by about a factor of 5 with respect to the total flux.

In the naive picture of Figure 68 the reaction will occur in two steps, the emission of the two nearly real photons and their collision with a cross section $\sigma_{\gamma\gamma}$ (equivalent photon approximation [97]). The order of magnitude of this cross section is given by the vector-meson dominance model. In the simplest version of this description the gamma-proton and proton-proton cross section can be related to the gamma-gamma cross section.

$$\sigma_{\gamma\gamma} = \frac{\sigma^2 (\gamma p)}{\sigma (pp)} \approx 300 \text{ nb} \quad .$$

A more refined description [98] yields even higher values in the asymptotic region above about 3 GeV invariant mass of the two photons. The energy dependence of the two-photon cross section is shown in Fig. 70 together with the annihilation cross section [99]. Already at moderate beam energies the two curves cross over and the two-photon cross section gains sizeable

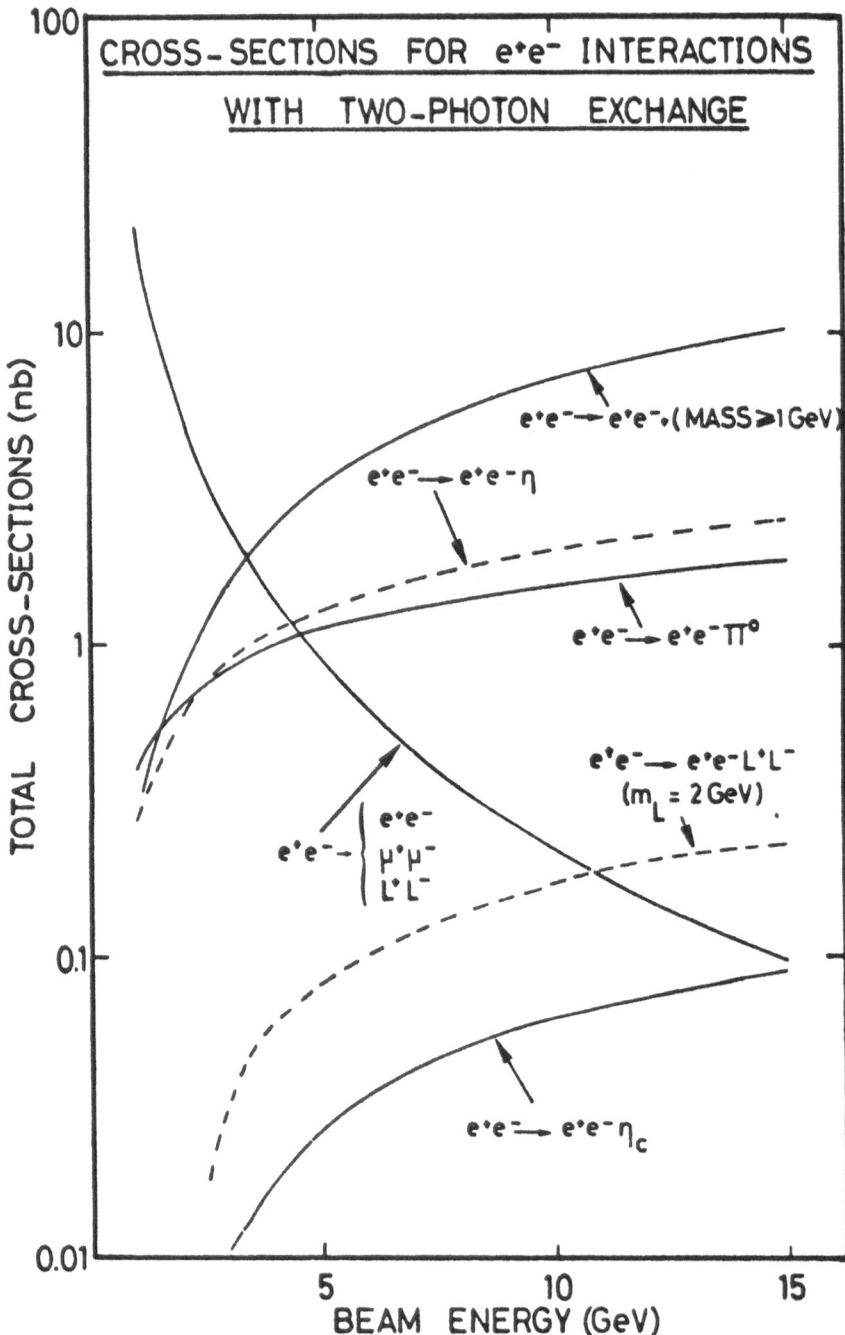

Fig. 70. The expectation for various two-photon cross sections compared to the annihilation cross section [99].

values compared to the annihilation cross section.

At first sight one might realize little interest in $\gamma\gamma$ reactions except getting rid of them in the e^+e^- annihilation data. In fact nearly real photons are supposed to behave very much like hadrons. Their collisions are much better studied, for instance at the ISR. There are, however, two aspects which show the topical importance of two-photon reactions.

C = +1 resonances

$C = +1$ resonances like η and η_c can be produced directly in $\gamma\gamma$ reactions. In view of the latest conflicting results on the η_c the importance of this does not need further advertisement. Apart from spotting a particle like the η_c a measurement of the cross section would yield the two-photon width $\Gamma_{\gamma\gamma}$ which provides a very sensitive test of the constituent quark charge. As one can see from Fig. 70 the estimated cross sections are not out of reach at 15 GeV.

Jet production

The most outstanding aspect of two-photon processes is, however, the direct production of quark jets [100,101].

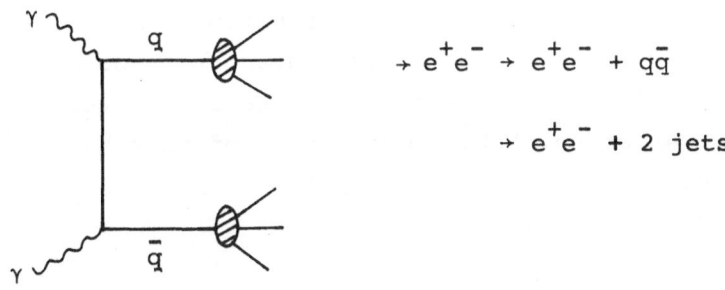

$$\to e^+e^- \to e^+e^- + q\bar{q}$$

$$\to e^+e^- + 2 \text{ jets}$$

236

Similar to the annihilation processes this mechanism
can be related to the QED process $e^+e^- \rightarrow e^+e^-\mu^+\mu^-$
defining a quantity

$$R^2\gamma = 3 \sum_q Qq^4 .$$

Evidence of hadronic γγ events

In 1974 two candidates for hadronic γγ events were reported at Frascati [102]. Since then this field remained
totally unexplored. To search for two-photon processes
the PLUTO group [103] looked for coincidences between
the inner detector and one or two of the forward
spectrometers. In the data presented here only the SAT
part of the forward spectrometer was used. A tagged
electron was defined by an energy deposition of more
than 3 GeV. To look for hadron production at least
three tracks or two tracks plus additional energy were
required in the central detector.

Single tag events

A vertex distribution of single tag events of the type

$$e^+e^- \rightarrow e^\pm + \text{hadrons}$$

is shown in Fig. 71. The distribution exhibits a clear
excess of events near to the interaction point. The
relatively large background can be attributed to beam-
gas scattering. After subtraction of this background
(15 events) a signal of 38 ± 11 events remains.

237

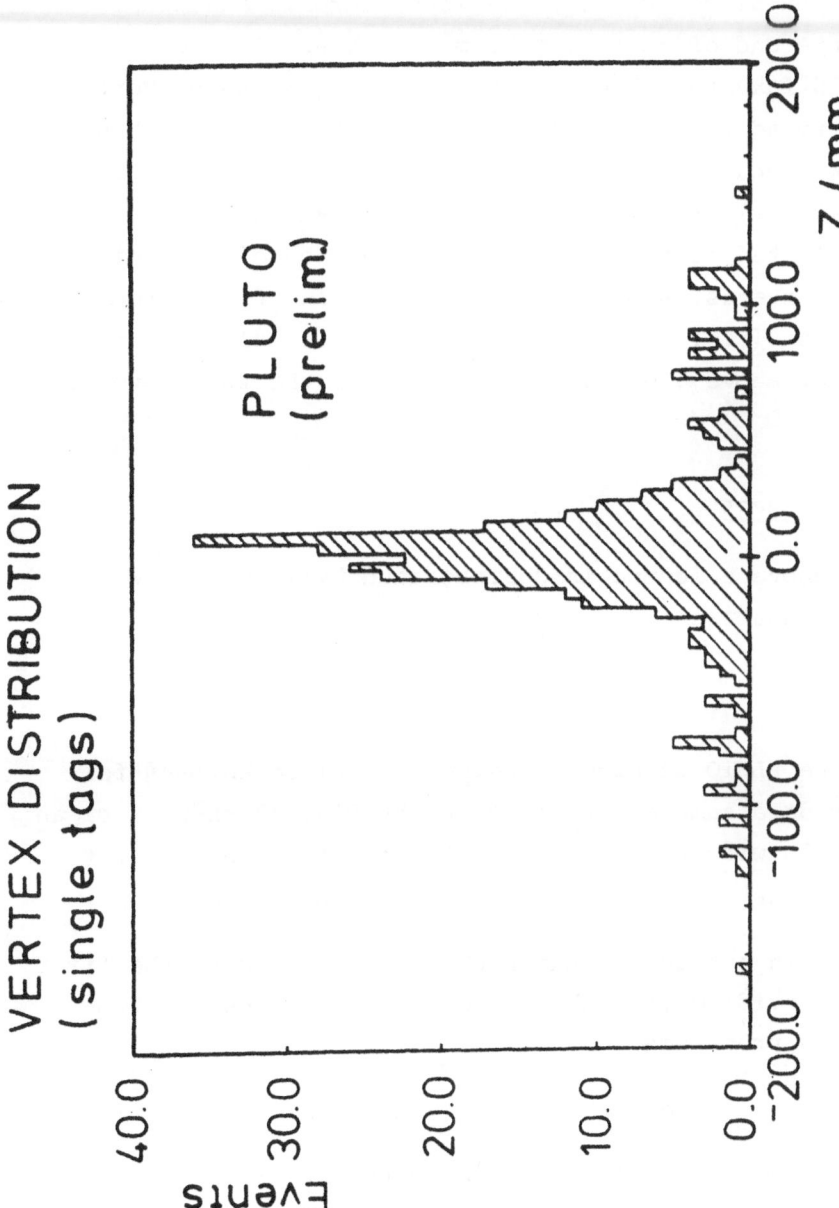

Fig. 71. PLUTO: (preliminary) vertex distribution of single-tag two-photon candidates.

Accidental coincidences between hadronic annihilation events and off-momentum beam particles hitting the SAT could fake the above signature. Since no such event was found in the hadronic annihilation events, this background is estimated to be less than 0.1%. Therefore, a clear signal remains which can only be attributed to the two-photon production of hadronic final states. A preliminary evaluation of the cross section (Fig. 72) yields an asymptotic value of about 500 nb above 3 GeV invariant mass of the two photons.

Double tags

The above evidence is corroborated by 12 events with the signature

$$e^+e^- \to e^+e^- + \text{hadrons}$$

in the PLUTO detector. The accidental background for this data sample is less than 1%. Fig. 73 shows a clean example with 2 electrons in the forward spectrometer and 4 non-showering tracks in the central detector.

In conclusion the PLUTO group has found the first clear evidence for the production of hadrons in two-photon processes:

$$e^+e^- \to e^+e^- + \text{hadrons} \quad .$$

This opens a new field of e^+e^- physics.

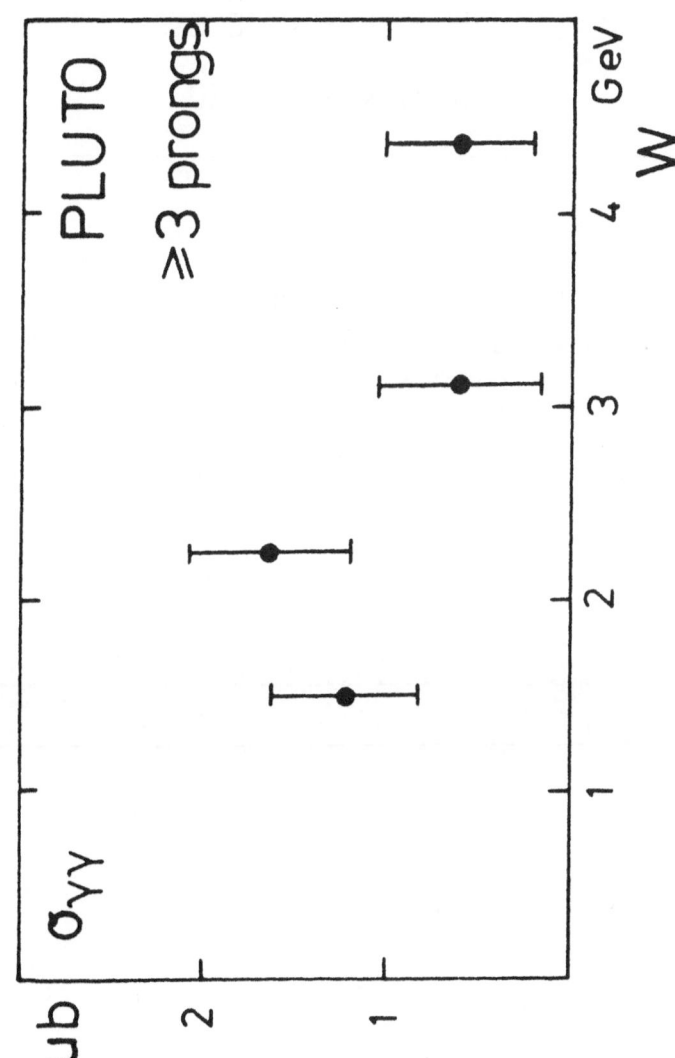

Fig. 72. PLUTO: (preliminary) invariant mass distribution of the hadronic system for single-tag two-photon events.

R U N ± 18994

EVENT ± 1867

X-VERSUS-Z AND
Y-VERSUS-Z VIEW
-7000. < Z < 7000.

-550. < X < 550.

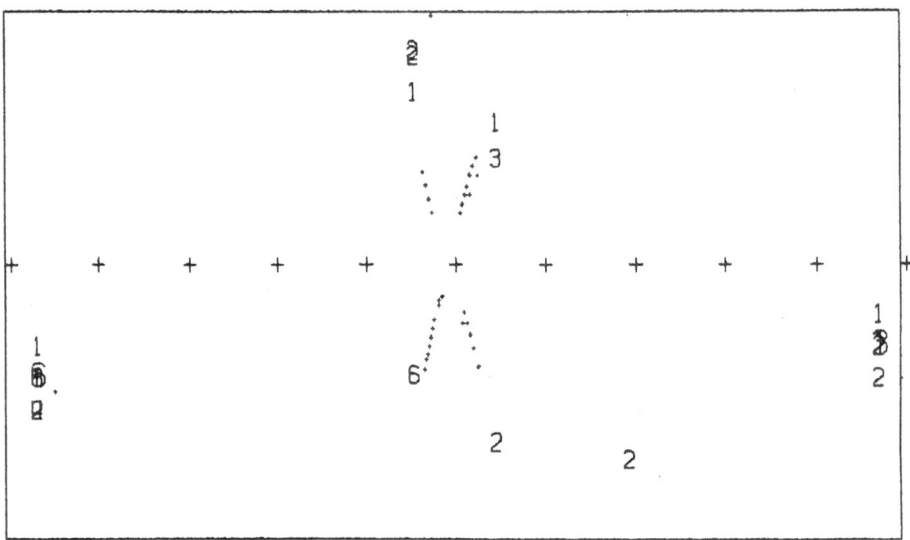

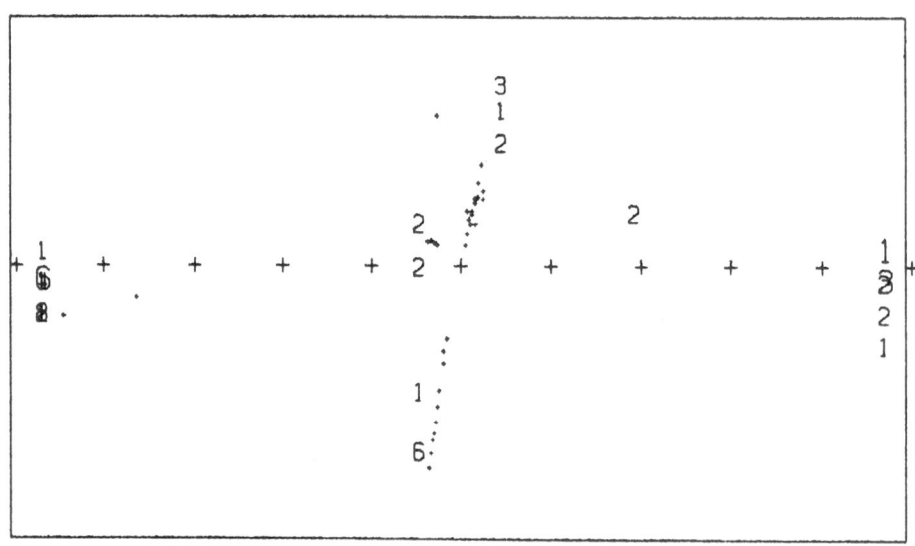

Fig. 73. PLUTO: a double-tag two-photon event.

4. Physics plans at PETRA

In this last section I would like to describe some of the physics plans for the immediate future at PETRA.

Search for the top quark

The search for the proposed t quark is of literally topical interest for the near future of PETRA. Therefore, PETRA has been prepared for energies of the order of 30 GeV c.m. by introducing further cavities. There are a number of handels and strategies to look for the new top quark. I will follow the common prejudice that this new quark has a charge of 2/3. The expected step in R will then be of the order of 1.5 including gluon corrections. Our measurements at 13 and 17 GeV have shown that the detectors at PETRA would be sensitive to such a threshold on the statistical basis of a few hundred events, provided the background can be controlled.

The ultimate aim of finding the toponium resonance will then probably be attacked in two steps. First the position of the threshold will be estimated from a sequence of R measurements at large energies. Once the threshold is roughly localized one could apply theoretical estimates to predict the position of the toponium resonances. A fine scan in steps of the machine resolution will then be applied to look for resonance structures in this region. To get an estimate of the relative height of the resonance (σ_p) compared to the continuum (σ_c) we can scale the relative cross sections from the T region to the toponium region (assuming $M_{t\bar{t}} = 25$ GeV):

$$\left(\frac{\sigma_p}{\sigma_c}\right)_{t\bar{t}} \approx \left(\frac{\sigma_p}{\sigma_c}\right)_T \cdot \frac{\Delta W(9.4)}{\Delta W(25)} \cdot \frac{\Gamma_{ee}(t\bar{t})}{\Gamma_{ee}(b\bar{b})} \approx 1.3 \cdot \left(\frac{\sigma_p}{\sigma_c}\right)_T \cdot$$

Since the energy resolution ΔW is about three times worse at the toponium and Γ_{ee} will scale like Q_q^2, the relative peak cross section at the toponium will be roughly as large as at the T. Therefore, from a total cross-section measurement the toponium could be found, if it exists, in the PETRA energy region.

Still there are at least 3 additional weapons which might turn out to be of great importance in the toponium search. The topology of the new events coming in above threshold will be drastically different from the dominant two-jet behaviour below threshold. Fig. 74 shows the expectation above top threshold [90]. If the differences are as large as expected one might even get a separation of top production by a simple cut in sphericity.

A further handle would then be to look for excessive lepton production [104]. If we follow the common believe that the top quark is cascading down the chain

$$t \rightarrow b \rightarrow c \rightarrow s$$

in successive weak decays we can estimate the total number of leptons per event. Assuming a branching ratio of 8% into each type of lepton in each branch of the decay chain we arrive at an average of about 1 lepton (μ or e) for each $t\bar{t}$ event. In addition spectacular events with up to 6 leptons may occur.

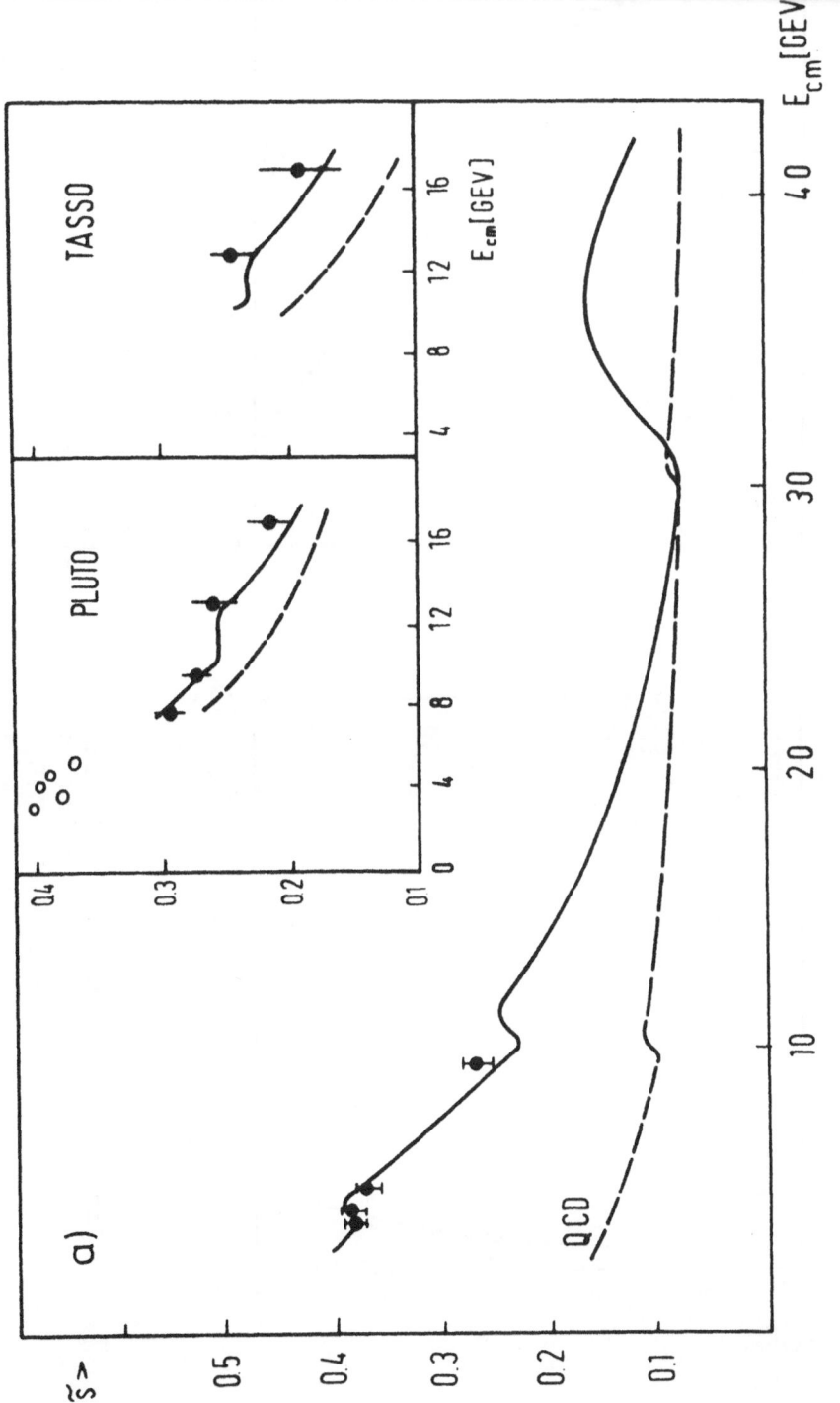

Fig.74a. Predicted change of sphericity and thrust above t threshold [90].
Mean sphericity as a function of energy.

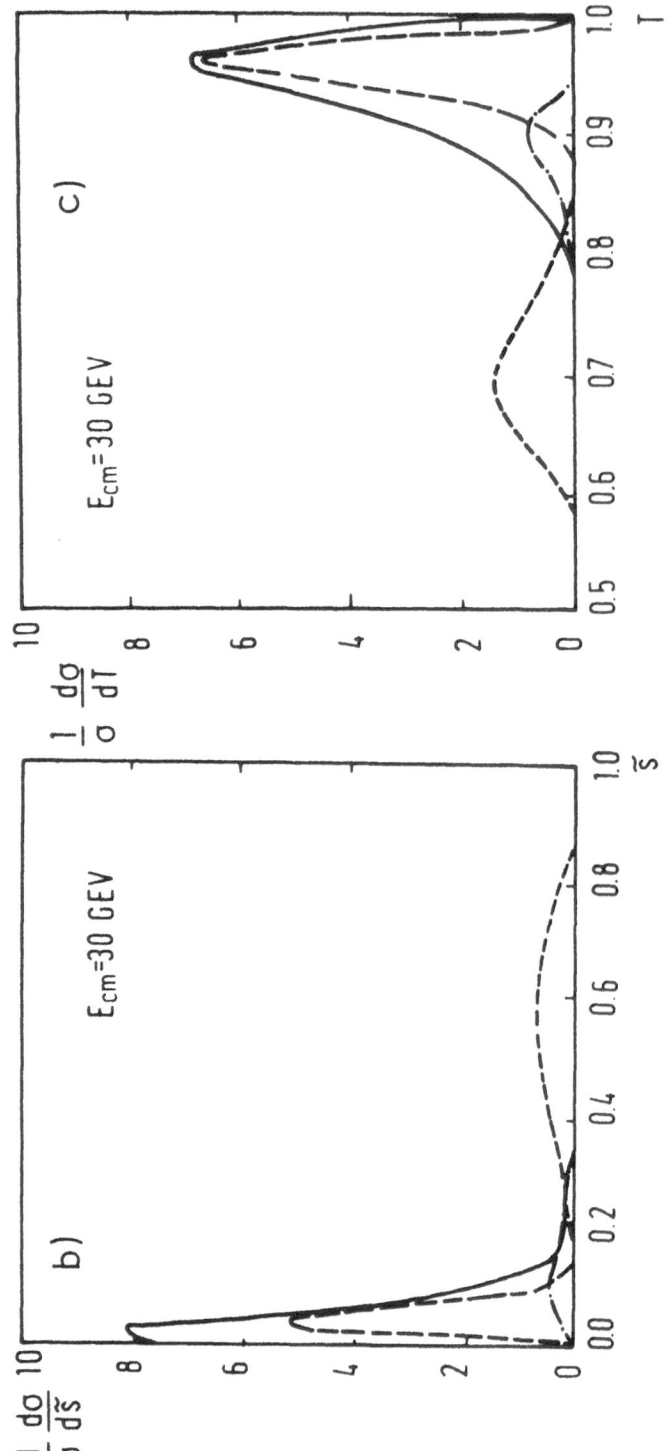

Fig. 74b. Sphericity distribution at the t threshold.
c. Thrust distribution. The "old" contribution (---) including b (-·-·-)
compared to the t part (......).

Search for heavy leptons

If further heavy sequential leptons exist, their branching
ratio into hadrons will be larger than for the τ. For
masses between 10 and 15 GeV the leptonic branching ratio
is expected to be of the order of 11% (for e, μ and τ)
whereas the remaining 67% will be hadronic decays [105].
The prominent signature for these events will, therefore,
be an isolated electron and a hadronic jet in opposite
directions with large missing energy. Of course, also
two-lepton events with large missing mass are expected,
however, on a much lower statistical level.

Two-photon physics

Last not least the wide field of two-photon physics will
be open in the high-energy region   looking for even C
resonances and jet production.

PETRA detectors

Let me close this lecture with a survey of the five
detectors which are either installed or in preparation
for the next running periods. A very short summary of
the physics abilities of the five detectors is given in
table 13. Since it is a hopeless attempt to describe five
detectors in the framework of this lecture I just try
to summarize the main physics goals of the different
experiments. Most detectors (except MARK J) provide
charged particle detection in large solenoidal magnetic
volumes over 80 to 90% of the solid angle. The typical
momentum resolution at 5 GeV is of the order of 5 %
for the drift chamber detectors whereas it is only 15 %

	Main Physics Goals	Charged Hadrons σ_p/p at 5 GeV	Electrons	Special Items
PLUTO	σ_{tot}	15 %	lead scintillator	$\gamma\gamma$ tagging 25 ÷ 250 mrad
MARK J	μ pairs	hadron calorimetry	lead scintillator + tubes	rotatable
TASSO	jets + hadron identification	4÷6 %	liquid argon	full Č-identification in 2 x 1.5 sr
JADE	jets + leptons	3÷5 %	lead glass	"jet chamber", 50 dE/dx samples
CELLO	leptons	3÷5 %	liquid argon	thin superconducting coil 0.5 X_0

Table 13. PETRA detectors.

for the proportional chamber-detector PLUTO. Myon
identification over a large solid angle is available
in all experiments. The same is true for electron and
photon detection. However, the method is quite different
for the various detectors as indicated in table 13. In
the last column I have listed a few items which are
special to the different devices. Fig. 75 and 76 show
the detectors JADE and CELLO which are now being
prepared for installation this year.

5. Summary

First results at 13 and 17 GeV c.m. energy have
been obtained at the new PETRA storage ring. The total
cross section and the event topology are consistent
with a fifth quark b and exclude that the $t\bar{t}$ threshold
is below 17 GeV. Statistics is too scarce yet to draw
any conclusion about QCD gluon-bremsstrahlung effects.
QED remains valid up to the highest energy. Hadronic
events from two-photon reactions have been found. The
race is on for top.

ACKNOWLEDGEMENT

I am indebted to all my colleagues at DESY who
helped me in preparing these lectures, in particular
the members of the PLUTO collaboration [106]. I want
to thank B. Koppitz and H. Spitzer for a critical
reading of the manuscript and R. Siemer for careful
typing. I am grateful to Prof.H. Mitter for his in-
vitation and the pleasant stay (and skiing) at Schladming.

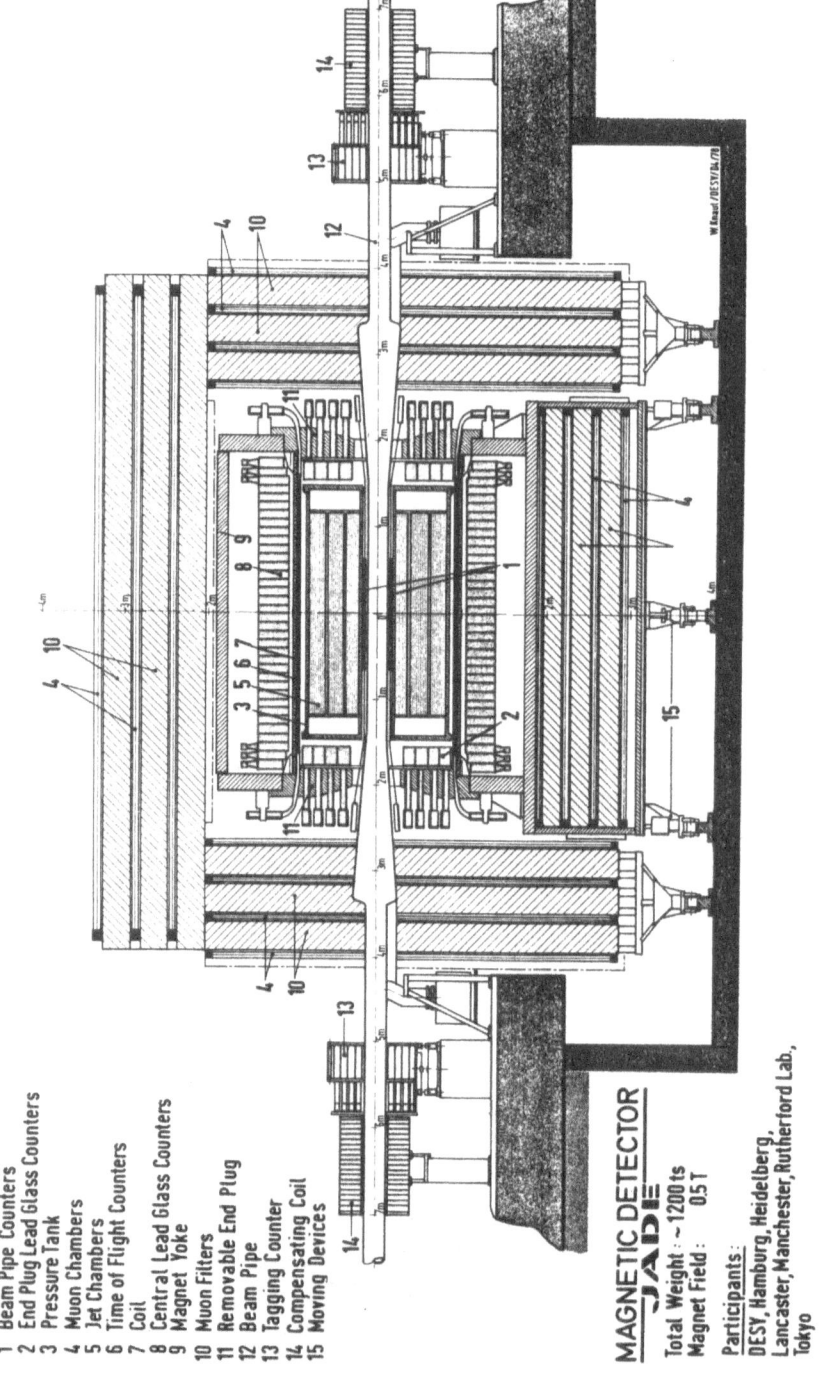

MAGNETIC DETECTOR
JADE

1 Beam Pipe Counters
2 End Plug Lead Glass Counters
3 Pressure Tank
4 Muon Chambers
5 Jet Chambers
6 Time of Flight Counters
7 Coil
8 Central Lead Glass Counters
9 Magnet Yoke
10 Muon Filters
11 Removable End Plug
12 Beam Pipe
13 Tagging Counter
14 Compensating Coil
15 Moving Devices

Total Weight : ~ 1200 ts
Magnet Field :   0.5 T

Participants :
DESY, Hamburg, Heidelberg,
Lancaster, Manchester, Rutherford Lab,
Tokyo

Fig. 75. The JADE detector (DESY–Hamburg–Heidelberg–Lancaster–Manchester–Rutherford–Tokyo Collaboration).

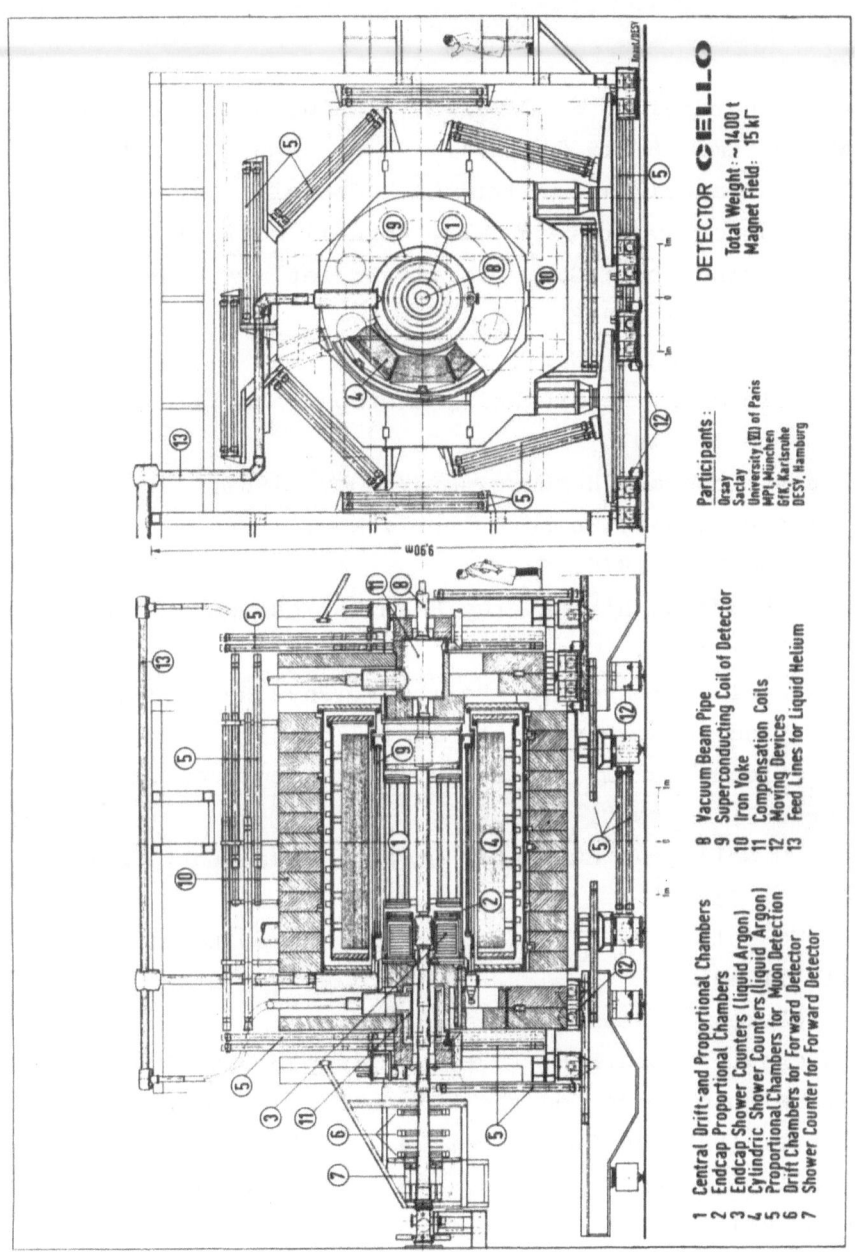

Fig. 76. The CELLO detector (DESY–Karlsruhe–München–Orsay–Paris Collaboration).

REFERENCES

[1] C. Bernardini, G.F. Corazza, G.Ghigo and B.Touschek, Nuovo Cim. 18 (1960) 1293.

[2] J. Le Francois, rapporteur talk, Cornell Conf. 1971. V. Sidorov, rapporteur talk, Cornell Conf. 1971. Pioneering work: G.K. O'Neill, Bulletin A. Phys. Soc. Ser. 23 (1958) 158.

[3] B.H. Wiik and G. Wolf, DESY 78/23. P. Waloschek, Kerntechnik, Isotopentechnik und -chemie 12 (1970) 525.

[4] Doris proposal, Hamburg, DESY, Oct. 1967.

[5] PETRA proposal, Hamburg, DESY, Nov. 1974. Updated Feb. 1976.

[6] K.G. Steffen, High Energy Beam Optics, John Wiley, New York 1965.

[7] T. Appelquist and H.D. Politzer, Phys.Rev. D12 (1975) 1404 and Phys. Rev. Lett. 34 (1975) 43.

[8] L. Criegee et al., Proc. 1973 Int. Conf. on Instrum. for HEP, Frascati 1973. PLUTO Coll., J. Burmester et al., Phys. Lett. 64B (1976) 369.

[9] PLUTO Coll., J. Burmester et al., Phys. Lett. 66B (1977) 395. A. Bäcker, Thesis, Siegen (1977). Internal report DESY F33-77/03 (unpublished).

[10] G. Feldman, Proc. of the XIX.Int. Conf. on HEP, Tokyo, 1978 and SLAC-PUB-2224 (1978).

[11] J. Siegrist et al., Phys. Rev. Lett. 36 (1976) 700.

[12] DASP Coll., R. Brandelik et al., Phys. Lett. 70B (1977) 125 and Phys.Lett. 73B (1978) 109.

[13] J. Kirz, Contrib. to the XIX.Int. Conf. on HEP, Tokyo, 1978.

[14] J.J. Aubert et al., Phys.Rev. Lett. 33 (1974) 1404.
J.E. Augustin et al., Phys.Rev.Lett. 33 (1974) 1406.

[15] Recent reviews:
G. Goldhaver, EPS conf., Budapest, 1977.
G. Feldman, Banff Summer Inst., Alberta (CA), 1977.
H. Schopper, DESY-report 77/79 (1977).
B.H. Wiik and G. Wolf, DESY-report 78/23 (1978).

[16] G. Flügge, loco cit. (ref.10) and DESY 78/55 (1978).

[17] M. Krammer and H. Krasemann, this school.

[18] DASP Coll., W. Braunschweig et al., Phys. Lett.67B
(1977) 243 and 249.
W.D. Apel et al., Phys.Lett. 72B (1978) 500.
The state X(2820) was questioned in recent experimen-
tal results obtained by the Cristal ball group
(E. Bloom, XIV. Rencontre de Moriond, Les Ares, 1979).

[19] W. Bartel et al., Phys. Lett. 79B (1978) 492.
C.J. Biddik et al., Phys.Rev. Lett. 38 (1977) 1324.
W. Tannenbaum et al., Phys.Rev.Lett. 35 (1975) 1323.
PLUTO Coll., V.Blobel, XII. Rencontre de Moriond,
Flaine, 1977.
S. Yamada, Hamburg Conf., 1977.
Recent results from the MARK II group do not confirm
the χ(3.45) (talk given at DESY by H. Taureg).

[20] G. Goldhaber et al., Phys.Rev. Lett. 37 (1976) 255.
I. Peruzzi et al., Phys. Rev. Lett. 37 (1976) 569.

[21] DASP Coll., R. Brandelik et al., Phys. Lett. 70B
(1977) 132 and 80B (1979) 412.

[22] M.L. Perl et al., Phys.Rev.Lett. 35 (1975) 1489
and 38 (1976) 117.

[23] S.W. Herb et al., Phys.Rev.Lett. 39 (1977) 252.
W.R. Innes et al., Phys.Rev. Lett. 39 (1977) 1240.

[24] PLUTO Coll., J. Burmester et al., Phys. Lett. 68B
(1977) 297 and 301.

[25] G.J. Feldman et al., Phys. Lett. 63B (1976) 466.
M.L. Perl et al., Phys. Lett. 70B (1977) 487.

[26] M. Rößler, Thesis, Hamburg (1978),
Internal report DESY F14-78/01 (unpublished).

[27] G. Flügge, Proc. of the Vth Int. Conf. on
Experimental Meson Spectroscopy, Boston(1977).
M.L. Perl, Hamburg Conf. (1977)

[28] DASP Coll., R. Brandelik et al., Phys.Lett. 73B
(1978) 109.

[29] W. Bartel et al., Phys. Lett. 77B (1978) 331.

[30] W. Bacino et al., Phys. Rev. Lett. 41 (1978) 13.

[31] PLUTO Coll., G. Alexander et al., Phys. Lett. 73B
(1978) 99.
W. Wagner, Thesis, Aachen 1978 (unpublished).

[32] Y.S. Tsai, Phys. Rev. D4 (1971) 2821.
H.B. Thacker, J.J. Sakurai, Phys.Lett. 36B (1971)
103.
J.D. Bjorken, C.H. Llewellyn-Smith, Phys.Rev. D7
(1973) 887.

[33] PLUTO Coll., G.Alexander et al., Phys.Lett. 78B
(1978) 162.

[34] A.M. Cnops et al., Phys. Rev. Lett. 40 (1978) 144.

[35] J.Kirkby, Summer Institute, Stanford 1978 (SLAC-PUB-
2231).
W. Bacino et al., Phys.Rev.Lett. 42 (1979) 6 and
42 (1979) 749.

[36] L. Lederman, XIX.Int. Conf. on HEP, Tokyo, 1978.

[37] K. Ueno et al., Phys. Rev. Lett. 42 (1979) 486.

[38] PLUTO Coll., Ch.Berger et al., Phys. Lett. 76B
(1978) 243.

[39] C.W. Darden et al., Phys.Lett. 76B (1978) 246.

[40] PLUTO Coll., Ch.Berger et al., DESY 79/19,
submitted to Zeitschr.f. Physik C.

[41] J.K. Bienlein et al., Phys.Lett. 78B (1978) 360.

[42] G.W. Darden et al., contribution to the XIX.Int. Conf. on HEP, Tokyo, 1978 and Internal report DESY F15-78/01 (1978).

[43] J.L. Rosner, C. Quigg, H.B. Thacker, Phys. Lett. 74B (1978) 350.

[44] D.R. Yennie, Phys.Rev.Lett. 34 (1975) 219.
F.E. Close, D.M. Scott, D.Sivers, Phys.Lett. 62B (1976) 213.
T. Walsh, DESY report 76/13 (1976).
G.J. Gounaris, Phys. Lett. 72B (1977) 91.
M. Greco, Phys. Lett. 77B (1978) 84.

[45] G. Sterman and S. Weinberg, Phys.Rev. Lett. 39 (1977) 1436.

[46] A. de Rujula, J. Ellis, E.G. Floratos and M.K. Gaillard, Nucl. Phys. B138 (1978) 387.

[47] J.D. Bjorken and S.J. Brodsky, Phys.Rev. D1 (1970) 1416.

[48] E. Fahri, Phys. Rev. Lett. 39 (1977) 1587. See also
S. Brandt, Ch.Peyrou, R. Sosnowski and A.Wroblewski, Phys. Lett. 12 (1964) 57.
S. Brandt and H. Dahmen, Zeitschr. f. Phys. C1 (1979) 61.

[49] H. Georgi and M. Machacek, Phys. Rev. Lett. 39 (1977) 1237.

[50] G. Hanson et al., Phys.Rev. Lett. 35 (1975) 1609.
G. Hanson, XIII.Rencontre de Moriond, Les Ares (1978); SLAC-PUB-2118 (1978).

[51] PLUTO Coll., Ch. Berger et al., Phys. Lett. 78B (1978) 176.

[52] R.D. Field and R.P. Feynman, Phys. Rev. D15 (1977) 2590 and Nucl. Phys. B136 (1978) 1.

[53] V.P. Sukhatme, XIII.Rencontre de Moriond, Les Ares, 1978.

254

[54] K. Koller, H. Krasemann, T.F. Walsh, Zeitschr. f. Physik C1 (1979) 71.

[55] K. Koller, T.F. Walsh, Phys. Lett. B72 (1977) 227, B73 (1978) 504, and Nucl. Phys. B140 (1978) 449.

[56] T.A. De Grand et al., Phys.Rev. D16 (1977) 3251.
S. Brodsky et al., Phys. Lett. B73 (1978) 203.
H. Fritzsch, K.H. Streng, Phys.Lett. B74 (1978) 90.

[57] E.g.: K. Shizuya, S.H.H. Tye, preprint Fermilab-PUB-79/16-THY with further references.

[58] PLUTO Coll., Ch. Berger et al., DESY report 78/71 (1978); Phys. Lett. B82 (1979) 449.

[59] J. Bienlein et al., to be published.

[60] A. Ore, J.L. Powell, Phys. Rev. 75 (1949) 1696.

[61] Compare also: K. Hagiwara, Nucl. Phys. B137 (1978) 164.

[62] G. Alexander, XIX. Int. Conf. on HEP, Tokyo, 1978.

[63] R.P. Feynman, Phys. Rev. Lett. 23 (1969) 1415.

[64] V. Blobel et al., Nucl. Phys. B69 (1974) 454.

[65] M. Breidenbach et al., Phys. Lett. 39B (1972) 654.
M. Breidenbach, G. Flügge, K.R. Schubert, E.G.H. Williams, contribution to XVI. Int. Conf. on HEP, Batavia, 1972.
J.C. Sens, Proc. 4th Int. Conf. on High Energy Coll., Oxford, 1972.
G. Giacomelli, NAL-PUB-73/74-EXP (1973).

[66] I.I. Bigi, S. Nussinov, Phys.Lett. 82B (1979) 281.

[67] PLUTO Coll., J. Burmester et al., Phys. Lett. 67B (1977) 367.
DASP Coll., R. Brandelik et al., Phys.Lett. 67B (1977) 363.
V. Lüth et al., Phys. Lett. 70B (1977) 132.

[68] H. Fritzsch, K.H. Streng, Phys.Lett. 77B (1978) 299.

[69] K. Wacker, talk at the DPG-Frühjahrstagung, Bonn, 1979.

[70] H. Schopper, Hamburg Conf., 1977.

G. Weber, Symp. on HEP and the Role of Quarks, Stockholm, 1978.

[71] PETRA proposal, Hamburg, 76/19 (1976).

[72] PETRA proposal, Hamburg, 76/15 (1976).

[73] PETRA proposal, Hamburg, 76/13 (1976).

[74] PETRA proposal, Hamburg, 76/16 (1976).

[75] PETRA proposal, Hamburg, 76/14 (1976).

[76] G.A. Voss, progress report at DESY, Febr. 1979.

[77] M. Kobayashi, K. Maskawa, Prog. Theor. Phys. 49 (1973) 652.

[78] T. Kinoshita et al., Phys. Rev. D2 (1970) 910.

[79] Recent experimental limit: T. Himel et al., Phys. Rev. Lett. 41 (1978) 449.

Recent review: J.J. Sakurai, III. Int. Symp. on HEP with Pol. Beams and Targets, Argonne, 1978.

[80] E. Iarocci, P. Waloschek, DESY report 72/13.

H. Mehrgardt et al., Int. Conf. on Instr. in HEP, Frascati, 1973.

[81] PLUTO Coll., Ch. Berger et al., Phys. Lett. 81B (1978) 410.

[82] H. Salecker, Z. Physik 160 (1960) 385.

[83] D. Barber et al., Phys. Rev. Lett. 42 (1979) 1110.

[84] PLUTO Coll., H. Spitzer, talk given at DESY, April 1979.

[85] TASSO Coll., H. Martyn, talk given at DESY, April 1979.

[86] J.E. Augustin et al., Phys. Rev. Lett. 34 (1975) 233.

L.H. O'Neill et al., Phys. Rev. Lett. 37 (1976) 395.

B.L. Beron et al., Phys.Rev. Lett. 33 (1974) 663.

[87] F.A. Berends et al., Nucl. Phys. B57 (1973) 381, B68 (1974) 541, B101 (1975) 234, and Phys. Lett. 63B (1976) 433.

256

[88] D. Barber et al., Phys. Rev. Lett. 42 (1979) 1113.

[89] TASSO Coll., R. Brandelik et al., DESY report 79/14
(1979) submitted to Phys. Lett.

[90] A. Ali, J.G. Körner, G. Kramer, J. Willrodt,
Z. Physik C1 (1979) 203 and DESY report 79/16
(1979), to be published.

[91] J. Ellis, M.K. Gaillard, G.G. Ross, Nucl. Phys.
B111 (1976) 253.

[92] C.L. Basham et al., Phys. Rev. D17 (1978) 2298.
F. Steiner, DESY 78/59 (1978).

[93] I.I.Y. Bigi, R.F. Walsh, Phys. Lett. 82B (1979) 267.
I.I.Y. Bigi, preprint Th 2626 CERN (1979).

[94] G. Kramer, G. Schierholz, Phys. Lett. 82B (1979)
108.

[95] S.J. Brodsky, T.Kinoshita, H. Terazawa, Phys.Rev.
Lett. 25 (1970) 972 and Phys.Rev. D4 (1971) 1532.
A. Jaccarini, N. Artega-Romero, J. Parisi, P.Kessler,
Lett. Nuovo Cim. 4 (1970) 933.

[96] H. Terazawa, Rev. Mod. Phys. 45 (1973) 615.

[97] C. Weizsäcker, E.J. Williams, Z. Physik 88 (1934)
612.

[98] M. Greco, Y. Srivastava, Nuovo Cim. 43A (1978) 88.

[99] G.J. Bobbink et al., PETRA proposal 76/18, Hamburg
(1976).

[100] S.J. Brodsky, T.A. De Grand, J.F. Gunion, J.H.Weis,
Phys. Rev. Lett. 41 (1978) 672.

[101] Compare also: K. Kajantie, this school.

[102] G. Barbiellini et al., Phys. Rev. Lett. 32 (1974)
385.
L. Paoluzzi et al., Lett. Nuovo Cim. 10 (1974) 435.

[103] PLUTO Coll., W. Wagner, XIV. Rencontre de Moriond,
Les Ares, 1979.

[104] J. Ellis, M.K. Gaillard, D.V. Nanopoulos,
S. Rudaz, Nucl. Phys. B131 (1977) 285.
A. Ali, Z. Physik C1 (1979) 25.
V. Barger, T. Gottschalk, R.J.N. Phillips,
Wisconsin preprint COO-881-82 (1979).
F. Bletzacker, H.T. Nieh, preprint ITP-SB-79-10,
Stony Brook, 1979.

[105] Y.S. Tsai, SLAC-PUB 2105 (1978).

[106] PLUTO Collaboration:
Ch. Berger, H. Genzel, R. Grigull, W. Lackas,
F. Raupach, W. Wagner;
I. Physikalisches Institut der RWTH Aachen, Germany.
A. Klovning, E. Lillestöl, E. Lillethun, J.A.Skard;
University of Bergen, Norway.
H. Ackerman, G. Alexander, F. Barreiro, J. Bürger,
L. Criegee, H.C. Dehne, R. Devenisch, G. Flügge,
G. Franke, W. Gabriel, Ch. Gerke, G. Horlitz,
G. Knies, E. Lehmann, H.D. Mertiens, B. Neumann,
K.H. Pape, H.D. Reich, B. Stella, U. Timm,
P. Waloschek, G.G. Winter, S. Wolff, W. Zimmermann;
Deutsches Elektronen-Synchrotron DESY, Hamburg,
Germany.
O. Achterberg, V. Blobel, L. Boesten, H. Kapitza,
B. Koppitz, W. Lührsen, R. Maschuw, R. van Staa,
H. Spitzer;
II. Institut für Experimentalpyhsik der Universität
Hamburg, Germany.
C.Y. Chang, R.G. Glasser, R.G. Kellog, K.H. Lau,
B. Sechi-Zorn, A.Skuja, G. Welch, G.T. Zorn;
University of Maryland, USA.
A. Bäcker, S. Brandt, K. Derikum, C. Grupen,
H.J. Meyer, M. Rost, G. Zech;
Gesamthochschule Siegen, Germany.

T. Azemoon, H.J. Daum, H. Meyer, O. Meyer,
M. Rössler, D. Schmidt, K. Wacker;
Gesamthochschule Wuppertal, Germany.

Acta Physica Austriaca, Suppl. XXI, 259—349 (1979)

Q U A R K O N I A[+]

by

M. KRAMMER and H. KRASEMANN

Deutsches Elektronen-Synchrotron DESY
Notkestraße 85, D-2000 Hamburg 52

CONTENTS

[+]Lectures given by M. Krammer at the
XVIII.Internationale Universitätswochen für Kernphysik
Schladming, Austria, February 28 - March 10, 1979.

1. INTRODUCTION

When the J/ψ and ψ' were discovered in fall 1974
[1] we all witnessed a dramatic and beautiful revival of
the quark model [2]. The predicted new quark flavour,
c = charm [3] could be added to the hadron spectroscopy
by the interpretation of the new particles as $c\bar{c}$ bound
states. It was argued that the new system is nonrelativistic: Charmonium [4]. Whereas the old mesons suffered
from the fact that the quarks are extremely relativistic
(mass differences are of the order of the masses themselves), in Charmonium the heavy ($\approx$ 1.5 GeV) c-quarks
should move relatively slowly, $\beta^2 = (v/c)^2 \approx 0.2$. The
well known powerful methods of exploring a nonrelativistic
system could be used. This was the source of real excite-
ment about the new particles.

In spring 1977 a still heavier meson family, the
$\Upsilon, \Upsilon', \Upsilon''$ [5] was discovered. From the measurement of
their leptonic decay widths in DORIS [6] we learned that
the charge of the fifth quark [7] is $|e_Q| = 1/3$: we found
the bottom quark. The $b\bar{b}$ system, which is supposed to be
even much more nonrelativistic than $c\bar{c}$, is thus called
Bottonium. We hope to discover a heavier sixth quark

flavour, t = top maybe, in the new machines PETRA and
PEP. In these lectures we will describe the dynamics
of a nonrelativistic Q$\bar{\text{Q}}$ bound system, Q = c,b,t,...,
called QUARKONIUM.

Quantumchromodynamics [8], QCD, turned out to be
the most promising candidate for the theory of quark
dynamics, i.e. the strong interactions. QCD is a non-
abelian gauge field theory of the interactions of quarks
and eight massless vector gauge bosons, the gluons. The
coupling constant α_s, renormalized at the relevant
momentum transfer q^2 or the corresponding distance R,
is a monotonously falling function of $|q^2|$. It tends
logarithmically to zero as $|q^2| \to \infty$ or $R \to 0$

$$\alpha_s(q^2) = \frac{12\pi}{33-2N} \cdot \frac{1}{\log(q^2/q_o^2)} + O(\log^{-2}) \qquad (1.1)$$

N = number of light quark flavours

this is called "asymptotic freedom" [9].

If α_s becomes small, perturbation theory is fine
and in Born approximation the quark interaction is just
one gluon exchange. The nonabelian selfinteraction of the
colour-charged gluons plays no rôle in lowest order graphs,
thus QCD is very similar to QED, the static potential for
short distances being essentially of the Coulomb type.

α_s becomes large for some large R of the order of
1/2 fm, which is the typical hadron radius. At present
this region, where perturbation theory does not work, is
subject to educated speculations only [10]: we believe
that the rising coupling "confines" the quarks. Models

which give a hint at this are lattice gauge theories or
the string model.

When QCD is in fact the underlying theory for
Quarkonia, we should be able to probe QCD features by
studying these systems. First we should be able to probe
the one gluon exchange at short distances. The static
Coulomblike potential gives rise to the spin-spin, spin-
orbit and tensor interactions known from Positronium,
since the quark gluon vertex has the same Dirac structure
as the electron photon vertex (γ_μ-coupling). Second, at
large distances the "confining" potential should be linear,
$V_C(R) = aR$. It should be flavour independent and the interquark force a at large distances should also be somehow
related to the inverse Regge slope of the low mass mesons.
The "confining" potential should be essentially spin independent [10].

We have no guess for the potential at intermediate
distances. In Charmonium a superposition of the spin
dependent one gluon exchange potential and the scalar
linear potential has worked out rather well [11]. However, this potential is not universal, as we have learned
from Bottonium. The intermediate region of the potential
has to be treated in a more sophisticated manner.

Besides the spectra, the fine and hyperfine
structure we will discuss radiative transitions in some
detail. Chapter 7 will be devoted to the puzzling states
X(2.83), χ(3.45) and χ(3.59/3.18).

As an application of QCD we will describe the
gluonic annihilation of Quarkonia, which leads to the
total hadronic decay width via a nonperturbative dressing

mechanism. With the experimentally accessible regime of
c.m. energies of 10 GeV or more, the gluons which govern
annihilations in QCD, might show up as hadron jets [12].
These jets should carry the directed momentum of the
initial gluon. In angular distributions of these jets
one should be able to measure gluon helicities [12,13].
One can further speculate on the existence of glueballs
[14] in Quarkonium decays. It is rather difficult, however, to find an easy test of the nonabelian gluon self
coupling. Nevertheless, finding the gluons is very
important: they are the gauge bosons of QCD and a proof
of their existence is as crucial for QCD as a proof of
the existence of Z° and $W^{\pm}$ for the Salam-Weinberg theory.

2. THE SPECTRA

Quarkonium is essentially nonrelativistic. The
perturbative Hamiltonian is then obtained from the Bethe-
Salpeter equation in nonrelativistic approximation or
from the exact relativistic scattering amplitude (Born
graph only). One obtains the Schrödinger equation in
zeroth order of β^2 and the Fermi-Breit Hamiltonian terms
up to order β^2. In 0th order

$$H^{\circ} = 2m_Q + \vec{p}^2/m_Q + V(R) + \text{const.} \tag{2.1}$$

a) Charmonium

For Charmonium the standard potential used is [11]

$$V(R) = V_{AF}(R) + V_C(R) = -\frac{4}{3} \cdot \frac{\alpha_s}{R} + a \cdot R \quad . \tag{2.2}$$

V_{AF} is the "asymptotically free" short distance part due to one gluon exchange. -4/3 is a group factor from $SU(3)_{colour}$ and α_s is the effective coupling. For this one can take two points of view. Either α_s is R-dependent [15] or α_s is a constant, different for each quark flavour mass: [9]

$$\alpha_s(M_2^2) = \alpha_s(M_1^2)[1 - \frac{33-2N}{12\pi} \alpha_s(M_1^2)\log(M_1^2/M_2^2)]^{-1} \qquad (2.3)$$

N = number of "light" quarks.

In the standard calculations [11,16] the second point of view is taken.

    At large distances the potential should be "confining". The linear potential is suggested by lattice gauge theories, or string models where the field lines are parallel and the force between two coloured constituents is constant:

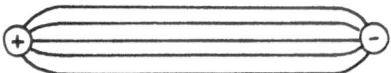

One example is the Meissner effect in superconductors of the second kind and another example is QED in one space dimension.

    The potential and the structure of the spectrum is shown in Fig. 2.1 and is compared with the experimental Charmonium.

    The parameters are determined as

a = 1 - 0.9 GeV/fm   from ψ' - J/ψ

$$\alpha_s = 0.3 - 0.4 \text{ from } \frac{|\psi_{\psi'}(0)|^2}{|\psi_{J/\psi}(0)|^2} = \frac{M_{\psi'}^2 \Gamma_{e\bar{e}}(\psi')}{M_{J/\psi}^2 \Gamma_{e\bar{e}}(J/\psi)} =$$

$$= \frac{(3.7 \text{ GeV})^2 2.2 \text{ keV}}{(3.1 \text{ GeV})^2 4.8 \text{ keV}} \qquad (2.4)$$

or from the center of gravity of the P waves. The quark mass only slightly influences this fit. It mainly influences the wave functions themselves, the dipole matrix elements and the velocity of the quarks. For the dipole matrix elements one would like a large quark mass, $m_c \simeq 2$ GeV. Fitting m_c to $|\psi(0)|^2$ from the naive[+] Van-Royen-Weisskopf formula [17]

$$\Gamma_{e\bar{e}}(V) = 16\pi\alpha^2 e_Q^2 \frac{|\psi_V(0)|^2}{M_V^2} = \alpha^2 e_Q^2 \frac{|R_V(0)|^2}{(M_V/2)^2} \qquad (2.5)$$

gives a rather small value, $m_c \simeq 1.1$ GeV. However, this does not destroy the nonrelativistic approximation. We found $\beta^2 = (v/c)^2 < 0.3$ in J/ψ and $\beta^2 < 0.4$ in ψ' for $m_c = 1.16$ GeV and $\alpha_s < 0.41$. Thus the quark masses are an open question here.

Is the large value of α_s reasonable? From the decay formulae to be described in Chapter 8 one finds α_s (annihilation at 3 GeV) $\simeq 0.2$. However, this refers to annihilation distances which are shorter than the average interquark distances. Furthermore the probability for

---

[+]There is a next order QCD correction to the Schrödinger wave function at the origin $|\psi(0)|^2 \rightarrow |\psi(0)|^2(1-16\alpha_s/3\pi)$ due to a gluon vertex correction [18]. This is large in Charmonium.

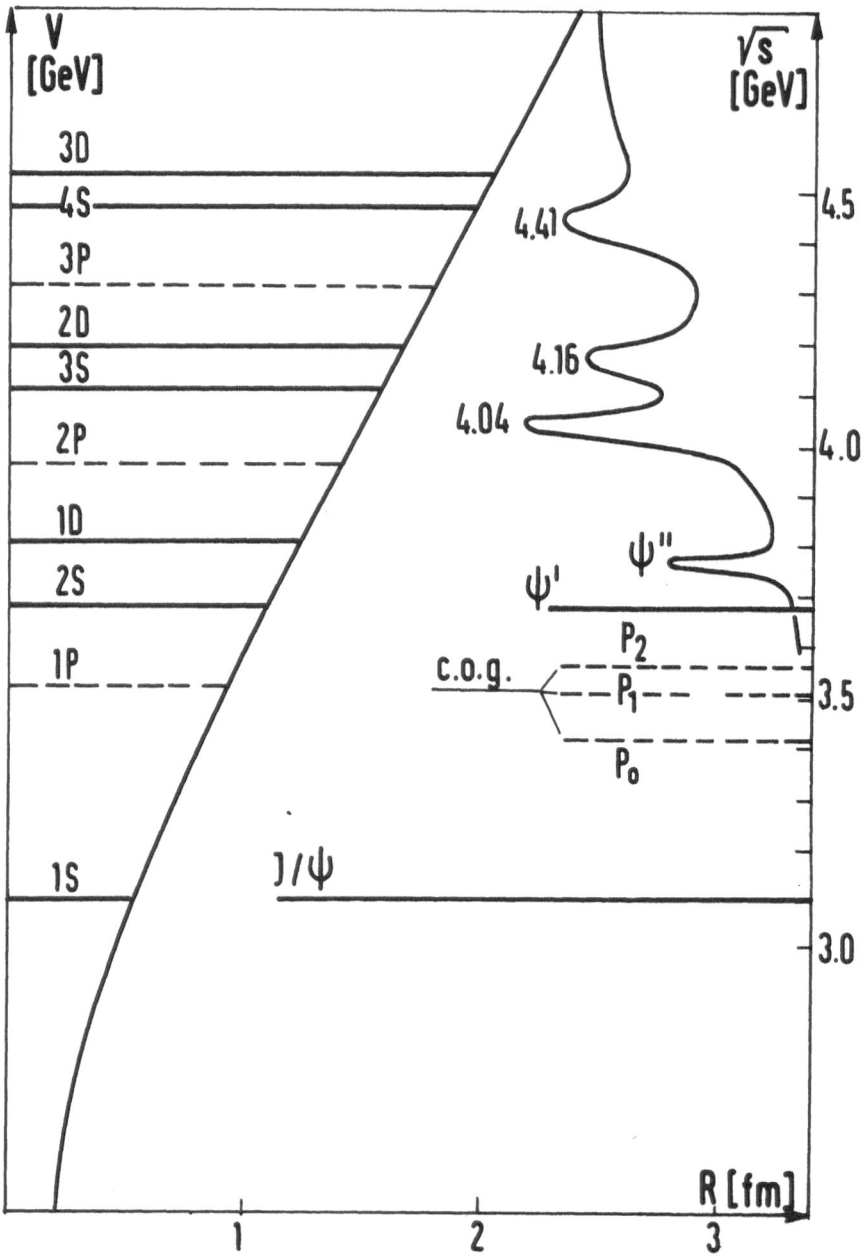

Fig. 2.1. The shape of the standard Charmonium potential
(2.2) with $\alpha_S = 0.41$, the spectrum and the experimentally observed Charmonium states. The
other parameters are a = 0.8665 GeV/fm and
$m_c = 1.6$ GeV.

three gluons → hadrons may be smaller than one. Then α_s (annihilation) is larger than 0.2. From deep in-elastic lepton scattering on the other hand we find α_s (3 GeV) $\simeq \alpha_s$ (0.07 fm) $\simeq$ 0.4 taking the scale para-meter λ = 0.5 GeV. From Fig. 2.1 we see that 0.07 fm is just in the middle of the range where V_{AF} dominates.

It is remarkable to see that the predicted D wave coincides with ψ'' (3.77), especially if the spin-orbit and tensor force splittings are taken into account! Amusingly also the higher S and D waves seem to coincide with the observed peaks in R above 4 GeV. Historical discrepancies in this respect are due to the choice of α_s.

The level sequence in Charmonium is 1S, 1P, 2S, 1D,... There are two theorems [19] about this ordering, let us quote them here:

i) For $V(R) = \frac{-g^2}{R} + V_C(R)$, where V_C is a confining potential non-singular at the origin, 2S is above 1P if V_C satisfies the following sufficient condition

$$(\frac{d}{dR})^2 [R(2V_C(R)+R\frac{dV_C(R)}{dR})] > 0 \qquad \forall\ R$$

(2.6)

$$[V_C(R) = R^\epsilon,\ \epsilon > 0]$$

ii) $V(R) = \frac{-g^2}{R} + V_C(R)$, where V_C is a non-singular confining potential. If $(d/dR)^3 (R^2 V_C(R))$ is positive, then the 1D state lies above the 2S state, provided

$$\frac{d}{dR}\ \frac{1}{R}\ \frac{d}{dR}(2V_C(R) + R\frac{dV_C(R)}{dR}) < 0 \qquad \forall\ R$$

(2.7)

$$[V_C(R) = R^\epsilon,\ 0 < \epsilon < 2]$$

b) Scaling the Schrödinger Equation

We now ask the question whether or not $V_c(R) = aR$ is unique for the next quark flavour (quark mass) as QCD suggests. For this purpose we first consider the scaling behaviour of the Schrödinger equation. The radial equation

$$[\frac{-d^2}{dR^2} + \frac{\ell(\ell+1)}{R^2} + 2\mu(V(R) - E)]u(R) = 0$$

with $V(R) = a \cdot R^\varepsilon$, $\varepsilon > -2$ \hfill (2.8)

can be brought into dimensionless form

$$[\frac{-d^2}{d\rho^2} + \frac{\ell(\ell+1)}{\rho^2} + \rho^\varepsilon - \xi] v(\rho) = 0 \hfill (2.9)$$

with

$$\rho = R(2\mu a)^{1/(2+\varepsilon)}, \qquad \xi = 2\mu E(2\mu a)^{-2/(2+\varepsilon)}. \hfill (2.10)$$

From (2.10) one reads off the scaling laws

$$R \sim m^{-1/(2+\varepsilon)}, \qquad E \sim m^{-\varepsilon/(2+\varepsilon)}. \hfill (2.11)$$

These scaling laws are also applicable for $\varepsilon = 0$, in which case the potential is logarithmic, $V(R) = a \cdot \log R/R_0$.

c) Bottonium[+] etc.

If the linear potential dominates, like in Charmonium,

[+] The justification for bottom, i.e. $e_Q = -1/3$, will be given in Chapter 3.

then $\Delta E \sim m^{-1/3}$. Eichten und Gottfried [20] predicted for the Υ system a spacing of $M_{\Upsilon'} - M_{\Upsilon} = 425$ MeV using $M_{\psi'} - M_{J/\psi} = 589$ MeV as input. However, experimentally [6] $M_{\Upsilon'} - M_{\Upsilon} = 556 \pm 3$ MeV which is much closer to $M_{\psi'} - M_{J/\psi}$. Thus the standard Charmonium potential is not the universal potential in QCD. Phenomenologically one is better off with a logarithmic potential [21] because of the constant spacings. However a pure log potential has no justification in QCD. But one can approximate the intermediate $Q\bar{Q}$ potential by a logarithmic one. This has been done by Bhanot and Rudaz [22] with the Ansatz (model I)

$$V(R) = \begin{cases} -4/3 \cdot \alpha_s/R & R < R_1 \\ b \cdot \log (R/R_o) & \text{for} \quad R_1 \leqslant R \leqslant R_2 \\ a \cdot R & R > R_2 \end{cases} \qquad (2.12)$$

and with the requirement of V(R) being continuously differentiable at R_1 and R_2. With a constant α_s for Charmonium and Bottonium, $\alpha_s = 0.31$, and a = 0.787 GeV/fm they obtained $M_{\Upsilon'} - M_{\Upsilon} = 560$ MeV. It is amusing to note that this value of a is even in agreement with what one would expect from the old meson spectroscopy.

For a study of very heavy Quarkonia, however, $\alpha_s = $ const is a bad approximation. One has to take the logarithmic variation of the effective α_s with R into account, as done by Ono and one of the authors, model II [23].In this approach the Coulomb potential is modified via Eq.2.3

$$-\frac{4}{3}\alpha_s(q^2,\Lambda^2)\frac{4\pi}{q^2} \xrightarrow{N=3} \approx -\frac{64\pi^2}{27}\frac{1}{q^2 \log(-q^2/\Lambda^2)} \qquad (2.13)$$

270

and the Fourier transform becomes [24]

$$V_{AF}(R) = \frac{1}{R} \left\{ \frac{-16\pi}{27} \frac{1}{\log(1/R^2\Lambda^2 e^{2\gamma})} + O(\log^{-3}) \right\} \qquad (2.14)$$

where γ = Euler's constant.

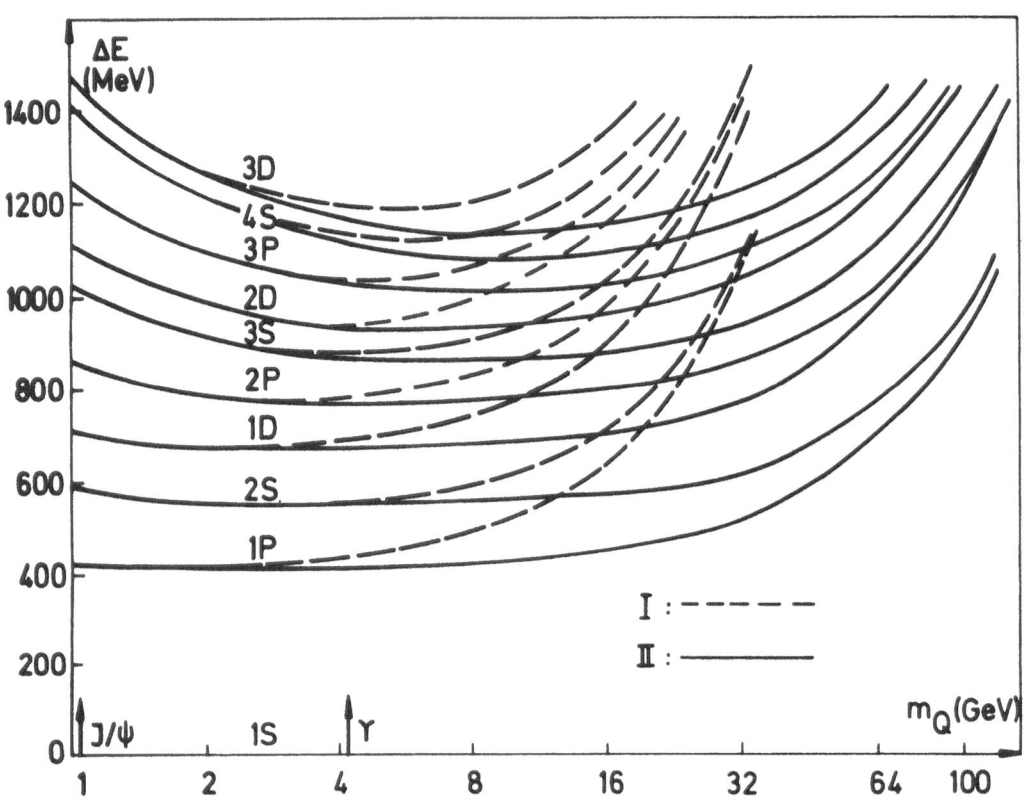

Fig.2.2. Mass difference to the ground state in model I
[23] (full line) and model II [22] (dashed
line).

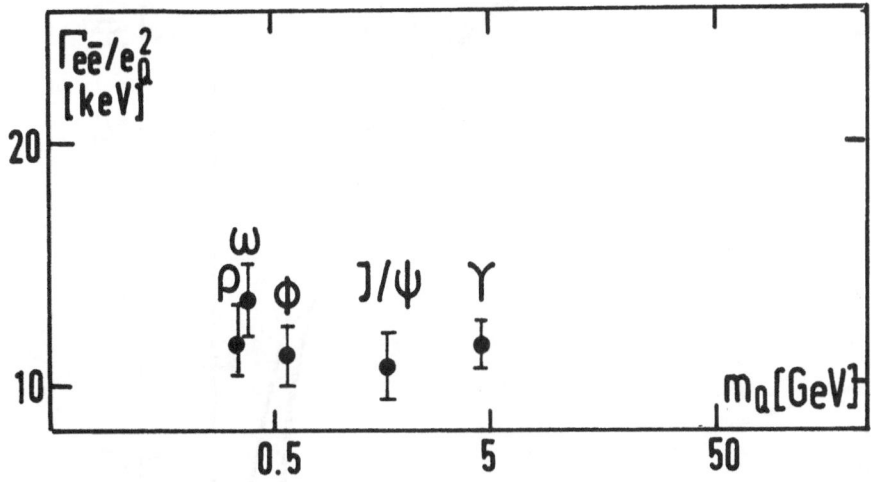

Fig. 2.3. Reduced leptonic decay widths.

For Charmonium and Bottonium the two potentials give
identical predictions. The differences will show up at
the next Onium, if its mass is high enough. In Fig.2.2
the level spacings are shown as functions of the quark
mass. In Fig.2.3 the reduced leptonic widths are shown.
The constancy of $\Gamma_{e\bar{e}}/e_Q^2$ from $\rho^\circ(770)$ to $\Upsilon(9.46)$ is a
rather interesting fact. But its explanation lies out-
side the scope of nonrelativistic potential models.
Whereas in model I (Ref.[22]) $\Gamma_{e\bar{e}}/e_Q^2$ increases for
Quarkonia heavier than Υ, Fig. 2.4, so that it will have
doubled for $m_{Q\bar{Q}} \simeq 30$ GeV, this increase in model II
(Ref.[23]) is much slower.

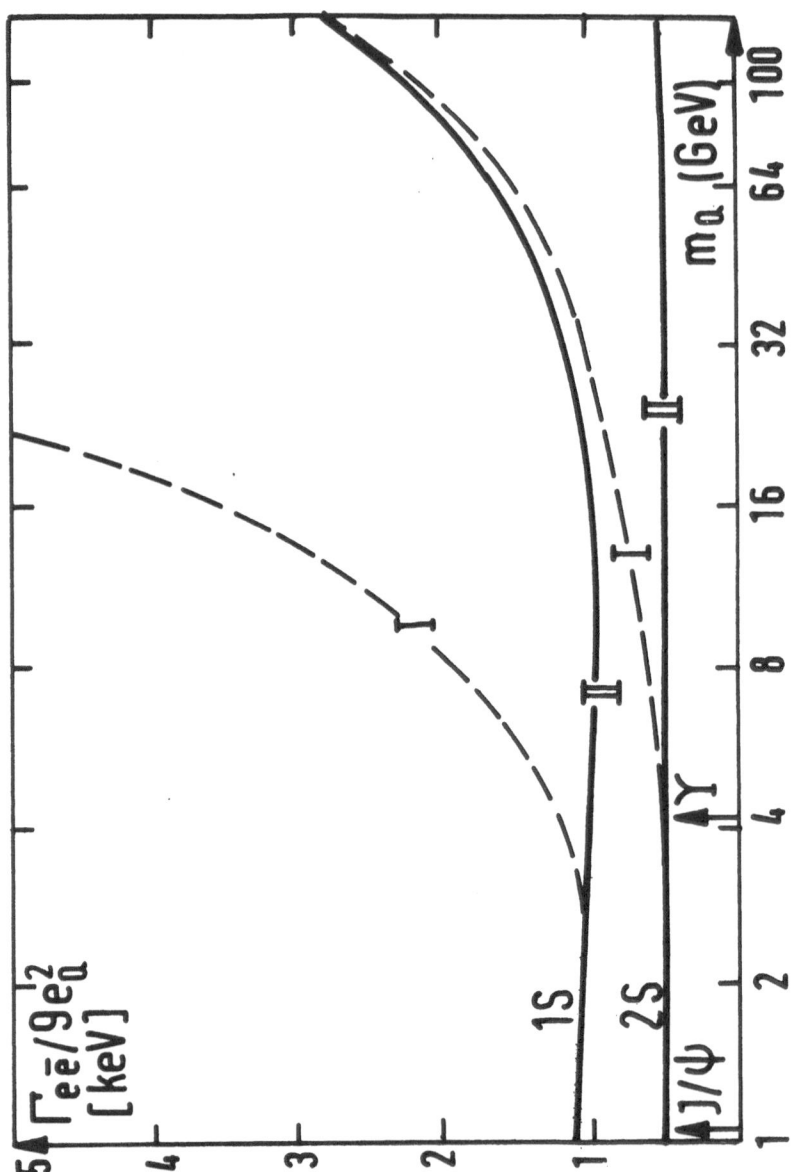

Fig. 2.4. $\Gamma_{e\bar{e}}/9e_Q^2$ as a function of m_Q for model I and II.

Threshold in e^+e^- due to the sixth quark:	$M_{threshold}$ = 21.3 GeV $M_{n+1} = M_n + 0.7\ n^2$ M_1 = 0.3 GeV	H.Lehmann [25]
Bound states of the sixth quark:	$M_{t\bar{t}}$ = 23.8 GeV	G.J.Aubrecht II and D.M.Scott [26]
	$M_{Q\bar{Q}}$ = 28-29 GeV by the 'factor 3 or π rule': $M_\phi:M_{J/\psi}:M_\Upsilon$ = = 1.02:3.1:9.46	every third expert [27]
	$M_{Q\bar{Q}}$ = 40 GeV	G.Preparata [28]
Current mass of the sixth quark (top quark):	$2m_t$ = 17.6, 25.8 GeV	W. Kummer [29]
	$2m_t$ = 22, 26, 100 GeV	M.-A. de Crombrugghe [30]
	$2m_t$ = 27.2 GeV	H. Georgi and D.V.Nanopoulos [31]
	$2m_t$ = 28 GeV	H. Harari, H. Haut and J. Weyers [32]
	$2m_t$ = 52 GeV	S.Pakvasa and H.Sugawara [33]
	$2m_t$ = 54 GeV	J.D. Bjorken [34]
	$2m_t$ = 80 GeV	T.F. Walsh [35]
Current mass of the fourth quark with e_Q = -1/3:	$2m_Q$ = 32-40 GeV	T.F. Walsh [35]

Table 2.1. Predictions for thresholds, bound state masses and current quark masses. We are aware of the fact that we have certainly missed the prediction of one or another of our colleagues.

The mass of the next quark is an interesting question. In Table 2.1 we give a list of numbers. This list is typical for the speculative character of the predictions.

d) Number of Bound States below Threshold

The higher the mass of the quark the more states are below threshold. Their number can be fairly easy estimated. The condition for lying below threshold is $M_{Q\bar{Q}} < 2M_{Q\bar{q}}$ or with $M_{q_1 q_2} = m_{q_1} + m_{q_2} + E_{q_1 q_2}$

$$E_{Q\bar{Q}} < 2m_q + 2E_{Q\bar{q}} \quad . \tag{2.15}$$

The binding of $Q\bar{Q}$ depends on the mass of the heavy quark. The states fall deeper into the potential well with increasing m_Q. For $Q\bar{q}$, on the other hand, the light quark mass (reduced mass) determines the properties of the system. In a very crude first approximation $E_{Q\bar{q}}$ can be taken as a constant and then the threshold is fixed.

The number of bound states is obtained semi-classically from the Bohr-Sommerfeld quantization condition

$$\int_0^{R_o} dR \sqrt{m_Q(E_{thr} - V(R))} = \pi(n - 1/4) \tag{2.16}$$

and one finds, since $E_{thr} = V(R_o)$ and R_o become independent of m_Q [36]

$$n = \frac{1}{4} + const \cdot \sqrt{m_Q} \quad . \tag{2.17}$$

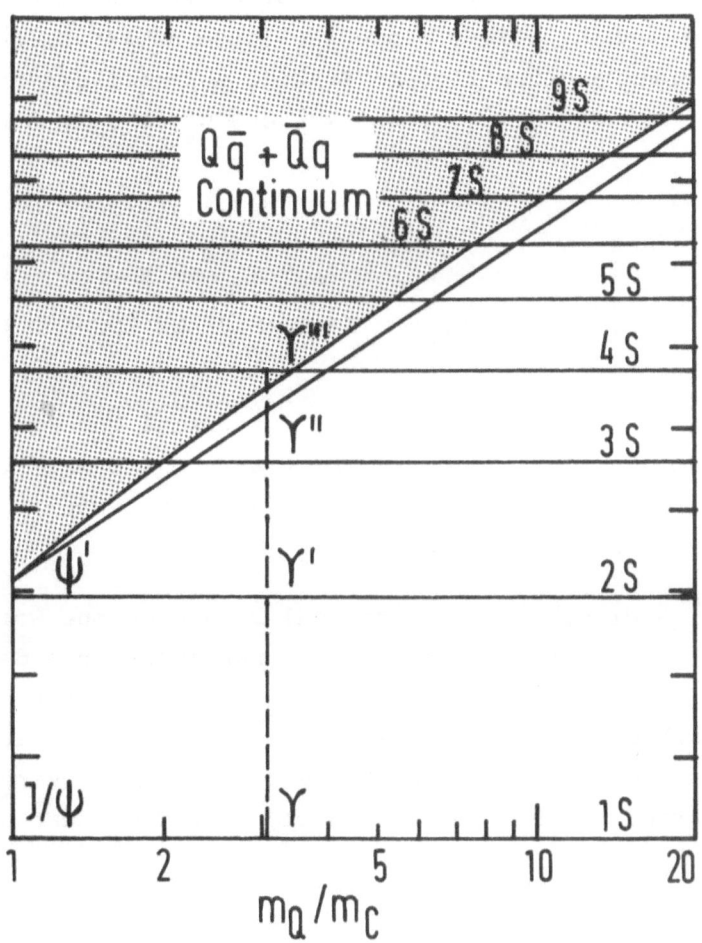

Fig. 2.5. Number of bound states below the strong-decay threshold [36].

Quigg and Rosner fixed the constant in Charmonium, where
n = 2. Fig. 2.5 shows their prediction for heavy quarks.
In Bottonium n = 3 is expected and Υ''' should already be
a factory for B mesons via $B\bar{B}$ and/or eventually $B\bar{B}^*$, $\bar{B}B^*$.
The width of Υ''' may be well below the machine width in
e^+e^- annihilation, since the large number of nodes in the
radial wave function of Υ''' will suppress its decay into
two slowly moving ground state S waves like B or B^*.

3. THE CHARGE OF THE QUARKS AND LOCAL DUALITY

The charmed quark and its charge were predicted in
1970. The GIM [3] mechanism fixes e_c to + 2/3. For the
charge of the quark constituing the Υ-system no such
strong prediction exists.

Without any theoretical prejudice one would have
to wait until one finds the B particles in order to
tell the charge. If the combinations with the strange
quark $Q\bar{s}$, $\bar{Q}s$ are neutral, e_Q = -1/3, if they are
charged, e_Q = 2/3. In this sense, the discovery of the
charged $F^{\pm}(2040)$ meson finally confirmed that e_c = 2/3.

From the parton point of view, the charge of the
new quark is measured by the step in $R = \sigma_{hadron}/\sigma_{\mu\bar{\mu}}^{QED}$
in e^+e^- annihilation above the new particle threshold:
ΔR = 1/3 (4/3) for e_Q = -1/3 (2/3). However, we need
not go into the continuum, the resonance spikes below
threshold already tell us the charges of the quarks.
The quantity

$$\frac{1}{\Delta M} \int\limits_{res} dM \; \sigma_M (res \rightarrow all) / \sigma_{\mu\bar{\mu}}^{QED} \; (q^2 = M^2) \equiv \Delta R_{res} \qquad (3.1)$$

measures a step ΔR for a given spacing ΔM. With help
of the relation

$$\int\limits_{res} dM \; \sigma_M (res \rightarrow all) = 6\pi^2 \Gamma (res \rightarrow e^+e^-)/M_{res}^2 \qquad (3.2)$$

one obtains in the Υ system from [6] $\Gamma(\rightarrow e^+e^-) =$
$= (1.32 \pm 0.09)$ keV and $\Delta M = M_{\Upsilon'} - M_{\Upsilon} = (556 \pm 3)$ MeV

$$\Delta R_{\Upsilon} = 0.63 \pm 0.04 \quad . \qquad (3.3)$$

This value of ΔR lies between the values for the two
possible charges. If we continue with $\Gamma(\Upsilon' \rightarrow e^+e^-) =$
$= 0.38 \pm 0.10$ keV and $\Delta M = M_{\Upsilon''}(10.41) - M_{\Upsilon'}(10.015) =$
$= 0.40$ GeV we obtain

$$\Delta R_{\Upsilon'} = 0.26 \pm 0.07 \qquad (3.4)$$

which clearly favours $|e_Q| = 1/3$, i.e. the bottom quark.

Originally specific models predicted specific
values for $\Gamma_{e\bar{e}}$ [21]. Another way to evaluate the charge
of the quark is studying the scaling properties of $\Gamma_{e\bar{e}}$
in possible potential models [37]. Also in this approach
the decisive evidence for a charge of $-1/3$ comes from
$\Gamma_{e\bar{e}}$ of Υ'. A simple continuation of the observed constancy
of $\Gamma_{e\bar{e}}/e_Q^2$ (Fig. 2.3) would give $(e_Q/e_c)^2 = \Gamma_{e\bar{e}}(\Upsilon)/\Gamma_{e\bar{e}}(J/\psi)$
and thus $|e_Q| = 1/3$. Potential models predict $\Gamma_{e\bar{e}}(\Upsilon'') \approx$
≈ 0.17 keV [38].

We shall now demonstrate that a parton type scaling

278

behaviour is a property of a rather general class of
potential models. We use Eq. 3.2 and

$$\Gamma(res \rightarrow e^+e^-) = 16\pi\alpha^2 \frac{3e_Q^2}{3} \frac{|\psi_{res}(0)|^2}{M_{res}^2} \tag{3.5}$$

and approximate the mass squared of the nth resonance by

$$M_n^2 \simeq 4m^2 + 4mE_n \tag{3.6}$$

where E_n is the Schrödinger eigenvalue. Semiclassically
one can relate the S wave functions at the origin to the
spectrum in a rather potential independent way. With

$$\psi_n(\vec{x}) = \frac{1}{\sqrt{4\pi}} \frac{u_n(R)}{R} \quad \text{and} \quad p_n^2 = 2\mu(E_n - V(R)) \tag{3.7}$$

the normalized WKB solution is for finite $V(0)$

$$u_n(R) = \frac{p_n^{-1/2}(R)}{R_{max}} \cdot \sqrt{2} \cdot \sin(\frac{1}{\hbar} \int_0^R d\hat{R} \, p_n(\hat{R})) . \tag{3.8}$$

By differentiating the Bohr-Sommerfeld quantization
condition

$$\int_0^{R_{max}} d\hat{R} \, p_n(\hat{R}) = \pi\hbar(n + const.) \tag{3.9}$$

with respect to n, one obtains the relation [39]

$$4\pi^2 |\psi_n(0)|^2 = \sqrt{2\mu E_n/\hbar^2} \cdot \frac{d(2\mu E_n/\hbar^2)}{dn} . \tag{3.10}$$

Insertion of this relation into Eqs. 3.5 and 3.2 leads to [40]

$$\frac{1}{\Delta M} \int_{res} dM \, \sigma_M (res \rightarrow all) = \frac{4\pi\alpha^2}{3M^2} \, 3e_Q^2 \cdot \frac{3}{2} \, \sqrt{1-4m^2/M^2} \, . \qquad (3.11)$$

For comparison the parton model cross section reads

$$\sigma_{parton}(q^2) = \frac{4\pi\alpha^2}{3q^2} \, 3e_Q^2 (1 + \frac{2m^2}{q^2}) \, \sqrt{1-4m^2/q^2} \, . \qquad (3.12)$$

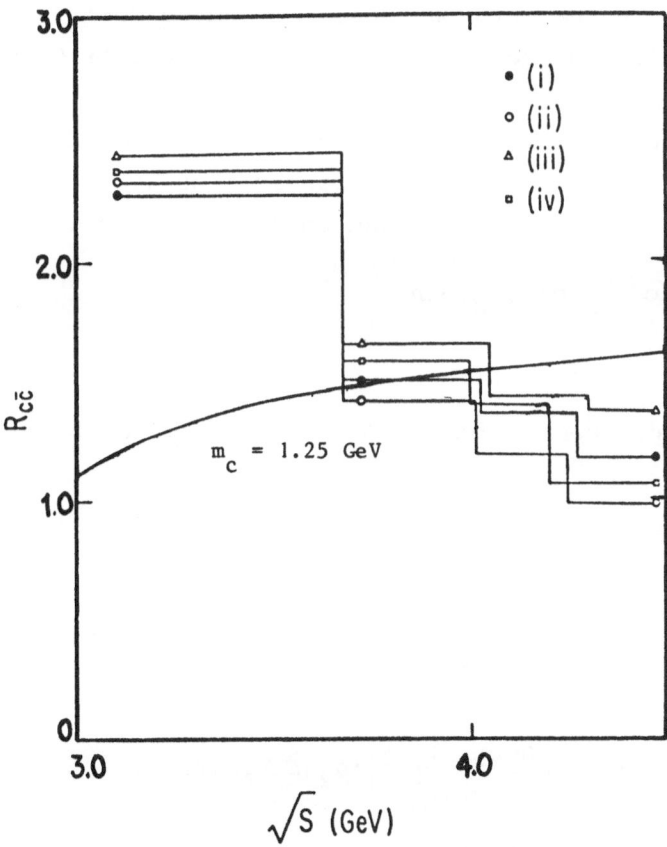

Fig. 3.1. From Ref.[40]. (i)-(iv) denote the four potential models of Ref.[41].

Thus with the nearly correct threshold factors
$(3/2 \simeq 1 + 2m^2/q^2)$ the resonances reproduce in an
approximate way the parton behaviour: this is called
local q^2 duality. For Charmonium Fig. 3.1 shows the
smoothened resonances for four different potential
models [41] and the curve corresponds to Eq. 3.11 with
the current quark mass $m_c = 1.25$ GeV.

4. LEVEL SPLITTINGS (FINE AND HYPERFINE STRUCTURE)

In the physical spectrum the Schrödinger states are
split up due to spin interactions. In this Chapter we want
to compare the magnitude of these splittings with the
simplest Ansatz we can imagine, the Fermi-Breit Hamiltonian
[42]. These higher order corrections are relativistic
kinematic corrections and spin corrections:

$$H = H^O + H^{rel} + H^{spin} \tag{4.1}$$

The spin corrections have three contributions:

spin-orbit: $\quad H^{LS} = \dfrac{2}{m_Q^2} \, \vec{L} \cdot \vec{S} [\dfrac{1}{R} \dfrac{d}{dR}] (V_{AF}(R) - \dfrac{1}{4}V(R))$

tensor: $\quad H^T = \dfrac{-1}{12m_Q^2} (3\vec{\sigma}_1 \cdot \hat{R}\vec{\sigma}_2 \cdot \hat{R} - \vec{\sigma}_1 \cdot \vec{\sigma}_2) [\dfrac{d^2}{dR^2} - \dfrac{1}{R} \dfrac{d}{dR}]V_{AF}(R)$

spin-spin: $\quad H^{SS} = \dfrac{1}{6m_Q^2} \, \vec{\sigma}_1 \cdot \vec{\sigma}_2 \, \Delta \, V_{AF}(R)$. $\tag{4.2}$

Here $\vec{\sigma}_1/2$ is the quark spin, $S = 1/2 \, (\vec{\sigma}_1 + \vec{\sigma}_2)$ the meson spin, $\vec{L}$ its angular momentum, R the interquark distance. For the potential V(R) we take the simplest ansatz $V = V_{AF} + V_C$ with only $V_{AF}(R)$ being spin-dependent. Nevertheless, the spin-independent V_C contributes to the spin orbit interaction due to the relativistic kinematic effect of the Thomas precession [43], represented by the term $-1/4 \, V(R)$ in H^{LS} (4.2). In Quarkonium this additional Thomas precession decreases the spin orbit splittings and it decreases $R_p = \Delta M(^3P_2 - {}^3P_1)/\Delta M(^3P_1 - {}^3P_0)$. While in Positronium, where $V(R) \sim -1/R \sim V_{AF}(R)$, $R_p = 0.8$, the experimental value in Charmonium is $R_p = 0.5$ [1].

We are confident that the Fermi-Breit Hamiltonian (4.2) is not a too bad approximation. As an example let us consider the part of the relativistic corrections due to the kinetic energy of the quarks. This correction is $<(\vec{p}^2)^2/4m_Q^3> \sim E_{kin} <\beta^2/4>$. Up to β^2 of 0.4 the relativistic kinetic energy correction is less than 10 %. The β^2 one obtains in Charmonium calculations is 0.2 to 0.3 for J/ψ and 0.27 to 0.4 for ψ' varying m_c from 1.6 to 1.16 GeV.

Let us now compare experiment with the predictions from (4.2). The three P waves of Charmonium shown in Table 4.1 are quite well established.

state	$^3P_2\,(2^{++})$	$^3P_1\,(1^{++})$	$^3P_0\,(0^{++})$	center of gravity
mass [GeV]	3.552	3.508	3.415	3.522

Table 4.1

The P wave splittings can be parametrized as

$$\langle H^{LS} \rangle = A \langle \vec{L} \cdot \vec{S} \rangle \ , \qquad \langle H^T \rangle = B \langle T \rangle \tag{4.3}$$

where the tensor operator $T \equiv 3\vec{\sigma}_1 \cdot \hat{R} \vec{\sigma}_2 \cdot \hat{R} - \vec{\sigma}_1 \cdot \vec{\sigma}_2$. The expectation values of $\vec{L} \cdot \vec{S}$ and T can be found in text-books on Quantum Mechanics [44]. They are displayed in Table 4.2. A Charmonium analysis with the experimental masses of Table 4.1

j	$\langle \vec{L} \cdot \vec{S} \rangle$	$\langle T \rangle$
$\ell+1$	ℓ	$-2\ell/(2\ell+3)$
ℓ	-1	2
$\ell-1$	$-(\ell+1)$	$-2(\ell+1)/(2\ell-1)$

Table 4.2

yields for A and B

$$A \simeq 34 \text{ MeV} \ , \qquad\qquad B \simeq 10 \text{ MeV} \ . \tag{4.4}$$

We obtain from (4.2) for the standard Charmonium potential

$$A = \frac{2}{m_Q^2} \langle \alpha_s R^{-3} - \tfrac{1}{4} aR^{-1} \rangle \ ,$$

$$B = \frac{1}{3m_Q^2} \langle \alpha_s R^{-3} \rangle \ . \tag{4.5}$$

We see that the spin dependence from the one gluon ex-change (V_{AF}) is governed by $\langle R^{-3} \rangle$ while the Thomas precession is governed by $\langle R^{-1} \rangle$. Taking our $\alpha_s = 0.4$,

$m_c^{-1} \langle R^{-3} \rangle \simeq 0.07$ GeV^2 and $\langle R^{-1} \rangle \simeq 0.4$ GeV from numerical fits we obtain the values of A and B given in Table 4.3 for two different values of m_c.

m_c [GeV]	1.6	1.1
A [MeV]	35-12=23	56-32=24
B [MeV]	6	9

Table 4.3. A and B from numerical fits. In row A the contribution from the Thomas precession is 12 and 32 MeV respectively.

By comparison of Table 4.3 with Eq.(4.4) we see that we are in the right ball park. We could not have expected a better agreement from our crude approximation!

Let us now try the spin-spin interaction which arises from the short range one gluon exchange V_{AF} alone. The relevant term in the Fermi-Breit Hamiltonian (4.2) was

$$H^{SS} = \frac{1}{6m_Q^2} \vec{\sigma}_1 \cdot \vec{\sigma}_2 \, \Delta V_{AF}(R) \quad . \tag{4.6}$$

Because $\Delta V_{AF}(R) \sim \Delta(\frac{-1}{R}) = 4\pi\delta(R)$ the integral over the wave functions becomes trivial and with $\vec{\sigma}_1 \cdot \vec{\sigma}_2 = 2\vec{S}^2 - 3$ we have

$$\langle H^{SS} \rangle = \frac{4}{3} \alpha_s \, 4\pi \, \frac{1}{6m_Q^2} \, (2\vec{S}^2 - 3) |\psi(0)|^2 \quad . \tag{4.7}$$

It is clear that (4.7) is an overestimate because the one gluon exchange potential has a weaker singularity at the origin than the $-1/R$ potential.

Taking $|\psi(0)|^2$ from $\Gamma_{e\bar{e}}$ via Eq.(2.5) and α_s from Eq.(2.4) gives us for the splittings

$$M(1^3S_1) - M(1^1S_0) \simeq 70 \text{ MeV } (M_{J/\psi}/2m_c)^2$$

$$M(2^3S_1) - M(2^1S_0) \simeq 35 \text{ MeV } (M_{\psi'}/2m_c)^2 \quad . \tag{4.8}$$

Trying to identify $\eta_c(1^1S_0) \equiv X(2.83)$ means $\gtrsim 70$ MeV $\equiv$ $\equiv 250$ MeV, $\eta_c'(2^1S_0) \equiv \chi(3.45)$ means $\gtrsim 35$ MeV $\equiv 230$ MeV, or $\eta_c'(2^1S_0) \equiv \chi(3.59)$ means $\gtrsim 35$ MeV $\equiv 80$ MeV. Many solutions have been proposed to solve this puzzle, among these are instanton effects [45] and an anomalous colour magnetic moment of the c-quark [46]. The simplest solution might be that the $|\psi(0)|^2$ in Eq.(2.5) and in (4.7) are different objects. The next order correction to $|\psi(0)|^2$ in (2.5) comes in through gluon exchange between the two quark lines before annihilation. It yields a factor [18]

$$(1 - \frac{16\alpha_s}{3\pi}) \tag{4.9}$$

which in no case is small. But before continuing this discussion let us wait for estimates of some decay rates involving the pseudoscalars. Then we will find that we have much more severe problems which question the identifications above.

Let us now discuss the P and D wave splittings for heavier Quarkonia in models I (Ref.[22]) and II (Ref.[23]). Fig. 4.1 shows the mass difference $\Delta M(^3P_2 - {}^3P_0)$ and the ratio $R_P = \Delta M(^3P_2 - {}^3P_1)/\Delta M(^3P_1 - {}^3P_0)$ as a function of the quark mass. The decrease of the overall splittings can be estimated from the scaling behaviour of the radius R

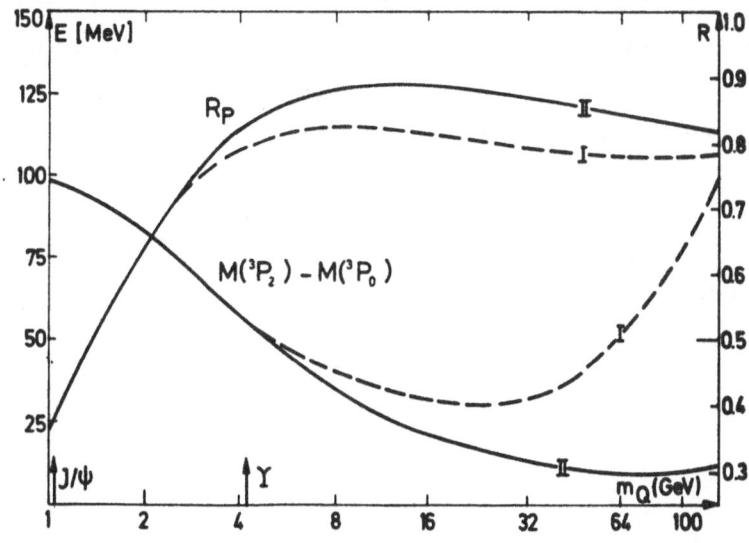

Fig. 4.1. The mass difference $M(^3P_2 - ^3P_0)$ and the ratio R_P as a function of m_Q in model I and II.

(2.11) in intermediate potential regimes, where the potential is logarithmic, $R^{-3}m_Q^{-2} \sim m_Q^{-1/2}$, compare Eqs. (4.5). The Thomas precession contribution from confinement, however, decreases much faster like $R^{-1}m_Q^{-2} \sim m_Q^{-3/2}$. The sharp rise of R_P between 1 and 8 GeV quark mass is due to this difference in the scaling behaviour. The increase of the splittings above $M(Q\bar{Q}) \approx 200$ GeV in model II is due to the Coulomb like potential at short distances for which all mass differences have to scale like m_Q, the better the smaller Λ/m_Q is. If QCD were not asymptotic free but pure Coulombic at short distances, the increase would show up much earlier, namely above $M(Q\bar{Q}) \approx 30$ GeV (model I). With asymptotic freedom the ratio R_P approaches the asymptotic Coulomb value of 0.8 from above, because the spin dependent potential is weaker than Coulomb.

For the D wave masses we also parametrize like in Eq. (4.3) with $\langle \vec{L} \cdot \vec{S} \rangle$ and $\langle T \rangle$ again given in Table 4.2. The coefficients A, B and $R_D = \Delta M(^3D_3 - ^3D_2)/\Delta M(^3D_2 - ^3D_1)$ are shown in Fig.4.2 as functions of the quark mass. The interesting thing is the change of sign of A for $m_Q \approx 1.4$ GeV which in case of small m_c leads to an inversion of the D wave multiplet in Charmonium. A similar inversion of the P wave multiplet of the charmed mesons, D_P^*, F_P^*, has been predicted by Schnitzer [47].

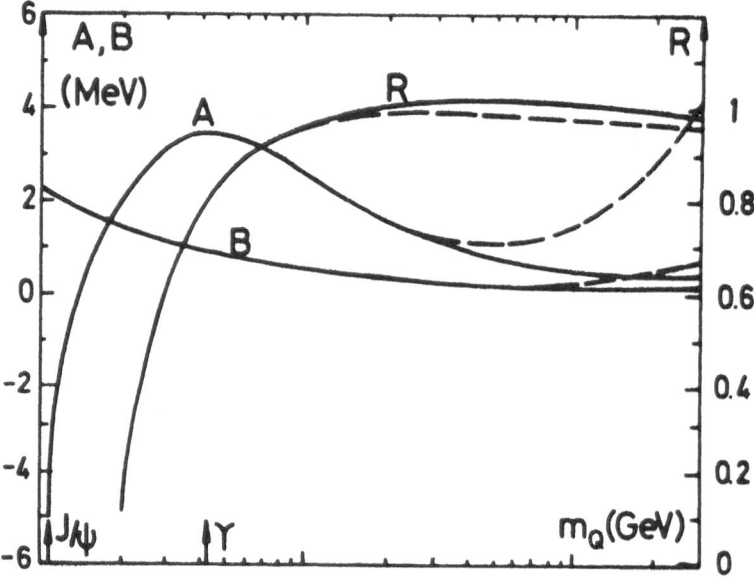

Fig. 4.2. The parameters A,B,C and the ratio R_D for Quarkonium D waves as functions of m_Q in model I (---) and II (———).

Effects of this Thomas precession also seem to show up in the baryon spectrum, where a phenomenological description of the splittings indicates that the net spin orbit splittings are small. In a potential model the $\vec{L}\cdot\vec{S}$ term and the Thomas precession have to cancel [48].

5. HOW TO FIND THE LEVELS

The best place to study the levels of Quarkonia is in e^+e^- annihilation. There the vector states are produced via one-photon annihilation. The cross section is given by

$$\sigma(e^+e^- \to V \to f) = \frac{12\pi\Gamma_{V\to e^+e^-} \cdot \Gamma_{V\to f}}{(q^2-M_V^2)^2 + M_V^2\Gamma_{tot}^2} \tag{5.1}$$

which yields for a narrow resonance

$$\int_{res} dM\ \sigma(e^+e^- \to V \to f) = 6\pi^2\ \frac{\Gamma_{V\to e^+e^-}}{M_V^2}\ B_{V\to f} \quad . \tag{5.2}$$

From the 1^{--} $Q\bar{Q}$ states one can reach the lower levels by photonic and/or hadronic transitions.

The $C = +$ levels can be reached by one photon transitions. In the next Chapter we shall discuss the electric and magnetic dipole transitions. The $C = -$ states can be reached from the vector mesons via two photon emission, preferentially via a $C = +$ state.

In Fig. 5.1 the hadronic transitions via $\pi\pi,\eta,3\pi$ are shown for Bottonium.

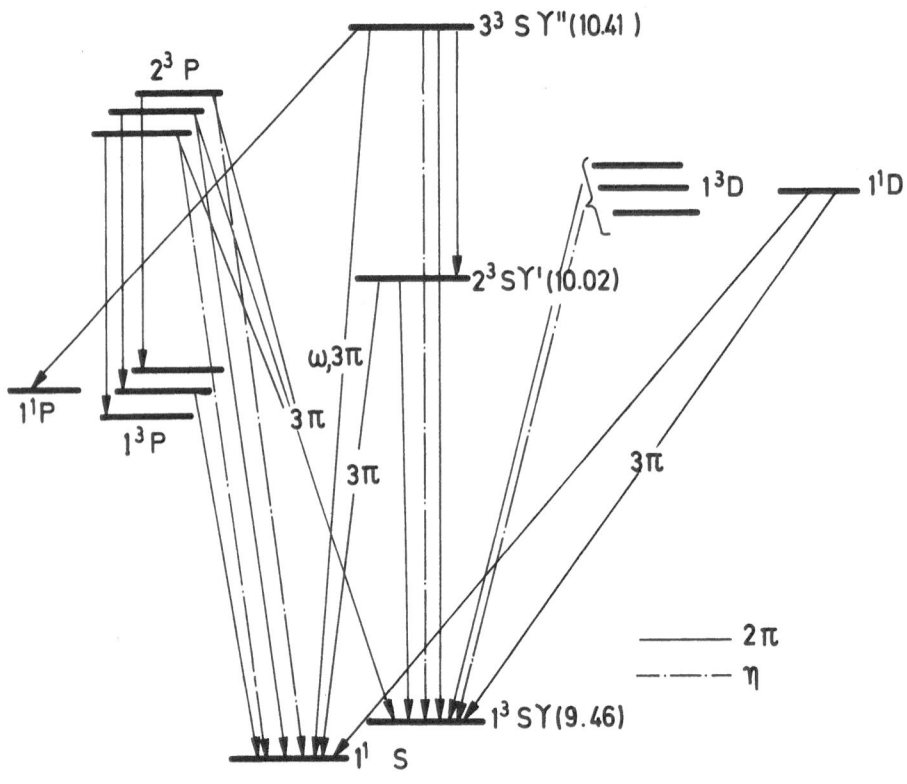

Fig. 5.1. Hadronic transitions in Bottonium. The Figure
is an updated version of that of Ref.[20].

A rather important transition to be looked for is the
decay $3^3S_1(\Upsilon") \rightarrow 1^1P_1(1^{+-}) + \epsilon(\pi\pi)$. It would reveal the
1^1P_1 state, whose existence is difficult to prove in
Charmonium, where one had to look for $\psi'(3.7) \rightarrow 1^{+-} + \gamma\gamma$.
Finding of the 1^1P_1 state is important because its mass
allows to determine, whether there are long range spin-
spin correlations or not. In our Ansatz for the Hamiltonian
we only had short range spin-spin forces. They do not act
on P waves and therefore the 1^1P_1 state is degenerate with

the c.o.g. of the 1^3P_j states. A long range spin-spin force [49], however, would act on the P waves and would lift this degeneracy.

According to Gottfried [50] the hadronic cascade decays can be understood as radiative gluon transitions which can be subject to a multipole expansion similar to electromagnetic radiation. While the expansion in $(kR/2)^2$ might converge, the expansion in α_s/π needs not. The distances involved in the process are of the order of the wave function radius and α_s will be large. This is the essential reason why we do not expect to be able to calculate absolute rates for gluon radiation. But we might be able to derive selection rules and the scaling behaviour. The typical example is [51]

$$\Gamma(Q\bar{Q}_2 \to Q\bar{Q}_1 + \lambda) \sim m_Q^{-2} \tag{5.3}$$

for any two $Q\bar{Q}$ states of the same flavour Q and λ any state of light quarks, provided $Q\bar{Q}_1$ and $Q\bar{Q}_2$ have the same parity, $\Delta L = 0$ or 2 and $\Delta S = 0$. If this scaling law is already valid in Charmonium, then $\Gamma(\psi' \to J/\psi\pi\pi) \simeq$ $\simeq 100$ keV implies $\Gamma(\Upsilon' \to \Upsilon\pi\pi) \simeq 10$ keV.

In a recent paper, Billoire, Lacaze, Morel and Navelet [52] have investigated the cascade $2^3S_1(Q\bar{Q}) \to$ $\to 1^3S_1(Q\bar{Q})$ + hadrons. They constrain themselves to the lowest order in α_s and $k \cdot R$ ("two gluon colour electric dipole emission") and project out the final spin states. They are able to roughly reproduce the experimentally observed $\pi\pi$ spectrum in $\psi' \to J/\psi\pi\pi$, see Fig. 5.2. However, in this approximation the transition $2^3S_1(Q\bar{Q}) \to$ $\to 1^3S_1(Q\bar{Q}) + \eta$ is zero because the gluon momenta are neglected. (η is emitted with ℓ = 1, but zero momentum gluons cannot carry angular momentum).

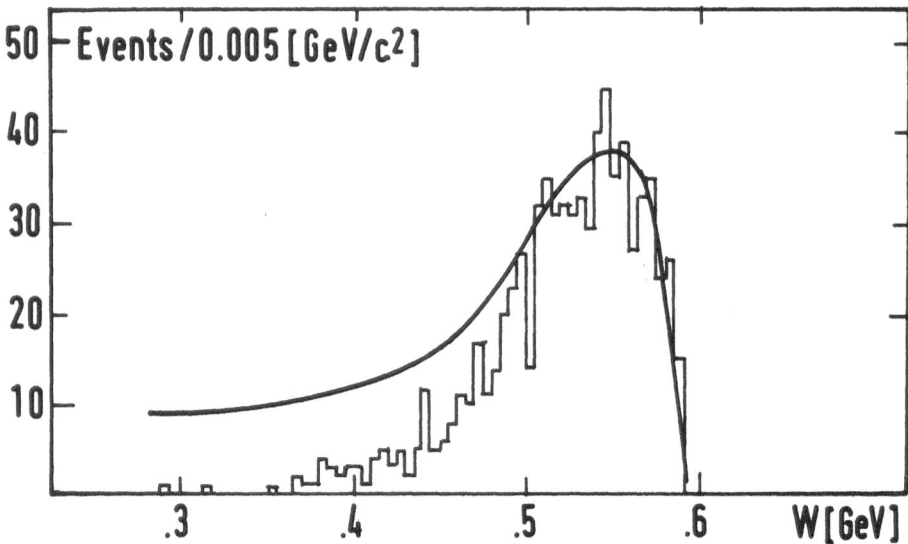

Fig. 5.2. Comparison of the 0^+ spectrum for $\psi' \to J/\psi + 2g$ to the experimental data for $\psi' \to J/\psi\pi\pi$. The spectrum corrected for acceptance is normalized to the observed events in the peak region. The disagreement on the low mass side is due to phase space and the absence of an Adler zero in the matrix element of Ref.[52]. [53].

6. PHOTON TRANSITIONS AND SUM RULES

For photon wave lengths long against the bound state size one can do a multipole expansion. The widths are [54]

$$\Gamma \sim \alpha\, e_Q^2 \begin{Bmatrix} k^3\ R^2 \\ k^3\ m_Q^{-2} \end{Bmatrix} (\frac{k \cdot R}{2})^{2(n-1)} \quad \text{for} \quad \begin{Bmatrix} E_n \\ M_n \end{Bmatrix} \text{transitions.} \quad (6.1)$$

Here k is the photon wave number, R the bound state
radius in the reduced system. The expansion parameter
is $(kR/2)^2$ and is roughly 1/4 to 3/100 in Charmonium,
thus justifying the multipole expansion.

a) Electric Dipole Transitions and El Sum Rules

In Quarkonium the formula for an electric dipole
transition (El) is [54]

$$\Gamma^{El}(Q\bar{Q} \to \gamma Q\bar{Q}) = \frac{4}{3}\, \alpha e_Q^2\, k^3\, |\vec{x}_{fi}|^2 \tag{6.2}$$

where $\vec{x}_{fi}$ is the matrix element of the dipole operator.
There are many corrections to the naive formula:

i) higher multipoles which are at most 5% in ψ' decays;
ii) interference of the finite wave length of the photon
     field e^{ikR} with the bound state wave function. Okun
     and Voloshin [55] have shown that these corrections
     amount to at most 5% in Charmonium;
iii) relativistic corrections, consisting of a) relativ-
     istic corrections to the wave functions, b) the
     interaction of the quark magnetic moment with the
     electric vector of the photon field, the last
     correction gives factors 1.0 to 0.6 [55]; c) the
     recoil corrections have been found to be + 20% in
     a relativistic model [56].

The radiative El widths of the standard Charmonium model
are shown in Fig. 6.1. The numbers in brackets are the
model widths with corrections of type iii b). All this
indicates that the model numbers are only correct up to
a factor two.

There are two kinds of electric dipole sum rules,

the so called Thomas-Reiche-Kuhn (TRK) sum rule (SR)
and the Wigner SR [57]. They were rediscovered for
Charmonium by Jackson [58]. Both SR's apply to the
dipole matrix element (6.2) without corrections. The
SR's are derived from Heisenberg's uncertainty relation

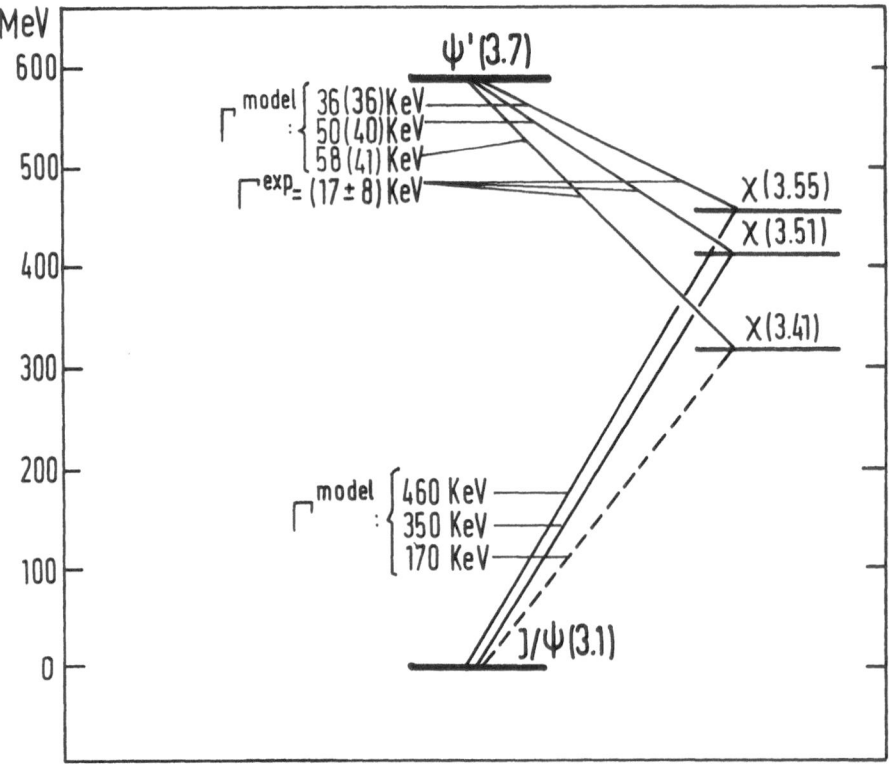

Fig. 6.1. E1 transitions in Charmonium. Model widths
are calculated via Eqs. (6.11, 12). The
numbers in brackets are the corrections of
type iii,b).

$$[\vec{x}, \vec{p}] = 3i \qquad\qquad (\hbar = c = 1) \qquad\qquad (6.3)$$

In a static $Q\bar{Q}$ potential without velocity dependent terms, we can replace $\vec{p}$ via the equation of motion

$$\vec{p} = i \; \frac{m_Q}{2} \; [H^O, \vec{x}] \quad . \qquad\qquad (6.4)$$

After taking the expectation value in a state $|i\rangle$ and inserting a complete set of states $|f\rangle$ the replacement of $\vec{p}$ leads to

$$\Sigma_f (E_f^O - E_i^O) \, |\vec{x}_{fi}|^2 = \frac{3}{m_Q} \quad . \qquad\qquad (6.5)$$

The number of final states is restricted by selection rules. In an arbitrary static potential $\Delta \ell = \pm 1$ for dipole transitions. In a harmonic oscillator potential, however, the number of final states is further restricted by the oscillator selection rule: The change of the number of radial nodes Δr is either O or $-\Delta \ell$. This fact is called saturation of the SR by the harmonic oscillator.

To derive the Wigner SR we recall Eq. (6.3) which can be written as

$$\Sigma_f \left(\langle i|\vec{x}|f\rangle \langle f|\vec{p}|i\rangle - \langle i|\vec{p}|f\rangle \langle f|\vec{x}|i\rangle \right) = 3i \quad . \qquad\qquad (6.6)$$

The angular selection rule now enables us to project out the final states with $\Delta \ell = + 1$ and those with $\Delta \ell = -1$. After some elaborate algebra one arrives at two SR's [57]:

$$\Sigma_{f,\ell-1}(E_f^o - E_i^o)|\vec{x}_{fi}|^2 = \frac{-\ell(2\ell-1)}{2\ell+1} \cdot \frac{1}{m_Q} \qquad (6.7a)$$

$$\Sigma_{f,\ell+1}(E_f^o - E_i^o)|\vec{x}_{fi}|^2 = \frac{(\ell+1)(2\ell+3)}{2\ell+1} \cdot \frac{1}{m_Q} \qquad (6.7b)$$

which of course add up to Eq.(6.5). We have gained two things: first the number of final states on the l.h.s. of (6.7) is smaller than in the TRK SR, and second, (6.7a) has a negative sign which will be very helpful.

b) Explicit Transition Rates and Bounds

To write down the rates it is convenient to express the dipole operator of Eq.(6.2) through the radial operator $R_{f,i}$ [59]

$$\Gamma(r,\ell,s,j \to r',\ell',s,j') = \frac{4\alpha e_Q^2}{3} k^3 (2j'+1) \left\{ \begin{matrix} \ell'j's \\ j\ \ell\ 1 \end{matrix} \right\}^2 |<r',\ell'||x||r,\ell>|^2 \qquad (6.8)$$

with

$$<r',\ell-1||x||r,\ell>=i\sqrt{\ell}\int_o^\infty R^2 dR\ R_{r',\ell-1}(R)\cdot R\cdot R_{r,\ell}(R) \equiv i\sqrt{\ell}\ R_{f,i}$$

$$<r',\ell+1||x||r,\ell>= -i\sqrt{\ell+1}\int_o^\infty R^2 dR\ R_{r',\ell+1}(R)\cdot R\cdot R_{r,\ell}(R) \equiv -i\sqrt{\ell+1}R_{f,i}. \qquad (6.9)$$

The matrix element $|\vec{x}_{fi}|$ of the sum rules Eqs. (6.5,7) does not involve quark spin. It is related to $<r',\ell'||x||r,\ell>$ by

$$\Sigma_{m'} \; |<r',\ell',m'|\vec{x}|r,\ell,m>|^2 \; = \; \frac{1}{2\ell+1}|<r',\ell'\|x\|r,\ell>|^2. \quad (6.10)$$

We can now write down rates and bounds. In Fig. 6.2 we show the quark spin triplet El transitions for Bottonium. Transitions for which there is an upper and/or lower bound are labelled by the formula numbers.

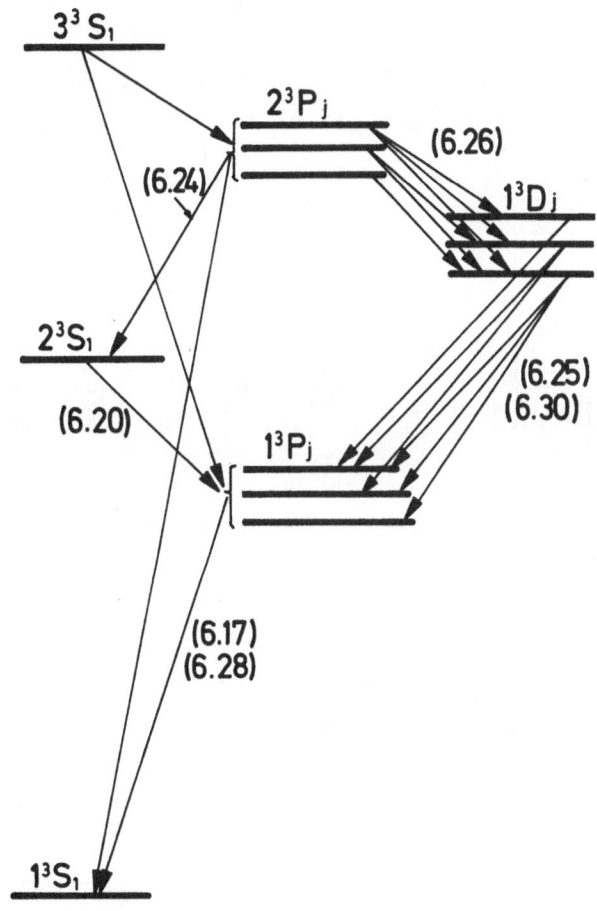

Fig. 6.2. El transitions in Bottonium. The numbers in brackets indicate formulae for upper and lower bounds.

The transition rates are:

$$\Gamma(^3P_j \rightarrow \gamma^3S_1) = \tfrac{4}{9} \, \alpha \, e_Q^2 \, k^3 |R_{S,P}|^2 \tag{6.11}$$

$$\Gamma(^3S_1 \rightarrow \gamma^3P_j) = \tfrac{4}{3} \, \frac{2j+1}{9} \, \alpha \, e_Q^2 \, k^3 |R_{P,S}|^2 \tag{6.12}$$

$$\Gamma(^3P_j \rightarrow \gamma^3D_{j'}) = \tfrac{8}{3} \, (2j'+1) \, \alpha \, e_Q^2 \, k^3 \begin{Bmatrix} 2 & j' & 1 \\ j & 1 & 1 \end{Bmatrix}^2 |R_{D,P}|^2 \tag{6.13}$$

$$\Gamma(^3D_j \rightarrow \gamma^3P_{j'}) = \tfrac{8}{3}(2j'+1)\,\alpha e_Q^2 \, k^3 \begin{Bmatrix} 1 & j' & 1 \\ j & 2 & 1 \end{Bmatrix}^2 |R_{P,D}|^2 \tag{6.14}$$

$$\begin{Bmatrix} 2 & j' & 1 \\ j & 1 & 1 \end{Bmatrix}^2 = \begin{Bmatrix} 1 & j & 1 \\ j' & 2 & 1 \end{Bmatrix}^2 = \frac{1}{900} \times$$

j\j'	1	2	3
0	100	0	0
1	25	45	0
2	1	9	36

$$. \tag{6.15}$$

The TRK SR (6.5) gives us as a bound

$$(E_{1P}^o - E_{1S}^o) \, |R_{1P,1S}|^2 \leq \frac{3}{m_Q} \tag{6.16}$$

which implies an upper bound on 1P → 1S

$$\Gamma(1^3P_j \rightarrow \gamma 1^3S_1) \leq \tfrac{4}{9} \, \alpha \, e_Q^2 \cdot \frac{k^3}{k^{(o)}} \cdot \frac{3}{m_Q} \, . \tag{6.17}$$

For $\ell = 1$ in the initial state the first two terms of (6.7a) give

$$(E^O_{2S} - E^O_{1P}) |R_{2S,1P}|^2 + (E^O_{1S} - E^O_{1P}) |R_{1S,1P}|^2 \leq \frac{-1}{m_Q} . \qquad (6.18)$$

With (6.16) this results in

$$(E^O_{2S} - E^O_{1P}) |R_{2S,1P}|^2 \leq \frac{2}{m_Q} \qquad (6.19)$$

and leads to the upper bound

$$\Gamma(2^3S_1 \to \gamma 1^3P_j) \leq \frac{4}{3} \frac{2j+1}{9} \alpha \; e^2_Q \frac{k^3}{k^{(o)}} \frac{2}{m_Q} . \qquad (6.20)$$

In a similar way we obtain

$$(E^O_{2P} - E^O_{2S}) |R_{2P,2S}|^2 \leq \frac{3}{m_Q} + (E^O_{2S} - E^O_{1P}) |R_{2S,1P}|^2 \leq \frac{5}{m_Q} \qquad (6.21)$$

$$(E^O_{1D} - E^O_{1P}) |R_{1D,1P}|^2 \leq \frac{5}{m_Q} \qquad (6.22)$$

$$(E^O_{2P} - E^O_{1D}) |R_{2P,1D}|^2 \leq \frac{-3}{m_Q} + (E^O_{1D} - E^O_{1P}) |R_{1P,1D}|^2 \leq \frac{2}{m_Q} \qquad (6.23)$$

and thus upper bounds on the widths:

$$\Gamma(2^3P_j \to \gamma 2^3S_1) \leq \frac{4}{9} \alpha \; e^2_Q \frac{k^3}{k^{(o)}} \frac{5}{m_Q} , \qquad (6.24)$$

$$\Gamma(1^3D_j \to \gamma 1^3P_{j'}) \leq \frac{8}{3} (2j'+1) \alpha e^2_Q \begin{Bmatrix} 1 & j' & 1 \\ j & 2 & 1 \end{Bmatrix}^2 \frac{k^3}{k^{(o)}} \frac{5}{m_Q} , \qquad (6.25)$$

$$\Gamma(2^3P_j \to \gamma 1^3D_{j'}) \leq \frac{8}{3} (2j'+1) \alpha e^2_Q \begin{Bmatrix} 2 & j' & 1 \\ j & 1 & 1 \end{Bmatrix}^2 \frac{k^3}{k^{(o)}} \frac{2}{m_Q} . \qquad (6.26)$$

Next, we write (6.18) as

$$(E^o_{1P} - E^o_{1S})\,|R_{1S,1P}|^2 \geq \frac{1}{m_Q} + (E^o_{2S} - E^o_{1P})\,|R_{2S,1P}|^2 \qquad (6.27)$$

and obtain by 'inverting' (6.20) the lower bound

$$\Gamma(1^3P_j \to \gamma 1^3S_1) \geq \frac{4}{9}\alpha e^2_Q \frac{k^3}{k^{(o)}} \frac{1}{m_Q} +$$

$$\qquad\qquad (6.28)$$

$$+ \frac{3}{2j+1} \frac{k^3_{1P,1S} k^{(o)}_{2S,1P}}{k^3_{2S,1P} k^{(o)}_{1P,1S}} \Gamma^{exp}(2^3S_1 \to \gamma 1^3P_j) \ .$$

Similarly we obtain from (6.23)

$$(E^o_{1D} - E^o_{1P})\,|R_{1P,1D}|^2 \geq \frac{3}{m_Q} + (E^o_{2P} - E^o_{1D})\,|R_{2P,1D}|^2 \qquad (6.29)$$

and thus the lower bound on the width

$$\Gamma(1^3D_j \to \gamma 1^3P_{j'}) \geq \frac{8}{3}(2j'+1)\alpha e^2_Q \begin{Bmatrix} 1 & j' & 1 \\ j & 2 & 1 \end{Bmatrix}^2 \frac{k^3}{k^{(o)}} \frac{3}{m_Q} +$$

$$\qquad\qquad (6.30)$$

$$+ \frac{(2j'+1)\begin{Bmatrix} 1 & j' & 1 \\ j & 2 & 1 \end{Bmatrix}^2}{(2j'''+1)\begin{Bmatrix} 2 & j''' & 1 \\ j'' & 1 & 1 \end{Bmatrix}^2} \frac{k^3_{1D,1P} k^{(o)}_{2P,1D}}{k^3_{2P,1D} k^{(o)}_{1D,1P}} \cdot \Gamma^{exp}(2^3P_{j''} \to \gamma 1^3D_{j'''}) \ .$$

In Fig. 6.3 we give the upper bounds for Bottonium using $m_b = 4.6$ GeV. We neglect the splittings and restrict ourselves to states with $j = 1$.

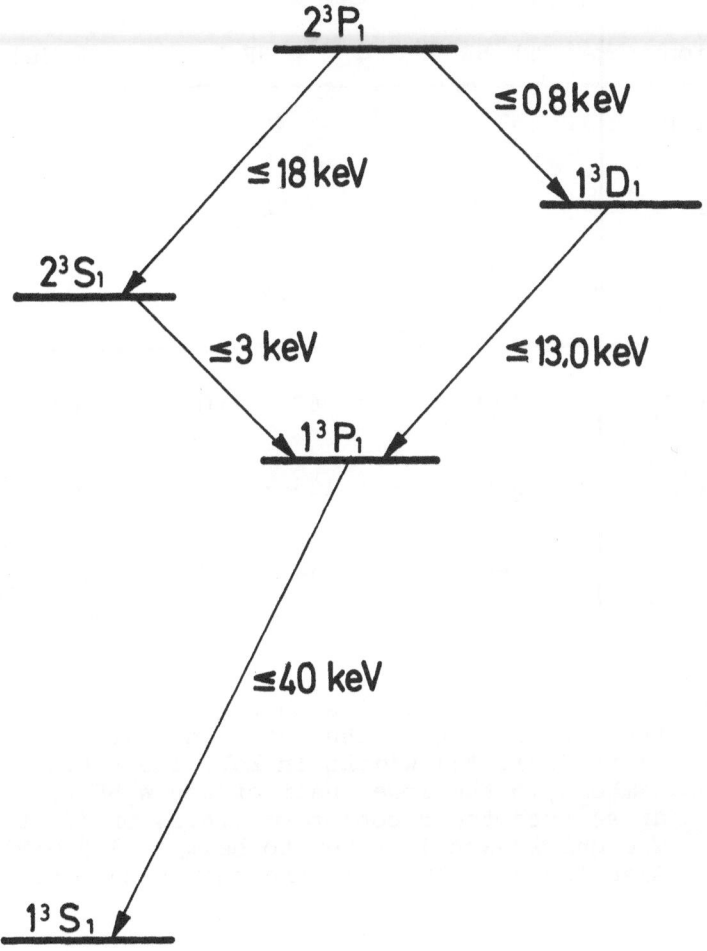

Fig. 6.3. Upper bounds for El transition widths in
Bottonium.

The bounds in Charmonium are of more practical use at
present. They are compiled in Table 6.1. Combining the
bounds of Table 6.1 and the experimentally measured
branching ratios for $P_c/\chi \to \gamma J/\psi$ one can obtain bounds
for the total widths, as shown in Table 6.2.

transition	TRK SR	W SR	model
$2^3S_1 \to \gamma 1^3P_2$		< 40	36
$2^3S_1 \to \gamma 1^3P_1$		< 56	50
$2^3S_1 \to \gamma 1^3P_0$		< 64	58
$1^3P_2 \to \gamma 1^3S_1$	< 490	> 160 + 140	460
$1^3P_1 \to \gamma 1^3S_1$	< 370	> 125 + 75	350
$1^3P_0 \to \gamma 1^3S_1$	< 180	> 60 + 30	170

Table 6.1. Upper and lower limits on El transitions from
the Thomas-Reiche-Kuhn and Wigner (W) sum
rules (SR). All widths in keV. The second
numbers in the lower half of the W SR column
arise from the second term r.h.s. of (6.28).
The quark mass is taken to be $m_c = 1.6$ GeV.
Also the model numbers are only sensitive to
m_c.

P states	$\chi(3.41)=0^{++}$	$P_c/\chi(3.51)=1^{++}$	$\chi(3.55)=2^{++}$
BR($\gamma J/\psi$),exp. [%]	3 ± 3	35 ± 7	14 ± 6
$\Gamma_{tot}(P_c/\chi)$,bounds[MeV]	3....6	0.57...1.05	2.15...3.5

Table 6.2. Bounds on $\Gamma_{tot}(P_c/\chi)$ derived from the sum rules,
Table 6.1, and the experimental BRs of $P_c/\chi \to \gamma J/\psi$.
The sum rules correspond to an uncorrected El
transition, this gives an additional theoretical
uncertainty of a factor 2.

The total widths of the P_c/χ states are calculable as the
sum of the radiative and the hadronic (gluon annihilation)
widths. A comparison of these total widths with the bounds
in Table 6.2 will be a comparison of theory with "experiment". This will be done in Chapter 8.

c) Magnetic Dipole Transitions

M1 decays are due to the interaction of the magnetic
photon field vector $\vec{m} = \vec{k} \times \vec{\epsilon}$ and the quark magnetic moment
$\mu_Q = e \cdot e_Q/2m_Q$. The matrix element reads

$$<f|\mu_Q \cdot \vec{\sigma} \cdot (\vec{k} \times \vec{\epsilon})|i> \qquad (6.31)$$

and involves the spin part of the states $|i>$ and $|f>$ only.
We have two graphs and therefore 4 times the rate as for
atomic M1 transitions [54]:

$$\Gamma(V_r \rightarrow \gamma PS_{r'}) = \frac{4}{3} \alpha \, e_Q^2 \, \frac{k^3}{m_Q^2} \, \delta_{rr'} \, ,$$

$$\qquad (6.32)$$

$$\Gamma(PS_r \rightarrow \gamma V_{r'}) = 4 \alpha \, e_Q^2 \, \frac{k^3}{m_Q^2} \, \delta_{rr'} \, .$$

An M1 transition requires $\Delta \ell = 0$ and the spatial overlap
between $|i>$ and $|f>$ is either $1(r = r')$ or 0 $(r \neq r',$
forbidden M1) in this approximation. Relativistic corrections of course modify the rate (6.32) and lead to small
transitions also between orthogonal $(r \neq r')$ states. In
allowed transitions $(r = r')$ the spatial overlap of one
should not be changed much by relativistic corrections
(but compare section 7).

d) Problems with M1 in Charmonium

In Fig. 6.4 possible candidates for the pseudoscalar partners of J/ψ and ψ' and the corresponding M1 transitions are shown. If the second χ is not at 3.59 GeV but at 3.18 GeV (second experimental solution) it can hardly be explained as a pseudoscalar. In the figure the calculated M1 widths are shown.

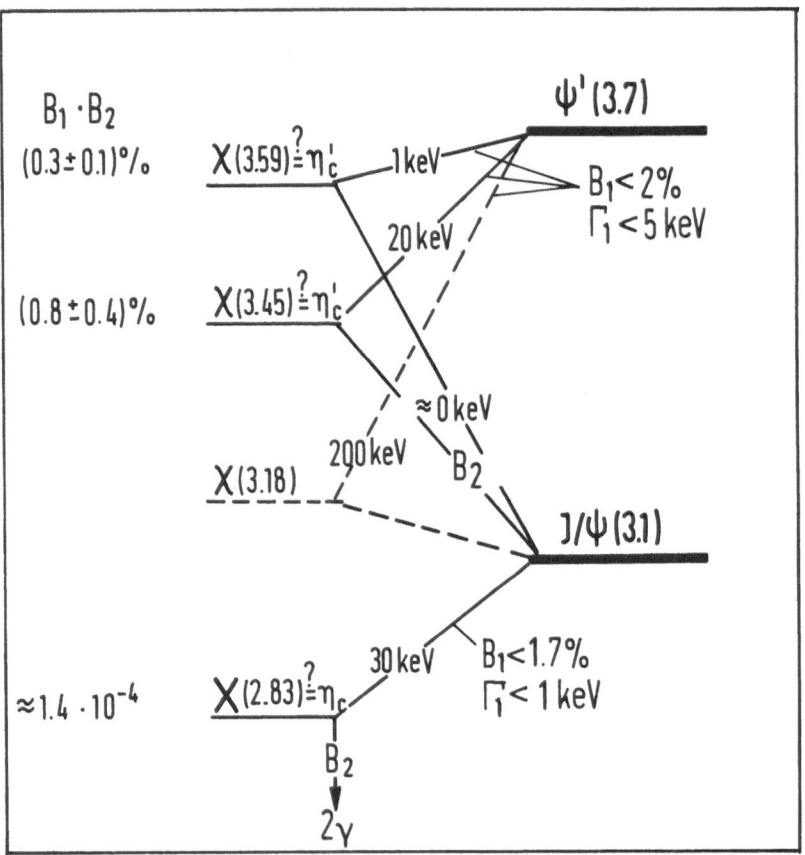

Fig. 6.4. M1 transitions in Charmonium. Theoretical widths, eq.(6.32) are indicated at the transition lines. $B_1(\Gamma_1)$ and $B_1 \cdot B_2$ are from experiment, Ref.[1], [60].

They have at first to be contrasted with the experimental
bounds on these transitions as indicated. Together with
the experimental product of branching ratios these bounds
allow to derive lower limits on the decay branching
fractions of these states, as shown in Table 6.3. There
is no way of assigning one of the experimental states to
a pseudoscalar without coming into conflict with a) absolute M1 widths, b) branching fractions for the decay of
this state. In the next Chapter we shall discuss alternatives which have been proposed in the literature.

State	$\chi(3.59)$	$\chi(3.45)$	$X(2.83)$
$B_1 \cdot B_2$ (Exp.) [%]	0.3 ± 0.1	0.8 ± 0.4	0.014 ± 0.004
B_1 (Exp.) [%]	< 2	< 2	< 1.7
B_1 (Theory) [%]	≈ 0.5	≈ 9	≈ 45
B_2 (Exp.) [%]	> 10	> 20	> 0.7
B_2 (Theory) [%]	< 1	< 1	≈ 0.1

Table 6.3. Experimental upper bounds on B_1 and lower
bounds on B_2 via $B_1 \cdot B_2$, and comparison with
theory. The kind of transition for B_1, B_2 is
indicated in Fig. 6.4. The theoretical numbers
arise from allowed and "forbidden" M1 transitions and the ratio of 2γ versus 2 gluon annihilation. For the latter see Chapter 8. The
forbidden M1 transition should lead to a B_2
not bigger than a few 10 keV/a few MeV $\approx 10^{-2}$.

e) Scaling of E1 and M1 transitions

For E1 transitions the scaling behaviour is most easily
obtained from the sum rules:

$$\Gamma \sim \frac{k^3}{k^{(o)}} \frac{1}{m_Q} \sim \frac{k^2}{m_Q} \tag{6.33}$$

M1 transitions, on the other hand, scale like

$$\Gamma \sim \mu_Q^2 k^3 \sim \frac{k^3}{m_Q^2} . \tag{6.34}$$

The ratio M1/E1 scales like k/m_Q. A comparison of related
radiative transitions in Charmonium, Bottonium and
eventually Toponium can thus help to distinguish E1
from M1 transitions.

7. THE $X(2.83)$, $\chi(3.45)$ AND $\chi(3.59/3.18)$ PUZZLES

In Chapter 4 we learned that the mass splittings $J/\psi-$
$X(2.83)$ and $\psi'-\chi$ are a factor three larger than naively
expected. But the really important point which disfavours
the interpretation of X and any χ as the pseudoscalar
partners of J/ψ and ψ' is the following: The radiative
M1 transitions, which should be allowed, are by factors
5 to 30 smaller than expected. The branching fraction
of the forbidden transition $\chi(3.45) \rightarrow \gamma J/\psi$ is at least
a factor 20 larger than estimated. Similarly the $B(X(2.83) \rightarrow$
$\rightarrow 2\gamma)$ is about one order of magnitude larger than predicted
in Charmonium. These numbers were shown in Fig. 6.4 and
Table 6.3. The discrepancies with the M1 transitions are
most puzzling, because related estimates in the 'old' meson

spectroscopy work within a factor 2 or better. We just like to recall the famous $\omega \rightarrow \gamma\pi^o$ transition [61].

The pleasant solution would be to find the true pseudoscalars much closer to J/ψ and ψ' respectively. Experimentally this is not ruled out. Then, however, the X and χ states are either not real or at least no simple $Q\bar{Q}$ states.

At the 1977 Hamburg Conference Gottfried [51] has enumerated possible explanations for the X and χ(3.45) states. We are going to repeat them and add some new speculations (the following also applies to χ(3.59/3.18).

There is the suggestion of 4 quark states $c\bar{c}q\bar{q}$ with all quarks in the same orbital state. De Rújula and Jaffe [62] estimate a $j^P = 0^+$ level at 3.6 GeV. This model suffers from a disease: The hadronic width must be large because it can decay Zweig allowed into $X\pi$ or $X\eta$ if it has I = 1 or 0.

Lipkin [63] has a similar idea. He argues that estimates of masses of 4 quark states are unreliable, but the level ordering is to be taken seriously. He further assigns the states δ(970) [I = 1, $j^P = 0^+$] and S^*(993) [I = 0, $j^P = 0^+$] to the configurations ($s\bar{s}q\bar{q}'$), both below the $\phi(s\bar{s})$. By analogy one would expect $c\bar{c}q\bar{q}'$ to lie below J/ψ. The states would be Zweig stable and possibly narrow enough to be assigned to an I = 0,1 doublet **X**(2.83). It is to be noted that the transitions $J/\psi \rightarrow X\gamma$, $\psi' \rightarrow \chi\gamma$ are now of E1 type and $\psi' \rightarrow X\gamma$ has lot of phase space and, maybe, a sizeable dipole matrix element! To test the idea, one should search for the transitions $\psi' \rightarrow X\rho$ and $\psi' \rightarrow \gamma\chi$(3.51) $\rightarrow \gamma X\pi$. Their observation would show that X is not a $c\bar{c}$ state. The χ(3.45) could be fitted into this scheme as a 0^+ $c\bar{c}s\bar{s}$ state. In a recent paper Eilam, Margolis and Rudaz [64] discuss the X(2.88) production in $\pi^- p \rightarrow \gamma\gamma n$ at 40 GeV/c [65].

Another explanation for $\chi(3.45)$ was given by the authors [66]: In the Bethe Salpeter framework with strong binding there are excitations also in the relative time coordinate, which have no analogue in the nonrelativistic Schrödinger equation. The first extra C = + state is a 'time like' P wave with $j^{PC} = 0^{++}$ or 1^{++}, depending on the Dirac structure of the model. The γ-ray transitions become of E1 type. However, this model offers no explanation for the X(2.83), besides being the 0^{-+} ground state.

Harari proposed that the $\chi(3.45)$ is the $c\bar{c}$ state $^1D_2$ [67]. However, there are problems with the magnetic transition to the J/ψ. For a detailed discussion see Ref.[55].

It could be that the $\chi(3.45)$ is a $c\bar{c}$-gluon state [68,69]. Purely gluonic states were conjectured e.g. by Jaffe and Johnson [70]. In a recent paper, Ishikawa [71] claims a glue ball at 2.81 GeV with $j^{PC} = 0^{-+}$ which mixes with the true η_c.

We already mentioned in Chapter 4 that the contribution to the hyperfine interaction induced by the presence of instantons has been considered as a solution to the large J/ψ-X splitting [45]. However, quantitative estimates are unreliable [72] and instantons do not cure the M1 problem.

Maybe the perturbative treatment of the hyperfine interaction is grossly misleading? In a recent paper,

Gromes [73] gives the following qualitative arguments:
if the hyperfine splitting is essentially due to one
gluon exchange, α_s must be large ($\alpha_s \simeq 0.5$). This then
leads to a strongly localized an relativistic wave
function of the η_c and a rather strange form of the
wave function of η_c', while nothing peculiar happens to
the triplet states. Overlap integrals are quite
different from the naive expectation and the problems
connected with radiative decays may well be solved.
The hadronic width of η_c may become $\simeq$ 20 MeV, that of
η_c' is much smaller. The branching ratio B($\eta_c \to 2\gamma$) remains
unaffected. The naive use of the Fermi-Breit Hamiltonian
for the Charmonium hyperfine splitting was already
critized by Ditsas, McDougall and Moorhouse [74]. There
is, however, one problem in the approach of Gromes:
$\psi' \to \gamma X$ becomes an almost allowed M1 transition. But
experimentally we do not see it [1].

From all these considerations we conclude that we
need additional experimental information. Maybe the
first piece of new experimental information already
rules out the $\chi(3.45)$: The Mark II collaboration does
not find the $\chi(3.45)$ at the expected level [75].

8. ELECTRO- AND CHROMOMAGNETIC ANNIHILATION

Quarkonium states may annihilate into photons
and/or gluons. Since for nonrelativistic bound states
annihilation is a pointlike process, the quarks must
come together to annihilate. Not only the annihilation
into photons is governed by a small coupling $\alpha = 1/137$
but hopefully also that into gluons by a small α_s (small R).
We can approximate the decay by the lowest order (Born-)

graph [4], i.e. we can apply the 'minimal gluon scheme'.
What do we mean by annihilation into gluons? Of course
Quarkonium annihilates into hadrons, not gluons. However, QCD suggests that this annihilation proceeds via
gluons. We can approximate the amplitude for hadronic
annihilation as the product $A(Q\bar{Q} \to$ hadrons$) = A_1(Q\bar{Q} \to$
$\to$ minimal number of gluons$) \cdot A_2$ (minimal number of
gluons $\to$ hadrons). A_1 is calculable in QCD and this is
what we will do. For A_2 we have to make assumptions.
The simplest assumption is $A_2^2 = 1$ [4]. This is an absolute upper bound. In practice A_2^2 may well be smaller.
First, the gluons may be in a state with zero overlap
to any known hadron system, e.g. two gluons in a spin
2 state at an invariant mass of M = 300 MeV. Second,
colour bleaching effects can only make A_2^2 smaller,
not larger. Third, in some regions of the three gluon
phase space higher orders become very important, because a 400 MeV decay gluon cannot be discriminated
against a 400 MeV confinement gluon. Realistic rates
might therefore be much smaller than what we will cal-
culate.Additionally, not only $A_2^2 = 1$ may be too
optimistic, but also the calculation of A_1 in lowest
order might be grossly misleading because of higher
order corrections. We do not know these higher order
corrections in QCD but we know the next order QED
correction to the ${}^3S_1$ positronium decay into three
photons [76]: With $\alpha = 1/3$ it would be - 100%. But
note, the more massive $Q\bar{Q}$ is the better our approximations should hold. For Charmonium they are most probably
wrong. This might explain a part of the discrepancy
between the α_s (annihilation), found from the calculation described above, and the α_s (spectrum). We will, however, ignore all these shortcomings and just calculate

hadronic widths as gluon annihilation widths.

We proceed as follows: first we collect well known formulae. Since in Born approximation there is no gluon selfinteraction yet, the conversion from photons to gluons is just done by redefining the charge. We will then discuss applications to Quarkonia. The results will also be relevant for the next Chapter on jets.

a) Annihilation Formulae

The vector state can decay into a lepton or quark pair (hadrons), Fig. 8.1.

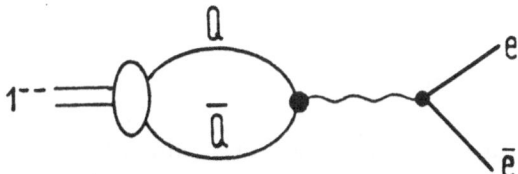

Fig. 8.1. Leptonic decay of 1^{--} $(Q\bar{Q})$. The electrons may be replaced by μ's, τ's or quarks lighter than Q.

For $M_V^2 \gg 4m_e^2$ and including colour the formula for a $^3S_1$ state is [17]:

$$\Gamma(^3S_1 \to e^+e^-) = \alpha^2 e_Q^2 |R(0)|^2 (M_V/2)^{-2} . \tag{8.1a}$$

Here $R(0)$ is the radial Schrödinger wave function at the origin. Quarks couple to the photon in the same way as leptons so that (8.1) is understood for each lepton or quark flavour separately: $\Gamma_{q\bar{q}} = 3e_q^2 \Gamma_{e^+e^-}$. For a $^3D_1$ state one has [55]:

$$\Gamma(^3D_1 \to e^+e^-) = 200\ \alpha^2 e_Q^2 |R_D''(0)|^2\ M_V^{-6} . \qquad (8.1b)$$

Numerical estimates of (8.1b) for Charmonium ($\psi''(3.77) \to$ $\to e^+e^-$) only give a few eV. The experimental value for $\Gamma_{e\bar{e}}(\psi''(3.77)) = 0.36$ keV [77]. So, there must be other decay mechanisms. The simplest would be an admixture of the $\psi'(3.7)$. Within the nonrelativistic bound state picture only the tensor forces, arising from one gluon exchange, mix the 2^3S_1 and 1^3D_1 wave, but numerical estimates again only give a few eV [55]. There must be another source of mixing. The Cornell group [78] proposed an S-D mixing via virtual $D\bar{D}$ states and they successfully predicted $\Gamma_{e^+e^-}$ of $\psi''(3.77)$.

The decay of 1^{--} into two photons as well as two gluons is impossible. In the two photon case this is just the photon C parity. Also two gluons in a symmetric colour singlet state have even C. This can be seen as follows [55]: gluons can be represented by matrices in colour space, A,B,C,... There is only one way to construct a colour singlet out of two gluons A,B: Tr(AB). But Tr(AB) is even under charge conjugation. For three gluons we have two ways to construct a colour singlet, Tr(ABC) and Tr(ACB). The symmetric combination has negative C parity (D-coupling), the antisymmetric one has C = +1 (F-coupling). Therefore a 1^{--} $Q\bar{Q}$ state can decay into any number of gluons larger than two. Remember that for electromagnetic decays only an odd number of photons is allowed: photons are C eigenstates.

In lowest order the 1^{--} can decay into three photons as well as three gluons or one photon plus two gluons. The three photon decay of $^3S_1$, Fig. 8.2 has been originally calculated by Ore and Powell [79] (here including the statistical colour factor):

$$\begin{vmatrix} \Gamma(^3S_1 \to 3\gamma) \ [79] \\ \\ \Gamma(^3S_1 \to 3g) \ [4] \\ \\ \Gamma(^3S_1 \to \gamma 2g) \ [80] \end{vmatrix} = \begin{vmatrix} \frac{4}{3}\alpha^3 e_Q^6 \\ \\ \frac{10}{81}\alpha_s^3 \\ \\ \frac{8}{9}\alpha e_Q^2 \alpha_s^2 \end{vmatrix} \cdot \frac{\pi^2-9}{\pi} \frac{|R(0)|^2}{m_Q^2} \tag{8.2}$$

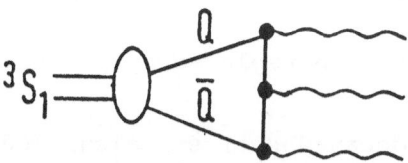

Fig. 8.2. 3γ decay of $^3S_1(Q\bar{Q})$. When the photons are replaced by gluons, this denotes the "direct" hadronic decay.

The conversion factor from the three photon to the three gluon decay is [55]

$$\Gamma_{3g}/\Gamma_{3\gamma} = \frac{\alpha_s^3}{\alpha^3 e_Q^6} \frac{1}{9} \Sigma_{a,b,c} \left| Tr\left(\frac{\lambda^a}{2}\frac{\lambda^b}{2}\frac{\lambda^c}{2}\right)_{sym.} \right|^2 \tag{8.3}$$

and has the following origin: $\alpha_s^3/\alpha^3 e_Q^6$ just converts the charges together with

$$\left| \text{Tr}\left(\frac{\lambda^a}{2} \frac{\lambda^b}{2} \frac{\lambda^c}{2}\right)_{\text{sym.}} \right|^2 .$$

The $\Sigma_{a,b,c}$ counts the number of coloured graphs in the 3g case, while the 3^{-2} counts the number of coloured graphs in the 3γ case.

The pseudoscalar $^1S_0$ ground state can decay into two photons or two gluons in lowest order, Fig. 8.3.

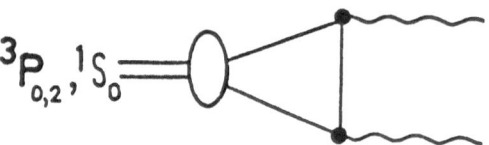

Fig. 8.3. 2γ decay of $Q\bar{Q}$. For the hadronic decay the photons are replaced by gluons.

The two photon decay was first calculated by Pomeranchuk [81]. Including colour one has

$$\left| \begin{array}{l} \Gamma(^1S_0 \to 2\gamma) \quad [81] \\ \\ \Gamma(^1S_0 \to 2g) \quad [4] \end{array} \right| = \left| \begin{array}{l} 3\alpha^2 e_Q^4 \\ \\ \frac{2}{3}\alpha_s^2 \end{array} \right| \frac{|R(0)|^2}{m_Q^2} \qquad (8.4)$$

with the conversion factor

$$\Gamma_{2g}/\Gamma_{2\gamma} = \frac{\alpha_s^2}{\alpha^2 e_Q^4} \frac{1}{9} \Sigma_{a,b} \left| \text{Tr}\left(\frac{\lambda^a}{2} \frac{\lambda^b}{2}\right) \right|^2 . \qquad (8.5)$$

We do not discuss the decay into more photons or gluons.

Assuming that the 2g decay is the basic process for the
dominant hadronic decay of the pseudoscalar, Eq.(8.5)
yielded the branching fraction for the 2γ decay (Table
6.3).

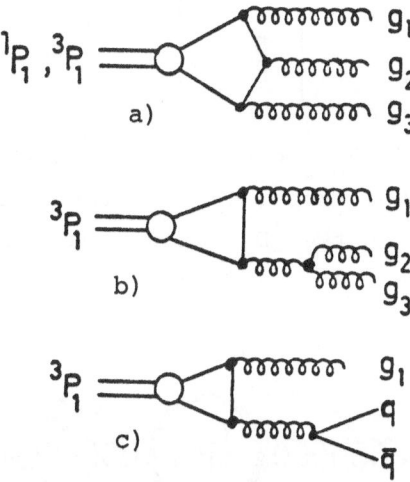

Fig. 8.4. The gluonic decay diagrams of spin-1 P waves.

We now turn to P wave annihilation, Figs. 8.3 and
8.4. In a P wave the wave function at the origin is zero.
That means that the quarks do not like to come together
to annihilate. The P waves, however, can annihilate when
the two quarks come near each other and simultaneously
have a relative velocity ≠ 0. This is a higher order
process in $\beta^2 = (v/c)^2$. It is governed by the spatial

derivative of the wave function. The annihilation widths of P waves will be smaller than that of the $^1S_0$ wave! The widths of the spin 0 and spin-1 P waves of Positronium have first been calculated by Alekseev [82]. The same calculation for Charmonium has been done by Barbieri, Gatto and Kögerler [83]. They yield

$$\left| \frac{\Gamma(^3P_0 \rightarrow 2g)}{\Gamma(^3P_2 \rightarrow 2g)} \right| = \left| \frac{6\alpha_s^2}{\frac{8}{5}\alpha_s^2} \right| \frac{|R_P'(0)|^2}{m_Q^4} . \tag{8.6}$$

The 2γ width of $^3P_{0,2}$ can be obtained from (8.6) by the conversion factor (8.5). It is interesting to note that the two gluons in the $^3P_2$ decay come out with opposite helicities ($\longrightarrow$), $J_z = \pm 2$ [84]

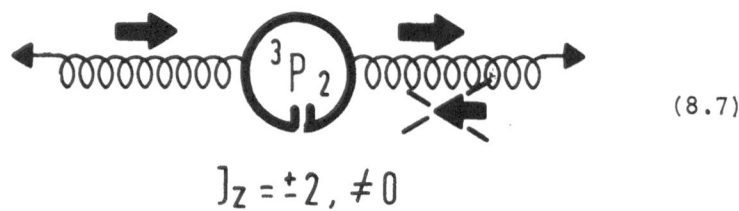

$$J_z = \pm 2, \neq 0 \tag{8.7}$$

whereas in the $^3P_0$ decay both gluons necessarily have the same helicity. The decays of the $j = 1$ P waves are more complicated. A spin 1 state cannot decay into two massless vector bosons, either photons or gluons in a colour singlet [85]. We therefore have to consider the next order (in α_s) diagrams, which for gluon annihilation are shown in Fig. 8.4. They bring up another complication. We now have a three body phase space and have to integrate

over all possible energies of, say, gluon 1. Gluon 1 is
allowed to be soft. It further is allowed to carry away
the angular momentum of the P wave. So it has all
characteristics of a bremsstrahlungs gluon. The same
is true for photon annihilation, except that in this
case diagram b) of Fig. 8.4 is absent. A bremsstrahlungs
gluon or photon in the annihilation of a _free_ $Q\bar{Q}$ pair
with $\ell = 1$ leads to the typical bremsstrahlungs singularity. The cross section factorizes into the bremsstrahlungs
part and the annihilation of an $\ell = 0$ $Q\bar{Q}$ pair into two
photons or gluons. For a bound state, however, the
annihilation amplitude cannot be singular, because the
quarks are _not_ on shell. Their virtuality is of the order
of the bound state dimensions. For a bound state annihilation we therefore may cut the amplitude at momenta of the
soft (bremsstrahlungs) photon or gluon which correspond
to the bound state radius. In diagram language, the
singularity will be cancelled by higher order graphs
like vertex corrections. For QED this procedure is well
defined [86]. We hope that it will work parallel for QCD.
As a cutoff momentum for QCD annihilation we take the
typical momentum for a soft "confinement" gluon, 400 MeV,
since in a QCD process higher order graphs will involve
such "confinement" gluons. For heavier Quarkonia one has
to take the minimum of R_{Bohr}^{-1} and 400 MeV. We will express
the cutoff in terms of a parameter $\Delta = 2M \cdot 400$ MeV [87],
M being the Quarkonium bound state mass. Let us first
discuss the $^1P_1$ decay. This state has $j^{PC} = 1^{+-}$ and
therefore only diagram a) of Fig. 8.4 can contribute,
in either photon or gluon annihilation. Its decay has
been calculated by Barbieri, Gatto and Remiddi [86].
They find

$$\Gamma(^1P_{1+-} \to 3g) \simeq \frac{20}{9} \alpha_s^3 \frac{|R'_P(0)|^2}{m_Q^4} \log \frac{M^2}{\Delta} \tag{8.8}$$

where the log arises from the bremsstrahlungs singularity of the diagram. For the decay of the $^3P_1$ state, $j^{PC}=1^{++}$, only diagram c) can contribute to the photon annihilation while in principle all three diagrams can contribute to the gluon annihilation. Barbieri, Gatto and Remiddi found that the singular parts of the diagrams a) and b) cancel each other. Okun and Voloshin [55] gave the general argument for this: The amplitudes a) and b) interfere, since they lead to the same final state. Since they can both be factorized into the bremsstrahlungs part times the corresponding annihilation diagram for the 2 gluon annihilation of a coloured $^3S_1$ state, also their sum can be factorized in this way. This sum, however, contains all graphs to this order for $^3S_1$ (coloured) $\to$ 2g, which must be zero from unitarity arguments [55]. Neglecting the non-singular parts of amplitudes a) and b) against the singular c) means that also for the gluon annihilation the calculation of graph c) is sufficient. It gives [86,87]

$$\Gamma(^3P_{1++} \to q\bar{q}g) \simeq \frac{N}{3} \frac{8\alpha_s^3}{3\pi} \frac{|R'_P(0)|^2}{m_Q^4} (\log \frac{M^2}{\Delta} - \frac{1}{2}) \tag{8.9}$$

where N is the number of light flavours q. The photon versions of (8.8) and (8.9) can be found in Ref.[55].

For completeness we note the formula for the decay of the spin 2 D wave into 2 gluons which is given by the second derivative of the wave function, this is the second order in an expansion of $\beta^2 = (v/c)^2$

and therefore even less reliable. Okun and Voloshin [55] calculated

$$\Gamma(^1D_{2^{-+}} \rightarrow 2g) = \frac{2}{3} \alpha_s^2 \frac{|R_D''(0)|^2}{m_Q^6} \qquad (8.10)$$

b) Applications

The ratio of Eqs. (8.2) and (8.1a) gives

$$\frac{\Gamma(^3S_1 \rightarrow 3g)}{\Gamma(^3S_1 \rightarrow e^+e^-)} = \frac{10}{81} \cdot \frac{\pi^2-9}{\pi} \cdot \frac{\alpha_s^3}{\alpha^2 e_Q^2} \simeq 1440 \frac{4}{9e_Q^2} \alpha_s^3 \ . \qquad (8.11)$$

If we interpret as usual the 3g annihilation as the total direct hadronic annihilation then this is a measurable quantity and we have e.g. in Charmonium

$$\frac{\Gamma(J/\psi \rightarrow hadr.)_{dir}}{\Gamma(J/\psi \rightarrow e^+e^-)} \simeq 10 \qquad (8.12)$$

from which follows that the α_s (annihilation) is $\alpha_s=0.19$. Because of the third power of α_s in (8.11) this value is quite stable even against large corrections to the widths. But it certainly is subject to many corrections like those discussed at the beginning of this Chapter or corrections to the wave function at the origin like (4.9).

A very interesting ratio is that of Eqs.(8.6) to Eq. (8.9):

$$\Gamma({}^3P_{0^{++}} \rightarrow 2g) : \Gamma({}^3P_{1^{++}} \rightarrow q\bar{q}g) : \Gamma({}^3P_{2^{++}} \rightarrow 2g)$$

$$\tag{8.13}$$

$$= \qquad 15 : \frac{20 \, N\alpha_s}{9\pi}(\log \frac{M^2}{\Delta} - \frac{1}{2}) : 4 \quad .$$

It leads to ratios of

$$15 : 2.1 \, \alpha_s : 4 \qquad \text{J}/\psi \text{ system}$$

$$15 : 5.7 \, \alpha_s : 4 \quad \text{in the } \Upsilon \text{ system} \qquad (8.14)$$

$$15 : 11 \, \alpha_s : 4 \qquad 30 \text{ GeV } Q\bar{Q} \text{ system} .$$

We can of course calculate more than these ratios, namely the total widths of the P waves, assuming that these are given by the gluon annihilation width and radiative transition width essentially. The result is shown in Table 8.1 for Charmonium, the Υ and the $t\bar{t}$ (30 GeV) system, and compared to the quasi-experimental bounds of Table 6.2. For the calculation of Eqs.(8.6, 8, 9) we need $|R'(0)|^2$. Numerical calculations give us $|R'_{c\bar{c}}(0)|^2 m_c^{-3} \simeq 0.024 \text{ GeV}^2$, $|R'_{b\bar{b}}(0)|^2 m_b^{-3} \simeq 0.012 \text{ GeV}^2$ and $|R'_{t\bar{t}}(0)|^2 m_t^{-3} \simeq 0.007 \text{ GeV}^2$. These quantities are relatively quark mass independent. Although the widths of Table 8.1 are very model dependent, we conclude that the pattern of (8.14) agrees very well with the observed branching fractions of the Charmonium P waves. This is one of the successful predictions of QCD within Charmonium.

We now give ratios of widths of the ${}^3S_1$ decay in Table 8.2. The decay channels of the vector ground state are: i) into lepton pairs, $e\bar{e}$, $\mu\bar{\mu}$, $\tau\bar{\tau}$, ii) into $\Sigma q\bar{q}$,

iii) the three gluon annihilation and iv) the annihilation into one photon and two gluons (Eqs. 8.2).

The radiative decays of J/ψ (or Υ) $^3S_1 \to \gamma +$ + hadrons are very interesting because they allow to study the Zweig suppression mechanism. QCD predicts the branching ratio γ2g/3g to be of the order of 1/10

		$\Gamma_{tot}(^3P_0)$ [keV]	$\Gamma_{tot}(^3P_1)$ [keV]	$\Gamma_{tot}(^3P_2)$ [keV]
$c\bar{c}$	theory	4000	500	1500
	"quasiexp."	6000±6000	1000±200	3200±1600
$b\bar{b}$	$\alpha_s = 0.15$	350	50	150
	$\alpha_s = 0.20$	600	80	200
$t\bar{t}$	$\alpha_s = 0.12$	85	53	60
	$\alpha_s = 0.15$	105	56	65

Table 8.1. Comparison of Charmonium "experiment" and theory for the P wave total widths, including the radiative transitions from Table 6.1. The "experiment" line is taken from Table 6.2, here only errors arising from BR(γJ/ψ) are shown. We also give the prospects for the Υ and $t\bar{t}$ systems of 30 GeV, where the radiative transitions are included, they are ≈ 40 keV for Υ P waves (Fig.6.3) and 50 keV for each $t\bar{t}$ P wave.

cc̄ decay channel:	eē+μμ̄ :	Σqq̄ :	3g :	γ2g
$e_Q = 2/3$	2 :	R :	$\dfrac{5(\pi^2-9)\alpha_s^3}{18\pi\alpha^2}$:	$\dfrac{8(\pi^2-9)\alpha_s^2}{9\pi\alpha}$
a) $\alpha_s = 0.19$	2 :	2.5 :	10 :	1.2
bb̄ decay channel:	eē+μμ̄ :	Σqq̄+ττ̄:	3g :	γ2g
$e_Q = -1/3$	2 :	R :	$\dfrac{20(\pi^2-9)\alpha_s^3}{18\pi\alpha^2}$:	$\dfrac{8(\pi^2-9)\alpha_s^2}{9\pi\alpha}$
b) $\alpha_s = 0.15$	2 :	5 :	20 :	0.8
$\alpha_s = 0.18$	2 :	5 :	34 :	1.1
tt̄ (30 GeV) decay channel:	eē+μμ̄ :	Σqq̄+ττ̄ :	3g :	γ2g
$e_Q = 2/3$	2 :	R :	$\dfrac{5(\pi^2-9)\alpha_s^3}{18\pi\alpha^2}$:	$\dfrac{8(\pi^2-9)\alpha_s^2}{9\pi\alpha}$
c) $\alpha_s = 0.12$	2 :	5 :	2.5 :	0.5
$\alpha_s = 0.15$	2 :	5 :	5 :	0.8

Table 8.2. Ratios of the ground state decay channels
a) in Charmonium, b) in the Υ system,
c) in a 30 GeV tt̄ system. For Charmonium
$\alpha_s = 0.19$ agrees with experiment (lowest
order formulae). For Υ decays the value of
α_s best compatible with experiment, $B_{\mu\bar{\mu}}$
[88], seems to be 0.18 at present.

in Charmonium, while in the old hadron spectroscopy
photon inclusive decays were usually down by a factor
of 200. For Charmonium three exclusive contributions
to $^3S_1 \rightarrow \gamma 2g$ have been seen so far, namely $J/\psi \rightarrow \gamma\eta$,
$\gamma\eta'$, γf with branching fractions 0.082 ± 0.01, $0.24 \pm$
± 0.07, 0.2 ± 0.07 % respectively. Billoire, Lacaze,
Morel and Navelet [89] have performed a spin parity
analysis of the two gluon system in $^3S_1 \rightarrow \gamma 2g$. Their
result is shown in Fig. 8.5. The remarkable feature
is the dominance of the 2^+ two gluon state, especially
at small invariant mass W. This can be seen from the
Ore-Powell matrix element:

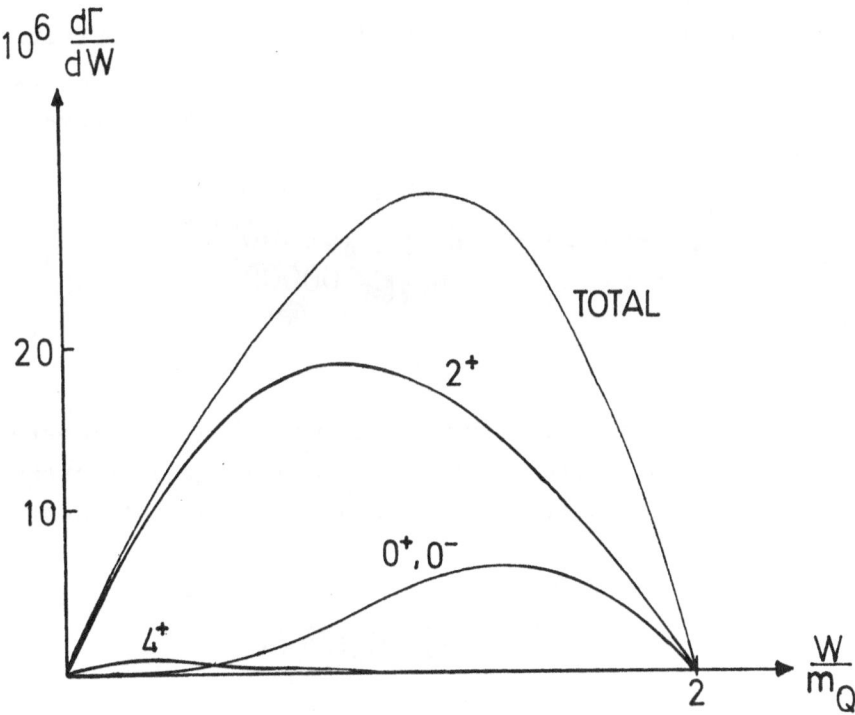

Fig. 8.5. The total hadronic mass spectrum and the
contributions of $0^\pm$, 2^+ and 4^+ states to it.
Other contributions are negligible [89].

$$M_1^*(k_1,k_2,k_3;\varepsilon_1,\varepsilon_2,\varepsilon_3;k,\varepsilon) \sim [(k_1+k_2)\cdot k_3 (k_2+k_3)\cdot k_1 (k_3+k_1)\cdot k_2]^{-1},$$

$$\cdot\{-\varepsilon_1\cdot\varepsilon_2[k_1\cdot k_3\varepsilon_3\cdot k_2\varepsilon^*\cdot k_1+k_1 k_3 k_2\cdot k_3\varepsilon^*\cdot\varepsilon_3+k_2\cdot k_3\varepsilon_3\cdot k_1\varepsilon^*\cdot k_2]+$$

$$+\varepsilon^*\cdot\varepsilon_3[k_3\cdot k_1\varepsilon_1\cdot k_2\varepsilon_2\cdot k_3-k_1 k_2\varepsilon_1\cdot k_3\varepsilon_2\cdot k_3+k_2\cdot k_3\varepsilon_2\cdot k_1\varepsilon_1\cdot k_3]+$$

$$+ 1 \leftrightarrow 3 + 2 \leftrightarrow 3\} \ . \tag{8.15}$$

In the limit of small gluon-gluon invariant mass $M_{1,2}^2 = (k_1 + k_2)^2 \to 0$, which for real quanta $(k_i^2 = 0)$ implies $k_2 \to \lambda k_1$, $\varepsilon_2\cdot k_1 \to 0$, $\varepsilon_1\cdot k_2 \to 0$, it is easily checked that in this limit only the first term $\varepsilon_1\cdot\varepsilon_2[\ldots]$ survives [90]. From this structure follows: the helicities of the two gluons are opposite, their transverse polarizations are parallel

$$J_z = 0, \neq \pm 2 \tag{8.16}$$

If we allow a finite but small two gluon mass (corresponding to a small angle between g_1 and g_2) the transformation into the two gluon rest frame yields the configuration

$$\tag{8.17}$$

i.e. $J_z = \pm 2$. The gluons are now polarized in such a way that they like to couple to the $^3P_2(q\bar{q})$ state (see Eq.8.7). Interpreting the f(1270) meson as a $^3P_2(q\bar{q})$ state the

diagram

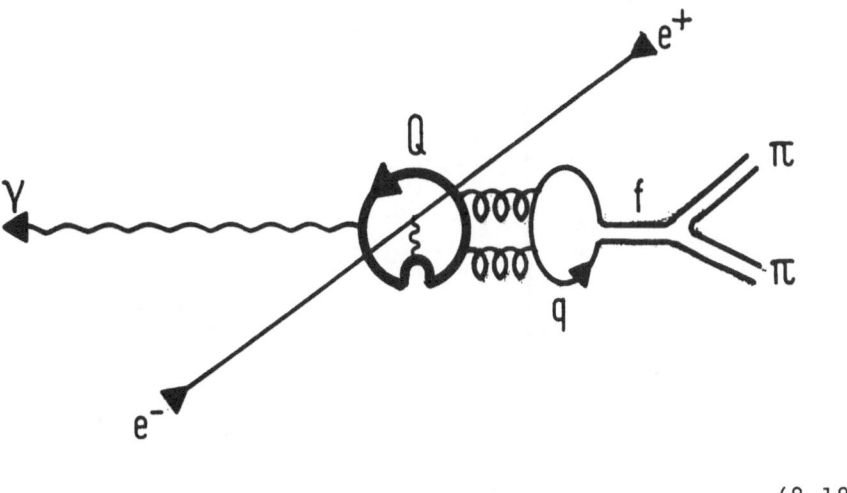

$$(8.18)$$

has been calculated [90], describing the production and decay of the f meson by three kinematically independent helicity amplitudes

$$1^{--}(1,0,-1) \rightarrow \gamma(1) + 2^{++}(0,1,2) \equiv (A_o, A_1, A_2) \qquad (8.19)$$

which can be measured via the decay kinematics. The result of the calculation is shown in Fig. 8.6 as function of the mass ratio $M(^3P_2(q\bar{q}))/M(^3S_1(Q\bar{Q}))$. At the point $J/\psi \rightarrow \gamma f$ these amplitude ratios have been measured [91] and agree with this QCD calculation, as shown in Fig. 8.7.

Billoire et al.[89] compare the $0^{\pm}$ channels where there is no 'width anomaly' at small W (see Fig. 8.5) with experiment, Fig. 8.8. The agreement with η and η' is rather satisfactory.

324

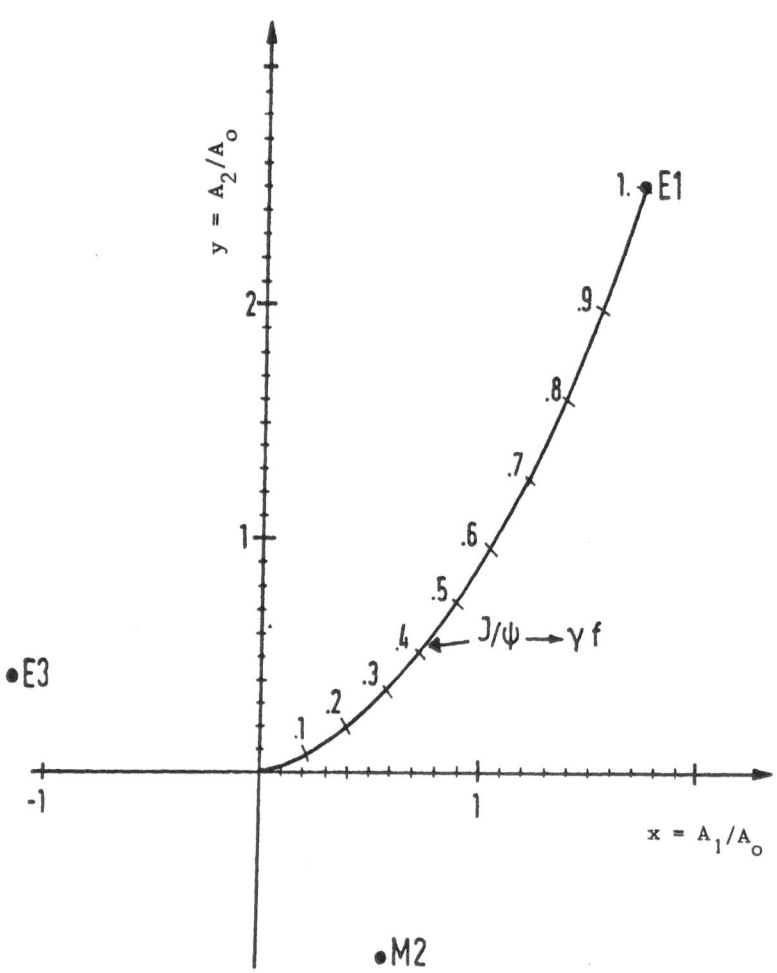

Fig. 8.6. Ratios of helicity amplitudes A_i in $^3S_1(Q\bar{Q}) \to$
$\to \gamma + {}^3P_2(q\bar{q})$ as a function of $\bar{M}(^3P_2)/M(Q\bar{Q})$.
E1, M2, E3 denote the familiar multipole
transitions, the (x,y) pair for $J/\psi \to \gamma f$ is
indicated.

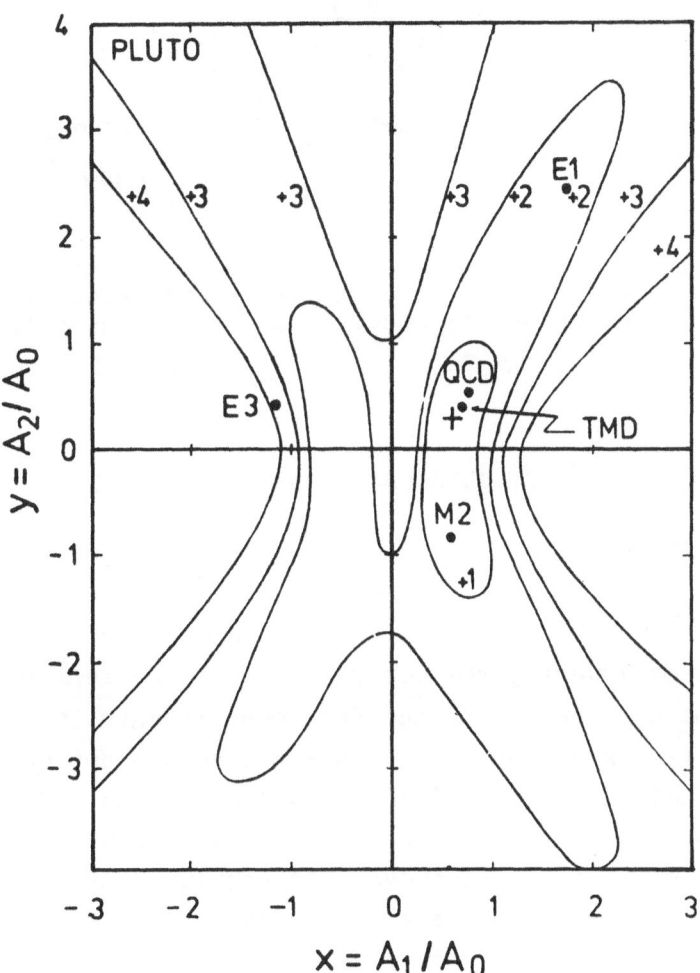

Fig. 8.7. A measurement of the predicted (x,y) pair of
Fig.8.6 for $J/\psi \to \gamma f$ by the PLUTO collaboration. The cross is the central value of the
experiment, the lines indicate standard
deviations. TMD denotes the prediction of
the tensor meson dominance model [92].

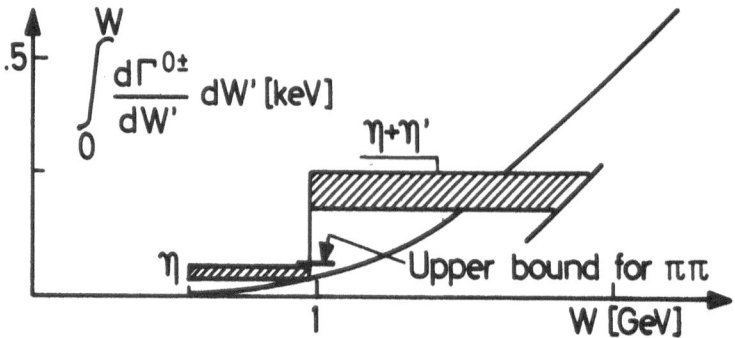

Fig. 8.8. The spectrum integrated up to W, versus W,
for $0^{\pm}$ final states. The shaded area is the
experimental result for $\Gamma(J/\psi \rightarrow \gamma\eta+\gamma\eta')$.

9. JETS FROM QUARKONIA

The exploration of QCD suffers from the fact that
its constituents, the quarks and gluons, cannot exist as
free particles because of the confinement. Their properties
cannot be investigated directly. But there is a surrogate
for the observation of the free constituents, that are
the jets. Experimentally jets are observed not only in
deep inelastic hadron-hadron and lepton-hadron scattering but especially in $e^{+}e^{-}$ annihilation, once the c.m.
energy of 5 GeV is exceeded. The angular distribution
of these jets is completely consistent with the production of two spin 1/2 (almost) massless particles [93,94],
the quarks, via photon vacuum polarisation. The fragmentation of quarks or gluons into hadrons is imagined as a
nonperturbative confinement effect, which conserves the
original directed momenta.

At present there is no way of calculating this
process, but there exists a very suggestive picture:
Inside a small space region with $\simeq$ 1/2 fm radius colour

can exist and within this region the $q\bar{q}$ pair (or gluon)
production is a short distance effect (see Fig. 9.1).

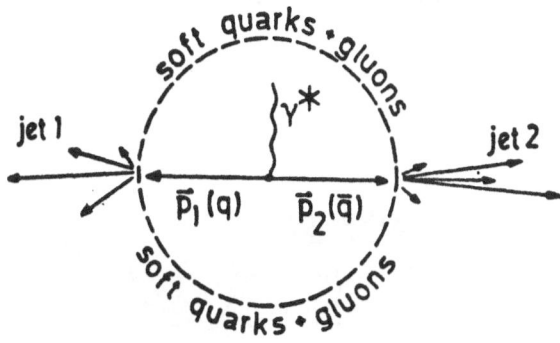

Fig. 9.1. Quark jets.

When hard coloured quanta (quarks or gluons) with momenta
p_i reach the confinement sphere they must fragment into
white hadrons since colour fields cannot exist outside
this sphere. The coloured quanta break up into hadrons
with a finite perpendicular momentum $p_\perp$. This breaking
up is energetically much favoured over a further existence
as coloured quanta. When the perpendicular momenta are
small compared to the longitudinal hadron momenta, which
add up to the momentum of the original quantum, we see
hadron jets. The confinement effects, however, are
assumed to be soft, carried by long wavelength quarks
and/or gluons. The wavelength corresponds to the colour
bag of 1/2 fm. Therefore the jet momenta equal the
original quantum momenta up to the order of 400 MeV.
This picture demands the production of the original jet
quanta to be a short distance effect ($\ll$ 1/2 fm). This
is certainly true for the (electromagnetic) quark pair

production in e^+e^-. It is also true for a hard gluon
bremsstrahlung process [95]. Resonance decays, however,
are not pointlike but involve propagators (Fig. 9.2
and 9.8). Here it is not so clear, how well the jet
picture will work. However, because the propagators are

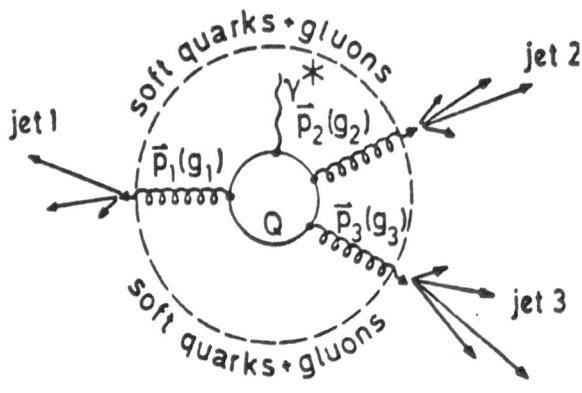

Fig. 9.2. $Q\bar{Q} \rightarrow 3$ gluon jets.

mass dependent the picture will work the better the
higher the mass of the decaying $Q\bar{Q}$ resonance is. For
a Q-mass of 5 GeV the propagator length in Fig. 9.2
is probably already short enough to apply the jet
picture and for the next new flavour (higher) $Q\bar{Q}$
resonance it will definitely be so.

The quark jets in e^+e^- annihilation became visible
above $s = (p_1 + p_2)^2 \gtrsim (5 \text{ GeV})^2$, i.e. a massless quark
needs $\gtrsim 2.5$ GeV of energy against the c.m. to be able to

form a jet. The first PETRA experiments with c.m. energies
up to 17 GeV show quark jets which can clearly be seen
by eye and which have a fixed transverse momentum of
350 to 400 MeV [94]. This experimental evidence is - up
to present energies - consistent with the nonperturbative
fragmentation picture described above. What can we guess
for gluon jets? For gluons the jet threshold[+] certainly
is not lower. It is equal to that of quark jets if the
gluons fragment like quarks. But it can also be higher,
since a gluon carries the colour indices of a quark
antiquark pair and each index may fragment separately.
Then the multiplicity of gluon jets may be higher and
the longitudinal hadron momenta may be lower than those
of quark jets. Physics will be somewhere in between. It
follows that a gluon jet of a certain longitudinal
momentum will have a somewhat higher multiplicity and
larger opening angle than a quark jet of the same
momentum. The threshold for gluon jet production will
be higher than that for quark jet production.

Some possible sources of gluon jets are shown in
Fig. 9.3. The pseudoscalars are omitted, they may also
lead to two jets from the two decay gluons. We begin
with the $^3S_1$ decay into 3 gluons. The three gluons of
this decay will form a plane. The angular distribution
of the normal $\hat{n}$ to this plane against the beam is [12]

$$\frac{d\Gamma}{d \cos \theta_{\hat{n}e}} \sim 3 - \cos^2 \theta_{\hat{n}e} \quad . \tag{9.1}$$

_____

[+]Speaking of a jet threshold we refer to the energy of
a single quark or gluon versus the center of mass of
the colour bag.

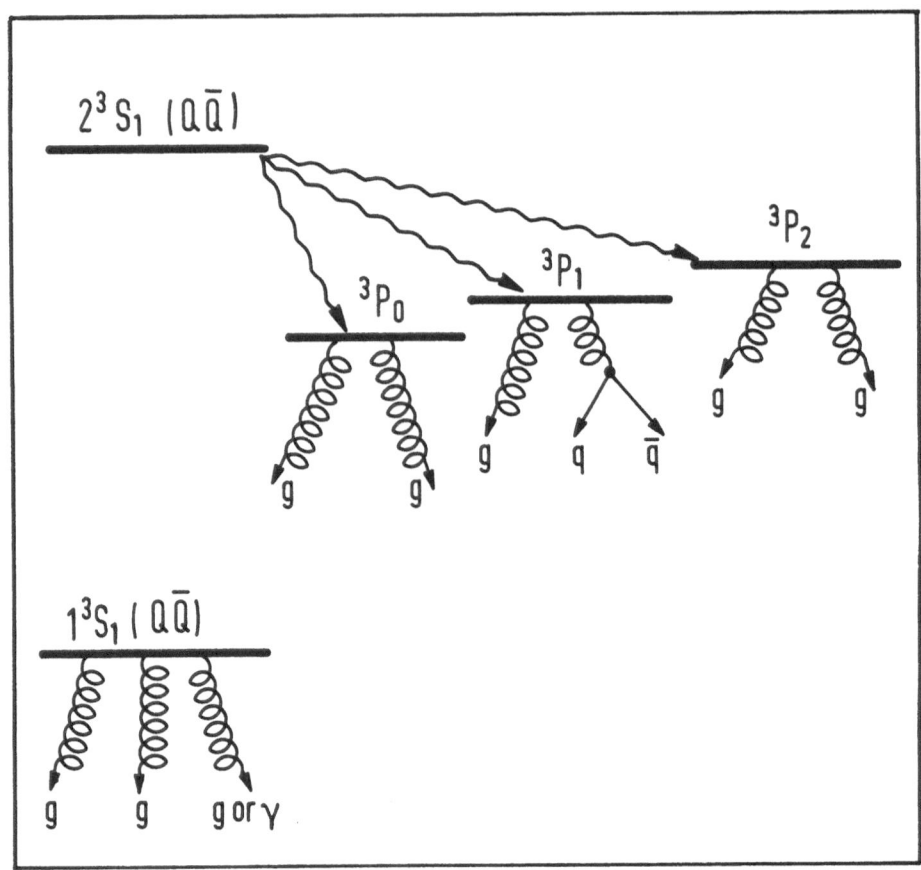

Fig. 9.3. Possible sources of gluon jets in heavy
          Quarkonia.

For these decays one defines a variable T, "Thrust",
[96] which in lowest order is just the scaled energy
of the most energetic gluon, $T = x_1 = 2p_{g1}/M_{Q\bar{Q}}$. The
direction of g_1 defines the thrust axis. The differential
rate of the 3 gluon decay together with the angular
distribution of this thrust axis as calculated from
the Born graph is shown in Fig. 9.4. While off

resonance the coefficient of the $\cos^2$ term, α, is
uniquely 1, it shows a T dependence for $Q\bar{Q}$ decays.
The average $\alpha(T)$ for $Q\bar{Q} \rightarrow 3g$ is 0.39 [97]. This
average can readily be compared to existing ex-
periments [98], in contrast to the distributions of
Fig. 9.4. The reason is, that experimentally we cannot measure the gluon momenta and thus the thrust as
we defined it. The corresponding experimental quantity,
also called thrust, is $T_E = \max\ (\Sigma|P_{\shortparallel}|/\Sigma|p|)$ where the
projection axis (the thrust axis) is chosen so that T_E
is maximal. If the jet would have zero transverse
momentum $x_1 = T = T_E$. But with the nonperturbative
$p_\perp \neq 0$ the relation between T and T_E is only statistical.
We do not know the momentum of the original jet quantum
on an event by event basis. But for averages we need not
know it, experimentally as well as theoretically we just
sum over all events. The experimental result for the
average $\alpha(T)$ in Y decays is shown in Fig. 9.5.
The authors of Ref.[97] have also calculated the opening
angle θ_{23} of the two gluons opposite the most energetic
gluon, as defined in Fig. 9.6. Also this calculation was
done in lowest order (Born graph only). Fig. 9.7 shows if
one remembers the differential rate of Fig. 9.4, that
the opening angle is rather small, on the average it is
only 75 degrees. If gluons fragment like quarks (which
for this discussion is a conservative assumption!), one
can extract the opening angle of a gluon jet from Y
decays from data on quark jets at $\sqrt{s}$ = 6 GeV: Half of
the hadronic energy lies inside a cone of half angle
$\delta = 30^\circ$ [94]. This means that an average Y event will
not show a three jet structure! Only a subsample of decays
with small T will show a three jet structure, e.g. for
$\theta_{23} \geq 90^\circ$ we need to cut at $T \leq 0.85$ and are left with at most
30% of all events. Another average to be compared to ex-

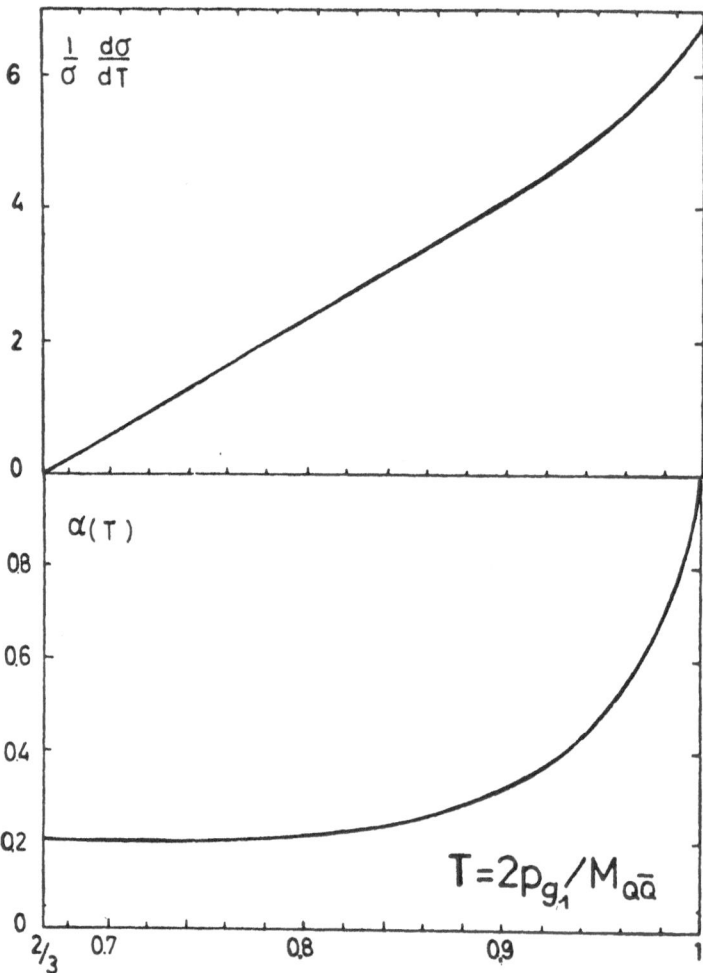

Fig. 9.4. The differential rate of $^3S_1(Q\bar{Q}) \to 3g$ and the thrust angular distribution $W \sim 1+\alpha(T)\cos^2\theta_{Te}$ as functions of T. θ_{Te} is the angle between the thrust axis and the beam.

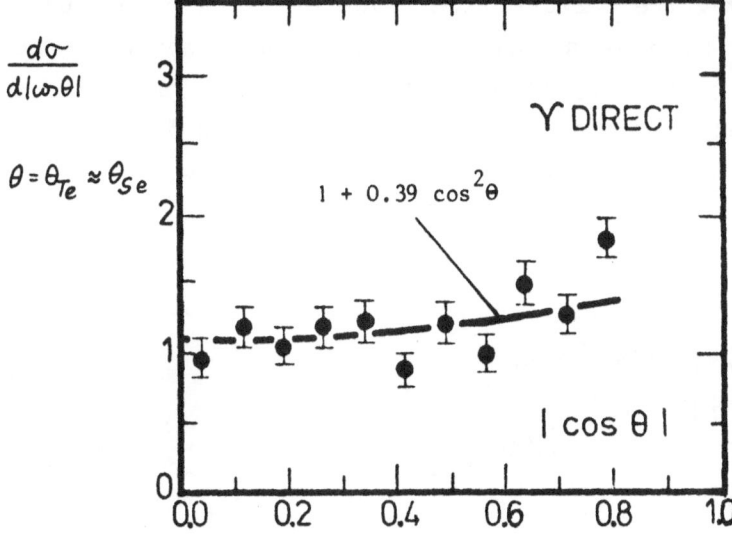

Fig. 9.5. The angular distribution of the most energetic gluon in $Q\bar{Q} \rightarrow 3g$ and the data points for $\Upsilon \rightarrow 3g$.

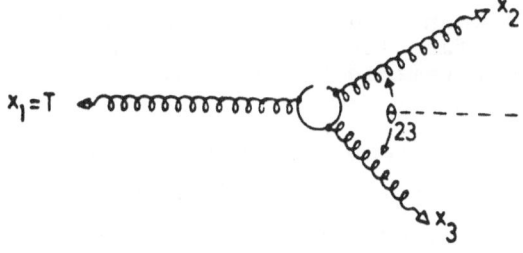

Fig. 9.6. Definition of θ_{23}.

periment is the average thrust <T> of all direct Υ
decays. The perturbative value of T can be calculated,
<T> = 0.89. Assuming jet cones of half angle $\delta \simeq 30^\circ$
simple geometry allows one to calculate the value of
the experimental thrust, $<T_E> \simeq <T> \cos \delta = 0.77$ [97].
Experimentally it is measured to be 0.76 ± 0.01 [98].

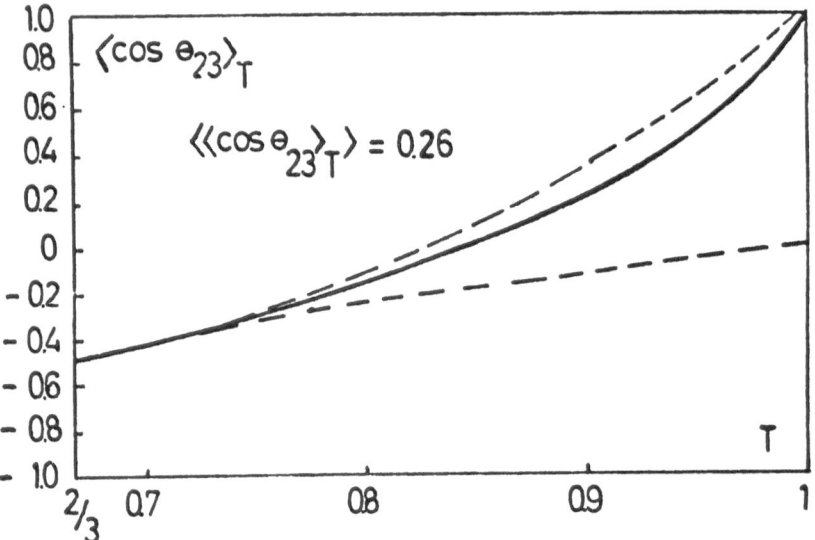

Fig. 9.7. The mean value of θ_{23}, as defined in Fig.9.6
as a function of T. The dashed lines show
the kinematic boundaries.

Higher order QCD calculations for Quarkonium
decays into hadrons may serve as a test for the cubic
(and quartic) self coupling of the gluons. De Rújula,
Lautrup and Petronzio [99] have found a measure for
the three gluon coupling in the next order of α_s: Interference terms like those of graphs 9.8 a plus 9.8b allow

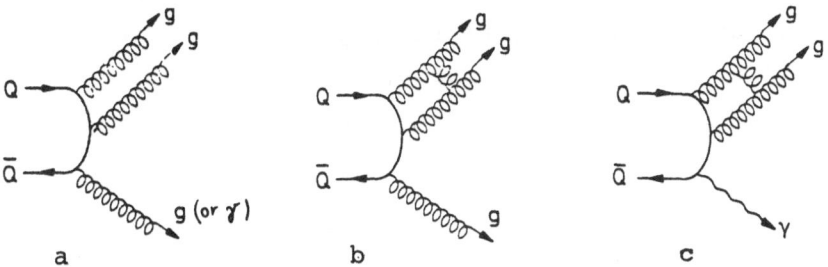

Fig. 9.8. The lowest and possible higher order (α_s)
diagrams for $Q\bar{Q}$ decays.

to calculate an asymmetry of the angular distribution
(9.1) when the $Q\bar{Q}$ state is produced with longitudinally
polarized beams. This asymmetry is intimately related
to the 3g self coupling of Fig. 9.8b. Unfortunately
cancellations occur in the calculation, so that the
effect is rather small, the asymmetry is only 0.3% in
a subsample of all $Q\bar{Q}$ decays (this sample with $0.8 \leq$
$\leq T \leq 0.9$ contains 20% of the events). It may therefore
be more favourable to go to the next order. There we
expect a change $\sim \alpha_s^2$ of the opening angle θ_{23} (Fig.9.6)
in $\gamma 2g$ or 3g decays of the Quarkonium ground state.
In events with high thrust (3g) or a large photon energy
($\gamma 2g$) the two gluons opposite the most energetic quantum
are nearby in phase space and interact with each other
according to graphs like Fig.9.8b and 9.8c (there are
many more graphs of course). But we see that the colour
combinations of $Q\bar{Q} \rightarrow 3g$ and $Q\bar{Q} \rightarrow \gamma 2g$ are different. In
the first case (hadronic decay) the two slower gluons
are in a colour octet, they repel. In the second case
(radiative decay) the two gluons are in a colour singlet,
they attract. The effects on $\theta_{23}(T)$ of Fig. 9.7, although

of order α_s^2, may well be larger than the asymmetry of
order α_s discussed before, because here no cancellations
should occur (we just square the amplitudes). If these
speculations turn out to be true, then the decays of
heavy Quarkonium ground states become a laboratory for
QCD.

We now turn to jets originating from Quarkonium P
wave decays. The lowest P waves can be reached from the
first radially excited S wave, e.g. Υ', via an E1
transition (compare Fig.9.3). Experimentally it will
be necessary to trigger on this monochromatic photon
to identify the P wave. The P state then can decay
into 2 gluons in case of the $^3P_0$ and $^3P_2$ states. We
will discuss the jet decay of the $^3P_1$ state later. These
two gluons have a distinct energy of half the P state
mass. This is the essential difference to the 3 jet
decay of Quarkonium. Here we have monochromatic jets!
In Υ' the jet energy is almost 5 GeV, this should be
sufficient to determine the original gluon direction
via the jet direction. A measurement of the gluon angular
distributions becomes feasible! For the decay of the $^3P_0$
state this angular distribution is trivial: no matter,
what the dynamics are, there is only one helicity
amplitude which can contribute. But in the $^3P_2$ decays
there are two independent helicity amplitudes for
massless gluons. The QCD matrix element for the $^3P_2 \to gg$
decay reads with $q \equiv k_1 - k_2$ [13]

$$\varepsilon_{\mu\nu}(\lambda)\,[\,4k_1 \cdot k_2 \varepsilon_1^{*\mu} \varepsilon_2^{*\nu} - \varepsilon_1^* \cdot \varepsilon_2^* q^\mu q^\nu + 2k_1 \cdot \varepsilon_2^* \varepsilon_1^{*\mu} \cdot q^\nu -$$

$$- 2k_2 \cdot \varepsilon_1^* \varepsilon_2^{*\mu} \cdot q^\nu\,] \qquad (9.2)$$

and for on shell gluons we find the selection rule (8.7),
i.e. the decay proceeds via the helicity ± 2 state. The
formula for the kinematics [13] gives us, integrated,
the distribution

$$W_{2g}^{2^{++}} (\theta_{\gamma j}) \sim 1 + \cos^2 \theta_{\gamma j} \qquad (9.3)$$

where $\theta_{\gamma j}$ is the angle between the trigger photon and
one of the jets, measured in the c.m.s. of the jets
(Fig. 9.9). If the $^3P_2$ would decay into two quark jets

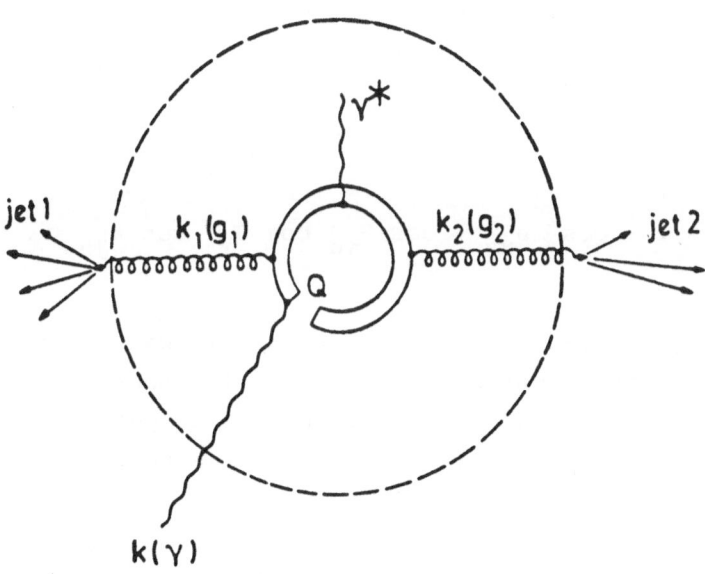

Fig. 9.9. $2^3S_1(Q\bar{Q}) \rightarrow \gamma + 1^3P_{0,2}(Q\bar{Q}) \rightarrow \gamma + 2g$ jets, as imagined within the colour bag.

by some arbitrary mechanism, the helicity of the two quarks can at most add up to $\lambda = \pm 1$. The kinematic formula then gives

$$W_{q\bar{q}}^{2^{++}}(\theta_{\gamma j}) \sim 1 - \frac{6 + 3A^2}{10+9A^2} \cos^2 \theta_{\gamma j} \qquad (9.4)$$

where A gives the weight of helicities $\lambda = \pm 1$ over helicity 0. The sign difference between (9.4) and (9.3) allows a clear test of the QCD mechanism. The rate for this process will be around 5% of all Υ' decays [13].

As we have discussed in Chapter 8 the $^3P_1$ decay proceeds via the complicated graph c) of Fig. 8.4. The decay is displayed again in Fig. 9.10. We will see two quark jets and a hadron cloud from the soft gluon from this decay. The quark jets should be easy to detect. Their angular distribution is given by

$$W_{q\bar{q}g}^{1^{++}} \sim 2 - \cos^2 \theta_{\gamma e} + \cos \theta_{\gamma e} \cos \theta_{\gamma j} \cos \theta_{je} \qquad (9.5)$$

already smeared over the important kinematic regime of small gluon momentum [100]. $\theta_{\gamma j}$ is the same angle as before, $\theta_{\gamma e}$ is the angle between the trigger photon and the beam, say e^-, and θ_{je} is the angle between the same jet arm as for $\theta_{\gamma j}$ and e^-. $\theta_{\gamma e}$ is measured in the lab frame, but θ_{je} as well as $\theta_{\gamma j}$ are in the c.m.s. of the jets. As an alternative process, the decay into two massless quarks would give [100]

$$W_{q\bar{q}}^{1^{++}} \sim 1 - \cos \theta_{\gamma e} \cos \theta_{\gamma j} \cos \theta_{je} \qquad (9.6)$$

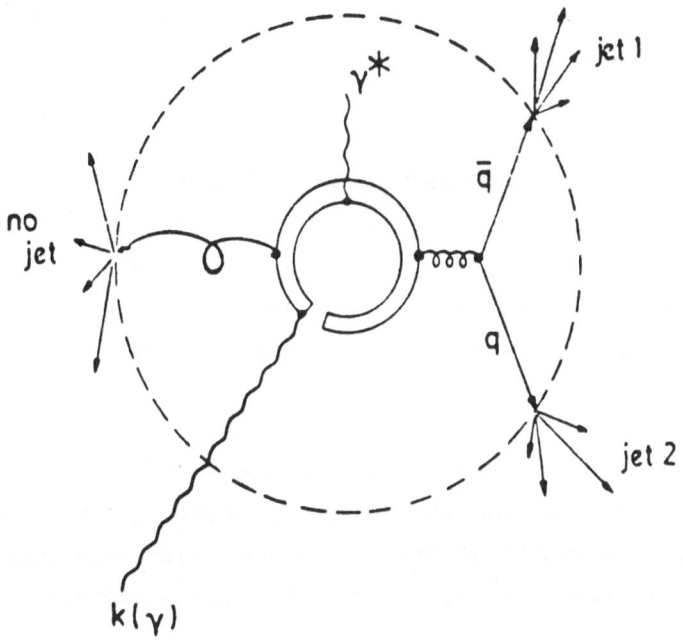

Fig. 9.10. $2^3S_1(Q\bar{Q}) \rightarrow \gamma + 1^3P_1(Q\bar{Q}) \rightarrow \gamma + 2$ quark jets. Here a soft gluon recoils against the two quark jets.

Here the $\cos^2\theta_{\gamma e}$ term is missing and the term linear in the cosines has a different sign. But the most important difference between the two alternatives (9.5) and (9.6) is the recoil of the soft gluon in the $gq\bar{q}$ decay of the $^3P_1$ state. This recoil will be much larger than the recoil of the trigger photon alone which of course is always present. But with the recoil of the trigger photon alone, the 2 jets would be collinear up to a 10° deviation at most. From the recoil of the soft gluon in the decay of the $^3P_1$ state, however, the angle between the two quark jets may be as small as 110°. This is true for the Υ system. For a heavier Quarkonium the "soft" gluon may even form a third jet in a small subset of all events.

10. CONCLUSIONS AND OUTLOOK

QCD gives us hints to the static $Q\bar{Q}$ potential at short distances and allows an educated guess for long distances. The simplest potential constructed this way, the "standard potential" $V(R) = -(4/3)\alpha_s/R+aR$ works astonishingly well in the Charmonium system. In first order perturbation theory (Breit-Fermi Hamiltonian) we are able to describe the fine structure reasonably well, while the application to the hyperfine structure might suffer from the lack of trustworthy experimental candidates. The Υ system is nonrelativistic to a much better approximation, $\beta^2 \simeq 0.08$, semirelativistic methods should give even better results here. But the attempt to describe Charmonium and Bottonium with the same static potential in the Schrödinger equation requires a modification of the "standard potential" at intermediate distances (QCD gives us no hints here). With this modification and a refinement of the short distance shape of the potential according to the distance dependence of the running coupling constant one can speculate about and predict items of still heavier Quarkonia yet to be discovered: Level spacings and Γ_{ee} will be as in Charmonium or

Bottonium, but the fine and hyperfine splittings will
decrease considerably. The ratio

$$R = \frac{\Delta M(^3P_2 - {}^3P_1)}{\Delta M(^3P_1 - {}^3P_0)}$$

is interesting. It will already be close to 0.8 in
Bottonium (experimentally it is 0.5 in Charmonium)
and exceed 0.8 for more massive Quarkonia to approach
its asymptotic value of 0.8 from above! The spectra
will nevertheless look very different from Charmonium.
Not only that the number of bound states below the
strong decay threshold will increase like $\sqrt{m_Q}$, also
the first states above threshold will be very narrow
because of the large number of radial nodes in the
wave function. E.g. Υ''', the first $b\bar{b}$ state above $B\bar{B}$
threshold will be very narrow, probably even below e^+e^-
machine widths, thus turning out as the ideal $B\bar{B}$
factory.

The confinement part of the potential may be
spin independent as suggested by lattice gauge theories.
Numerical fits to Charmonium and Bottonium are consistent with a complete spin, flavour and mass independence of the long range interquark forces. Even
the inverse Regge slope of the "old" mesons fits into
this picture. We seem to understand parity changing
photon transitions in terms of El radiation. This only
means that we understand the "size" of Charmonium. El
transitions of heavier Quarkonia therefore allow to
test their size, too. We also seem to understand the
branching fractions of P wave decays via the special
QCD annihilation mechanism into two gluons. This is
a short distance phenomenon. We further seem to under-

stand radiative decays of J/ψ into γf and γη as well as γη' via simple gluon spin arguments.

Up to now we do not know any Quarkonium pseudo-scalar state definitely. The experimental candidates X(2.83), χ(3.45) and χ($^{3.59}_{3.18}$) can hardly be understood. Especially their M1 transitions and gluon annihilation properties should be much different from what is observed for these states.

Hadronic decays of Υ indicate the existence of gluon jets. Maybe we caught the first glimpse of the gauge bosons of the strong interactions. Our hopes for the future are that these gluon jets can finally be proven. Then Quarkonium states, S and P waves, become a laboratory for QCD. We can measure the gluon spin and verify certain QCD processes like $^3P_1 \rightarrow g q \bar{q}$. Studying the three jet decays of Quarkonium ground states we can learn about the gluon selfinteraction either by finding asymmetries or better by comparison of angular distributions of the γgg decay with the ggg decay.

ACKNOWLEDGEMENT

We thank Professor H. Mitter and his team for the provision of an excellent atmosphere during the school, at the Stammtische and on top of the mountains. We acknowledge many discussions on the subjects of the lectures which we had with our colleagues in Hamburg.

REFERENCES

[1] J.J. Aubert et al., Phys.Rev.Lett. 33 (1974) 1404;
    J.E. Augustin et al., Phys.Rev.Lett. 33 (1974) 1406;
    G.S. Abrams et al., Phys.Rev.Lett. 33 (1974) 1453;
    For reviews see:
    B.H. Wiik and G. Wolf, DESY 78/23 (May 1978);
    G.J. Feldman and M.L.Perl, Phys.Rep. 33C (1977) 285.

[2] M. Gell-Mann, Phys.Lett. 8 (1964) 214;
    G. Zweig, CERN Preprints TH 401, 412 (1964).

[3] J.D. Bjorken and S.L.Glashow, Phys.Lett.11 (1964)255;
    S.L.Glashow, J.Iliopoulos and L.Maiani, Phys.Rev.D2
    (1970) 1285,
    M.K. Gaillard, B.W.Lee and J.L. Rosner, Rev.Mod.
    Phys. 47 (1975) 277.

[4] T.Appelquist and H.D.Politzer, Phys.Rev.Lett. 34
    (1975) 43;
    Phys. Rev. D12 (1975) 1404;
    A.De Rújula and S.L.Glashow, Phys.Rev.Lett. 34
    (1975) 46;
    T.Appelquist et al., Phys.Rev.Lett. 34 (1975) 365;
    E.Eichten et al., Phys.Rev.Lett. 34 (1975) 369.

[5] S.W.Herb et al., Phys.Rev.Lett. 39 (1977) 252;
    W.R. Innes et al., Phys.Rev.Lett. 39 (1977) 1240;
    K.Ueno et al., Phys.Rev.Lett. 42 (1979) 486.

[6] Ch.Berger et al., Phys.Lett. 76B (1978) 243;
    C.W.Darden et al., Phys.Lett.76B (1978) 246;
    Phys.Lett. 80B (1979) 419;
    J.K. Bienlein et al., Phys.Lett. 78B (1978) 360;
    C.W. Darden et al., Phys.Lett. 78B (1978) 364;
    For a review see: G. Flügge, this school.

[7] M.Kobayashi and K.Maskawa, Progr.Theor.Phys. 49
    (1973) 652;
    Y.Achiman, K.Koller and T.F.Walsh, Phys.Lett. 59B
    (1975) 261;

344

For a review see:

J. Ellis et al., Nucl. Phys. B131 (1977) 285.

[8] H.Fritzsch, M.Gell-Mann and H.Leutwyler, Phys.Lett.
B47 (1973) 365;

D.J. Gross and F. Wilczek, Phys. Rev. D8(1973) 3497;

S. Weinberg, Phys.Rev.Lett. 31 (1973) 494.

[9] G. 't Hooft, unpublished;

D.J. Gross and F. Wilczek, Phys. Rev. Lett. 30
(1973) 1343;

H.D. Politzer, Phys.Rev.Lett. 30 (1973) 1346.

[10] For a review see: H.Joos, this school.

[11] E. Eichten et al., Ref. 4 .

[12] J.Ellis, M.K.Gaillard and G.G.Ross, Nucl.Phys.B111
(1976) 253;

T.A. DeGrand, Y.J.Ng and S.H.H. Tye, Phys.Rev.D16
(1977) 3251;

K.Koller and T.F. Walsh, Phys.Lett. 72B (1977) 227
and 73B (1978) 504;

S.J. Brodsky, D.G.Coyne, T.A.DeGrand and R.R.Horgan,
Phys.Lett. 73B (1978) 203;

H.Fritzsch and K.H.Streng, Phys.Lett. 74B (1978) 90;

K. Hagiwara, Nucl.Phys.B137 (1978) 164;

A. De Rújula et al., Nucl. Phys. B138 (1978) 387;

K. Koller and T.F. Walsh, Nucl. Phys. B140 (1978) 449;

K. Koller, H.Krasemann and T.F. Walsh, Z. Physik
C1 (1979) 71.

[13] M. Krammer and H. Krasemann, Phys.Lett. 73B (1978)58;
H.Krasemann, Z. Physik C1 (1979) 189.

[14] D.Robson, Nucl. Phys. B130 (1977) 328 and references
therein;

P. Roy and T.F. Walsh, Phys. Lett. 78B (1978) 62.

[15] V.B.Berestetski, Sov. Phys. Uspekhi 19 (1976) 934.

[16] R. Barbieri et al., Nucl. Phys. B105 (1976) 125.

[17] H.Pietschmann and W.Thirring, Phys.Lett. 21
     (1966) 713;
     R. Van Royen and V.F. Weisskopf, Nuovo Cim. 50
     (1967) 617; ibid. 51 (1967) 583.

[18] R.Barbieri et al., Phys. Lett. 57B (1975) 455;
     R. Karplus and A. Klein, Phys. Rev. 87 (1952) 848;
     We thank A. Billoire for pointing out an error and
     drawing our attention to the paper by: W. Celmaster,
     Phys. Rev. D19 (1979) 1517.

[19] A. Martin, Phys. Lett. 67B (1977) 330;
     H. Grosse, Phys. Lett. 68B (1977) 343.

[20] E. Eichten and K. Gottfried, Phys.Lett. 66B (1977)286.

[21] C. Quigg and J.L. Rosner, Phys. Lett. 71B (1977) 153.

[22] G. Bhanot and S. Rudaz, Phys. Lett. 78B (1978) 119.

[23] H. Krasemann and S. Ono, DESY 79/09 (Nucl.Phys.,

[24] W. Celmaster, H. Georgi and M. Machacek, Phys.Rev.
     D17 (1978) 879.

[25] H. Lehmann, private communication.

[26] G.J. Aubrecht II and D.M. Scott, COO-1545-247.

[27] See various note books.

[28] G. Preparata, Phys. Lett. 82B (1979) 398.

[29] W. Kummer, this school.

[30] M.-A. de Crombrugghe, Phys.Lett. 80B (1979) 365.

[31] H. Georgi and D.V. Nanopoulos,Phys.Lett.82B(1979)392.

[32] H. Harari, H. Haut and J. Weyers, Phys. Lett. 78B
     (1978) 459.

[33] S. Pakvasa and H. Sugawara, Phys. Lett. 82B (1979) 105.

[34] J.D. Bjorken, SLAC-PUB 2195.

[35] T.F. Walsh, DESY 78/58.

[36] C. Quigg and J.L. Rosner, Phys.Lett.72B (1978) 462.

[37] J.L. Rosner, C.Quigg and H.B. Thacker, Phys. Lett.
     74B (1978) 350.

[38] R. Bertlmann, H.Grosse and A. Martin, Phys. Lett.
     81B (1979) 59.

[39] M. Krammer and P. Leal Ferreira, Rev. Brasil.
     Fis. 6 (1976) 7.

[40] K. Ishikawa and J.J. Sakurai, UCLA-78-TEP-2;
     Z. Physik C1 (1979) 117.

[41] M. Machacek and Y. Tomozawa, Ann. Phys. (N.Y.)
     110 (1978) 407.

[42] L.D. Landau and E.M. Lifshitz, Relativistic Quantum
     Theory (Pergamon Press, 1971); A.I. Achieser and
     W.B. Berestezki, Quantenelektrodynamik (Teubner,
     Leipzig, 1962); J. Pumplin, W.Repko and A. Sato,
     Phys. Rev. Lett. 35 (1975) 1538 and references
     therein;
     D. Gromes, Nucl. Phys. B131 (1977) 80.

[43] For a derivation on the classical level see the
     text books, e.g. A. Sommerfeld, Atombau und
     Spektrallinien (Vieweg, Braunschweig, 1950) or
     J.D. Jackson, Classical Electrodynamics (J. Wiley
     and Sons, New York, 1962). In the framework of the
     Bethe Salpeter equation for Charmonium see
     A.B. Henriques, B.H. Kellett and R.G. Moorhouse,
     Phys. Lett. 64B (1976) 85.

[44] See e.g.: A.I. Achieser and W.B. Berestezki,
     ref.[42], § 39.

[45] F. Wilczek and A. Zee, Phys. Rev. Lett. 40 (1978)
     83.

[46] H.J. Schnitzer, Phys. Lett. 65B (1976) 239.

[47] H.J.Schnitzer, Phys. Lett. 76B (1978) 461.

[48] N. Isgur, Erice Lectures 1978 and references therein,
     Oxford Univ. Ref. 67-78;
     U. Ellwanger, Nucl. Phys. B139 (1978) 422.

[49] H. Leutwyler and J. Stern, Phys. Lett. 73B (1978)
     75.

[50] K. Gottfried, Phys. Rev. Lett. 40 (1978) 598.

347

[51] K. Gottfried, in "Proc. 1977 International Symposium on Lepton and Photon Interactions at High Energies" (DESY, F. Gutbrod ed.).

[52] A. Billoire et al., preprint DPh-T/78/111.

[53] J. Ellis, communication at this school.

[54] See the text books, e.g. W. Heitler, The Quantum Theory of Radiation (Oxford Univ. Press, London, 1947);
J.M. Blatt and V.F. Weisskopf, Theoretical Nuclear Physics (J.Wiley and Sons, New York, 1952); or
S.A. Moszkowski in Beta and Gamma Ray Spectroscopy, K. Siegbahn ed. (North Holland, Amsterdam, 1955).

[55] L.B. Okun and M.B. Voloshin, preprint ITEP 152 (Moscow 1976);
V.A. Novikov et al., Phys.Rep. 41C (1978) 1.

[56] H. Krasemann, Thesis, Hamburg 1978.

[57] See e.g.: H.A. Bethe and E.E. Salpeter, Quantum Mechanics of One- and Two-Electron Atoms, Springer Verlag (Berlin, Göttingen, Heidelberg, 1957).

[58] J.D. Jackson, Phys.Rev. Lett. 37 (1976) 1107.

[59] L.D. Landau and E.M. Lifschitz, Relativistische Quantentheorie (Akademie-Verlag, Berlin, 1975).

[60] W. Bartel et al., Phys. Lett. 79B (1978) 492.

[61] C.Becchi and G.Morpurgo, Phys. Rev. 140B (1965) 687;
W.Thirring, Schladming Lectures 1965, Acta Phys. Austr. Suppl. II (1966) 205.

[62] A. De Rújula and R.L. Jaffe, MIT preprint CTP 658 (August 1977).

[63] H.J. Lipkin, Fermilab Conf.-77/93 THY (October 1977).

[64] G. Eilam, B. Margolis and S. Rudaz, McGill Univ. preprint.

[65] W.D. Apel et al., Phys.Lett. 72B (1978) 500.

[66] H. Krasemann and M. Krammer, Phys.Lett. 70B (1977) 457.

348

[67] H. Harari, Phys. Lett. 64B (1976) 469.

[68] R.C. Giles and S.-H.H. Tye, Phys.Rev. D16 (1977) 1079.

[69] D. Horn and J. Mandula, Phys. Rev. D17 (1978) 898.

[70] R.L. Jaffe and K. Johnson, Phys. Lett. 60B (1976) 201.

[71] K. Ishikawa, UCLA/78/TEP/21, September 1978.

[72] C.G. Callan et al., Phys.Rev. D18 (1978) 4684.

[73] D. Gromes, HD-THEP-79-1.

[74] P.Ditsas, N.A. McDougall and R.G. Moorhouse, Nucl.Phys. B146 (1978) 191.

[75] H. Taureg, talk given at DESY.

[76] W.E. Caswell, G.P. Lepage and J.R. Sapirstein, Phys. Rev. Lett. 38 (1977) 488.

[77] Particle Data Group, Phys. Lett. 75B (1978).

[78] E. Eichten et al., Phys. Rev. Lett. 36 (1976) 500.

[79] A. Ore and J.L. Powell, Phys. Rev. 75 (1949) 1696.

[80] M. Chanowitz, Phys. Rev. D12 (1975) 918.

[81] I. Ya. Pomeranchuk, Doklady Akademii Nauk SSSR 60 (1948) 263.

[82] A.I.Alekseev, Sov. Phys. JETP 34 (1958) 826.

[83] R.Barbieri, R.Gatto and R.Kögerler, Phys. Lett. 60B (1976) 183.

[84] M. Krammer and H. Krasemann, Ref.[13].
The dominance of the f decay amplitude into two photons with opposite helicities has already been discussed in a different context, e.g. by:
B. Schrempp-Otto, F. Schrempp and T.F. Walsh, Phys. Lett. 36B (1971) 463;
P. Grassberger and R. Kögerler, Nucl. Phys. B106 (1976) 451.

[85] C.N.Yang, Phys.Rev. 77 (1950) 242; compare also L.D. Landau and E.M. Lifshitz, Ref.[42].

[86] R.Barbieri, R.Gatto and E. Remiddi, Phys. Lett. 61B (1976) 465.

[87] H. Krasemann, Ref.[13].

[88] C.W. Darden et al., Internal Report DESY F15-78/01 (Aug. 1978); G. Flügge, this school.

[89] A. Billoire et al., Phys.Lett. 80B (1979) 381.

[90] M. Krammer, Phys. Lett. 74B (1978) 361.

[91] G. Alexander et al., Phys. Lett. 76B (1978) 652.

[92] See H. Genz, this school.

[93] R.F. Schwitters et al., Phys.Rev.Lett. 35 (1975) 1320; G. Hanson et al., Phys. Rev. Lett. 35 (1975) 1609; G. Hanson, SLAC PUB 2118 (1978).

[94] Ch.Berger et al., Phys. Lett. 78B (1978) 176; Phys. Lett. 81B (1979) 410; see also G. Flügge, this school.

[95] J. Ellis, M.K. Gaillard and G.G. Ross, Ref.[12].

[96] S. Brandt et al., Phys. Lett. 12 (1964) 57; E. Farhi, Phys. Rev. Lett. 39 (1977) 1587; A. De Rújula et al., Ref.[12].

[97] K. Koller, H. Krasemann and T.F. Walsh, Ref.[12].

[98] Ch. Berger et al., Phys. Lett. 82B (1979) 449.

[99] A. De Rújula, B.Lautrup and R.Petronzio, Nucl. Phys. B146 (1978) 50.

[100] A. De Rújula et al., Ref.[12]and H. Krasemann, Ref.[13].

Acta Physica Austriaca, Suppl. XXI, 351—406 (1979)
© by Springer-Verlag 1979

MANIFESTATIONS OF COLOUR IN HADRON SPECTROSCOPY[+]

by

CHAN HONG-MO[*]

CERN

CH-1211 Genève 23

ABSTRACT

The spectroscopy of multiquark hadron states is
shown to be a very suitable area for testing colour
dynamics recent work on which is reviewed with
particular emphasis on the simplest examples $(QQ)(\bar{Q}\bar{Q})$
or diquoniums which have considerable overlap with
experiments.

[+]Lectures given at the
 XVIII. Internationale Universitätswochen für Kern-
 physik,Schladming,Austria,February 28-March 10,1979.
[*]On leave of absence from the Rutherford Laboratory,
 Oxford, England

CONTENTS

I. INTRODUCTION - COLOUR CHEMISTRY

The object of these lectures is to seek evidence of colour in spectroscopy. The concept of colour is so popular these days that it needs no introduction. However, it is good to remember that its popularity is due not so much to empirical evidence as to aesthetic appeal. Indeed, to anyone but a particle physicist the whole idea of colour must sound suspiciously like political propaganda: it is advertised as a new degree of FREEDOM, but its main visible effect is CONFINEMENT. There are indeed some empirical checks such as the values of:

$$R = \sigma(e^+e^- \to \text{hadrons})/\sigma(e^+e^- \to \mu^+\mu^-)$$

and

$$\Gamma(\pi^0 \to \gamma\gamma) \, ,$$

which count the number of colours, but these cannot be regarded as sufficient for such a fundamental concept. One ought to expect as evidence a whole new class of phenomena qualitatively different from those in a colour-less world.

Now one obvious place to seek evidence for colour is in spectroscopy where a new degree of freedom necessarily introduces an additional complexity in the spectrum. (One familiar example is the effect of the electron spin in atomic spectroscopy.) The effect of colour in hadron spectroscopy, however, is not so obvious for the following reasons:

(i) hadrons are colour singlets so that colour is visible only at subhadronic levels.

(ii) in ordinary $\bar{Q}Q$ mesons and QQQ baryons one has only subhadronic parts of colour 3. (A diquark QQ can in principle be in colour $3 \times 3 = \bar{3}$ or 6, but only $(QQ)^{\bar{3}}$ can combine with $(Q)^3$ to form a colour singlet since $6 \times 3 = 8 + 10$ does not contain the singlet representation.)

Hence to see effects of colour one must study hadrons with more quarks or gluons. The simplest example is $QQ\bar{Q}\bar{Q}$. Here the diquark QQ can be in colour $\bar{3}$ or 6 and the antidiquark $\bar{Q}\bar{Q}$ in colour 3 or $\bar{6}$. One can then form colour singlets in two ways:

$$(QQ)^{\bar{3}} \ (\bar{Q}\bar{Q})^{3}: \qquad 3 \times \bar{3} = \underline{1} + 8 \tag{1}$$

$$(QQ)^{6} \ (\bar{Q}\bar{Q})^{\bar{6}}: \qquad 6 \times \bar{6} = \underline{1} + 8 + 27 \ . \tag{2}$$

The number of states is doubled because of colour which, if experimentally verified, would be a clear direct evidence for the colour degree of freedom.

Unfortunately such a test is not easy to achieve. It is practically impossible simply to count the number

354

of states in experiment. Because of the complex spectrum
in a $Q Q \bar{Q} \bar{Q}$ system the difficulty will be many times that
in baryon spectroscopy which has taken us twenty years
and is still not settled. One must seek instead to
identify directly the different colour configurations.
In general, however, colour configurations such as
$(QQ)^3 - (\bar{Q}\bar{Q})^3$ and $(QQ)^6 - (\bar{Q}\bar{Q})^6$ mix since the Hamiltonian
has no reason to be diagonal in the diquark colour. The
resulting states have then very similar properties and
will be hard to distinguish experimentally. The question
is thus: are there situations in which mixing between
colour configurations become negligible? If so, we may
have a chance to identify the colour configurations.

As an example of how this may happen [1] let us
consider explicitly a qq system[+] in a relative s-wave
state, where q = u,d. The diquark can have spin S = 0,1
and isospin I = 0,1, which we denote by subscripts.
Thus: $(qq)^x_{2S+1, \, 2I+1}$. Because of the Pauli principle,
the only diquark states are:

$$(qq)^{\bar{3}}_{1,1} \, , \quad (qq)^{\bar{3}}_{3,3} \, , \quad (qq)^{6}_{1,3} \, , \quad (qq)^{6}_{3,1} \qquad . \tag{3}$$

Combining these with the corresponding antidiquarks we
obtain our two sets of colour singlets which we denote
by:

$$(T): \; (qq)^{\bar{3}} - (\bar{q}\bar{q})^3, \quad \text{and} \quad (M): \; (qq)^6 - (\bar{q}\bar{q})^{\bar{6}} \, ,$$

respectively.

[+]We shall denote a quark of any flavour by Q. The small
letter q is reserved for u or d quarks.

The diquark and antidiquark interact by exchanging
gluons, which may had to mixing of the two colour con-
figurations.Since gluons are flavourless, this can mix
only states with the same flavour. From (3) then one
sees that in order to change the colour of a diquark by
gluon exchange one must also change the diquark spin [1].

Now, spin-dependent forces normally decrease like
same power of the distance. That this is probably true
also for interactions between quarks can be demonstrated
as follows [1]. In the quark model π and ρ are both s-
wave $q\bar{q}$ states differing only by the total quark spins:
for π, $S = 0$ (singlet), and for ρ, $S = 1$ (triplet).
Therefore, the mass difference $\Delta m(0) = m_\rho - m_\pi$ measures
the strength of the spin-dependent interaction between
q and $\bar{q}$ in relative s-wave, namely $\ell = 0$. Similarly, B
and A_2 are p-wave $q\bar{q}$ states differing only by the quark
spin S. $\Delta m(1) = m_{A_2} - m_B$ measures the strength of the
spin-dependent interaction between q and $\bar{q}$ for $\ell = 1$.
Hence, plotting the mass difference $\Delta m(\ell)$ between the
triplet and singlet states along Regge trajectories gives
the variation of spin-dependent forces as functions of ℓ,
or equivalently of the distance between q and $\bar{q}$. Fig. 1
shows such a plot for the $q\bar{q}$ and $q\bar{s}$ mesons. In both
cases, $\Delta m(\ell)$ is seen to decrease rapidly with ℓ.

Assuming that this is true also for interactions
between diquarks would imply that the mixing between
colour configurations is suppressed if the diquarks are
kept apart by some 'high' angular momentum; thus:
$(qq) \overset{L}{\longrightarrow} (\bar{q}\bar{q})$. The amount of mixing at a given L can
actually be estimated as follows [1]. The mixing angle θ
is given by perturbation theory as:

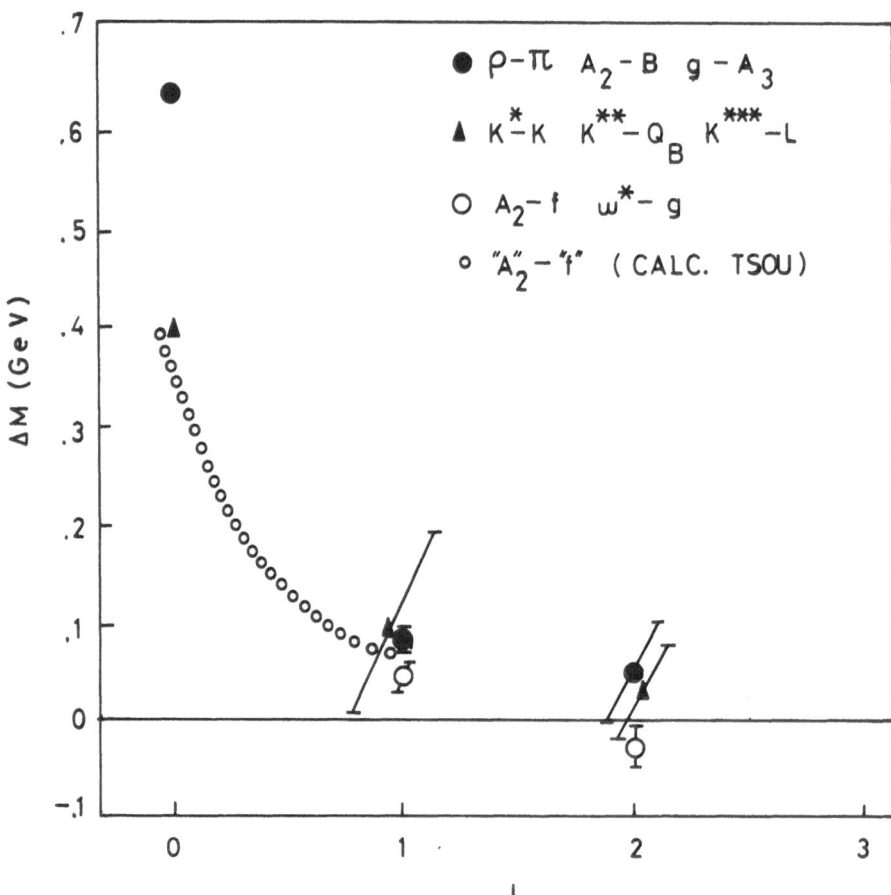

Fig. 1. Spin-dependent and annihilation forces between quark constituents as functions of the orbital angular momentum L.

$$\tan \theta \sim \frac{<T|V|M>}{E_T - E_M} \qquad . \qquad (4)$$

The numerator $<T|V|M>$ represents the mixing matrix element as measured by the strength of the spin-dependent forces at angular momentum L, e.g. by $\Delta m(1) = m_{A_2} - m_B$ at L = 1 and by $\Delta m(2) = m_g - m_{A_3}$ at L = 2. The denominator $E_T - E_M$ is the mass difference between the two colour configurations before mixing. This is at least the mass difference between the coloured $\bar{3}$ and 6 diquarks, which are themselves split by gluon exhcanges between these two quark constituents. Now, the two quarks in a diquark being in a relative s-wave by assumption, this mass-splitting must be of the same order as $\Delta m(0) = m_\rho - m_\pi$. Hence we obtain the estimate:

$$\tan \theta (L) \sim \frac{\Delta m(L)}{\Delta m(0)} \sim .2 \text{ at } L = 1, \ .1 \text{ at } L = 2 \ . \qquad (5)$$

One sees that the mixing is expected to decrease quite rapidly with L. For sufficiently high L the colour configurations then became approximately pure [1,2]. This is exactly the situation which interests us.

Similar considerations relying on spin-dependence apply to several other cases of interest. Even in cases where they do not apply one may still expect colour mixing forces to decrease with separation. To illustrate this point but $V(\underline{r})$ represent the interaction potential between two quarks or antiquarks at distance $\underline{r}$ [3]. Consider now a cluster of quarks of colour charges (matrix) λ_i at nearby positions $\underline{x}_i$. The potential they exert on another quark at the origin is then proportional to $\sum_i \lambda_i V(\underline{x}_i)$. We make now an expansion about the centre-of-mass of the cluster $\underline{r} = \sum_i \underline{x}_i$, thus:

$$\sum_i \lambda_i V(\underline{x}_i) = (\sum_i \lambda_i) V(\underline{r}) + \sum_i \lambda_i (\underline{x}_i - r)\frac{\partial V}{\partial \underline{r}} + \ldots \quad . \qquad (6)$$

Notice that the leading term is proportional to $\sum_i \lambda_i$ which is the total colour charge of the cluster; it can therefore not change the cluster charge. It follows then that colour mixing forces between two quark clusters are 'dipole' forces which devease like some power of the clusters' separation as compared with the confining potential V. This may well be sufficient to suppress the mixing between colour configurations when the clusters are spatially separated. Unfortunately this conjecture cannot be checked empiricically in $Q\bar{Q}$ or QQQ spectroscopy.

We are thus led to consider multiquark hadrons with the elongated structure shown in Fig. 2.

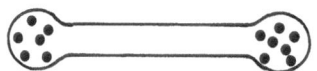

Fig. 2. A chromionic molecule.

They consist of two portions of opposite colour charges x and $\bar{x}$ separated by a 'high' angular momentum L, where the constituents in each portion are close together, say, in a relative s-wave state. The two portions are held together by colour flux lines which are confined inside a narrow tube. We are interested in these states primarily because they may have approximately pure colour configurations which may hence be distinguishable, but it terms out, as we shall see later, that we have an additional bonus in that they also have the following properties:

(i) Their spectrum is largely predictable based on the experience gained in $Q\bar{Q}$ and QQQ spectroscopy [1,2].

(ii) Many of them have unusual decay characteristics which make them particularly suitable for experimental observation [1].

To investigate the spectrum and properties of this class of hadrons, it is convenient to borrow the language of chemistry [4]. They have some similarity to chemical ionic molecules, e.g.: Na^+-Cl^-, which consists of two oppositely charged ions neutralized by an ionic band. Here we have a 'chromionic' molecule: $(A)^x-(B)^x$, where two 'chromions' of opposite colour charges are linked together and neutralized by a tube of colour flux lines which we may call a 'chromionic band'. The similarity is of course only superficial since the dynamics are very different. In contract to ordinary ionic bands which break easily at large distances chromionic bands are supposed even to increase in stability for increasing separation. Nevertheless, the chemical language is convenient for the following reason. The properties of ordinary molecules are largely controlled by the properties of their component ions and bands. This fact allows us to reduce the study of molecules to that of their components resulting in saving of labour and simplification of concept. It seems that a similar reduction is possible also in 'colour chemistry'. In what follows, therefore, we shall examine our colour molecules in terms of their component 'chromions' and 'chromionic bands'. Once sufficient is known about their components, we can construct our own molecules and predict a fair amount of their properties.

II. CHROMIONS

In analogy to (3) we introduce the following notation for chromions in general:

$$(\text{constituents})^{\times}_{2S+1}$$

where $\times$ is the total colour representation and S is the total quark spin. This will be sufficient to distinguish most of the simpler ions we are interested in. For example, some familiar hadrons ('chromatons') in this notation are

$$\rho^+ = (u\bar{d})^1_3; \quad \Delta^{++} = (uuu)^1_4; \quad \Lambda = (uds)^1_2 \; (I = 0) \; .$$

First we discuss the spectrum of chromions. Notice that the absolute value for the mass of a chromion has no meaning since it cannot exist free; it may only be assigned an effective mass inside a hadron but even this requires some detail knowledge of the confining forces which at present means constructing a potential or bag model. The estimates so obtained, however, are insufficiently accurate for practical purposes. We shall therefore leave them eventually to be determined by phenomenology.

On the other hand, mass splittings between ions of similar construction are better known. The experience gained in $Q\bar{Q}$ and QQQ spectroscopy suggests that to a good approximation:

(i) Hadron masses are linear in the masses of their constituents, e.g.

$$m_{K^*} - m_\rho \sim m_\Lambda - m_p \sim m_s - m_p \sim 175 \text{ MeV}^+$$

(ii) Hadrons with the same constituents but different S are split by one-gluon exchange between their constituents [5,6].

We shall illustrate three points by considering again the diquark ions (QQ). To leading order in the strong coupling α_s, the gluonic interaction between the constituents is given by the one-gluon-exchange diagram of Fig. 3.

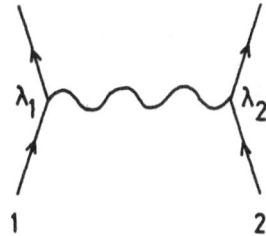

Fig. 3. The one-gluon-exchange diagram.

$^+$These relations are better satisfied when spin-dependent forces are properly taken into account.

Following the procedure adopted in charmoniums, we make
a Breit reduction of this diagram leading to a potential
with spin-spin, spin-orbit and tensor terms [5]. Since
by assumption, our diquark is in s-wave, there are no
contributions from the spin-orbit or tensor forces. We
are then left with only the spin-spin term which may be
written as:

$$H_1 = -C \sum_a \sum_r (\lambda_1^a \sigma_1^r)(\lambda_2^a \sigma_2^r) \qquad (7)$$

λ^a (a = 1,...,8) are 3x3 Gell-Mann matrices corresponding
to the colour charges of the quarks, and σ^r (r = 1,2,3) are
2x2 Pauli matrices corresponding to their spins. The
coefficient $C \sim \alpha_s |\psi(0)|^2$ depends via the wave function
ψ on the total colour $\times$ and on the flavours of the two
quarks 1 and 2.

To find the masses of the diquark states we need
to diagonalise (7). This is readily done since:

$$\sum_a \lambda_1^a \lambda_2^a = \frac{1}{2} \{ \sum_a (\lambda_1^a + \lambda_2^a)^2 - \sum_a (\lambda_1^a)^2 - \sum_a (\lambda_2^a)^2 \}$$

$$= \frac{1}{2} \{ b_\times - b_3 - b_3 \} \qquad (8)$$

where $b_\times$ is the value of the quadratic Casimir operator
for colour $\times$. Similarly,

$$\sum_r \sigma_1^r \sigma_2^r = \frac{1}{2} \{ 4S(+1) - 4s_1(s_1+1) - 4s_2(s_2+1) \} \qquad (9)$$

where S is the diquark spin and $s_i = \frac{1}{2}$ are the individual quark spins, respectively. We obtain then that
(7) is diagonal already in the diquark states $(Q_1 Q_2)^x_{2S+1}$.
The eigenvalues are listed in Table I [1,2].

Ion	$<H_1>$
$(Q_1 Q_2)^{\overline{3}}_2$	$-8C$
$(Q_1 Q_2)^{\overline{3}}_3$	$\frac{8}{3}C$
$(Q_1 Q_2)^{6}_1$	$4C$
$(Q_1 Q_2)^{6}_3$	$-\frac{4}{3}C$

Table I. Diquark mass shifts due to the spin-spin interaction from one-gluon-exchange.

We are not particularly interested in calculating the
value of C theoretically and shall try therefore to
estimate them phenomenologically from $Q\bar{Q}$ and QQQ
spectroscopy. Colour $x = 3$ diquarks exist already
in ordinary baryons where we have three quarks in a
relative s-wave state with a spin-spin interaction
similar to (7) between each pair, thus

$$H_1 = -C \sum_{i>j} \sum_a \sum_r (\lambda^a_i \, \sigma^r_i)(\lambda^a_j \, \sigma^r_j) \quad . \qquad (10)$$

Diagonalising (10) leads to a mass difference between
N and Δ as:

$$m_\Delta - m_N = (qqq)^1_4 - (qqq)^1_2 = 16C \qquad (11)$$

which when equated to the empirical value of about
300 MeV gives [1]:

$$C_{qq} \sim 20 \text{ MeV} \tag{12}$$

for x = 3   and   q = u,d .

One may object to the estimate (12) on the grounds that
the wave function ψ of the two quarks in a baryon may
be affected by the presence of a third quark nearby
and therefore does not simulate sufficiently an isolated
diquark in a colour molecule. To answer this objection
we evaluate the mass difference between the excited
states of N and Δ on the leading trajectories with
$J^P = \frac{5^+}{2}$ and $\frac{7^+}{2}$, respectively [1]. In a string or bag
model these correspond to diquark-quark states with the
diquark in a s-wave and separated from the third quark
by an orbital angular momentum L = 2. We may then
neglect the spin-dependent interactions between the
diquark and the third quark as argued before. We then
obtain from the spin-spin interaction between the
quarks in the diquark:

$$m_{\Delta(\frac{7^+}{2})} - M_{N(\frac{5^+}{2})} = \frac{32}{3} C \tag{13}$$

as compared with the experimental value of 245 $\pm$ 30 MeV.
This gives C_{qq} = 23 $\pm$ 3 MeV, with larger errors but
consistent with the previous estimate (12). The same
can be repeated for higher resonances on the same
trajectories giving similar results. Indeed, the
close agreement of (13) with (12) suggests that the

parameter C governing the spin-spin interaction in a
quark pair may not depend two much on other quarks
nearby, and that the same estimate may be used for
multiquark ions in general.

The parameter C may depend on the flavours of the
interacting quark pair. To estimate this dependence we
write the spin-spin interactions between quark pairs in
a QQQ baryon as:

$$H_I = - \sum_{i>j} C_{ij} \sum_{a,r} (\lambda_i^a \sigma_i^r)(\lambda_j^a \sigma_j^r) \quad . \tag{14}$$

We thus fit (14) to the masses of the baryon octet and
baryon[*] decuplet with C_{ij} and $m_s = m_s - m_u$ as
parameters [7]. One obtains, apart from $\Delta m_s = 175$ MeV
as quoted above,

$$C_{qq} = 20 \text{ MeV}, \quad C_{qs} = 12.5 \text{ MeV}, \quad C_{ss} = 8.5 \text{ MeV} \tag{15}$$

quite close to the dependence $C_{ij} \sim 1/m_i m_j$ as obtained
from the Breit reduction.

Finally, the parameter C may depend on the total
colour × of the diquark. This dependence we have unfortunately no way to estimate in QQQ spectroscopy. As
a working hypothesis, we may suppress this dependence
on ×, which seems in rough agreement with the
information available at present from multiquark
spectroscopy, as we shall see later. This assumption,
however, has to be subjected to scrutiny when the
experimental situation improves.

With these observations one can now calculate the masses of all diquark ions $(Q_1 Q_2)_{2S+1}$ as:

$$\Delta M = \Delta m_1 + \Delta m_2 + <H_1>^x_{2S+1} \tag{16}$$

with $<H_1>$ given in Table I, C estimated in (12), (15) and $m_s - m_u = 175$ MeV, the masses of all diquarks formed from u,d and s are then fixed up to one overall normalising constant which we leave to be determined later phenomenologically.

The same considerations can be applied to ions with any number of quarks (or antiquarks) of any flavours provided we assume, as observed after (13), that the parameter C does not depend on the presence of other quarks nearby. We have now, however, to diagonalize (14) summed over all quark pairs in the ion [4]. This may be technically a messy affair especially when several quark flavours are involved, but is otherwise straightforward and of no particular theoretical interest.

More interesting are ions containing both quarks and antiquarks, e.g. $(Q\bar{Q})$ and $(QQ\bar{Q})$, which introduce two new features [4,8,9,10]. First, the 'colour magnetic' interaction as represented by Fig. 3 depends now on a new parameter $C_{Q\bar{Q}}$. For $x = 1$, this may be estimated from $Q\bar{Q}$ mesons, for example:

$$m_\rho - m_\pi = \frac{64}{3} C_{q\bar{q}} \tag{17}$$

which gives for $m_\rho - m_\pi \sim 630$ MeV:

$$C_{q\bar{q}} \sim 29 \text{ MeV} \qquad . \qquad\qquad (18)$$

Similarly from $m_{K^*} - m_{K'}$, and $m_{D^*} - m_{D'}$, we have

$$C_{q\bar{s}} = 18.5 \text{ MeV}, \qquad C_{q\bar{c}} = 6.5 \text{ MeV} \qquad\qquad (19)$$

again quite close to the dependence $C_{ij} \sim 1/m_i m_j$ from the Breit reduction. The fact that these estimates for $x = 1$ are similar to those of C_{QQ} in (15) for $x = 3$ is suggestive that these parameters C may indeed be insensitive to the total colour x of the interacting pair x.

Second, to leading order in α_s there is now also an annihilation diagram [10] as shown in Fig. 4.

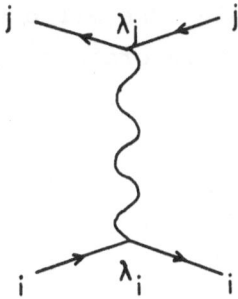

Fig. 4. The one-gluon annihilation diagram.

Clearly, this contributes only to a $Q\bar{Q}$ pair in a flavour-zero, colour-octet, and spin-triplet state. It does not occur for example in ordinary $Q\bar{Q}$ mesons which are necessarily colour singlets. We have here then an

occasion for verifying a typically quantum field
theoretical effect in chromodynamics which is readily
calculable. Indeed, we may write the diagram as:

$$H_2 = 6C \; p^i \, p^j \tag{20}$$

where

$$p^i = \frac{1}{18} \{16 + 3 \sum_a \lambda_i^a \, \lambda_{\underline{i}}^a \} \cdot \frac{1}{4} \{3 + \sum_r \sigma_i^r \, \sigma_{\underline{i}}^r \} \tag{21}$$

is the projection operator into the colour octet spin
triplet. The coefficient C is again essentially
$\alpha_s |\psi(0)|^2$ and is expected to be of the same order as
(18) and (19). Notice that H_2 is repulsive and tends
in general to push up masses of ions with low flavours
typically by about 100 MeV. Moreover, H_2 mixes $Q\bar{Q}$ states
with different flavoured quarks Q, leading thus to
violations of the OZI rule. The effects should, however, be much larger than in ordinary $Q\bar{Q}$ mesons which
are of order α_s^2 or α_s^3. Indeed, in $(Q\bar{Q})_3^8$ ions of colour 8
and spin triplet, the estimated mixing is so big that
there is no OZI rule to speak of. One sees therefore
that the consequences of H_2 should be clearly observable
once multiquark hadrons containing such ions are experimentally identified.

To find the masses of ions containing both quarks
and antiquarks, we have now to diagonalize simultaneously
both H_1 in (14) and H_2 in (20) each summed over all
appropriate pairs of constituents [10]. This is again
technically messy but theoretically uninteresting. It
suffices to note that a list now exists for the masses

of all ions with up to three quark constituents, namely, $Q Q$, $Q \bar{Q}$, $Q Q Q$, and $Q Q \bar{Q}$, as well as some 'atoms' with even more [4,10,11]. In other words, the stage here is all set for a lot of phenomenology if and when experimental data become available.

With some knowledge of the chromions' masses, we can now also discuss their stability. Chromions can be unstable in two ways, whose general properties are summarised in Table II. What we are most interested in are those ions which are stable against both 'dissociation' and 'emission', since they are likely to lead to metastable molecules with narrow widths which are experimentally easy to identify. A list of such stable chromions has also been compiled.

III. CHROMIONIC BONDS

Chromionic bonds are tubes of confined colour field lines which bind together chromions and neutralize their colour charges. Because of the field energy confined inside a chromionic bond also carries a mass [12]. Fig. 5 is an illustration of the colour field lines due to the chromionic constituents at one and of the tube.

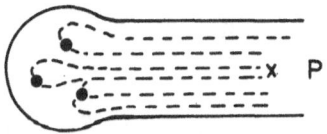

Fig.5. Colour field lines in a chromion and a chromionic bond.

Type	Dissociation	Emission
Illustration		
Characteristic	No change in number of constituents	$Q\bar{Q}$ pairs created
Example	$\varepsilon \to \pi\pi?$ $(q\bar{q}\bar{q}q)_1^{\,1} \to (q\bar{q})_1^{\,1}(q\bar{q})_1^{\,1}$	$\Delta \to N\pi$ $(qqq)_4^{\,1} \to (qqq)_2^{\,1}(q\bar{q})_1^{\,1}$
Typical width	Hundreds of MeV	$\sim$ 100 MeV

Table II. Instabilities of chromions.

The field of a point P inside the tube may be written as:

$$F^a \propto \sum_i \lambda_i^a / A_x \qquad (22)$$

where λ_i^a represent the colour charges of the chromionic constituents and A_x is the cross sectional area of the flux tube. The field energy density at P is proportional to the square of the field F^a:

$$E \propto \sum_a |F^a|^2 \propto \sum_a |\sum_i \lambda_i^a|^2 / A_x^2 = b_x / A_x^2 \qquad (23)$$

where b_x is the value of the quadratic Casimir operator for colour x.

To estimate the cross sectional area A_x for the flux tube, we turn now to the MIT bag model [12]. Here equilibrium is obtained by balancing the pressure of the colour fields inside against the external bag pressure B. We obtain then:

$$A_x \propto \sqrt{b_x} \qquad . \qquad (24)$$

The chromionic bond has then a linear mass density or energy per unit tube length of the form:

$$\rho_x = E \cdot A_x \propto \sqrt{b_x} \qquad . \qquad (25)$$

This result is more conveniently stated in terms of the angular momentum L. Asymptotically for large L Regge trajectories in the bag model are linear in M^2, thus

$$L = L_o + \alpha'_x M^2 \qquad (26)$$

where the slope α_x' is inversely proportional to linear mass density ρ_x

$$\alpha_x' \propto 1/\rho_x \propto 1/\sqrt{b_x} \qquad . \qquad (27)$$

Now experimentally from ordinary $Q\bar{Q}$ and QQQ spectroscopy which involves only colour 3 bonds we know that:

$$\alpha_3' \sim .9 \text{ Gev}^{-2} \qquad\qquad (28)$$

from which using (27) we can calculate the Regge slope for any colour $\times$, e.g.

$$\alpha_6' = \sqrt{\tfrac{2}{5}} \ .9 \ \sim .57 \ \text{Gev}^{-2}$$

$$\alpha_8' = \tfrac{3}{5} \ .9 \ \sim .54 \ \text{Gev}^{-2} \qquad . \qquad (29)$$

Next we discuss the various possible modes of chromionic band instability [4]. A band may for example "rupture' as illustrated in Fig. 6.

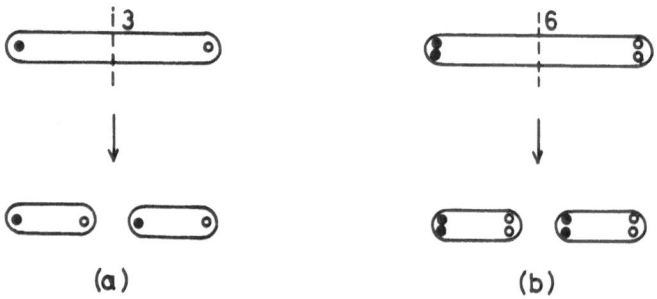

(a)   (b)

Fig. 6. The decay of a colour bond by rupture.

Since colour is confined, however, quark or gluon pairs
will have to be created to neutralise the colour charges
which are thereby exposed. For a colour 3 bond only one
quark pair needs to be created to neutralise the colour.
Such a rupture is easy and indeed represents the common
decays of most high-spin hadrons in $Q\bar{Q}$ and QQQ spectros-
copy, e.g. $f \rightarrow \pi\pi$, $A_2 \rightarrow \rho\pi$ etc. Typical widths for such
decays are of the order of about 100 MeV. For a colour
6 bond on the other hand the creation of one quark pair
is insufficient, since a quark with colour 3 cannot
neutralise the exposed colour 6 and at least two quark
pairs are required. The necessity to create an extra
quark pair presumably means that its decay width is
suppressed [1]. The suppression factor may in principle
be estimated empirically by comparing the branching
ratios of high-mass mesons decaying into $p\bar{p}$ and $\pi\pi$,
respectively. Unfortunately, so such data are as yet
available. Estimates from the Regge model yields
typically a suppression by about an order of magnitude,
and if so it makes chromionic molecules with $\times = 6$
interestingly narrow for experimenters. For a colour 8
bond with triality zero, there are two possibilities:
either we create to quark pairs as for $\times = 6$ or we
create a gluon pair to neutralise the exposed colour
charges. The second possibility is probably easier and
may prove to be an interesting starting point for in-
vestigating the gluon degree of freedom in spectroscopy.

Further a chromionic bond may perhaps be also split
along its length, as illustrated in Fig. 7. Whether this
is possible or not kinematically will depend on the
linear mass densities of the initial bond and of the
products of the split. If one believes the bag model

374

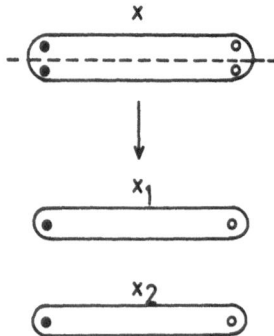

Fig. 7. The decay of a colour bond by splitting.

result of (25), then it can be shown in general as a consequence of a triangular inequality between Casimir operators that [13,14]:

$$\rho_x \leq \rho_{x_1} + \rho_{x_2} \qquad (30)$$

so that a split is always kinematically forbidden[+].

The properties of the lowest colour bonds are summarised in Table III.

[+]This result is in direct contrast to a naive string picture in which the total linear mass density of two parallel strings is always just the sum of the two densities. The difference is clearly due to a defect of the naive string picture which neglects the interference between the fields generated by nearby charges [14].

Bond x	b_x	ρ_x	α'_x (GeV^{-2})	Rupture		Split	
				Creating	Width (MeV)	Into	$\Delta\rho$
3	$\frac{16}{3}$	1 (Norm.)	.9	$q/\bar{q}$	~ 100	3,3	1.
6	$\frac{40}{3}$	1.58	.57	$qq/\bar{q}\bar{q}$	~ 10?	3,3	.42
8	12	1.50	.54	g/g	?	3,3	.50

Table III. Properties of chromionic bonds.

IV. COLOUR MOLECULES

Having now studied both chromions and chromionic bonds we shall combine them to form colour molecules. For large angular momentum L spin-dependent and colour-mixing forces are both negligible. We may then imagine the mass of a colour molecule to be made up of the chromionic masses and the mass of the bond plus some confinement energy which we cannot calculate. The masses of the chromions are given by diagonalising the matrices $\Delta M = H_1 + H_2$ for each of the ions as already discussed in II. The mass of the bond depends on its length and is more conveniently stated in terms of the angular momentum L. We shall assume that Regge trajectories are linear even at finite masses as seems true approximately in experiment. From (26) we then obtain a mass for the bond:

$$M_{BOND} = \sqrt{(L-L_o)/\alpha'_x} \tag{31}$$

where the free parameter L_o may be taken to represent the unknown confinement energy. The mass of the molecule is then

$$M = \Delta M + \sqrt{(L-L_o)/\alpha'_x} \tag{32}$$

depending on one free parameter L_o for each family of molecules with the same chromionic constituents and the same colour bond ×.

For small L (i.e. $L \sim 1$) spin-orbit and tensor interactions between the ions are not entirely negligible.

They can, however, the estimated from $Q\bar{Q}$ spectroscopy
[5] and taking them into account leads to mass splittings
in spin multiplets of the order $\sim$ 100 MeV. We shall report
later as examples same calculations on the diquark-antidiquark molecules.

Next we consider the decay of colour molecules.
First a colour molecule can decay via any instability of
its component ions and bonds. It is best to illustrate
this with again our simplest example, namely, diquark-
antidiquark molecules or diquoniums. The masses of diquark
ions were listed already in Table I. Substituting the
values of C in (15) one sees that apart from $(qq)\frac{\bar{3}}{3}$ which
may decay marginally by π-emission, thus:

$$(qq)\frac{\bar{3}}{3} \rightarrow (qq)\frac{\bar{3}}{1} + \pi \tag{33}$$

all other (QQ) ions are stable. The decay (33), in which a
π is emitted just by flipping the constituent spins, is
very similar to $\Delta \rightarrow N\pi$ through with a reduced phase space.
This suggests then that all diquoniums containing a $(qq)\frac{3}{3}$
ion will have such a mode with an estimated width of a few
tens of MeV. Further, due to the instability of the bond,
a colour 3 banded molecule, or "T"-diquonium can decay
readily (width $\sim$ 100 MeV) by rupture creating one $Q\bar{Q}$ pair
and resulting in a $B\bar{B}$ final state, thus:

$$(QQ)^{\bar{3}} \not\xrightarrow{} (\bar{Q}\bar{Q})^{\bar{3}} \rightarrow (QQQ)' + (\bar{Q}\bar{Q}\bar{Q})' \quad . \tag{34}$$

Whereas, a colour 6-bonded molecule or "M"-diquonium cannot easily decay in this way; here bond rupture requires
creating two quark pairs leading to two diquoniums of the
same type:

$$(QQ)^6 \not\leftarrow (\bar{Q}\bar{Q})^{\bar{6}} \to (QQ)^6 - (\bar{Q}\bar{Q})^{\bar{6}} + (QQ)^6 - (\bar{Q}\bar{Q})^{\bar{6}} \qquad (35)$$

with a greatly suppressed width ~ 10 MeV.

Of course the molecule may also decay via a molecular transition involving the molecule as a whole. For example one may imagine a quark transfer between the two ions, thus:

$$(QQ)^x - (\bar{Q}\bar{Q})^x \to (Q\bar{Q})' + (Q\bar{Q})' \qquad (36)$$

resulting in a mesonic final state. Because of the angular momentum barrier separating the diquarks, which the quark and antiquark exchange will have to overcome, this transition is expected also to decrease rapidly with increasing L. That this is indeed the case can be demonstrated phenomenologically in $Q\bar{Q}$ spectroscopy as follows [1]. The transition (36) may be represented by the quark diagrams (a) and (b) in Fig. 8, which apart from the 'spectator' quarks are similar to diagrams known to give violations of exchange degeneracy in the dual Regge model [16]. For example the diagram (c), which is the analogue to (a) with the spectator quarks removed, is responsible for breaking the degeneracy between the f and ρ trajectories in the dual topological unitarisation scheme. Hence, the amount of exchange-degeneracy violations along a Regge trajectory is a measure of the quark interchange force as a function of the angular momentum L. In Fig.1 the breaking of degeneracy is shown between the f and ρ trajectories and is seen to decrease rapidly as L increases. For this reason we believe that mesonic decays of diquoniums by quark interchange are also strongly suppressed when L is large.

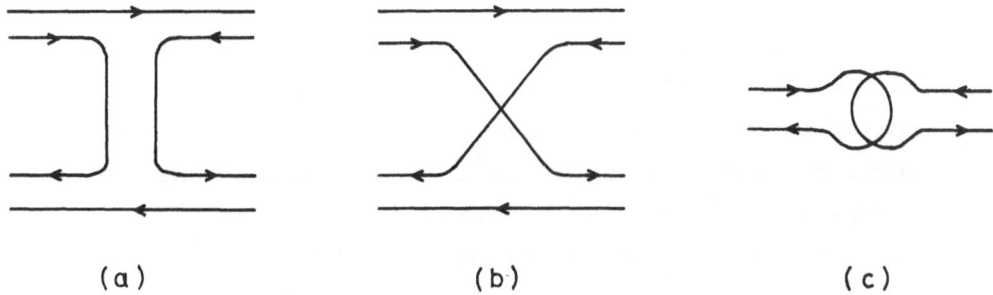

Fig.8. Quark-pair annihilation and quark interchange.

Another type of molecular transitions which may
contribute significantly to the decay of molecules with
a stable colour bond × is the following. The molecule
first acquires, by colour mixing, a component in an
isomesic configuration with an unstable colour bond, ×'
say, and then decays by one of the favoured modes of the
latter molecule [1]. For example an "M"-diquonium may
decay as follows:

$$(QQ)^6 - (\bar{Q}\bar{Q})^6 \rightarrow (QQ)^3 \not\!\!/ \ (\bar{Q}\bar{Q})^3 \rightarrow (QQQ)' + (\bar{Q}\bar{Q}\bar{Q})' \quad (37)$$

resulting in a $B\bar{B}$ final state. Obviously such a mode

is important only where L is not too large in view of
our basic assumption that colour mixing forces are
strongly L-dependent. For $L \sim 1$, however, it may lead
to $B\bar{B}$ widths for "M"-diquoniums of the order

$$\tan^2\theta \, . \, \Gamma(\text{"T"} \to BB)$$

which by (5) yields an estimate of several MeV.

A more likely molecular transition, however, for
stable-bonded molecules such as "M"-diquoniums is the
following [1]. A constituent partially overcomes the
angular momentum barrier (by quantum fluctuations for
example) moving slightly inwards so that for some
instant the molecule exists as a triatomic molecule,
thus:

$$(QQ)^6 \, \underline{\quad L \quad} \, (\bar{Q}\bar{Q})^{\bar{6}} \to (QQ)^6 \, \underline{\quad L' \quad} \, (\bar{Q})^{\bar{3}} \, \underline{\quad\quad} \, (\bar{Q})^{\bar{3}} \, . \tag{37'}$$

This then decays by rupturing the colour 3-bond so
exposed resulting in a cascade decay into a meson and
a shorter diquonium state of the same type:

$$(QQ)^6 \, \underline{\quad L' \quad} \, (\bar{Q})^{\bar{3}} \, \underline{\quad\not\quad} \, (\bar{Q})^{\bar{3}} \to (QQ)^6 \, \underline{\quad L' \quad} \, (\bar{Q}\bar{Q})^{\bar{6}} + (Q\bar{Q})' . \tag{38}$$

Although we believe this transition to be suppressed
compared with the normal cascade decays of for example
high-spin baryons because of the barrier effect against
the transition (37'), we have no way of reliably
estimating its width. We think, however, that it may
well be the dominant mode in molecules such as "M"-
diquonium for which all other normal channels are
effectively closed. Now like all cascade decays of high-
spin hadrons these should follow very regular patterns

being strongly governed by kinematics [17,18]. In order to minimize the angular momentum barrier it has to penetrate; the decay prefers a high Q-value and a small change in spins between the decaying resonances and the decay products. As a result the emitted meson has to be light being thus most frequently a pion, while the product resonance prefers to lie on the highest Regge trajectory which is consistent with selection rules as illustrated in Fig. 9.

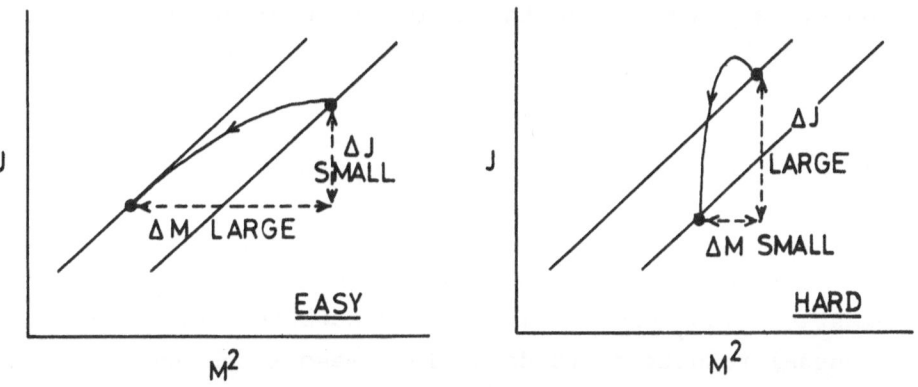

Fig.9. The kinematics of cascade decays for high-spin resonances.

Because of this, such cascade decay modes may surve as good signatures for "M"-diquoniums, as we shall demonstrate later.

Clearly the considerations we have given here for diquoniums can readily be generalized to more complicated molecules. Given our lists of chromions and chromionic bonds one can construct molecules and qualitatively predict their decay properties [4]. Many of them will have unusual decay characteristics which surve for this identification. One need not be restricted to the diatomic

molecules we have so far considered. With a stretch of
the imagination we may be tempted to build some large
linear chains or even the colour equivalent of the
benzene ring. However, as physicists we are not interested in complications for complications' sake. We are
more interested in probing and resolving the mysteries
of colour dynamics and for this it is more profitable
to study the simplest molecules: $(QQ) - (\bar{Q}\bar{Q})$, $(Q\bar{Q}) - (Q\bar{Q})$,
$(Q\bar{Q}) - (QQQ)$, $(QQ) - (QQ\bar{Q})$ etc. In particular, we shall
concentrate on diquoniums which are the colour chemists'
equivalents of the hydrogen atom for the atomic
physicists and the deuteron for the nuclear physicists.

V. DIQUONIUM PHENOMENOLOGY

We begin by summarising our conclusions in the
last section. Diquoniums are characterised by their
unusual reluctance to decay into mesonic channels in
spite of the absence of any obvious selection rules
and by their preference instead to decay into $B\bar{B}$
channels. They are therefore associated with the so-
called 'baryonium' states observed experimentally which
have just these characteristics [19]. There are two
types of diquoniums [1]. The colour 3-bonded molecules
or "T"-diquoniums are strongly coupled to $B\bar{B}$ and are
therefore broad. They are for this reason best studied
in the direct channel in $B\bar{B}$ formation experiments as
illustrated in Fig.10(a). The colour 6-bonded molecules
or "M"-diquoniums on the other hand are coupled to $B\bar{B}$
only via colour mixing. They prefer whenever possible
to decay by cascade via pion emissions into a lower

resonance of the same type. Because of this narrower
widths and suppressed coupling to B$\bar{\text{B}}$ they are best
looked for in the final states of production experiments,
as indicated in Fig. 10(b).

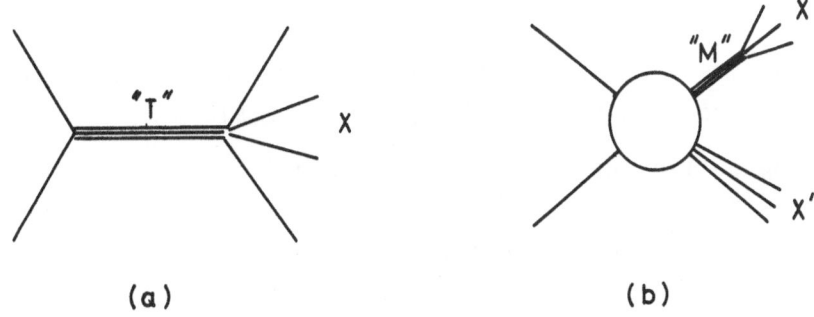

(a)   (b)

Fig.10. Formation and production of diquoniums.

Neglecting the small spin-oribt and tensor coupling
between the diquark ions the spectrum of "T"-diquoniums
is calculable in terms of one parameter L_o which represents the unknown confinement energy [1]. We may
now fix L_o by normalizing to for example the so-called
U-meson observed experimentally at 2.15 GeV with
$I^G(J^P) = 0^+(4^+)$. The masses of all "T"-diquoniums are
then determined. It should be emphasized that neither
the existence nor the quantum numbers of the U-meson
and indeed of all 'baryonium' states can as yet be
considered definitive so that our normalization may
have to be revised in view of future experimental
development. This should not affect our basic considerations but may alter the detail features of the
following phenomenology, which should therefore be
regarded as being tentative.

Consider first those diquoniums with only u,d quarks as constituents. The calculated spectrum [1] is shown in Fig. 11. One notices first a clustering of states at certain energies. This is due partly to the smallness of spin-orbit and tensor forces between di-quarks which were neglected in Fig. 11 and partly to the particular value of C as determined in (12). Since "T"-diquoniums are strongly coupled to $B\bar{B}$, these bunches of states are expected to show up as structures in $N\bar{N}$ cross sections. Existing information on these come from these main groups of experiment [19]:

(i) $\bar{p}p$ collisions above $\bar{p}p$ threshold.

(ii) $\bar{p}$ "n" collisions for off-mass-shell neutron "n" from deuterons close to the $N\bar{N}$ threshold.

(iii) $\bar{p}p \rightarrow \gamma + X$ for X below $N\bar{N}$ threshold.

The result obtained from all these groups are summarized in Fig.11. One observes that structures are seen indeed at or near the positions where the calculated spectrum clusters. In some cases, (S,T,U,V), the branching ratios of these structures into mesons have been measured and are found to be small, e.g. at T and U [19]

$$\frac{\sigma(\bar{p}p \rightarrow \pi\pi)}{\sigma(\bar{p}p \rightarrow \bar{p}p)} \lesssim .1$$

which is as one should expect for "T"-diquonium states.

Our models of course predicts not only the masses but also the quantum numbers of diquoniums. To check these it is insufficient just to measure the total cross sections. Fortunately, some good data exist in a number of special channels where not only the differential cross sections but also the polarisations are measured

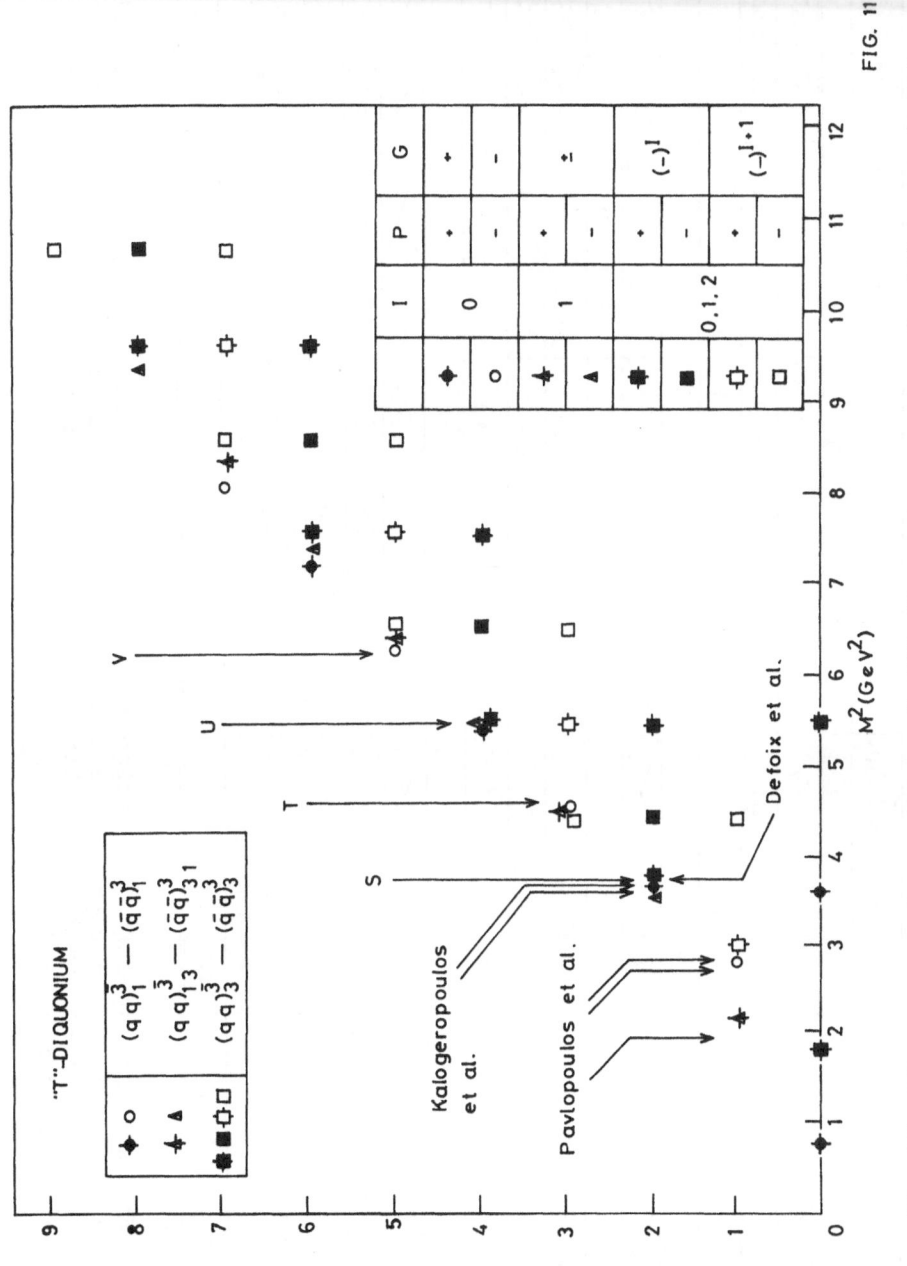

FIG. 11

Fig.11. The proposed spectrum of "T"-diquonium compared with experiment.

Energy Region →		**L**	**S**		**T**			**U**			**V**		
Traj. $I^G(J^P)$	$(q\bar q)^{\bar3}-(\overline{q\bar q})^{\bar3}$	$^{\bar3}_1{-}^{\bar3}_1$ (2)	$^{\bar3}_3{-}^{\bar3}_3$ (0)	$^{\bar3}_1{-}^{\bar3}_3$	$^{\bar3}_1{-}^{\bar3}_1$ (3)	$^{\bar3}_1{-}^{\bar3}_3$	$^{\bar3}_3{-}^{\bar3}_3$ (1)	$^{\bar3}_1{-}^{\bar3}_1$ (4)	$^{\bar3}_1{-}^{\bar3}_3$	$^{\bar3}_3{-}^{\bar3}_3$ (2)	$^{\bar3}_1{-}^{\bar3}_1$ (5)	$^{\bar3}_1{-}^{\bar3}_3$	$^{\bar3}_3{-}^{\bar3}_3$ (3)
Traj. $I^G(J^P)$		$0^+(2^+)$	$0^+(2^+)$	$1^-(2^+)$	$0^-(3^-)$	$0^-(3^-)$	$1^+(3^-)$	$0^+(4^+)$	$0^+(4^+)$	$1^-(4^+)$	$0^-(5^-)$	$0^-(5^-)$	$1^+(5^-)$
Traj. Mass		1.92	1.94	1.94	2.13	2.11	2.11	2.33	2.34	2.34	2.51	2.56	2.56
σ_{el} I,G,J,P		$J^{PC}{=}2^{++}$	$J^{PC}{=}2^{++}$ or 4^{++}		$I=0,1$			$I=0,1$			$I=0,1$		
$\sigma_{ch.ex.}$ Mass		1.936	1.9–2.0		2.155			2.345			2.5		
$pp{\to}\pi^+\pi^-$ $I^G(J^P)$		—	$[0^+(2^+)]$	×	×	×	$1^+(3^-)$	$0^+(4^+)$	×	×	×	×	$1^+(5^-)$
$pp{\to}\pi^+\pi^-$ Mass		—	$[2.0]$	×	×	×	2.150	2.310	×	×	×	×	2.5
$\bar p p{\to}\pi^0\pi^0$ $I^G(J^P)$		×	—	—	×	×	×	×	$0^+(4^+)$	×	×	×	×
$\bar p p{\to}\pi^0\pi^0$ Mass		×	—	—	×	×	×	×	2.35	×	×	×	×
$\bar p p{\to}K^+K^-$ $I^G(J^P)$		×	—	—	$0^-(3^-)$	×	$1^+(3^-)$	×	$?(4^+)$	×	$0^-(5^-)$	×	$1^+(5^-)$
$\bar p p{\to}K^+K^-$ Mass		×	—	—	2.15	×	2.15	×	2.34	×	2.5	×	2.5
$pp{\to}\pi^0\eta$ $I^G(J^P)$		×	—	—	×	×	×	×	×	$1^-(4^+)$	×	×	×
$pp{\to}\pi^0\eta$ Mass		×	—	—	×	×	×	×	×	2.32	×	×	×
$\bar p p{\to}\rho^0\pi^0$ $I^G(J^P)$		×	—	$1^-(2^+)$	×	×	×	×	×	×	×	×	×
$\bar p p{\to}\rho^0\pi^0$ Mass		×	—	1.940	×	×	×	×	×	×	×	×	×
Sum $I^G(J^P)$		$0^+(2^+)$	$0^+(2^+)$	$1^-(2^+)$	$0^-(3^-)$		$1^+(3^-)$	$0^+(4^+)$		$1^-(4^+)$	$0^-(5^-)$		$1^+(5^-)$
Sum Mass		1.936	~2.0	1.940	2.15		2.15	2.34		2.32	2.5		2.5
Total Mass		1.936	~2.0	1.940	2.15		2.11	2.34		2.32	2.5		2.5

Table IV. The proposed spectrum of peripheral "T"-diquonium states compared in detail with experiment. A cross denotes a state inaccessible to that particular channel because of selection rules. A line means that no experimental information is at present available.

so that some amplitude analysis can be made [20,21,22].
The total information available at present is summarized
in Table IV [23], most of which unfortunately cannot be
regarded as established. The theoretical spectrum is
rather rich and cannot at present be fully unravelled.
We concentrate therefore on the so-called 'peripheral'
states with $kl \sim m^{-1}$ which are expected to be most strongly
excited. These are listed also in Table IV. One notices a
close correspondence between theory and experiment both in
the masses and in the quantum number of the resonances
apart from several cases where theory predicts two
resonances with the same quantum numbers which experiment
is not yet able to resolve. In some cases there are further
checks on the resonance's preference for decay into certain
channels [23]. We note in particular an interesting
apparent isospin degeneracy of both the observed and
theoretical states in Table IV. Theoretically this is a
special feature of the diquark-antidiquark colour
molecule. Combining an $I = 1$ diquark with an $I = 1$
antidiquark we obtain molecules with $I = 0,1$ and 2 (of
which only $I = 0,1$ can be seen in $N\bar{N}$ channels). Now because the colour forces binding them together are
flavour-independent these isospin states are necessarily
degenerate. In contrast, if we were to consider these
resonances as nucleon-antinucleon bound states via nuclear
forces (in our present colour chemical language these
may perhaps be thought of as 'covalent molecules'), there
will be no reason for isospin degeneracy, since nuclear
forces,due to e.g. pion exchange, are strongly dependent
on isospin [24].

Next we turn to "M"-diquoniums which are more
interesting theoretically since their very existence is
a direct consequence of colour. Again we restrict our-

selves for the moment to molecules with only u,d quarks
as constituents, since most experimental information is
in this sector. The spectrum now depends on C_{qq} for
x = 6 for which we have no direct estimate. Let us
assume, however, as mentioned before, that C is weakly
dependent on x and use for C_{qq} the same value [12],
namely, $C_{qq} \sim 20$ MeV as for x = 3. We can then calculate
the whole spectrum again up to one constant parameter L_o.
Our predicted spectrum [1] is shown in Fig. 12, which
historically was obtained by normalising L_o to the ill-
fated state observed experimentally at 2.95 GeV [25]
about which you will hear more later. However, we could
have chosen some other of the existing experimental
states to normalise L_o without much change to the
spectrum, which is in any case only tentative.

Now "M"-diquoniums are supposed to be narrow
'baryonium' states found mainly in production. Experimentally there have been several such objects
reported in the literature [19] and these are
summarised in Fig. 12. Again one sees a pair correspondence between the observed states and the theoretical
predictions. One should emphasize,however, that most
of these states have been seen each in only are experiment under far less than ideal conditions and all
of them have yet to be confirmed. Further, none of them
(except perhaps the state at 2.204 GeV) has been
assigned quantum numbers so that no conclusion can yet
be drawn concerning the validity of our spectrum. None-
theless, let us proceed further with our analysis
assuming this spectrum. It will at least surve as an
exercise to illustrate how one can proceed when the
correct spectrum is known.

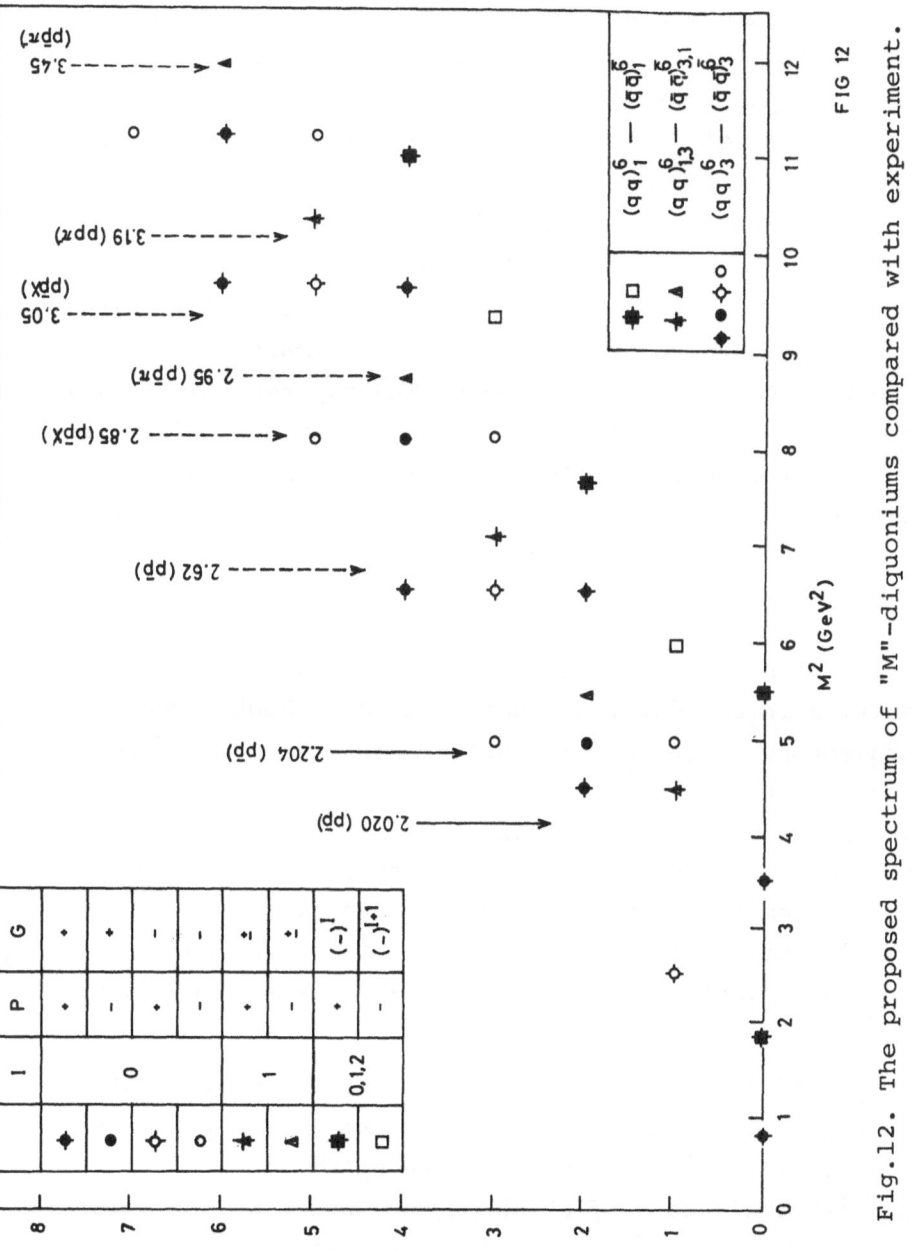

Fig.12. The proposed spectrum of "M"-diquoniums compared with experiment.

The spectrum in Fig.12 was calculated ignoring the
spin-orbit and tensor forces between the diquark and the
antidiquark which are believed indeed to be negligible
when L is large. For L $\sim$ 1,however, (and especially for
"M"-diquoniums because of some colour factors) these
forces may well become sizeable leading to splittings
of the order 100 MeV between diquoniums differing only by
the alignments of the diquarks' spins. In Fig.13 is shown
the detailed structure of the $(qq)_3^6 - (\bar{q}\bar{q})_3^{\bar{6}}$ (L = 1)
multiplet as calculated by Fukugita-Hanssen [15] using
a strength of the spin-orbit force estimated from the
spectrum of ordinary $q\bar{q}$ mesons. The apparent similarity
to the observed spectrum [26] may perhaps be fortuitous
but still deserves a closer examination by future experiment.

Next we consider the decay of "M"-diquoniums
assuming the spectrum of Fig.12. As discussed already
in section IV cascade decays prefer pion emissions resulting in another resonance lying on a high Regge
trajectory on the Chew-Frautschi plot. It may happen,
however, that the quantum numbers of an "M"-diquonium
are such that these cascade decays are unfavourable. In
that case the most likely mode (especially for low L)
would seem to be by mixing with "T"-diquoniums and decaying into $B\bar{B}$ final states. Take now the colour
molecules in turn according to the diquarks from which
they are formed:

(i) $(qq)_3^6 - (\bar{q}\bar{q})_3^{\bar{6}}$ which includes the leading trajectory
in Fig. 12 and has I = 0. In order to emit a pion such
a molecule must cascade to others with I = 1, the
highest trajectory of which on the Chew-Frautschi plot
being $(qq)_{1,3}^6 - (\bar{q}\bar{q})_{3,1}^{\bar{6}}$, as seem in Fig. 12. The latter,

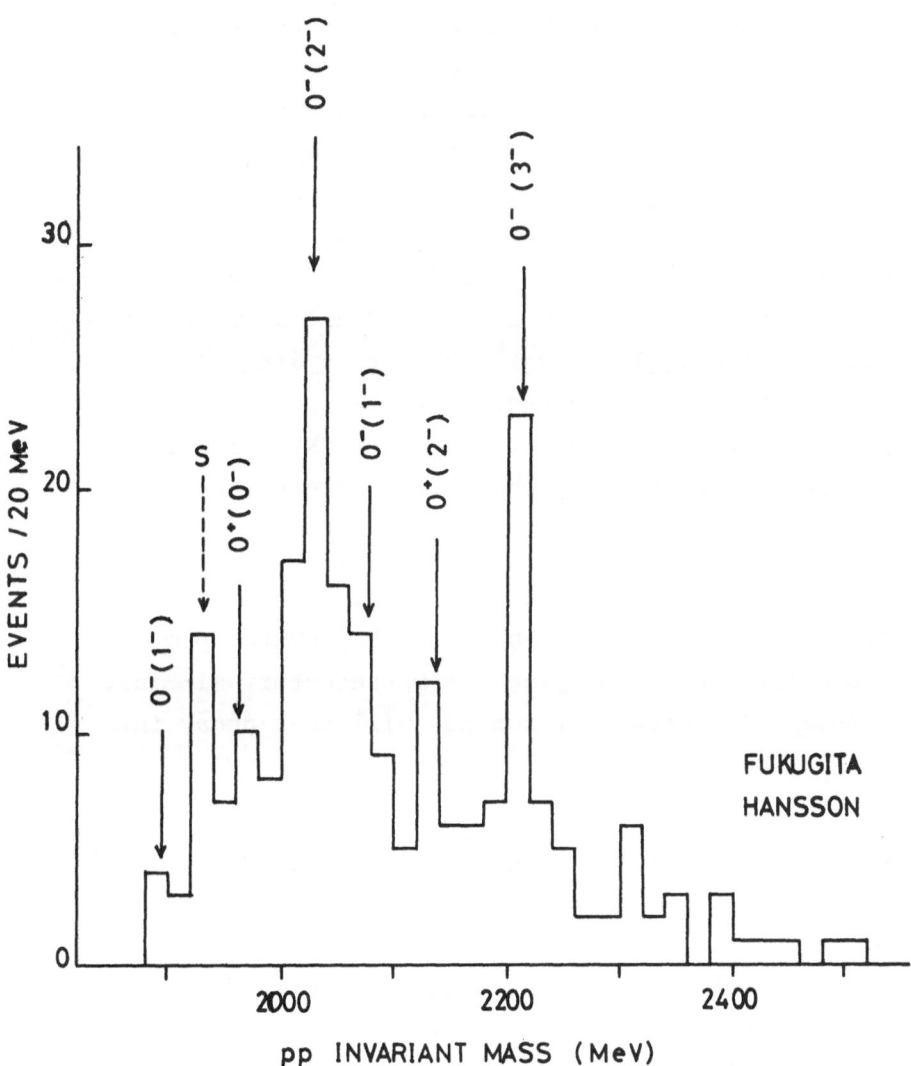

Fig. 13. The splitting of a diquonium spin multiplet
compared with experiment.

however, is considerably lower than that containing the decaying molecules. Therefore, by the arguments summarised in Fig.9, such cascade modes are suppressed easily by an order of magnitude as estimated from existing cascade models. Hence the lower L members of $(qq)_3^6 - (\bar{q}\bar{q})_3^6$ are likely instead to decay by the colour mixing mechanism [1]. In particular, there are two states reported in a $-p$ backward production experiment [26] with masses 2.020 and 2.204 GeV (see Fig.13). These were assigned tentatively in Fig. 12 and 13 to the L = 1 members of the $(qq)_3^6 - (\bar{q}\bar{q})_3^6$ trajectory with $I^G(J^P) =$ $= 0^-(2^-)$ and $0^-(3^-)$, respectively [23]. If so they are expected to decay preferentially into $N\bar{N}$, which is indeed the only channel in which they have yet been experimentally observed.

(ii) $(qq)_{1,3}^6 - (\bar{q}\bar{q})_{3,1}^{\bar{6}}$ with I = 1 and G = ±1. These molecules, in contrast, can favourably emit a pion and cascade into the high-lying I = 0 trajectory (i) just discussed. The product resonance will then decay into $N\bar{N}$, thus:

$$(qq)_{1,3}^6 - (\bar{q}\bar{q})_{3,1}^{\bar{6}} \rightarrow \pi + (qq)_3^6 - (\bar{q}\bar{q})_3^{\bar{6}} \tag{39}$$
$$\phantom{(qq)_{1,3}^6 - (\bar{q}\bar{q})_{3,1}^{\bar{6}} \rightarrow \pi + } \searrow N\bar{N}$$

resulting mostly in an $N\bar{N}\pi$ final state [1]. Moreover, existing models permit the estimation of branching ratios into the various cascade modes. Take for example the $(pp\pi^-)$ state (I ≥ 1) with mass 2.95 GeV reported in a $\pi-p$ forward production experiment [25]. This was assigned tentatively in Fig. 12 to the L = 3 number of this $(qq)_{1,3}^6 - (\bar{q}\bar{q})_{3,1}^{\bar{6}}$ trajectory with $I^G(J^P) = 1^+(4^-)$ [1,23].

Cascade models then suggest that its dominant modes
(39) will have the product resonance as either the
(2.020) or (2.204) discussed in (i) above or some
other states with L = 2 and mass ∿ 2.60 MeV [1,27].
In Fig.14 is shown the experimental p̄p spectrum from
the decay of the 2.95 state, which does seem to show
the structure expected.

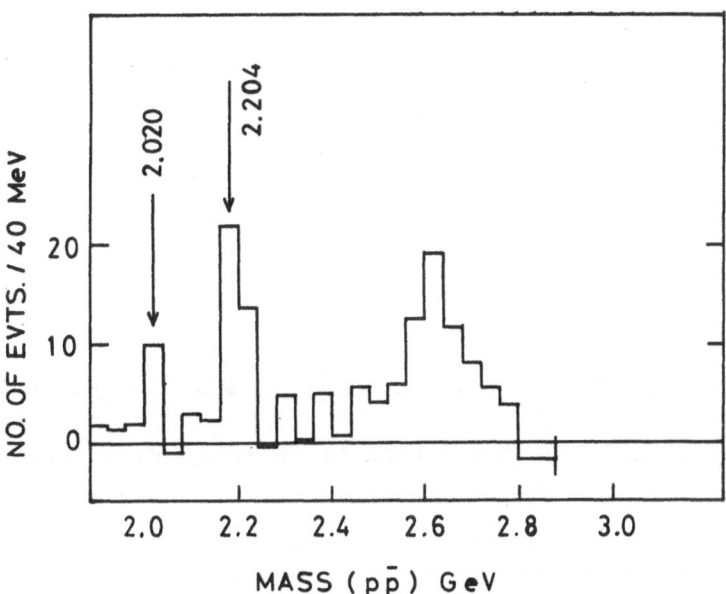

Fig. 14. The experiment mass spectrum of p̄p from the
decay of the (ppπ⁻) state at 2.95 GeV.

Therefore, if this were confirmed, it would have been a
dramatic verification of our theoretical considerations.
Unfortunately a later experiment with much higher
statistics but under somewhat different conditions [28]
have not found any signal at the mass value 2.95 GeV so

that even the existence of this state is at present
very much in doubt. Nevertheless, its purported behaviour is a good illustration of what one may expect.

(iii) $(qq)^6_1 - (\bar{q}\bar{q})^{\bar{6}}_1$ with I = 0,1,2. These will most
likely cascade first by π-emission into a resonance
of type (ii) above which then decays according to (39)
leading to the decay chain:

$$(qq)^6_1 - (\bar{q}\bar{q})^{\bar{6}}_1 \rightarrow \pi + (qq)^6_{1,3} - (\bar{q}\bar{q})^{\bar{6}}_{3,1} \qquad (40)$$

$$\hookrightarrow \pi + (qq)^6_3 - (\bar{q}\bar{q})^{\bar{6}}_3$$

$$\hookrightarrow N\bar{N}$$

and an $N\bar{N}\pi\pi$ final state. The fact that this trajectory
contains an exotic I = 2 component gives it added
interest. Experimentally these are some claims for
the possible existence of a $(p\bar{p}\,\pi^-\pi^-)$ state at
2.67 GeV [29] but these are not yet closely investigated.

Concerning the production of "M"-diquoniums one
question which naturally arises is that,if they are
indeed so weakly coupled to ordinary particles as
indicated by their narrow widths, then how can they
be produced? The answer [1] lies in the fact that this
suppression in coupling is not due to any selection
rule but only to kinematics in the form of an angular
momentum barrier inhibiting their decay. Their couplings
to or mixing parameters with ordinary hadrons are therefore strongly dependent on L, or on $M^2 = t$, being
small for large L (i.e. t large > 0), but sizeable for
L $\sim$ 0 (or equivalently t $\sim$ 0). Now the production of
"M"-diquoniums involves their mixing with ordinary hadron

trajectories at t > 0 and is therefore not subjected
to the same suppression as are their decays.

As an example, consider the production of "M"-
diquoniums from meson-baryon collisions in the forward
direction. This can be effected by the exchange of a
$Q Q \bar{Q} \bar{Q}$ trajectory as also for the production of a "T"-
diquonium state as illustrated in Fig.15.

(a)   (b)

Fig.15. The production of "M"-diquoniums.

The antidiquark in the diquonium is seen there to come
from a baryon in which it must have colour $\bar{3}$; hence if
its colour remains unchanged during the production
process only "T"-diquoniums can be produced. The colour
of the diquarks can be changed by exchanging a gluon
between them, for example. If this has to be effected
after the diquonium is produced i.e. on the diquonium
mass shell with $M^2 = t > 0$, as illustrated in Fig.15(a),
then the mixing is very small as argued in Section IV
above. However, the mixing can occur also in the exchanged trajectory, as depicted in Fig. 15(b), which has
t < 0, and there the mixing between colour $\bar{3}$ and colour 6

diquarks have no reason at all to be small. One way to see this is to remember that the exchange of a Regge trajectory may be considered as the exchange of a series of particles with varying spins [30] including in particular L = 0 states. For these our arguments against colour mixing in Section I breaks down completely. Indeed, in our language a L = 0 diquonium state is just another neutral chromion (or chromatom) and using the colour-magnetic interaction (7) one can calculate the spectrum as in other chromions [31,1]. One finds then that the mixing between the $(QQ)^3 (\bar{Q}\bar{Q})^3$ and $(QQ)^6 (\bar{Q}\bar{Q})^6$ states is indeed so large in general that it makes no sense at all to distinguish them. Now, a large colour mixing in the exchanged $QQ\bar{Q}\bar{Q}$ trajectory means of course that "M"-diquoniums can be produced by the process in Fig.15 with about the same abundance as "T"-diquoniums [1]. It is easily seen also that an analogous situation will prevail for the backward production of diquoniums via $QQQQ\bar{Q}$ exchange.

The arguments given above may be new to some and therefore sound suspect. It may perhaps help to quote as reminder a more familiar parallel case [16]. The decay of J/ψ into ordinary hadrons, e.g. ρπ , is strongly suppressed being of order only keV; yet J/ψ scatters from ordinary hadrons with typically hadronic cross sections of order mb, which may also appear at first sight to be mutually contradictory. The simple explanation is just that J/ψ decay involves the mixing between $c\bar{c}$ and $q\bar{q}$ pairs on the J/ψ mass shell $(M^2=t>>0)$, whereas its scattering concerns the same mixing but now for $t \leq 0$, as illustrated in Fig.16. The apparent contradiction then between J/ψ decay and J/ψ scattering merely

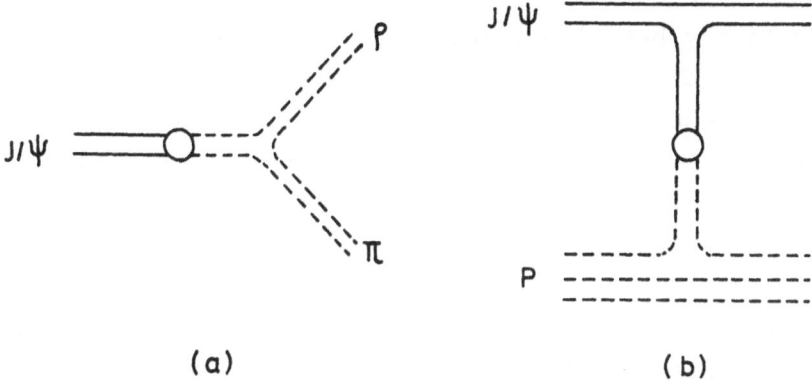

Fig. 16. The mixing of $c\bar{c}$ and $q\bar{q}$ states in J/ψ decay
and scattering.

says that the mixing angle has a strong t-dependence
and such a dependence can be deduced theoretically both
from the dual unitarised version of the Regge model [16]
and from the asymptotic freedom of QCD [32]. Our considerations outlined above for "M"-diquoniums, though
different in dynamical details, is conceptually very
similar.

Of course, if "M"-diquoniums can be produced only
by $QQ\bar{Q}\bar{Q}$ or $QQQ\bar{Q}\bar{Q}$ exchange, then their cross sections
will be very small at high energies because of the low
intercepts. However, these multiquark trajectories can
themselves mix with ordinary $Q\bar{Q}$ meson and QQQ baryon
trajectories via an OZI-violating mechanism very similar
to J/ψ scattering discussed above [1]. Again this mixing
is strongly t-dependent and becomes sizeable for t $\geq$ 0.
In particular, the mixing of the $QQ\bar{Q}\bar{Q}$ trajectory to the
Pomeron at t = 0 was estimated in dual unitarisation [33]

to give tan θ ∿.1 to .2. This means that diquoniums
(both "T"- and "M"-type) can be produced diffractively
[1] with cross sections roughly constant with respect
to beam energy but with values suppressed by a factor
$tan^2 θ$ ∿ 1 to 4% compared with the cross sections for
ordinary mesons of similar mass. No trustworthy estimate exists for the mixing of $QQQQ\bar{Q}$ to QQQ trajectories.
It appears,however, that because of certain colour-
magnetic effects due to (7) some $QQQQ\bar{Q}$ trajectories
may have comparable intercepts to ordinary QQQ
trajectories [9] and hence acquire large mixing to the
latter. In that case backward production of "M"-di-
quoniums may become comparable to that for "T"-di-
quoniums even for QQQ exchange.

Our arguments here for sizeable "M"-diquonium
cross sections do not of course apply to its formation
in the direct channel from $N\bar{N}$ collisions, Fig. 10(a).
The diquarks coming from the nucleons have colour 3
and have now problem combining to form "T"-diquonium.
However, the formation of "M"-diquoniums requires colour
mixing which must now be effected on its mass shell and
therefore remains small. One predicts then that "M"-
diquoniums have the very unusual property [1] of having
reasonable production cross sections but being very hard
to form. This property may be a valuable supplement to
their decay characteristics as a means for their identification.

It is encouraging to note that all narrow states
reported so far from experiment which we may wish to be
associated with "M"-diquoniums have been seen only in
production. In particular, for the (2.204) state there
exists an empirical upper limit to its formation cross

section from $N\bar{N}$ of < .4 mb, which is to be compared with
formation cross sections of 3-7 mb for S,T, and U (Fig.11)
which we associated with "T"-diquonium states [19]. Whereas the production cross section for (2.204) in the experiment where it was discovered was quoted as 21 ± 5 nb
as compared with 9 ± 5 nb for S which was also indicated
[26].

One serious practical problem in diquonium
spectroscopy, and in fact in the spectroscopy of all
multiquark systems, is the extremely richness of the
spectrum which makes it very hard to resolve the
quantum numbers of individual states. In this, important
advantages are gained by studying channels with restrictive
quantum numbers as already amply illustrated in Table IV.
In this respect one particularly interesting sector of
the diquonium spectrum is the states with $J^{PC} = 1^{--}$
which can couple directly to the photon. They can be
examined by formation in e^+e^- collisions or by diffractive
photoproduction of $p\bar{p}$ pairs. Since, if one belives Fig.12,
the whole diquonium spectrum with u,d quarks is now known,
it is an easy matter to select those states with $J^{PC} = 1^{--}$ [23,37]. These are shown in Fig.17 together with
all the relevant experimental information available [23].
One notices that the two (or three) narrow states observed by DCI and Frascati [34] below the $N\bar{N}$ threshold
seem unlikely to be radial excitations of ϕ and ω because of their narrow widths (ρ' has width $\sim$ 200 MeV).
They have, however, about the properties expected of
diquoniums below the $N\bar{N}$ threshold. On the other hand
above the $N\bar{N}$ threshold diquoniums exhibit themselves
as $N\bar{N}$ resonances. The $p\bar{p}$ spectrum from the scanty data
[35] on $\gamma p \rightarrow ppX$ available at present does show some

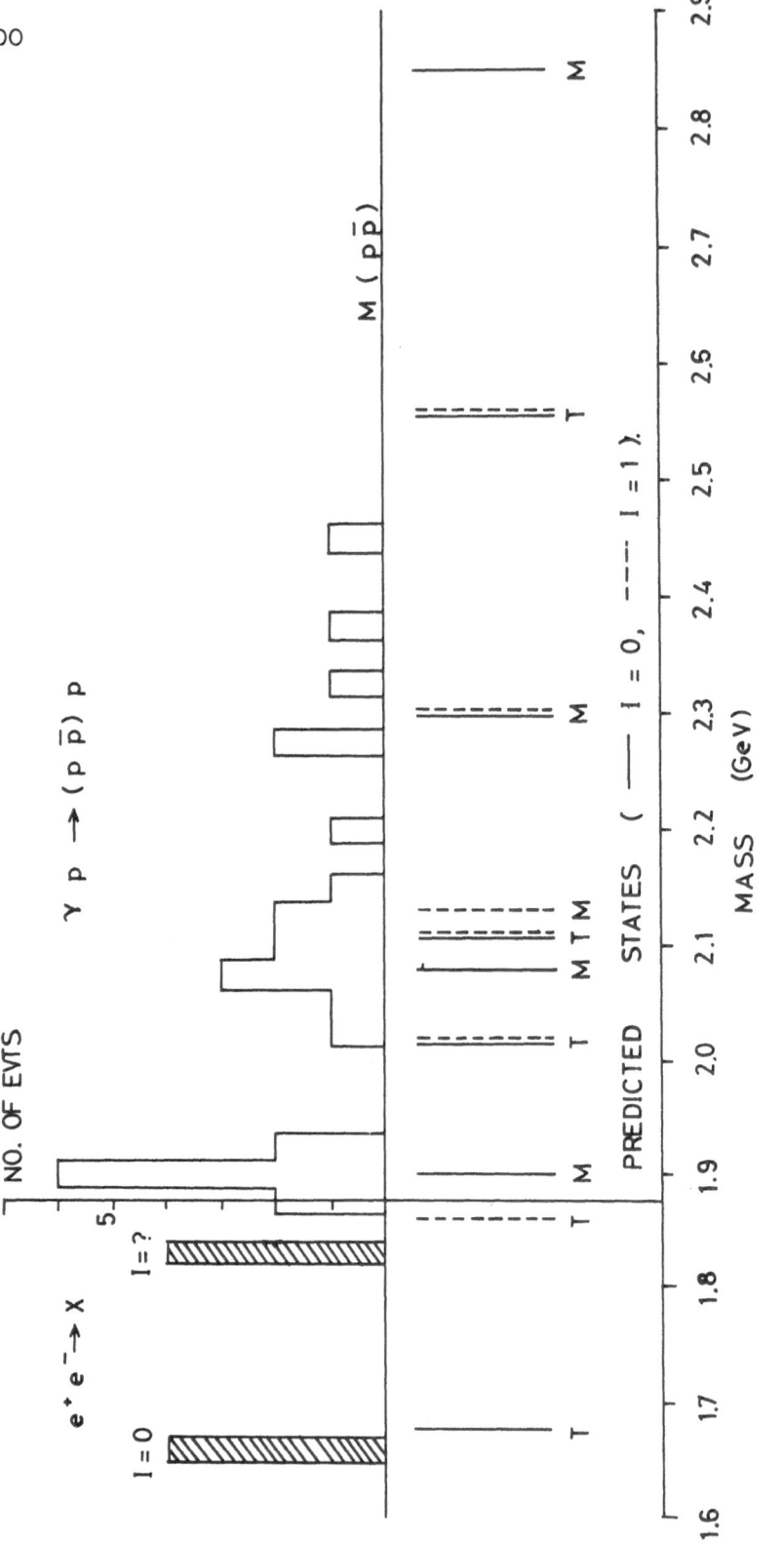

Fig. 17. The proposed spectrum of diquoniums with $J^{PC} = 1^{--}$ compared with experiment.

interesting correspondence with the predicted spectrum.

The extension of our discussions to diquoniums containing s and c quarks poses no theoretical problem. Given the parameters Δm_s, C_{qs}, C_{ss} in (15) and similar quantities for C the whole spectrum is completey predictable with no freedom [7], although because of the large extrapolation required the prediction for the charmed sector will be quite unreliable in normalisation. There are, however, little experimental data for comparison. Two exotic states with I = 3/2 and S = -1 were reported [36] whose masses fit well with predictions [7,38]; see Fig.18. We mention also that the lowest $(qs)-(\bar{q}\bar{s})$ diquonium state with $J^{PC} = 1^{--}$ is predicted at 2.15 GeV [37], while Frascati has seen a peak in $e^+e^- \to K^*X$ at 2.13 GeV [34] with $\Gamma \sim 30$ MeV, which is too narrow for a radial excitation of ϕ but reasonable for a diquonium below $\Lambda\bar{\Lambda}$ threshold.

VI. OTHER COLOUR MOLECULES

It is clear that similar considerations apply to other colour molecules but very little data exist as yet for phenomenology. There are some half-dozen tentative peaks seen experimentally which are possible candidates for $(QQ) - (Q\bar{Q}\bar{Q})$ or $(Q\bar{Q}) - (QQQ)$ [9]. Among these the most interesting is a $(K^+K^0_s\Sigma^-)$ state reported at 2.58 GeV [39] which was apparently not seen in the $\Lambda\pi^+\pi^-$ channel. The fact that it prefers to decay into $K^+K^0_s\Sigma^-$ instead of $\Lambda\pi^+\pi^-$ means presumably that it contains two s quarks and one $\bar{s}$ antiquark which implies at least five constituents, i.e. $qqss\bar{s}$. When data improve the spectroscopy of such $QQQQ\bar{Q}$ molecule may also be quite instructive.

402

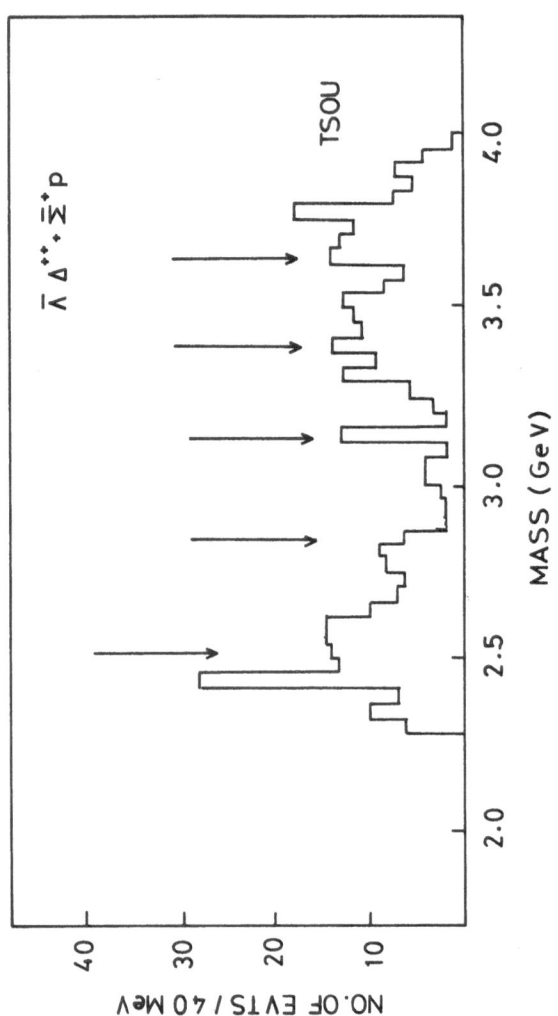

Fig. 18. Exotic diquoniums.

VII. CONCLUSION?

If all our predictions were verified by experiment, we would presumably have concluded as follows:

(a) From the existence of different types of molecules with very different properties (e.g. "T"-and "M"-di-quoniums) we would have deduced colour as a genuine degree of freedom.

(b) From the details of the spectrum, e.g. the dependence of the Regge slope on colour and the pattern of hyperfine splittings one would have found indications not merely for the number of colours as in $R(e^+e^-)$ but that the colour group is indeed (non-abelian) SO_3.

In addition one would have opened up a whole new branch of hadron spectroscopy of considerable interest and intricacy.

On the other hand, if our predictions were all wrong, one should not pronounce immediately that colour dynamics is dead, since in order to arrive at our results further assumptions were made which themselves may well be wrong. However, one has perhaps to regard to colour scheme with some suspicion, since similar assumptions have been made, though in a milder form, to claim successes for colour dynamics elsewhere.

Unfortunately the experimental situation is such that a firm conclusion either way is premature but, to my perhaps biased mind, the situation does appear at present to be somewhat encouraging for colour enthusiasts.

REFERENCES

[1] Chan Hong-Mo and H. Høgaasen, Phys. Lett. 72B (1977) 121; Nucl. Phys. B136 (1978) 401.

[2] R.L. Jaffe, Phys. Rev. D17 (1978) 1444.

[3] This is a paraphrase of an argument given by R.L. Jaffe in a conversation with the author over coffee.

[4] Chan Hong-Mo et al., Phys.Lett. 76B (1978) 634; see also Chan Hong-Mo, Proc. of the XIX. International Conference on High Energy Physics, Tokyo (1978).

[5] A. de Rujula, H. Georgi and S.L. Glashow, Phys. Rev. D12 (1975) 147.

[6] T. De Grand, R.L. Jaffe, K. Johnson and K. Kiskis, Phys. Rev. D12 (1975) 2060.

[7] Tsou Sheung Tsun, Nucl. Phys. B141 (1978) 397.

[8] M. Fukugita, T.H. Hansson, and K. Konishi, Phys. Lett. 74B (1978) 261.

[9] H. Høgaasen and P. Sorba, Nucl. Phys. B145 (1978) 119; M. De Crombrugghe, H. Høgaasen and P. Sorba, CERN preprint TH 2537 (1978), to appear in Nucl.Phys.B.

[10] H. Høgaasen and Chan Hong-Mo, in preparation.

[11] Chan Hong-Mo, M. Fukugita, T.H. Hansson, H. Høgaasen, unpublished.

[12] K. Johnson and C. Thom, Phys. Rev. D13 (1976) 1934.

[13] K. Konishi, private communication.

[14] Chan Hong-Mo and H. Høgaasen, Phys. Lett. 72B (1978) 400.

[15] M. Fukugita and T.H. Hansson, Rutherford Laboratory preprint RL-78-101 (1978).

[16] Chan Hong-Mo and Tsou Sheung Tsun, in "Many Degrees of Freedom in Particle Theory" ed. H. Satz, Plenum Publishing Corporation (1978) 83.

[17] Chan Hong-Mo and Tsou Shung Tsun, Phys. Rev. D4 (1971) 156.

[18] C. Quigg and F. von Hippel, Phys. Rev. D5 (1972) 624.

[19] L. Montanet, Vth International Conference on Experimental Meson Spectroscopy, Boston (1977); CERN preprint EP/PHYS 77-22; Proc. XIII. Rencontre de Moriond "Phenomenology of Quantum Chromodynamics" ed. J. Tran Thanh Vanh, Editions Frontières, France (1978).

[20] A.A. Carter et al., Phys. Lett. 67B (1977) 117.

[21] A.A. Carter, Proc. XIII. Rencontre de Moriond, ibid.; and private communication (1978).

[22] R.S. Dulude et al., Phys. Lett. 79B (1978) 329, 335.

[23] Chan Hong-Mo, CERN preprint TH 2540 (1978); and Rutherford Laboratory preprint RL-78-089 (1978), to appear in Proc. IV European Antiproton Symposium, Barr (Strasbourg) 1978.

[24] J.-M. Richards, contributed paper to appear in Proc. IV. European Antiproton Symposium, Barr (Strasbourg) 1978.

[25] C. Evangelista et al., Phys. Lett. 72B (1977) 139.

[26] P. Benkhieri et al., Phys. Lett. 68B (1977) 483; 81B (1979) 380; J. Six, Proc. XIII. Rencontre de Moriond, ibid. (1978).

[27] M. Fukugita and T.H. Hansson, Rutherford Laboratory preprint RL-79-034 (1979).

[28] T.A. Armstrong et al., CERN preprint EP/PHYS 79- (1979).

[29] A. Rodgers, V. International Conference on Experimental Meson Spectroscopy, Boston (1977).

[30] L. van Hove, Phys. Lett. 24B (1967) 183.

[31] R.L. Jaffe, Phys. Rev. D15 (1977) 267, 281.

[32] T. Appelquist and H.D. Politzer, Phys. Rev. Lett. 34 (1975) 43.

[33] Chan Hong-Mo and Tsou Sheung Tsun, Nucl. Phys. B118 (1977) 413.

[34] S.D. Protopopescu, Brookhaven Laboratory preprint BNL-23612 (1978); G. Cosme et al., Orsay preprint LAL-78/32 (1978).

[35] D. Aston et al., contribution to XIX.International Conference on High Energy Physics, Tokyo (1978).

[36] T.A. Armstrong et al., Phys. Lett. 77B (1978) 447.

[37] Tsou Sheung Tsun, in preparation as Oxford Mathematical Institute preprint (1979).

[38] Tsou Sheung Tsun, contributed paper to appear in Proc.IV. European Antiproton Symposium, Barr (Strasbourg).

[39] M. Mazzucato et al. CERN preprint EP/PHYS 78-24 (1978).

Acta Physica Austriaca, Suppl. XXI, 407—483 (1979)
© by Springer-Verlag 1979

INTRODUCTION TO QUARK CONFINEMENT IN QCD[+]

by

H.JOOS

Deutsches Elektronen-Synchrotron DESY

Notkestraße 85, D-2000 Hamburg 52

CONTENTS

[+]Lectures given at the
XVIII. Internationale Universitätswochen für Kernphysik,
Schladming, Austria, February 28 - March 10, 1979.

3. WILSON CRITERION

    3.1. Wilson's criterion for quark confinement

    3.2. Strong coupling approximation

    3.3. The continuum limit; renormalization group

    3.4. Semi-classical approximation

4. HAMILTONIAN APPROACH

    4.1. Hamiltonian in gauge theories

    4.2. Linear potential and string breaking in
        2-dimensional models

1. INTRODUCTION

1.1. Quantum Chromo-Dynamics [1]

Because of

(i)   the explanatory power of the quark model in the
phenomenological analysis of experimental elementary
particle physics,

(ii)   the unprecedented precision of the predictions of
quantum electrodynamics (QED)

(iii) the esthetic appeal of the geometry of gauge fields

we consider nowadays a gauge theory of interacting quarks
as the most promising design of the dynamics of hadrons.

In this "quantum chromo-dynamics" QCD, which is
very similar to QED, the interaction of the quarks with
the gauge field - the gluons - is introduced in the
Dirac equation

$$(D_\mu \gamma^\mu + iM) \psi = 0 \tag{1}$$

via the covariant derivative

$$D_\mu = \partial_\mu - ig\, A_\mu^a\, T_a \ , \qquad T_a = \frac{1}{2}\, \lambda_a \quad . \qquad (2)$$

The colour currents of the quarks $j_\mu^a(x) = \frac{1}{2}\bar{\psi}\lambda^a\gamma_\mu\psi(x)$
act as a source term for the gluon-field tensor $F_{\mu\nu}^a(x)$
in a Maxwell-type field equation

$$D^\mu\, F_{\mu\nu}^a(x) = G j_\nu^a(x)$$

$$F_{\mu\nu}^a(x) = \partial_\mu A_\nu^a(x) - \partial_\nu A_\mu^a(x) + g f_{bc}^a\, A_\mu^b(x) A_\nu^c(x) \qquad (3)$$

(for complete definitions, see sect.2.1).

These classical field equations can be derived via
the action principle from a Lagrange density

$$L = -\frac{1}{4}\, F_{\mu\nu}^a\, F_a^{\mu\nu} + \bar{\psi}(i\gamma^\mu D_\mu - M)\psi \qquad (4)$$

which allows also the most compact formulation of the
quantum theory of these fields: The generating functional
of the Green's functions: $T\{J_\mu;\bar{n},n\} = Z\{J_\mu,\bar{n},n\}/Z\{0\}$ is
formally represented as a functional-Fourier transform
of the "action exponential" with respect to the classical
field configurations

$$Z\{J_\mu;\bar{n},n\} = \int\dots\int D[A_\mu]D[\psi]D[\bar{\psi}]e^{i\int dx\,(L+L_{ga}+J_\mu A^\mu+\bar{\psi}n+\bar{n}\psi)} \qquad .(5)$$

The renormalizability of QCD [2] allows an evaluation of
(5) as a formal power series in the strong coupling
constant $\alpha_s = g^2/4\pi$, where the different terms are
described by the well-known Feynman diagrams of QCD.
Since at large momenta the effective coupling constant

is small ("asymptotic freedom"), this perturbation
expansion is meaningful and it describes successfully
gluonic radiative corrections to the parton model. The
success of these applications of QCD is discussed in
other Lectures of this school. On the other hand the
understanding of quark confinement requires probably
a non-perturbative evaluation of Equ.(5). Methods of
statistical mechanics for the evaluation of "partition
sum" $Z\{O\}$ seem to be promising [3].

1.2. The problem of "Quark and Gluon Confinement"

In contrast to QED with its electrons, myons, and
photons quanta with the properties of the fundamental
fields of QCD: ψ, $\bar{\psi}$, A_μ^a are not seen. There are no
indications on

(i)   quarks, i.e. particles with fractional flavour
        charges.
(ii)   gluons, i.e. massless, strongly interacting
        (flavour neutral) vector particles.
(iii) open colour, i.e. colour multiplets of particles.

The experimental evidence for these three statements is
of different quality - varying in time! The explanation
of these facts is the general problem of "Colour
screening and Quark confinement".

The problem of using the quark model without
quarks is nearly as old as the model itself. On the
phenomenological level the following physical notions
were used more or less successfully:

(iv) a large free quark mass, which in the hadron bound

states is compensated by a strong potential energy. (In Schladming it might be allowed to mention our model [4]; based on this idea - in a relativistic field-theoretical framework - which was presented here in 1973),

(v) a confining potential [5], which is increasing at large distances. The bound states of quarks with large "effective masses" (charm quark, bottom quark) allow the phenomenological determination of a non-relativistic potential of this type [6]:

$$V(r) = \begin{cases} - \dfrac{8\pi}{25r}(\log\dfrac{\mu}{r})^{-1} + c_1 \\ b \, \log(r/c) \\ a \cdot r \end{cases} \quad \text{for} \quad \begin{cases} r < R_1 = 0.072 \text{ f} \\ \\ R_1 < r < R_2 \\ \\ r > R_2 \approx 1 \text{ f} \end{cases} \quad \mu = 05 \text{ f.}$$

$$a = 0.8 \text{ GeV/f} .$$

A comparison of charmonium and bottonium allows a check on the flavour independence of $V(r)$, i.e. the universality of a. (For details see the lecture by M. Krammer).

(vi) The "Jet hypothesis" [7], states that the quark and gluon quanta seen as partons in inclusive deep-inelastic lepton scattering, show up as jets in the final states. There is good evidence for quark jets [8], barely an indication of gluon jets (see the lecture by G. Flügge), no information on the existence of "glue balls" as leading particles in gluon jets. Here we denote by a glue ball the lowest mass configuration which is flavourless and is showing indications of a large gluonic component.

Assuming the validity of the fundamental equations of
QCD <u>quark confinement is almost a mathematical problem.</u>
It is certainly the most crucial and challenging problem
in theoretical physics in these days. It is not yet
solved!

1.3. The chromoelectric Meissner effect [9]

There is a general folklore about the field-
theoretical picture of quark confinement in QCD.

For introducing this picture we remark that in
a gauge theory the statements 1.2 (i) - (iii) are not
completely independent! Because of the generalized
Gauss law (contained in 1.1. (3)) the gluonic flux
through a closed surface is equal to the colour charge.
Hence the absence of long-range (zero mass) coloured
fields (gluons) implies the non-existence of coloured
particles [10]. This effect is called "octet-colour
screening" by a gluonic mass gap [11]. The - yet unknown - mechanism, which provides the mass gap is
sometimes called <u>"gluon-dynamic Higgs mechanism"</u>, because it is supposed to do the job of the well-known
Higgs mechanism without introducing explicitly Higgs
fields (which probably would destroy asymptotic freedom).

The non-existence of coloured particles does not
imply that there are no quarks. The screening mechanism
may bleach the quarks [12] (make them colourless),
without destroying their fractional flavour charges.

However, an "Higgs-ersatz" made of octet-charged
objects cannot screen colour charges linked to

triality = 0 [13]. There are still long-range forces
between quarks. General energetic arguments [14] show
that this long-range gluon fields between quarks form
flux tubes of a diameter twice the Compton-wave length
of the mass-gap. This mechanism is called the "chromoelectric Meissner effect". The constant energy per
string length leads to a linear potential between
quarks and thus explains the confinement of quarks.

The conjecture of the chromoelectric Meissner
effect in QCD is nourished by the following theoretical
and phenomenological experience:

(i)   The crude calculations of gauge-invariant
      correlation length ("Wilson-loop") in the strong
      coupling limit of the lattice approximation to
      QCD [15].
(ii) The screening of magnetic charge in super-
      conductors of 2. kind, which leads to the
      formation of observed vortex strings [16].
(iii) The string interpretation of the dual resonance
      model [17].
(iv) The relative success of the MIT-bag model [18].

In order to get a feeling for the physical dimensions
of the string we interpret the linear potential
constant

$$a = 0.8 \text{ GeV/f} \approx 4 \text{ f}^{-2} \approx 0.16 \text{ GeV}^2 \tag{1}$$

in a classical consideration as field energy of unit
length. If the gluon flux between the quarks is
concentrated in a tube of diameter d, then a classical
estimate results in a string diameter between .5 and 1 f.

414

This order of magnitude of d should also give an estimate of the glue-ball mass m between .5 and 1 GeV.

Of course, this estimate should only give a rough orientation. In this physical string picture one should not forget that strings can break due to pair creation of (light) quarks.

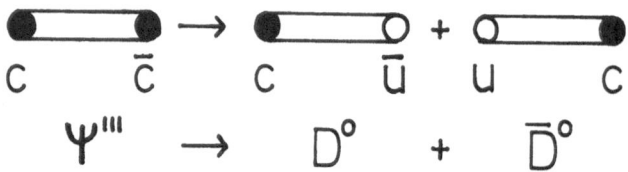

From the extension of the wave functions of the radial excitations of charmonium, which become unstable, we infer that the strings are stable up to a length of 2-3 fermi (see Fig. 21 by M. Krammer). - A string of a diameter of 1 f and a length of 3 f does not look essentially different from a bag!

For the theoretical study of string breaking one might consider 1+1-dimensional models. Here gauge fields form strings by dimensional restrictions! The abelian gauge theory of massless quarks (Schwinger model) [19] gives an explicit example where quark and antiquark pairs are confined in small string bits. These colour-less configurations are the only physical particles in this model, (Sect. 4.2).

1.4. The physical space-time continuum - a dynamical problem

Technically many of the theoretical discussions
of the non-perturbative effects considered with regard
to quark confinement are based on lattice approximations.
I mention the applications of methods of statistical
mechanics. Since I give much weight on this aspect,
I would like to make some general remarks explaining
why I consider this procedure also appropriate from the
physical point of view.

It is well-known that the quantum theory of
fields in continuum space-time leads to the so-called
"classical divergence problem". This is solved by the
renormalization procedure of quantum field-theory [20],
which consists of the following steps

(i)   regularization, i.e. mutilating the continuum
         by a momentum cut-off
(ii)   subtraction, i.e. combining bare action and
         reaction to a physically determined quantity
(iii)   renormalization, going to the continuum by
         keeping the subtracted quantities finite and
         fixed.

The invention of renormalizable field theory is the
discovery of the fact that the general topology of
physical space-time at small distances is a dynamical
problem. This discovery should be compared with that
of general **relativity,** namely that the structure of
the space-time continuity at large distances is also
a dynamical problem.

From this point of view the following coincidence
appears significant.

(iv) Classical non-abelian gauge theories describe a
     <u>differential</u> geometry of continuous space-time
     manyfold equipped with a local internal symmetry
     structure, ("coloured space-time") [21]
(v) Quantized non-abelian gauge theories have a
     physical continuum limit, which is particularly
     homogeneous at small distances: they are
     asymptotic free [22].

Is this a hint on a not yet completely understood
relation between differential geometric field theories
and the dynamical continuum-structure of physical
space-time? It would impose that the regularization
procedure should preserve the essential geometric
structure of gauge-field theories. (Usually one says
the regularization procedure should be gauge invariant).
Precisely this is the outstanding feature of the lattice
approximation that it provides a regularization procedure
preserving the differential geometric structure in the
spirit of a smooth simplex approximation to a smooth
manifold.

    It is the **merit** of K. Wilson [23] to emphasize
this point of view by constructing geometric lattice
gauge theories and introducing the general renormalization
group approach by which an "uncut-off" field theory is
obtained as a limit of cut-off theories - independent
of perturbation theory.

1.5. Some methodological remarks

    Quark confinement in QCD is an unsolved problem!
The central question is to understand the physical
vacuum as a medium in which a "gluon – dynamic Higgs

mechanism" leads to octet-gluon screening via a gluonic
mass gap, and in which consequently quarks are confined
by the chromoelectric Meissner effect.

To an unsolved problem one can give only a
guided tour through the theoretical environment in
which one searches for the solution and one can try
to show the point, which is closest to it. In this
spirit we shall consider other geometric field theories
similar to QCD, possibly in lower dimensions, possibly
of a simpler structure, possibly in lattice approximation.
The path we follow in these lectures might be seen from
the table of contents.

2. CLASSICAL FIELDS

2.1. Geometry of gauge fields [24]

Gauge-field theories are the most prominent members
of a class of field theories which describe an intimate
connection between the geometries of space-time and of
internal symmetry. In the following we review shortly
the geometric meaning of gauge theories in order
... to reveal the neighbourhood of classical gauge fields
and general relativity,
... to prepare the geometric formulation of lattice
gauge theories,
... to introduce a concise notation for the gauge-field
equations and the special solutions which play possibly
a rôle in quark confinement.

(i) Coloured space-time

(a) Usually we consider space-time as a 4-dimensional

manifold R consisting of points x depending analytically
on local coordinates: $x^1, \ldots, x^4$. (We hope the use of
the same letter for the point and its coordinates does
not lead to confusion). The points x(x') are equipped
with tangent spaces T_x spanned by the tangent vectors
$e_i(x)$ of the local coordinate lines.

Other differential geometric objects are based
on the tangent spaces. Such objects are: line elements
dx^i, differential p-forms $\omega = \omega_{i,\ldots,k} dx^i \wedge \ldots \wedge dx^k$ (i.e.
area elements, volume elements, etc.), symmetric tensor
fields $V = v^{i,j,\ldots k}(x) e_i(x) \times \ldots \times e_k(x)$ etc. In a manifold R with a metric $g_{ik}(x)$ vector fields and anti-
symmetric tensor fields are covariantly related to
differential forms: $\omega^v_{i_1 \ldots i_p}(x) = g_{i_1 j_1}(x) \ldots g_{i_p j_p}(x) \cdot v^{j_1 \ldots j_p}(x)$ (summation convention!). Under these
circumstances the calculus of differential forms re-
presents a particularly concise form of tensor analysis
which we want to use in the treatment of gauge-field
equations. For this we repeat the basic definitions and
formulas on exterior differentials and integration of
differential forms [25]:

The exterior differential dω of a p-form ω is
the (p+1)-form

$$d\omega = \frac{\partial}{\partial x^i} \omega_{i_1 \ldots i_p}(x) dx^i \wedge dx^{i_1} \wedge \ldots \wedge dx^{i_p}$$

with the rules for the application of d to a wedge
product of forms:

$$d(\omega \wedge \xi) = d\omega \wedge \xi + (-1)^p \omega \wedge d\xi$$

$$dd\omega = 0 \quad . \tag{1}$$

The <u>integral</u> $r_L\int^r\omega$ <u>of a r-form</u> $^r\omega$ on a r-dimensional submanifold ^rL is a generalization of the path integral on the path $^1L = \{x(s)\,|\,0 \leq s \leq 1\}$:

$$\int_0^1 {}^1\omega_i(x)\,\frac{dx^i(x)}{ds} = {}_1\int_L {}^1\omega \tag{2}$$

to oriented area integrals, 3-volume integrals etc.
For this integral we have the general Stokes' theorem

$$\int_{\Delta L}\omega = \int_L d\omega \tag{3}$$

where ΔL denotes the boundary of the submanifold L.

(b) We may consider <u>internal symmetry</u> as an enlargement of the geometric structure of space-time by adjoining a local "charge space H_x" to each point x. In the most interesting cases, the symmetry group $G = \{g\}$ of the charge spaces is a compact Lie group like SU(2), SU(3), generated by "charges" $\{T_a\} \equiv G$ with the commutation relations

$$[T_a, T_b] = i\,f_{ab}{}^c\,T_c \quad . \tag{4}$$

Under a symmetry operation the elements $\psi(x) \in H_x$ get transformed by a linear unitary representation D[G] of G: $\psi' = g\psi$. A field $\psi(x)$ in such a space-time with internal symmetry is then a function of the points x with values $\psi(x) \ni H_x$, and the most general symmetry operation on a field, a gauge transformation, is

$$\psi'(x) = g(x)\psi(x) \tag{5}$$

where $g(x) \in D[G]$ depends smoothly on x.

The quark fields in QCD are fields in such a space-time with internal colour symmetry: "coloured space-time". Each point x is provided with a tangent space T_x and a local 3-dimensional colour space H_x in which the colour charges T_a of $SU_c(3)$ are represented by the Gell-Mann matrices $T_a \rightarrow \frac{1}{2}\lambda_a$.

In the investigation of special problems of QCD one also studies models where G is not a Lie group; example: Z^n-gauge models [26], $G \approx Z^n$ (the cyclic group of order n) is the symmetry group of a space-time lattice. In another class of models the local "charge space" is not a linear space but a homogeneous space, on which G acts transitively [27]; example: The SO(3)-non-linear δ-model, where the fields $\psi(x)$ are points on a sphere and (5) denotes a rotation of this sphere. In later sections we shall use also such "simplifying generalizations" in order to illustrate special points of the confinement problem.

(ii) Connection and curvature

In coloured space time each point x is provided with a tangent space T_x and a local charge space H_x. It is familiar how a linear connection Γ^i_{mk} describes the <u>parallel displacement</u> of a tangent vector $v_x \in T_x$ to the tangent space of an infinitesimal close point x': the components of the displaced vector $v^i + \delta v^i$ in $T_{x'}$ are expressed by the components v^i of the vector in T_x and by the coordinates δx^i of the displacement $x(x^i) \rightarrow x'(x^i + \delta x^i)$

$$\delta v^i = \Gamma^i_{mk} \, \delta U^m \, v^k \quad . \tag{6}$$

The parallel displacement of a vector ψ_x of the colour space is described in the same way by a gauge connection

$$\delta\psi^a = A^a_{\mu b}\ \delta\mathbf{x}^\mu\ \psi^b \qquad (7)$$

where $A^a_{\mu b}$ is an infinitesimal symmetry transformation connected to the infinitesimal displacement $\delta\mathbf{x}$. It can be expanded with respect to the elements T_ℓ of the Lie algebra G of charges, Equ.(4),

$$A^a_{\mu b} = g.A^\ell_\mu(x)\ (T_\ell)^a_b \qquad (7')$$

$A^\ell_\mu(x)$ is the gluon potential in QCD, g the coupling constant. In a vector field we may compare the vectors $v(x)$, $v(x')$ at neighbouring points x,x' by forming the difference in T_x, between $v(x')$ and $v^*(x')$, the parallel transported $v(x)$. In a formula for components:

$$v^i(x+\delta x) - v^{i*}(x+\delta x) = (\frac{\partial v^i}{\partial x^k} - \Gamma^i_{k\ell}v^\ell)\delta x^k \equiv (D_k V)^i \delta x^k \quad . \qquad (8)$$

$D_k V$ is called the <u>covariant derivative</u> of the vector field. The covariant derivative of a charged field $\psi(x)$:

$$(D_\mu\psi)^a = (\partial_\mu - iA^a_{\mu b})\psi^b \qquad (9)$$

has the corresponding meaning with respect to the gauge connections of the charge spaces. Here we assume the space-time coordinates as cartesian ($\Gamma^i_{mn} = 0$) otherwise ∂_μ must be substituted by D_k of Equ.(9).

We may bring the definition (9) in a coordinate-independent differential form

$$D\psi = (D_\mu \psi) dx^\mu = d\psi - iA\psi \tag{10}$$

$A \equiv A_\mu dx^\mu$ is called the connection form. We have to consider A as a matrix-valued form, which transforms the charge vectors; or because A is a linear combination of charges, Equ.(7'), we may also consider A as a Lie-algebra-valued differential form. The appearance of such forms is characteristic for the differential geometry of coloured space-time, in which the structural elements of the space-time manifold and of the internal symmetry group get strongly intertwined. The general expression of a Lie-algebra-valued p-form is

$$^p\Omega = \omega^\ell_{\mu_1 \cdots \mu_p}(x) \; T_\ell dx^{\mu_1} \wedge \ldots \wedge dx^{\mu_p} \equiv \omega^\ell(x) T_\ell \tag{11}$$

the bracket product is defined as

$$[\Omega,\Xi] = i f^\ell_{ik} \; T_\ell \; \omega^i \wedge \xi^k \quad . \tag{12}$$

Calculation with such forms is straight-forward. We mention the product rule for exterior differentiation

$$d[\Omega,\Xi] = [d\Omega,\Xi] + (-1)^q [\Omega,d\Xi] \tag{13}$$

and the generalized Jacobi identity:

$$[\Omega,\Xi] = (-1)^{1+pq} [\Xi,\Omega] \tag{13'}$$

$$(-1)^{ps}[\Omega,[\Xi,\Psi]]+(-1)^{qp}[\Xi,[\Psi,\Omega]]+(-1)^{sq}[\Psi,[\Omega,\Xi]]=0 \quad (13'')$$

Ω,Ξ,Ψ are p,q,s forms, respectively.

By a linear representation of the Lie algebra, f.i. $T_a \rightarrow \frac{1}{2}\lambda_a$ in QCD, the Lie-algebra-valued forms become matrix-valued forms; for these $(\omega^a T_a) \wedge (\xi^b T_b) =$ $= {}^a\omega \wedge {}^b\xi \cdot T_a T_b$, where $T_a T_b$ is matrix multiplication.

The parallel transport of a vector is path dependent, the covariant derivatives do not commute

$$([D_\ell,D_m]V)^i = R^i_{k,\ell m} V^k \quad (14)$$

$R^i_{k,\ell m}$ is the curvature tensor. Similarly we have for the commutators of the covariant derivatives (9) of the charged fields

$$[D_\mu,D_\nu]\psi = \frac{g}{i} F^a_{\mu\nu}(x) T_a \psi = -iF_{\mu\nu}(x)\psi \quad (15)$$

where $F^a_{\mu\nu}(x)$ in QCD is the field-strength tensor of the gluon field. $F^a_{\mu\nu}$ determines a Lie-valued 2-form which is according to Equs.(10), (15)

$$F = \frac{g}{2}F^a_{\mu\nu} T_a dx^\mu \wedge dx^\nu = dA - \frac{i}{2}[A,A] \quad . \quad (16)$$

From the rules (13) follows easily the Bianchi identity

$$DF \equiv dF - i[A,F]$$

$$= dd A - \frac{i}{2} d[A,A] - i[A,dA] - \frac{1}{2}[A,[A,A]] \tag{17}$$

$$= 0$$

$$\left[dd A = 0, (Equ. (1)); \ d[A,A] = -2[A,dA], (Equ. (13),(13')); \right.$$

$$\left. [A,[A,A]] = 0 \ (Equ. \ (13")) \right] .$$

Let us now consider the parallel transport of charged vectors along a path L: $\{x(s) \mid 0 \le s \le 1\}$ of finite length:

$$\psi(x(1)) = U(L)\psi(x(0)) \qquad . \tag{18}$$

The U(L) will be generated infinitesimally from the infinitesimal parallel translations:

$$U(L) = \lim_{N \to \infty} \prod_{n=0}^{N-1} (1 + i A_\mu(x_n)(x^\mu(\tfrac{n+1}{N}) - x^\mu(\tfrac{n}{N})))$$

$$= P \ e^{\ i \int_L A_\mu \ dx^\mu} \qquad . \tag{19}$$

P: Path ordering. Less formally is U(L) defined by the solution of the matrix differential equation

$$\frac{dU(s)}{ds} = i A_\mu(x(s)) \frac{dx^\mu(s)}{ds} U(s)$$

$$U(0) = 1, \qquad U(1) = U(L) \qquad . \tag{19'}$$

Parallel vectors ψ at different points should remain parallel when we apply a general gauge transformation $g(x)$ on the coloured space-time: $\psi'(x) = g(x)\psi(x)$ Equ.(5). This implies the transformation law for the gauge connection

$$A'_\mu(x) = g(x)A_\mu g^{-1}(x) + ig(x)\partial_\mu g^{-1}(x) \tag{20}$$

from which follows

$$D'\psi'(x) = g(x)D\psi(x) \ , \qquad F'(x) = g(x)Fg^{-1}(x)$$

$$U'(L) = g(x(1))U(L)g^{-1}(x(0)) \ . \tag{21}$$

The inhomogeneous transformation (20) of the gauge connection corresponds to the transformation of the linear connection Γ^k_{mn} under general coordinate transformation.

(iii) Metric and duality

We already mentioned that a metric $g_{ik}(x)$ permits to use the differential calculus for tensor analysis. The Ricci tensor $e_{ik\ell m}$ $(= \sqrt{|g|}\ \varepsilon_{ik\ell m}$, $g = \det(g_{ik})$, $\varepsilon_{ik\ell m}$ antisymmetric, $\varepsilon_{1234} = 1)$ allows to define a duality transformation between r-forms and (d-r)-forms, which plays an important rôle in the dynamics of gauge theories. In d = 4 dimensions the duality maps

426

scalar $\leftrightarrow$ 4-form

$$\phi \rightarrow {}^*\phi = \frac{1}{4!} \, \phi \cdot e_{ik\ell m} \, dx^i \wedge dx^k \wedge dx^\ell \wedge dx^m$$

1-form $\leftrightarrow$ 3-form

$$j = j_k dx^k \rightarrow {}^*j = \frac{1}{3!} \, j^k \, e_{k\ell mn} \, dx^e \wedge dx^m \wedge dx^n$$

2-form $\leftrightarrow$ 2-form

$$F = \frac{1}{2} F_{k\ell} \, dx^k \wedge dx^\ell \rightarrow F^* = \frac{1}{2!} \, F^{k\ell} \, e_{k\ell mn} \, dx^m \wedge dx^n$$

etc.

$$({}^{**}{}^r\omega = (-1)^{r-1} \, {}^r\omega) \quad . \tag{22}$$

The Equ.(22') relates the volume integral of a scalar ϕ to an integral on a 4-form, Equ.(2), $\int {}^*\phi = \int \phi \sqrt{|g|} d^4x$. The integral $\int \alpha \wedge {}^*\beta = (\alpha,\beta)$ defines an inner product among r-forms ${}^r\alpha$, ${}^r\beta, \ldots$. This allows one to introduce a dual of the exterior differential operator d, denoted by δ, acting on dual forms: $(d\alpha,\beta) = (\alpha,\delta\beta)$ or $\delta\alpha = -{}^*d{}^*\alpha$ (in even dimensions).

In a manifold with metric invariance of length under parallel displacement leads to the well-known relation

$$\Gamma^r_{ki}(x) = -\frac{1}{2} g^{r\ell} \left(\frac{\partial g_{k\ell}}{\partial x^i} + \frac{\partial g_{\ell i}}{\partial x^k} - \frac{\partial g_{ik}}{\partial x^\ell} \right) \tag{23}$$

which implies the following symmetries for the

curvature tensor:

$$R_{k\ell;mn} = -R_{\ell k;mn} = -R_{\ell k;nm} = R_{mn;\ell k} \quad .$$

(iv) Dynamics

(a) It is the basic idea of <u>general relativity</u> that the geometry of space-time, i.e. the metric $g_{ik}(x)$ is determined by a dynamical equation from the matter distribution in the universe. The realization of this idea is Einstein's equation

$$R^{mn} - \frac{1}{2} g^{mn} R = - \frac{8\pi\kappa}{c^2} T^{mn} \quad . \tag{24}$$

Here $R_{mn} = R^k_{m,kn}$ is the contracted Riemann tensor, $R = R^m_n$ the Riemann scalar, T_{mn} the energy-momentum tensor, κ the gravitation constant. In view of Equs. (14) and (23) Einstein's equation is a non-linear differential equation for the metric.

(b) In the same spirit we should consider the field equation of QCD, Equ.(1.1.3),

$$D^\mu F^a_{\mu\nu}(x) = gj^a_\nu(x) \tag{1.1.3}$$

as the dynamical equation which determines the geométric structure of coloured space-time from the colour-current distribution of quark matter. With help of the differential calculus, introduced under (ii) and (iii), this and the other equations in Sect. 1.1 become a, particularly geometric, compact form:

$$D^*F = g^2 \, {}^*j \ , \qquad (\not{D} - iM)\psi = 0 \qquad (25)$$

where F is the curvature form of the gauge connection A: $F = DA = dA - i/2[A,A]$, which satisfies the Bianchi identity: $DF = dF - i[A,F] = 0$. $j = j^a_\mu(x) T_a dx^\mu$ is the quark current 1-form. Equ. (25) can be derived from the action

$$S = - \frac{1}{g^2} \text{ trace } \int F \wedge {}^*F + \int dx \bar\psi \, (\not{D}-iM)\psi \qquad (26)$$

(remember trace $T_a T_b = \frac{1}{2} \delta_{ab}$).

2.2. Lattice geometry

The renormalization procedure for gauge fields and the desirability of a strong coupling approximation to QCD in the "confinement region" suggest to con-sider gauge theories on a space (-time) lattice. For this we consider a 3- or 4-dimensional lattice embedded in Euclidean 3-space, or in Euclidean space-time, and we regard the lattice approximation to gauge theories as a sort of "triangulation" of the physical coloured space-time. This implies that we associate to the elements of the lattice: points, links, plaquettes, cubes, and super-cubes; geometric objects of the internal symmetry which correspond to integrals of differential forms over the lattice elements. This correspondence

between geometric elements and their relations in
continuum and lattice gauge theories becomes most
transparent if one uses the simplest notions of
algebraic topology.

(i) The lattice as a cell complex [28]

For simplicity we restrict our considerations to
a cubic lattice. Embedded in Euclidean space the
coordinates of the lattice points would be $x = \varepsilon(n_1,..,n_d)$
n_i integers: $-N \leq n_i \leq N$, ε = lattice constant. The
lattice may be finite or infinite, i.e. $N = \infty$. In the
terminology of algebraic topology a lattice is a cell
complex, its elements are r-cells: $^r c_i$; i.e. in a cubic
lattice: sites, links, plaquettes, cubes, super-cubes
are 0-, 1-, 2-, 3-, 4-cells, respectively. The cells
are provided with an orientation: direction of a link,
etc. $-^r c_i$ is the cell with opposite orientation. Incidence of a (r-1)-cell with an r-cell is indicated by
an incidence number $(^r c_i : ^{r-1} c_k)$:

$$(^r c_i : ^{r-1} c_k) = \begin{cases} 1 : ^{r-1} c_k \text{ face of } ^r c_i \\ -1 : -^{r-1} c_k \text{ face of } ^r c_i \\ 0 : \pm^{r-1} c_k \text{ not face of } ^r c_i \end{cases} \quad (1)$$

The incidence number determines the boundary of a cell

$$\Delta^r c = \sum_j (^r c : ^{r-1} c_j) \cdot ^{r-1} c_j \quad . \tag{2}$$

Example:

$$(^2C : {}^1C_2) = 1, \quad (^2C : {}^1C_3) = -1 \text{ etc.}$$

In describing the lattice approximation of the geometric objects of a manifold - Sect. 2.1. i.a. - we rely mainly on our intuition.

A path ${}^1L$ is approximated by a sum of 1-cells: ${}^1L = \sum_i {}^1C_i;$ we have used this notation already in the definition (2) of the boundary. The consideration of multiple paths leads to the concept of

"chains": $\qquad {}^1L = \sum_i \alpha^i \, {}^1C_i$ $\qquad\qquad\qquad$ (3)

where in the most general case the α^i are not restricted to integers, but may be elements of a general field (or even additive group). In a similar way a r-dimensional submanifold is generalized to r-chains. The r-chains form a linear space over the r-cells. This linear extension of the boundary operator (2) to chains: $\Delta^r L = \sum_i \alpha^i \, \Delta \, {}^r C_i$ coincides in simple cases with the intuitive notion of a boundary of a manifold. Δ has the property:

$$\Delta\Delta = 0 \qquad . \qquad\qquad\qquad\qquad (4)$$

We may consider the links incident in a point as the lattice analogue of the vectors $e_i(x)$ spanning the tangent space T_x. The chain of this links may be expressed with help of the incidence numbers: $\nabla^0 C =$ $= \sum_i {}^1 c_i \, ({}^1 c_i : {}^0 c)$. This definition of the "coboundary" operator can be generalized for r-cells and linearly extended to r-chanins:

$$\nabla^r C = \sum_i {}^{r+1} c_i \, ({}^{r+1} c_i : {}^r c)$$

$$\nabla^r L = \sum_i \alpha^k \cdot \nabla^r c_k \quad . \tag{5}$$

∇ maps r-chains on (r+1) chains. It has the property $\nabla^2 = 0$. Examples of its action are:

$$\nabla^0 C = \quad \text{in } Z_N^2; \quad \text{in } Z_N^3.$$

Linear functions on the space of r-chains are called r-cochains ω, these are defined by their values on r-cells: $\omega({}^r c_i)$. Differential r-forms ω' on a manifold define a r-cochain on a lattice embedded in this manifold [29]:

$$\omega' \xrightarrow{\text{La}} \omega : \omega({}^r c_i) = \int_{{}^r C} {}^r \omega' \quad . \tag{6}$$

The exterior differential of cochains is defined as

$$d\omega(^r L) = \omega(\Delta^r L) \tag{7}$$

$dd\omega = 0$ because of (4) .

Then it follows from Stokes' theorem (2.1.3) that $\omega' \xrightarrow{\text{La}} \omega$ implies $d\omega' \xrightarrow{\text{La}} d\omega$. The exterior differential of a r-cochain is a (r+1)-cochain, as might be illustrated by the following:

Example: A is 1-cochain, then $-\Delta^2 C = {}^1C_1 + {}^1C_2 + {}^1C_3 + {}^1C_4 -$

$$F(^2 C) = A(^1C_1) + A(^1C_2) + A(^1C_3) + A(^1C_4) \tag{8}$$

is a 2-cochain.

If $^r L = \sum \alpha^k \, {}^r C_k$ is an approximation to the sub-manifold L in a sufficient fine lattice, then (6) implies that $\omega(L)$ is an approximation to $\int_L \omega' : \omega(L) \sim \int \omega'$.

These considerations should illustrate how the correspondence (6) induces a natural translation of the tensor analysis on a manifold into the algebraic operations on chains and cochains in a lattice.

(ii) Connections in a coloured space time lattice

Now we have to apply the ideas of lattice approximation to the geometry of coloured space-time. This means: There is a local charge space $H_x \equiv H(^0C_k)$ associated with the sites, o-cells ^0C_k of the lattice. The symmetry group G acts on $\psi(x) \in H_x$ like in (2.1. 5). If $H(^0C_k)$ is a linear representation space of the group G,

like in QCD, one may consider the lattice approximation
to the charge field $\psi(x) \in H_x$ as vector valued 0-cochain
defined on the base ${}^{0}c_k$ of the 0-chains by $\psi({}^{0}c_k) \in H({}^{0}c_k)$.
According to the definition Equ.(7) the exterior
differential of ψ is an approximation to the partial
derivative along the "${}^{1}c_k$ direction":

$$(d\psi)\,({}^{1}c_k) = \psi({}^{0}c_k) - \psi({}^{0}c_k') \approx \epsilon \partial_k \psi(x)$$

$$\Delta^1 c_k = {}^{0}c_k - {}^{0}c_k' \ . \tag{8'}$$

The gauge connection A, Equ.(2.1. 7), determines on the
links a finite parallel transport - Equ.(2.1. 10) -

$$U({}^{1}c_k) = P \exp i \int_{{}^{1}c_k} A \approx e^{iA({}^{1}c_k)} \ . \tag{9}$$

It defines the parallel transport between charge spaces
at neighbouring points

$$\psi({}^{0}c_2) = U({}^{1}c)\psi({}^{0}c_1)$$

$$\Delta^1 c = {}^{0}c_2 - {}^{0}c_1 \qquad \text{with } U(-c) = U^{-1}(c) \ . \tag{9'}$$

The parallel transport of charge vectors along a path L,
Equ.(2.1. 9), is approximated by

$$Pe^{i\int_L A} \approx U(L) = P \prod_{{}^{1}c_i \in L} U({}^{1}c_i) \equiv \prod_{{}^{1}c_i \in L} U({}^{1}c_i) \ . \tag{10}$$

If G is non-abelian, U(L) is not a "linear" function on a chain; it cannot be considered a group-valued cochain. Only for abelian groups, when path ordering is irrelevant, we have $U(L) = e^{iA(L)}$, and the Lie-algebra-valued cochain A(L) is direct the lattice approximation, Equ.(6), of the connection form A: Equ.(2.1. 7).

It follows from Equs.(2.1. 10);(2.2. 8), (9), that we have to consider

$$(D\psi)\,(^{o}c_{1}) = \psi(^{o}c_{1}) - U(^{1}c)\psi(^{o}c_{2})$$

$$\Delta^{1}c = {}^{o}c_{1} - {}^{o}c_{2} \tag{11}$$

as the covariant derivative of ψ at $^{o}c_{1}$ along the link ^{1}c.

The lattice approximation to the field-strength 2-form, Equ.(2.1. 16), is related to the 2-cell ^{2}c. It corresponds to the logarithm of the path-ordered parallel transport, Equ.(10), around the boundary $\Delta^{2}c$, starting from a site $^{o}c_{1}$:

$$P(^{2}c;^{o}c_{1}) = U(^{1}c_{4})U(^{1}c_{3})U(^{1}c_{2})U(^{1}c_{1})$$

$$= \hat{\prod} U(^{1}c_{i}) \tag{12}$$

(see figure). The trace of $P(^{2}c;^{o}c_{1})$ is independent of $^{o}c_{1}$; it is called the "plaquette term":

$$P(^{2}c) = \text{trace } P(^{2}c;^{o}c_{i}). \tag{12'}$$

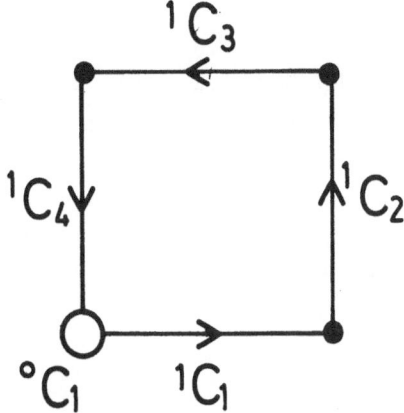

We add a short remark on the relation of the plaquette
terms $P[^2c;^0c_k]$, $P[^2c]$ to the field strength in the
continuum limit. For this we generate the lattice
connection infinitesimally: $U(^1c_k) = \exp i\, A(^1c_k)$,
Equ. (9). Using the Baker-Hausdorff formula

$$e^A\, e^B = e^{A + B + \frac{1}{2}[A,B] + \ldots}$$

repeatedly we get up to second order

$$P(^2c;^0c_1) = \exp i\, \{dA - \frac{i}{2}P_{^0c_1}[A,A] + \ldots$$

$$\equiv \exp i\, F(^2c;^0c_1) \tag{13}$$

where $P_{^0c_1}[A,A]$ denotes the "path-ordered Lie product"

$$P_{^0c_1}[A,A] = ([A(^1c_4),A(^1c_3)]+[A(^1c_4),A(^1c_2)]+$$

$$+[A({}^{1}C_{4}),A({}^{1}C_{1})]+[A({}^{1}C_{3}),A({}^{1}C_{2})]+\ldots[A({}^{1}C_{2}),A({}^{1}C_{1})]. \quad (14)$$

Comparing $F({}^{2}C;{}^{0}C_{1})$ with Equ.(2.1. 16) one can see immediately that $F({}^{2}C,{}^{0}C_{1})$ is a lattice approximation to the field-strength 2-form F.

Finally we summarize the transformation properties of the different geometric objects on the lattice under general gauge transformations:

$$\psi'({}^{0}C_{1}) = g({}^{0}C_{1})\psi({}^{0}C_{1})$$

$$U'({}^{1}C) = g({}^{0}C_{2})U({}^{1}C)g^{-1}({}^{0}C_{1}) \qquad \Delta^{1}C = {}^{0}C_{2} - {}^{0}C_{1} \qquad (15)$$

from which follows

$$(D\psi)'({}^{0}C_{1}) = g({}^{0}C_{1})D\psi({}^{0}C_{1})$$

$$P'[{}^{2}C;{}^{0}C_{1}] = g({}^{0}C_{1})P[{}^{2}C;{}^{0}C_{1}]g^{-1}({}^{0}C_{1})$$

$$P'[{}^{2}C] = P[{}^{2}C] \quad . \qquad (16)$$

(iii) Lattice dynamics

In our discussion of the dynamics of gauge theories on a lattice we want to consider only pure gauge fields. The lattice approximation to spinors, i.e. quark fields, has its peculiar difficulties, which we do not want to consider in this lecture.

Guided from the approximation of the field strength F by a plaquette term, Equ.(13), K.G. Wilson [30] suggests as substitute for the gauge field action

$$S = - \frac{1}{g^2} \text{ trace } \int F \wedge^* F, \quad \text{Equ.(2.1. 26)}$$

$$S_{La} = \frac{1}{g_o^2} \sum_{2C_k} P(^2C_k) . \qquad (17)$$

Calculating the trace Equ.(12') of the approximation (13) to the plaquette term one sees immediately that up to an additive constant (17) is in leading order an approximation to the classical action Equ.(2.1. 26).

Bearing in mind that $P(^2C_k)$ is the trace of a representation of the group element associated to the plaquette, i.e. a character $\chi_q(\hat{\Pi} U(^1C))$ of the quark representation of colour $S(U_3)$, one sees immediately generalizations to (17), with respect to arbitrary groups, arbitrary characters of group elements associated to more general lattice elements.

Such generalized actions

$$S[U] = \frac{1}{g^2} \sum_{r\text{-cells}} \chi(\prod_{C' \in \Delta^r C} U_{C'}) \qquad (18)$$

describe the dynamics of lattice approximations of all kinds of geometrical theories mentioned in Sect.2.1.i.

2.3. Classical solutions with topological charges

There are solutions of gauge theories, where the topology of the internal symmetry group conspires with the topology of space-time in such a way that particularly stable structures appear.

(i) A simple but very important example is the "Nielson-Oleson" string [17] solution of the field equations of the abelian Higgs model

$$(\partial_\mu - ieA_\mu)^2 \phi = m^2\phi - \lambda|\phi|^2\phi$$

$$\partial^\nu F_{\mu\nu} = ej_\mu = -\frac{ie}{2}(\phi^*\partial_\mu\phi - \partial_\mu\phi^* \cdot \phi) + e^2 A_\mu\phi^*\phi \quad . \tag{1}$$

These field equations are invariant under gauge transformations of the abelian group $G = U(1) = \{e^{i\alpha}|0 \le \alpha \le 2\pi\}$:

$$\phi'(x) = g(x)\phi(x) = e^{i\alpha(x)}\phi(x)$$

$$eA'_\mu(x) = eA_\mu(x) + ig(x)\partial_\mu g^{-1}(x) = eA_\mu(x) + \partial_\mu\alpha(x) \tag{2}$$

and form thus a simple example of gauge geometry.

There are cylindersymmetric, time-independent solutions of the form

$$\phi(x) = e^{i\theta} G(\rho) = e^{i\theta} (\frac{m}{\sqrt{\lambda}} + \bar{G}(\rho))$$

$$A_\theta(x) = \frac{1}{e\rho} + \bar{A}(\rho) \quad . \tag{3}$$

Here ρ, θ are the usual cylinder coordinates, $A_\mu dx^\mu = A_\theta d\theta$.
The requirement that the energy/length E_2:

$$E_2 = \frac{1}{2}\int d^2x \ \{B^2 + |(\partial - ieA)\phi|^2 + \frac{\lambda}{2}(|\phi|^2 - \frac{m^2}{\lambda})^2 \qquad (4)$$

is finite implies $|\phi| \to \frac{m}{\sqrt{\lambda}}$, $|(\partial - ieA)\phi| \to 0$ and is thus
responsible for the characteristic behaviour of the
solutions (3) at infinity:

$$\phi \to e^{i\theta}\frac{m}{\sqrt{\lambda}} = g(\theta)\frac{m}{\sqrt{\lambda}}$$

$$A \to igdg^{-1} \quad . \qquad (5)$$

The solutions (3) are asymptotically a pure gauge. This
asymptotic pure gauge is characterized by the 1-1 mapping
of the circle at infinity $S^2 = \{\theta | 0 \leq \theta \leq 2\pi\}$ on the
group G: $\theta \to \alpha$. There are solutions of finite energy
with asymptotic gauges $g(\theta) = e^{in\theta}$ which describe an
n-fold mapping of S^2 on G. This "winding number" n of
the mapping of the infinite circle S^2 of 2-space on the
group manifold G characterizes topologically the
boundary conditions of the cylinder symmetric solutions.
This topological quantum number for general time-
dependent, cylinder-symmetric solutions is conserved.
It can be expressed by the following integral

$$n = \frac{e}{2\pi}\int_{S^2} dA = \frac{e}{2\pi}\int_{R^2} B \quad . \qquad (6)$$

This shows that the topological structure described
above leads to the quantization of magnetic flux: B=dA.

The non-asymptotic part $\bar{G}(\rho)$ and $\bar{A}(\rho)$ cannot be calculated explicitly. For $\rho \to 0, \infty$, we get:

$$\bar{G}(\rho) \sim \begin{cases} \rho^{-1/2}\, e^{-\sqrt{2m}\,\rho} & \\ \\ -\dfrac{m}{\sqrt{\lambda}} & \end{cases} \quad ; \quad \bar{A}(\rho) \sim \begin{cases} \rho^{-1/2}\, e^{-\frac{em}{\sqrt{\lambda}}\cdot \rho} & \rho \to \infty \\ \\ -\dfrac{1}{e}\dfrac{1}{\rho} & \rho \to 0 \end{cases} \qquad (7)$$

Therefore the solution will have the form of

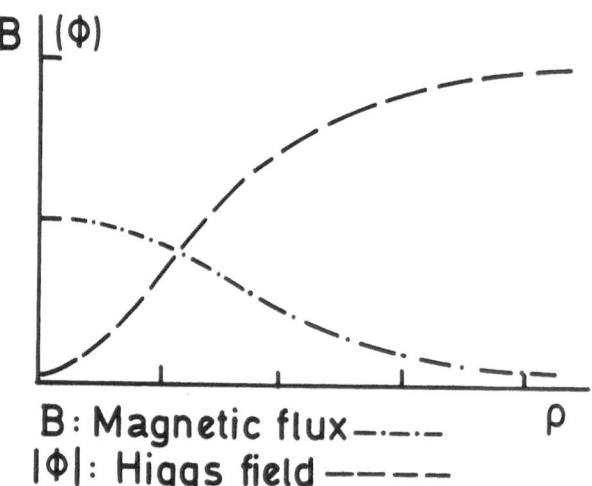

B: Magnetic flux $-\cdot-\cdot-$
$|\Phi|$: Higgs field $----$

The magnetic field $B(\rho)$ gets screened by the Higgs vacuum with a penetration length $\xi = \frac{em}{\sqrt{\lambda}}$, the compton wave length of the "physical" vector particle in the Higgs model. The Higgs vacuum $|\phi| = \frac{m}{\sqrt{\lambda}}$ gets displaced by the string of magnetic flux. The Higgs model is the Landau-Ginsberg theory [30] of superconductors of second kind, where Cooper pairs are phenomenologically described by the Higgs field. The formation of strings mathematically discussed above initiated the ideas of the chromoelectric Meissner effect. Since the coupling of gauge fields to Higgs scalars destroys asymptotic freedom, a Higgs model for the chromoelectric Meissner effect cannot be taken too seriously.

(ii) Instantons [32]

The discovery of the exact solution [33]

$$_1A = \frac{ix^2}{x^2 + \rho^2}\, g^{-1}\, dg \tag{8}$$

$$g(x) = \frac{1}{|x|}\, (x^4 - i\vec{x}\vec{\tau}) \tag{9}$$

of the SU(2)-gauge-field equations

$$D^*F = 0 \tag{10}$$

(in Euclidean space-time, τ_i: Pauli matrices) had a strong impact on our understanding of gauge fields. We mention some of these developements.

Asymptotically, $|x| \to \infty$, the solution (1) approaches a pure gauge $ig^{-1}dg$. This guarantees that the Euclidean action $S_e = -S = g^{-2}$ trace $\int F \wedge \overset{*}{F}$ is finite. The pure gauge $g^{-1}dg$, as asymptotic boundary value of solution (8), is topologically different from a trivial gauge $A = 0$, i.e. $g(x) \equiv g_o = $ const.: $g(x)$ of Equ.(2) maps the (infinite) sphere S^3 continuously on the group space G of $SU(2)$, which is topologically also a sphere. Opposed to that, the $A = 0$ gauge is related with a mapping of S^3 on a single point in G. These considerations suggest that one can classify the solutions of the gauge field equations of finite action by the winding number of their asymptotic pure gauges. The winding number κ of the asymptotic gauge counts how often the asymptotic gauge maps the infinite sphere S^3 continuously on G. The relation

trace $F \wedge F = d$ trace $\{A \wedge (dA + A \wedge A)\}$

allows an integral representation of the winding number κ of the asymptotic pure gauge of a gauge field with finite action:

$$\kappa = \frac{1}{8\pi^2} \text{ trace } \int F \wedge F \equiv \frac{1}{8\pi} (F, \overset{*}{F}) ; \tag{11}$$

here we used the notion for inner products of forms, Equ.(2.1. 22). One should compare this with our discussion above 2.3.(i).

The field strength $_1F = d_1A - \frac{i}{2}[_1A,_1A]$ of solution (8) is self-dual

$$\overset{*}{_1}F = {_1}F . \tag{12}$$

From Equ. (11) follows an interesting relation between
the action and the winding number:

$$S(A) = \frac{1}{g^2} (F,F) = \frac{1}{g^2} (F,{}^*F) = \frac{8\pi^2}{g^2} \kappa \qquad . \qquad (13)$$

Therefore $S({}_1A) = \frac{8\pi^2}{g^2}$, as we have shown above that
$\kappa = 1$ for solution Equ. (1). For general gauge fields
we have the Schwarz inequality

$$(F,F) = \sqrt{(F,F)({}^*F,{}^*F)} \geq (F,F^*)$$

which implies

$$S(A) \geq \frac{8\pi^2}{g^2} \kappa \qquad (14)$$

where the lower bound is reached by self-dual - or anti-
self-dual F. Self-dual or anti-self-dual gauge fields
minimize absolutely the action within a class of gauge
fields with given winding number. Because of the action
principle self-dual or anti-self-dual gauge fields

$$F = \pm {}^*F \qquad (15)$$

are solutions of the field equations (3). These self-
dual (anti-self-dual) gauge fields are called instantons
(anti-instantons).

The self-duality equations (15) for instantons
are simpler than the general dynamical equations of
gauge fields (1). They admit the successful application

of methods of algebraic geometry to find the manifold
of their solutions [34].

The manifold M_1 of instantons with winding number
$\kappa = 1$ is 5-dimensional corresponding to the parameter
ρ and a possible translation $x \rightarrow x + a$ of solution (8).
The general methods give for general κ the dimensionality
of the instanton manifold

$$\dim M_\kappa = \begin{cases} 5, & \kappa = 1 \\ \\ 8\kappa-3, & \kappa \geq 2 \end{cases} \qquad . \qquad (16)$$

There are relatively simple expressions for $(5\kappa+3)$-
parametric solutions [35] and there is progress in the
explicite description of the remaining ones [36].

The existence of topological inequivalent pure
gauges has implications on the classical description
of the empty space. It is conjectured that the instantons play a certain role for the vacuum of QCD
by describing "tunneling between classical vacua" [37].

3. WILSON CRITERION

3.1. Wilson's criterion for quark confinement

Wilson's criterion for quark confinement generalizes
the idea of a confining potential to the situation in a
gauge-field theory [39].

Having in mind QED one might expect that the

Green's function of the gauge field $<TA_\mu^i(x)A_\nu^k(x')> =$
$= D_{\mu\nu}^{ik}(x-x')$ might describe the long-range part of the
interaction of infinitely heavy quarks. Remember, in
free QED in (d+1) space-time dimensions, we have in
Feynman gauge $D_{\mu\nu}(x) = -(2\pi)^{-(d+1)}g_{\mu\nu}\int dp e^{-ipx}(p^2+i\varepsilon)^{-1}$.
Calculating $W(j) = e^2\int\int j_\mu(x)D_{\mu\nu}(x-x')j_\nu(x')dxdx'$ for
conserved currents, in particular static sources
$(j_\mu(x)) = (\delta(\vec{y}-\vec{x}),\vec{0})$, leads to

$$W = -\frac{e^2}{(2\pi)^{d+1}}\int_{-T}^{+T}dx^0\int_{-T}^{+T}dx^{0'}\frac{e^{i\vec{p}(\vec{y}-\vec{y}')-ip_0(x_0-x_0')}}{p_0^2-\vec{p}^2}$$

$$\approx 2T\cdot V(\vec{y}-\vec{y}') \quad \text{for} \quad T\to\infty \tag{1}$$

where $V(\vec{y})$ is the well-known Coulomb potential in d
space dimensions:

$$V(\vec{y}) = \frac{e^2}{(2\pi)^d}\int d\vec{p}\,\frac{e^{i\vec{p}\vec{y}}}{\vec{p}^2}$$

$$= \frac{e^2}{4\pi^{d/2}}\Gamma(\tfrac{1}{2}d-1)|\vec{y}|^{-d+2} \quad\quad \text{for} \quad d\geq 3$$

$$= \frac{e^2}{2\pi}\log|\vec{y}| \quad\quad \text{for} \quad d=2$$

$$= \frac{e^2}{2\pi}|y| \quad\quad \text{for} \quad d=1 \;. \tag{2}$$

This argumentation depends strongly on the gauge-fixing

procedure, which in QCD is rather complex. We think
about the unphysical states of the Fadeev-Popov ghosts,
or the questions related to the Gribov problem. In a
consistent quantum mechanical Hilbert space treatment
with a physical, gauge-invariant vacuum, the vacuum
expectation values of gauge non-invariant operators
like $A_\mu^i(x) A_\nu^k(x')$ would even vanish. In order to overcome these gauge problems K. Wilson suggested to consider instead of the 2-point correlation function of
the gauge fields the following expectation value:

$$W(L) = <\text{trace } Pe^{i\int_L A}>; \qquad L: \qquad (3)$$

$A_\mu ds^\mu = \frac{g}{2}\lambda_i A_\mu^i ds^\mu$ is the connection form of quark fields,
Equ.(2.1. 4'), P denotes path ordering, Equ.(2.1. 19),
trace P exp(i∫_L A) with closed path L is gauge invariant,
Equ. (2.1.(21). The loop L describes the history of a
(heavy, external) quark-antiquark pair. If the history
L has the shape of the figure, with large T and R, the
quantum mechanical interpretation of the expectation
value W(L) is related to the effective static potential
V(R) fór a colour-singlet pair of quark-antiquarks [40]:

$$V(R) = \lim_{T \to \infty} - \frac{1}{T} \log W(L) \quad . \qquad (4)$$

For illustration we may calculate the Wilson loop for

free QED in d+1 dimensions. With a Feynman gauge-fixing term $L_{ga} = -\frac{1}{2}(\partial_\mu A^\mu)^2$ the functional representation Equ.(1.1. 5) of W(L) is of the general form

$$W[L] = \frac{1}{Z}\int D[A_\mu] e^{\int dx[\frac{1}{2}A_\mu(x)\Box A^\mu + A_\mu(x)J_\mu(x)]}$$

$$= e^{-\frac{1}{2}\int J_\mu(x)D_{\mu\nu}(x-x')J_\nu(x')}$$

$$= e^{-\frac{e^2}{2}\int_L dx^\mu \int_L dx^{\mu'} D_{\mu\nu}(x-x')} \qquad . \qquad (5)$$

[Here we put $\int A_\mu J^\mu(x)dx = e\int_L A_\mu dx^\mu$; and we have evaluated (5) as a non-degenerate Gaussian integral in the Euclidean region $x_4 \to ix_o$, $A_4 \to iA_o$]. In evaluating the exponential in Equ.(5) we see that it consists of two parts

(i) the double line $\vec{x} = \vec{x}' = \vec{y}$ or $\vec{y}'$; this contributes an R-independent, infinite self-energy term (which should be compensated by a mass counter-term)

(ii) the double line $\vec{x} = \vec{y}$, $\vec{x}' = \vec{y}'$ or $\vec{x} = \vec{y}'$, $\vec{x}' = \vec{y}$; this gives according to Equ.(2): $I \cdot V(\vec{y}-\vec{y}')$; thus confirming Equ.(4) in the case of free QED. For a more careful discussion we refer to the literature.

Based on the interpretation of W(L) according to Equ.(4) K.G. Wilson formulates as a sufficient criterion for quark confinement

$$W(L) \sim e^{-\kappa \cdot arL} \qquad (6)$$

for a large area ar∟ enclosed by the loop ∟. The constant κ^{-1} determines the correlation between the gauge field at large distances (f.i. at space like distances in a temporal gauge A_o). Thus a $\kappa \neq 0$ indicates a mass gap $\Delta \sim \kappa$ between the vacuum and gluonic states of low energy.

The Wilson criterion is formulated for the confinement of static quarks. Because a mass gap leads to a chromoelectric Meissner effect, Sect.1.3., we expect that if the Wilson criterion is satisfied for a pure gauge theory, it also confines light physical quarks coupled to this gauge field.

The evaluation of the functional integral in the euclidean region, f.i. Equ.(5),allows the application of methods of statistical mechanics and is therefore for the following of greatest importance.

Finally we want to emphasize that it is the expectation value of a geometrical object, the colour connection along a closed path, which decides on an important physical property of coloured space-time.

3.2. Strong-coupling approximation [41]

The Wilson criterion for quark confinement is satisfied in the strong-coupling approximations to the lattice gauge theory. Therefore we want to describe shortly this method - following a representation due to J.M. Drouffe [42]. It gives a general treatment of all geometric lattice models described in 2.2 and thus emphasizes the geometrical aspects of our problem.

The starting points are the geometric Euclidean lattice actions of 2.2 with symmetry group G, which may be written in a compact form:

$$S[R] = \beta \sum_{\substack{r\text{-cells} \\ r_c}} \chi \left(\prod_{c' \in \Delta^r c} R_{c'} \right). \tag{1}$$

χ character of G, $\prod$ ordered product along the boundary, $\beta = \frac{1}{g^2}$ the inverse of the "dimension-free" bare-coupling constant - the inverse temperature in statistical language.

$r = 2, \; R = U = e^{iA} =$ link connection

in gauge thoeries.

$r = 1, \; R = \delta(x) = \delta(^{\circ}c)$

in Ising-type theories.

The expectation value of any functional $\Phi[R]$ of the basic fields R (in the Euclidean region) is calculated - Equ.(1.1. 5) - by

$$<\Phi(R)> = z\{\Phi\}/z\{1\} \tag{2}$$

with

$$z\{\Phi\} = \int D[R]\Phi(R)e^{S[R]} \tag{3}$$

where D[R] denotes the measure in the function space in basic field-configurations:

$$D[R] = \prod_{(r-1)\text{-cells}} d\mu(R(^{r-1}C)); \quad d\mu(g) \text{ Haar measure} \quad (4)$$
$$\text{of } G.$$

Of course, the most interesting functional for the question of quark confinement is the Wilson loop in a gauge theory, Equs.(2.2. 10),(3.1. 3):

$$W[L] = \text{trace } Pe^{i\int_L A} = \text{trace } {}_1\hat{\prod}_{C \in L} U(^1C) . \quad (5)$$

In this framework we have embedded the problem of quark confinement in QCD in a wide neighbourhood of geometric Euclidean theories in lattice approximation. This demonstrates the conceptional power of our notions in-troduced in chapt. 2. However, this scheme allows also for strong simplifications. A very drastic one is to regard as values of the basic fields R not group elements of the colour $SU(3)$, but simply $R = \pm 1 \in Z_2$. (Z_n-models: $G = Z_n$ = cyclic group of order n). For $r = 1$ we have the traditional Ising model. We shall often use this simplified model for illustration.

The evaluation of the multidimensional integral (3) in the strong-coupling limit, $\beta \to 0$, is simplified by expanding $\exp[\beta \cdot \chi(g)]$ in primitive characters of the symmetry group $\chi^k(g)$

$$e^{\beta\chi(g)} = K \sum_k \beta_k \chi^k(g)$$

$$\beta_k = \frac{1}{K} \int d\mu(g) e^{\beta\chi(g)} \chi^{k*}(g)$$

$$K = \int d\mu(g) e^{\beta \chi(g)} \quad , \tag{6}$$

This is a particular application of the orthogonality (and completeness) relations for the Fourier transformation on groups:

$$\int d\mu(g) D_{\alpha\beta}^{k}(g) D_{\gamma\delta}^{*k'}(g) = \frac{1}{\nu_k} \delta_{kk'} \delta_{\alpha\gamma} \delta_{\beta\delta} \quad . \tag{7}$$

$D^k(g)$ irreducible, ν_k-dimensional representation of G. From (7) follows for the primitive characters $\chi^k(g) =$ = Trace $D^k(g)$ the convolution formula:

$$\int d\mu(f) \chi^k(g'f^{-1}) \chi^{k'}(f,g^{-1}) = \frac{\delta_{kk'}}{\nu_k} \chi^k(g'g^{-1}) \quad . \tag{8}$$

In the simple case of $G = Z_2 = \{\sigma | \sigma = \pm 1\}$ there are two representations: $\chi^0(\sigma) = 1$, $\chi^1(\sigma) = \sigma$; the expansion (6) is

$$e^{\beta\sigma} = K(1 + X\sigma)$$

$$K = \cosh \beta \qquad \beta_1 \equiv X = \tanh \beta \quad . \tag{9}$$

For SU(3) the expansion (6) can be derived iteratively from the generating functional:

$$\int\limits_{SU(3)} d\mu(g) \exp(B^{\alpha\beta} g_{\alpha\beta}) = \sum_n \frac{2}{n!(n+1)!(n+2)!} (\det B)^n$$

$$g \in SU(3)$$

(M. Creutz [41]; approximations are given by J.M.Drouffe [42]).

The representation Equ.(6) gives for the exponential of the action

$$e^{S[R]} = K^N \prod_{r\text{-cells } C_\ell} (1+\sum_k \beta_k \chi^k (\hat{\prod}_{C \in \Delta^r C_k} R_c)), \tag{10}$$

N = number of r-cells,

because in our normalization the "inverse coupling constant" β of the trivial representation is $\beta_k = 1$. The expansion of (10) in powers of β_k can be re-presented by strong-coupling diagrams. A diagram D is a sum of a finite number $|D|$ of r-cells (links, plaquettes):

$$D = \sum_i^{|D|} {}^r C_i \quad \text{with a mapping} \quad {}^r C_i \to \beta_k \chi^k (\hat{\prod}_{C \in \Delta C_r} R_c).$$

The value of a diagram is

$$E(D) = \prod_{i=1}^{|D|} \beta_k \chi^k (\hat{\prod}_{C \in \Delta^r C_i} R_c) . \tag{11}$$

With this notation we get

$$K^{-N} e^{S[R]} = \sum_{\text{all } D} E(D) \tag{11'}$$

and the Equ.(3) becomes a sum of the contributions from all diagrams:

$$\int D[R]\,\phi(R)\,e^{S[R]} = K^N \sum_{\text{all } D} \int D[R]\,\phi(R)\,E(D) \qquad . \tag{12}$$

Let us consider first the contribution of a diagram to the partition function

$$\int D[R]\,E(D) = \int \ldots \int \prod_{\substack{(r-1)\text{-cells} \\ C_\ell}} d\mu\,(R_{C_\ell}) \prod_{i=1}^{|D|} \beta_{k}\chi^{k}(\hat{\prod}_{c\,\in\,\Delta^{r}C_{i}}\,r_{c}\,R_{c}) \qquad . \tag{13}$$

In the case of a gauge theory we may visualize the set of plaquettes of a diagram as a surface. The integration over regular links, incident with only two plaquettes, can be performed with help of the orthogonality relations Equ. (8). The integral is only different from zero, if the two plaquettes incident in the link are associated with the same irreducible representation. The integration over boundary lines vanishes. Hence diagrams contributing to the l.h.s. of Equ. (12) - "admissible diagrams" - must be closed: If D contains only regular links, we have

$$\int D[R]\,E(D) = \beta_{k}^{|D|} \tag{14}$$

where χ^{k} is the character of the irreducible representation of G uniformly attributed to all plaquettes of D. In order that the integration over irregular links does not vanish the multiple direct product of the representations associated with all incident plaquettes must contain the trivial representation. This leads to selection rules for "admissible, irregular graphs". We refer to the literature for further discussion.

There is no difficulty in extending this procedure to the computation of the Wilson loop. The contribution to (12) of a single graph is

$$\frac{Z}{K^N} W_D(L) = \int D[R] \chi_q \left(\prod_{C \in L} \hat{R}_{1_C} \right) E(D) .$$

As we perform the integrations, we see that the orthogonality relations, Equ. (8), require for a graph with non-vanishing contribution that the product of the representations of the plaquettes of D incident on L must contain the quark representation χ_q of G. The lowest-order graph consists of the simple surface with minimal area $|D|$ bounded by L:

$$W(L) \sim \beta^{|D|} + \ldots .$$

This is the only graph in 2 dimensions. We have rigorously confinement with a linear potential

$$V(r) = (\log g_o^2 + \ldots) \frac{r}{\epsilon} .$$

For higher dimensions, in particular for d = 4, one can show that after cluster decomposition the sum over the strong-coupling diagrams converges [43] for sufficiently high coupling constants: $g_o^2 = \frac{1}{\beta}$. Thus the Wilson criterion for quark confinement is satisfied for the SU(3) gauge theory on a lattice.

3.3. The continuum limit; renormalization group

The Wilson criterion is satisfied in the strong coupling limit of lattice gauge theories. Does this fact persist, if one applies the limiting procedure, which leads to the continuum? In the following we want to discuss some problems and notions related to this question.

We assume that the continuum gauge theory will be the zero lattice spacing limit of the lattice gauge theory described by the action

$$S[U(^1C);g] = \frac{1}{g^2} \sum_{2C} \text{trace } P[^2C] \quad . \tag{1}$$

In Equ.(2.2. 13) we indicated how one might understand the continuum limit for classical fields. Thus our assumption makes sense. However, because of renormalization the quantum-mechanical continuum-limit poses a completely different problem.

A renormalized continuum limit [44] is determined by a sequence of coupling constants $g^2(\varepsilon)$, $\varepsilon \to 0$, such that the expectation values of lattice approximations to physical quantities have a limit. In order to determine the singular behaviour of the "cut-off" dependence of the unrenormalized coupling constant $g(\varepsilon)$ we have to normalize a particular physical quantity to experiment. Let us fix these notions by considering as observables traces of path ordered connections along closed lines $O(L) = \text{trace } P \exp(i \cdot \int_L A)$. Their lattice approximations are, Equ.(2.2. 10),

$$O_\varepsilon(L) = \prod_{{}^1C_i \in L}^{\wedge} U({}^1C_i) \tag{2}$$

where $L \xrightarrow{La} L_\varepsilon$ is an approximation to L in a lattice with lattice constant ε, Sect.2.2.i. Of course the number of 1-cells contained in L_ε increases linearly with $\frac{1}{\varepsilon}$. Now we consider

$$\lim_{\varepsilon \to 0} <O_\varepsilon(L)> = \lim_{\varepsilon \to 0} \frac{1}{Z(g(\varepsilon))} \int D[U({}^1C)] O_\varepsilon(L) e^{S[U({}^1C);g(\varepsilon)]} \tag{3}$$

where $g(\varepsilon)$ is determined by the renormalization condition

$$<O_\varepsilon(L_o)> = e^{-a\ ar(L_o)} \underset{\sim}{\sim} e^{-g} \qquad . \tag{4}$$

L_o is the boundary of the unit square of 1.5f length, a the string constant, Equ.(1.3. 1), Equ.(3.1. 4). (R = 1.5f is in the linear region of the potential below the string breaking point, Equ.(1.2. 1)). We expect from the renormalizability of QCD that the renormalized continuum limit, Equ.(3) and (4), exists.

In order that such a limit exists with a finite continuum correlation length $\kappa \underset{\sim}{\sim} a$ the coupling constant must approach a critical value, for which the correlation length in lattice units becomes in-finite. Now we can pose our original question in the following form: Assume we start with a coupling constant g_o, for which the strong coupling approximation converges, and a lattice spacing ε such that the re-

normalization condition (4) can be satisfied.
(H.G. Dosch and V.F. Müller [45] give an estimate
$g_o^2 \sim 3.1 \times 20^4$, $\varepsilon \sim 1.5f$). Do we approach with
further refinement of the lattice the critical point
which describes asymptotic-free QCD? This means a
critical point at $g_{crit} = 0$.

It is known that the existence of phase
transitions depends on the space-time dimensions d.
A typical phase diagram looks like:

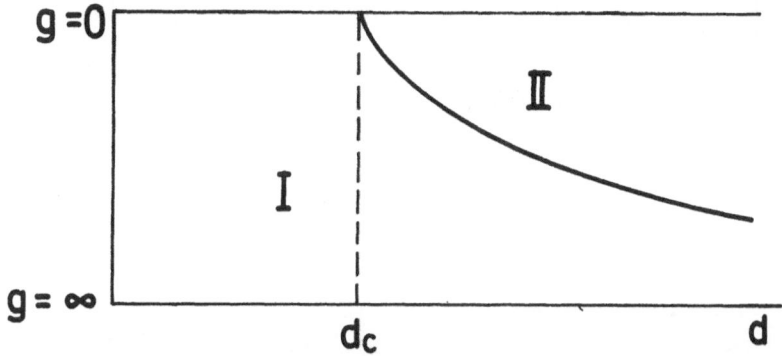

I: disordered, confining, strong-coupling phase.
II: ordered, quasi-free, weak-coupling phase.

There is no phase transition below the critical
dimension, there are phase transitions above the
critical dimension. At the critical dimension there
is a critical point at $g = 0$, the boundary of the
parameter space.

Hence the confinement of quarks and asymptotic
freedom would be properties of QCD, as a limit of a
lattice gauge theory, if the following could be
proven:

<u>Migdal's conjecture [46]:</u> The critical dimension of non-abelian gauge theories - in particular of $SU(3)$- gauge theory - is $d_c = 4$.

In the following table we give the critical dimensions of the geometrical models defined in Equ.(2.2. 18):

r \ G	abelian finite	abelian Lie group	non-abelian Lie group
$r = 1$ Ising type	$d_c = 1$	$d_c = 2$	$d_c = 2$
$r = 2$ gauge type	$d_c = 2$	$d_c = 3$ [47]	$d_c = 4?$

We draw two conclusions from this table:

(a) The conjecture on QCD fits nicely in the general systematics.
(b) The 2-dimensional non-linear σ-model and CP^n models are similar to the conjectured structure of QCD.

The evaluation of the partition function in the neighbourhood of the critical point is complicated because the large correlation length requests a large number of terms of the strong-coupling approximation. At the critical point this expansion diverges. The renormalization-group approach by Kadanoff and Wilson [48], f.i. in the form of a block-spin calculation, was invented to meet this problem.

The renormalization group defines iteratively a
sequence of actions $\{S_\ell\}$, which describes the behaviour
at lattice spacings $\{2^\ell \varepsilon_o\}$. This iteration procedure
is based on the following idea for the evaluation of
the partition function:

$$Z = \int D[\Phi] \, e^{S[\Phi]} \quad .$$

It is suggested that the integrations over the field
configurations on the lattice sites are arranged in
such a way   that one first averages over blocks and
then integrates over the block averages. In formulas:

$$S^{\ell+1} = T[S^\ell]:$$

$$e^{S_{\ell+1}[\Phi^{\ell+1}]} = \int D[\Phi^\ell] K [\Phi^{\ell+1}, \Phi^\ell] e^{S_\ell[\Phi^\ell]}$$

$$\int D[\Phi_n^{\ell+1}] K[\Phi_n^{\ell+1}, \Phi_n^\ell] = 1$$

$$Z = \int D[\Phi^{\ell+1}] e^{S^{\ell+1}[\Phi^{\ell+1}]} = \int D[\Phi^\ell] e^{S^\ell[\Phi^\ell]} \quad . \qquad (5)$$

Here we denote $\Phi^{\ell+1}$ the block-spin average associated
with the block lattice point (x), whereas Φ^ℓ is defined on the lattice points (●). (See Fig.)

Because the lattice size gets doubled, the
correlation length ξ measured in lattice units of
the block lattice is

$$\xi^{\ell+1} = \frac{1}{2} \xi^\ell \quad . \qquad (6)$$

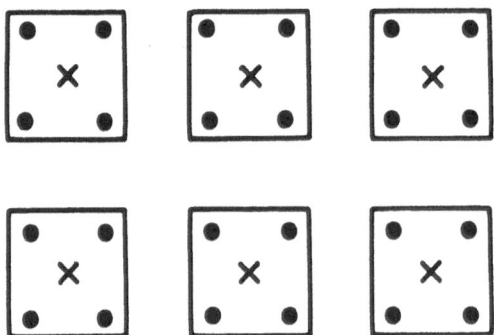

In general the expression of the iterated action becomes more complex than the original one. Nearest
neighbour interactions induce diagonal neighbour
interactions etc. However, under favourable circumstances and with an intelligent choice of the Kadanoff
kernel the general recursion relation

$$S^{\ell+1} = TS^{\ell} \tag{7}$$

might be expressed in recursion relations for a few
coupling parameters a^{ℓ} of the action. Particularly
interesting are the "fix-points" actions of the
renormalization group

$$S^{Fix} = TS^{Fix} \quad . \tag{8}$$

The correlation length of these interactions must be
$\xi^{Fix} = 0$ or $\xi^{Fix} = \infty$, because of Equ.(6). Hence
actions at critical points are fix points of the renormalization group.

The application of the block-spin method [48] to non-abelian gauge theories is up to now only possible with very crude approximations. They support and gave rise to Migdals conjecture. However, since in other examples this Migdal approximation does give wrong results, this evidence of the conjecture is not too strong. A numerical evaluation of the block-spin transformation in Migdal approximation was performed by Baaquie [49]. It supports the expectation for SU(2)-gauge theory that weak coupling at small distances leads after some iterations to strong coupling at large distances.

The discussion above is necessarily somewhat vague. Therefore we want to add a simple example, which allows to perform explicitely all the steps mentioned above: the 1-dimensional Ising model [50]. It even shares with QCD the critical dimension.

The action is

$$\sigma_n = \pm 1, \quad n = 0,1,\ldots N-1$$

$$S[\sigma] = \beta \sum_n \sigma_n \sigma_{n+1}$$

$$\sigma_o = \sigma_N \quad . \tag{9}$$

The strong coupling expansion of the partition function follows directly from the method of Sect. 3.2.Equ. (9).. ..(13). The admissible diagrams are closed lines, hence - Equ.(3.2. 9)

$$Z = \int D[\sigma] e^{\beta S[\sigma]} = \frac{1}{2^N} K^N (1 + X^N)$$

$$K = \cosh \beta, \quad X = \tanh \beta \quad . \tag{10}$$

In evaluating the correlation function $<\sigma_{n_1} \sigma_{n_2}>$ only diagrams which connect the two points contribute:

$$<\sigma_{n_1} \sigma_{n_2}> = (1+X^N)^{-1} (X^{|n_1-n_2|} + X^{N-|n_1-n_2|})$$

$$\rightarrow X^{|n_1-n_2|} \equiv \exp-(|n_1-n_2|/\xi) \tag{11}$$

for the thermodynamical limit $N\rightarrow\infty$.

Hence the correlation length of the 1-dimensional Ising model is

$$\xi(\beta) = -(\log X)^{-1} = \log \tanh \beta \tag{12}$$

$\xi(\beta) \rightarrow \infty$ for $\beta \rightarrow \infty$, hence $g = \beta^{-1} = 0$ is a critical point.

We get the same result with help of the block-spin method. For the lattice size we have to choose $N = 2^M$; to simplify the formulas we add a constant $-2^M \log \cosh \beta$ to the action Equ.(9).

Then the renormalization-group transformation $S' = TS$:

$$e^{S'[\sigma',\beta']} = \int D[\sigma]K(\sigma',\sigma)e^{S[\sigma,\beta]}$$

with

$$K[\sigma',\sigma] = \delta_{\sigma'_n,\sigma_{2n}} \qquad \begin{matrix} n = 1,2,\ldots 2^M \\ n'= 1,2,\ldots 2^{M-1} \end{matrix}$$

can be evaluated with help of the formula

$$\frac{1}{2} \sum_{\sigma_1 = \pm 1} e^{\beta(\sigma_o \sigma_1 + \sigma_1 \sigma_2)} = e^{\beta' \sigma_o \sigma_2 + \beta'}$$

with

$$\beta' = \frac{1}{2} \log \cosh 2\beta \qquad . \tag{13}$$

S' has the same form as S with β substituted by β'.
Hence Equ.(13) represents the recursion relation in
the parameter space induced by the block-spin transformation. $\beta' = \beta$ gives the fixed points $\beta = 0$, ∞.
The recursion relation can be iterated n-times

$$\tanh \beta^{(n)} = (\tanh \beta)^{2n} \qquad . \tag{14}$$

The partition function is invariant under the block-
spin transformation Equ.(5):

$$Z(\beta, 2^M) = Z(\beta^{(M)}, 2) \qquad . \tag{15}$$

Therefore the summation over 2^M lattice sites is
reduced to a summation over only 2 lattice sites
(and substituting β by $\beta^{(M)}$). We get again the
result Equ.(10).

Finally we consider the continuum limit. As
renormalization condition we fix the continuum value
of the correlation length:

$$\kappa = \xi(\beta) \cdot \varepsilon = 1 \quad . \tag{16}$$

Equs. (13) and (16) imply for the cut-off dependence of the renormalized coupling constant:

$$\beta = \text{artanh } e^{-\varepsilon} \quad .$$

Then we put $x = \varepsilon n$, and get for the continuum limit of the 2-point functions

$$<\sigma(x_1)\sigma(x_2)> = \lim_{\varepsilon \to 0} e^{-\frac{|x_1 - x_2|}{\varepsilon \xi(\beta)}} = e^{-|x_1 - x_2|} \quad . \tag{17}$$

The higher correlation function can be treated in a similar way [51].

3.4. Semi-classical approximation

In the continuum limit of the lattice gauge theory, the partition function and the expectation value of the Wilson loop

$$Z(g) = \int D[U] e^{S[U,g]} \tag{1}$$

$$<W(L)> = \frac{1}{Z(g)} \int D[U] \prod_{L} U(^1 c_a) e^{S[U,g]} \tag{1'}$$

must be evaluated near the critical point. At the critical point the strong coupling expansion is diverging. (However, we mentioned the improvement of the strong-coupling approximation by the block-spin method in Sect. 3.3). On the other hand, the conjecture of critical dimension $d_c = 4$ for QCD suggests the absence of a free phase and hence leaves only a limited significance to the usual perturbative weak-coupling expansion. This means, one does not yet know, what are the statistical significant contributions in the evaluation of the partition function of the lattice gauge theory. For a supplementary viewpoint on this problem one might look at the functional Integral (1), i.e. Equ.(1.1. 5) directly in the continuum limit. The discussion of this approach was not selected as a main topic of this lecture. Only for completeness we mention some investigations along this line.

In order to evaluate approximately the functional integral

$$Z[A] = \int D[A]\phi[A]e^{-S_\ell[A]}$$

in the Euclidean region one applies the saddle-point method. For this one has to look for the minimum of the action $S_e = -S > 0$. Field configurations which minimize S_e are solutions of the Euclidean field equations. In particular we discussed in Sect. 2.3.(ii) the instantons, for which the $S_e(A^{in})$ has an absolute minimum for a given topological quantum number. Expanding the field configuration around the classical solution - in the following we consider for definiteness instantons with topological quantum number κ, Equ.(2.3. 11), (16)

$$A_\mu = A_\mu^{cl} + \eta_\mu$$

$$S_\ell = S_\ell[A_\mu^{cl}] + (\eta\Delta\eta)$$

allows the evaluation of Z as a sum of Gaussian integrals:

$$Z[1] = \sum_\kappa e^{-\frac{8\pi^2}{g^2}\kappa} \int_{M_\kappa} d\mu (\det \Delta)^{-1/2} \quad .$$

M_κ instanton manifold with topological quantum number κ. The instanton gives non-perturbative contributions! The evaluation of the determinant of the fluctuation matrix Δ implies the consideration of some particularly important details:

(a) For every instanton parameter there will exist a zero-eigenvalue of Δ, which makes det Δ = 0. This zero energy modes can be separated with help of collective coordinates.

(b) Because the instantons are solutions of the field equations on the compactified Euclidean space-time, the infinite-volume limit ("thermodynamic limit") must be treated carefully.

(c) One considers Z as the grand canonical partition function of an "instanton gas". Its properties depend not only on the contribution from the action: $e^{-S(A^{cl})}$ but also on the instanton phase-space integral

$$\int_{M_\kappa} d\mu (\det \Delta)^{-1/2} \quad ,$$

which is not yet completely known.

Calculations along this line were performed

(i) in SU(2)-gauge theory with Higgs particles in (2+1)-dimensions. The contributions from the Polyakoff-'t Hooft monopoles lead to quark confinement [52].

(ii) The fluctuation determinant in the 1-instanton sector of SU(2)-gauge theory was calculated first by 't Hooft [53]. It was applied to the discussion of confinement in (3-1)-dimensional pure SU(2)-gauge theory in the "dilute-gas approximation" for the multiinstanton sectors. In this treatment confinement could not be shown [54].

(iii) Exact evolution of the instanton contributions in two dimensional non-linear σ-model and in the CP^n model (their critical dimension is $d_c = 2$) demonstrate that the "dilute-gas approximation" does not give correct results [5]. The instanton gas is dense! This is particularly interesting, because "confinement" is shown for the CP^n models in 1/n-expansion [56].

Hence the question, if instanton configuration of gauge fields lead to confinement in QCD, must be considered as open.

4. HAMILTONIAN APPROACH

4.1. Hamiltonian in gauge theories

In the discussion of quark confinement entirely based on the consideration of the Wilson criterion

in pure euclidean gauge field theories the connection
to physical intuition and possible problem solving gets
somewhat lost. For this reason we want to add some
complementary considerations of the quantum theory of
gauge fields, which uses more familiar methodological
notions like Hamiltonian, energy eigenstates, Schrödinger
equation etc.[54].

This approach embeds QCD in a canonical formalism.
It works strictly in the temporal gauge $A_o = 0$. One
considers as Hamilton operator of QCD [58]

$$H = \frac{1}{2} \int d^3x \sum_a \{\vec{E}^{a^2} + \vec{B}^{a^2}\}\tag{1}$$

with the equal-time commutation relations of the
canonical fields

$$A_i^a(x), E_i^a(x) = \frac{\partial}{\partial t}A_i^a(x), B_i^a(x) = \varepsilon_{ik\ell}(\partial_k A_\ell^a + \frac{g}{2}f_{bc}^a A_k^b A_\ell^c)$$

$$[E_i^a(\vec{x}), A_k^b(\vec{x}')] = -i\delta(\vec{x}-\vec{x}')\delta^{ab}\delta_{ik} .\tag{2}$$

The Hamiltonian is invariant under time-independent
infinitesimal gauge transformations $\lambda^a(\mathbf{x})$

$$\delta A_i^a = \frac{1}{g}\partial_i \lambda^a + f_{bc}^a A_i^b \lambda^c ,\tag{3}$$

which are generated by

$$Q[\lambda] = \int d^3x (\frac{1}{g}E_i^a \partial_i \lambda^a + E_i^a f_{abc} A_i^b \lambda^c)$$

$$\delta A_i^a = i[Q, A_i^a] \tag{4}$$

as a consequence of the canonical C.R.(2). Do the Heisenberg equations of motion reproduce the field equation of QCD? Only under this condition we could consider H as the correct Hamiltonian of QCD. We calculate

$$\frac{dA_i^a}{dt} = i [H, A_i^a] = E_i^a$$

$$\frac{dE_i^a}{dt} = i[H, E_i^a]$$

$$= -\epsilon_{ikj}(\partial_k B_j^a + gf_{bc}^a A_k^b B_j^c) \quad . \tag{5}$$

We see that the part of the field equations, which does not contain time derivatives

$$\partial_i E_i^a - gE_i^b f_{bc}^a A_i^c = 0 \tag{6}$$

is not a consequence of the Heisenberg equations. However, since the r.h.s. of Equ.(6) is the generator of the gauge transformations $Q^i(\lambda)$ which commute with H, condition (6) - for all times - is compatible with the Heisenberg equations of motion! We may define physical states as locally gauge-invariant states

$$Q[\lambda]|phys> = 0 \tag{7}$$

The field equations of QCD are then satisfied on
stationary physical states $H|phys> = E|phys>$.

This is the somewhat involved structure of the
canonical quantization formalism for gauge theories.
Of course, the manipulations above are purely formal.
There are many hidden inconsistencies characteristic
of a continuum-field theory. According to our statements of Sect. 1.4, we should start with this
formulation on the lattice with the intention to
perform later the continuum limit with renormalization.
We do not pursue further this line of approach for QCD.
We would meet at the end again the unsolved problem of
confinement - possibly under even less favourable
conditions. Instead of it, we study some two dimensional
gauge models for which confinement is trivial. Therefore these models allow to analyze other physical
features, which should be related to confinement like
string breaking, or the derivation of a non-relativistic
confining potential from a lattice gauge theory.

4.2. Linear potential and string breaking in 2-
dimensional models

This will be an easy exercise for the illustration
of the physical concepts related to the Hamiltonian
approach to Lattice gauge theories.

(i) As the model we choose a scalar charged field $\phi(x)$
in 2 dimensions which is coupled gauge-invariant to
the electric field F. The Hamiltonian of this system
is in the temporal gauge $A_o = 0$, $A_1(x) = A(x)$:

$$H = \int dy \{ \tfrac{1}{2}F^2 + \pi^*\pi + (\partial_y + ieA)\phi^*(\partial_y - ieA)\phi + \mu^2\phi^*\phi \} \quad . \tag{1}$$

The pairs of canonical coordinates are

$$\{F, A\}, \quad \{\pi = \partial_t\phi^*, \phi\}, \quad \{\pi^* = \partial_t\phi, \phi^*\} \quad .$$

Hence we have the commutation relations

$$[F(y), A(y')] = [\pi(y), \phi(y')] = [\pi^*(y), \phi^*(y')] = -i\delta(y-y')$$

$$[F, \pi] = [F, \phi] = \dots = 0 \quad . \tag{2}$$

The time-independent infinitesimal gauge transformations $\lambda(y)$:

$$\delta A = \frac{1}{e}\partial_y\lambda, \quad \delta F = 0, \quad \delta\phi = i\lambda\phi, \quad \delta\pi^* = i\lambda\pi^* \quad \text{etc.} \tag{3}$$

leave the Hamiltonian invariant. They are formally generated by

$$Q[\lambda] = \int dy \{ i\lambda(y)(\pi\phi - \phi^*\pi^*) - \frac{1}{e}F(y)\partial_y\lambda(y) \}$$

$$\delta A = i[Q, A], \quad \delta\phi = i[\Omega, \phi] \quad \text{etc.} \tag{4}$$

It is now an easy calculation to show that the Heisenberg equation of motion generates the field equations.

$$\frac{dA}{dt} = i[H,A] = F \quad ; \quad \frac{d\phi}{dt} = i[H,\phi] = \pi^* \quad , \quad \ldots$$

$$\frac{d\pi^*}{dt} = i[H,\pi^*] = +(\partial_y - ieA)^2\phi - \mu^2\phi, \quad \ldots$$

$$\frac{dF}{dt} = i[H,F] = -ie(\phi^*\partial_y\phi - \phi\partial_y\phi^*) - 2e^2A\phi^*\phi \qquad (5)$$

which are completed by the condition of local gauge invariance on physical states

$$(\partial_y F - ie(\pi\phi - \phi^*\pi^*))|\text{phys}> = 0 \qquad . \qquad (6)$$

This is Gauss' law not contained in the Heisenberg equations (5).

(ii) Now we consider a lattice approximation to our Hamiltonian. According to our considerations of Section 2.2.(ii), we have to set $y = \varepsilon n$, and

$$\phi(y) = \sqrt{\varepsilon}\,\phi(n) \quad , \quad \pi(y) = \frac{1}{\sqrt{\varepsilon}}\,\pi(n)$$

$$F(y) = \frac{1}{\varepsilon}F(n) \quad , \quad \exp(ie\int_{C_n} A(y)\,dy) = U(n).\,(7)$$

The lattice fields are dimensionless; we put also $\bar{e} = \varepsilon e$, $\bar{\mu} = \varepsilon\mu$. The canonical C.R. are:

$$[\pi(n),\phi(n')] = [\pi^*(n),\phi^*(n')] = -i\delta_{nn'}; [\pi(n),\phi^*(n)] = \ldots = 0 \quad .$$

$$[F(n),U(n)] = \bar{e}U(n), \quad [F(n),\phi(n')] = ... = 0 . \tag{8}$$

According to Equ.(2.2. 11) we have to set for the covariant derivative

$$\partial_y - ieA(y) \rightarrow U(n)\phi(n+1) - \phi(n) . \tag{9}$$

Thus the Hamiltonian becomes

$$H = \frac{1}{\varepsilon}\sum_n \{\pi^*(n)\pi(n) + (U^*(n)\phi^*(n+1) - \phi^*(n))(U(n)\phi(n+1) - \phi(n))$$

$$+ \bar{\mu}^2\phi^*(n)\phi(n) + \frac{1}{2}F^2(n)\} . \tag{1o}$$

The gauge properties the basic quantities $O(n)$: $\phi(n)$,
$\phi^*(n)$, $\pi(n)$, $\pi^*(n)$, $U(n)$, $F(n)$ are

$$\phi'(n) = e^{+i\theta(n)}\phi(n) , \qquad \pi'(n) = e^{-i\theta(n)}\pi(n)$$

$$U'(n) = e^{i\theta(n+1)}U(n)e^{-i\theta(n)} , \qquad F'(n) = F(n) . \tag{11}$$

These get generated, $O'(n) = e^{iQ}O(n)e^{-iQ}$, by $Q[\theta]$

$$Q[\theta] = \sum_n (\frac{1}{e}(\theta(n+1) - \theta(n))F(n) + i\theta(n)(\phi^*(n)\pi^*(n) - \pi(n)\phi(n)) .$$

$$\tag{12}$$

In order to simplify the discussion of gauge-invariant
states we introduce charge creation and annihilation
operators:

$$a(n) = \sqrt{\tfrac{\mu}{2}}\phi(n) + i\sqrt{\tfrac{1}{2\bar{\mu}}}\pi^*(n) \quad , \quad a^+(n) = \sqrt{\tfrac{\mu}{2}}\phi^*(n) - i\sqrt{\tfrac{1}{2\bar{\mu}}}\pi(n)$$

$$b(n) = \sqrt{\tfrac{\mu}{2}}\phi^*(n) + i\sqrt{\tfrac{1}{2\bar{\mu}}}\pi(n) \quad , \quad b^+(n) = \sqrt{\tfrac{\mu}{2}}\phi(n) - i\sqrt{\tfrac{1}{2\bar{\mu}}}\pi^*(n)$$

$$\rho(n) = a^+(n)\,a(n) - b^+(n)\,b(n) \quad . \tag{13}$$

The generator of the gauge transformations reads now

$$Q[\theta] = \frac{1}{e}\sum_n \theta(n)\,[F(n)F(n-1) - \bar{e}(a^+(n)a(n) - b^+(n)b(n))] \quad .\tag{14}$$

On gauge invariant states $Q(\theta)|phys> = 0$ we have Gauss' law:

$$(F(n) - F(n-1) - \bar{e}\rho(n))|phys> = 0 \quad .$$

The Hamiltonian gets expressed in $a^{\pm}$, $b^{\pm}$:

$$H = \frac{1}{\epsilon}\sum_n \{\tfrac{1}{2}F^2(n) + \bar{\mu}(a^+(n)a(n) + b^+(n)b(n))$$

$$-\frac{1}{2\bar{\mu}}(U(n)(a^+(n)a(n+1)+b^+(n+1)b(n)) + h.c.)$$

$$-\frac{1}{2\bar{\mu}}(U(n)(b(n)a(n+1)+a^+(n)b^+(n+1)) + h.c.)\} \quad . \tag{15}$$

We consider a Fock representation generated from the "vacuum" $|0>$:

$$F(n)|0> = 0, \quad a(n)|0> = 0, \quad b(n)|0> = 0 \quad . \tag{16}$$

This vacuum is gauge invariant $Q[\theta]|0> = 0$. Because of the pair-creation part in H, last line Equ.(15), $|0>$ is not an eigenstate of the Hamiltonian, i.e. $|0>$ is not the true vacuum. However, in the limit of heavy particles: $\mu/e >> 1$ we may neglect this part; we call the remaining part H_0.

Now we define gauge-invariant "string-creation operators" [59]

$$T(n_0,r) = a^+(n_0+r)\exp(-i\bar{e} \sum_{n_0-r}^{n_0+r+1} A(m))b^+(n_0-r)$$

$$[Q(\theta),T(n_0,r)] = 0 \quad . \tag{17}$$

These are approximations to quantized gauge connections between charged fields: $\phi^*(x)\exp(ie\int_{x'}^{x} A(x)dx)\phi(x')$, which formed the basis of our geometrical considerations of coloured space-time.

From the string states

$$|n_0,r>=T(n_0,r)|0>,F(n)|n_0,r>\approx e[\theta(n_0-r)\theta(n_0+r)]|n_0,r>$$

$$\tag{18}$$

we may get such with a definite momentum $p = 2\pi\alpha/\varepsilon$:

$$|P,r> = \sum_{n_0} e^{2\pi i n_0 \alpha}|n_0,r>$$

$$-\frac{1}{2} < \alpha < \frac{1}{2} \quad . \tag{19}$$

A direct calculation using Equ.(8), (15), (16) yields

$$H_0|\alpha,r> = \frac{1}{\varepsilon}(2\bar{\mu} - \frac{\cos2\pi\alpha}{\bar{\mu}}\Delta_r^2 + \frac{\bar{e}^2}{2}|r| - \frac{2}{\bar{\mu}}\cos2\pi\alpha)|\alpha,r>$$

$$= E|\alpha,r> \tag{20}$$

the Schrödinger equation with linear potential on the 1-dimensional lattice. The continuum limit can be easily performed in setting

$$P = \frac{2\pi\alpha}{\varepsilon}, \qquad \mu = \frac{\bar{\mu}}{\varepsilon}, \qquad e = \frac{\bar{e}}{\varepsilon}$$

constant and letting $\varepsilon \to 0$. For the difference operator $\Delta_r|r> = |r+1> - |r>$ we use $\frac{1}{\varepsilon}\Delta_r \to \partial_y$.

This example should mainly illustrate how practical calculations on a lattice may look like. The main simplifications come from the neglection of particle creation in H; one might expect that this is a reasonable approximation as long as $e/\mu \ll 1$. This condition is never fulfilled if the mass $\mu = 0$. Then the linear potential will play no rôle, and we have a completely different situation. Fortunately there is a model with mass $\mu = 0$, which is exactly solvable in the continuum; namely QED in 2 dimensions with massless quarks - the "Schwinger model":

$$i\gamma^\mu(\partial_\mu - ieA_\mu)\psi(x) = 0$$

$$\partial^\nu F_{\nu\mu}(x) = -e\bar{\psi}\gamma_\mu\psi \qquad . \tag{21}$$

The physical states consist of free neutral particles of mass $\mu = \frac{e}{\sqrt{\pi}}$. If $\Phi(x)$ denotes the field describing these particles

$$\partial_\mu \partial^\mu \Phi + \mu^2 \Phi = 0$$

$$[\Phi(x), \Phi(x')] = i\Delta(x-x'; \frac{e}{\sqrt{\pi}}) \tag{22}$$

then the gauge-invariant quantities derived from $\bar{\psi}, \psi, A_\mu$, namely, the electric field F and the "string operator" T can be expressed in $\Phi(x)$:

$$F_{\mu\nu}(x) = -\frac{e}{\sqrt{\pi}} \epsilon_{\mu\nu} \Phi(x) \tag{23}$$

$$T_{\alpha\beta}(x,x';L) = \psi_\alpha(x) \exp(ie\int_{xL}^{x'} dt^\mu A_\mu(t)) \bar{\psi}_\beta(x)$$

$$=: e^{i\sqrt{\pi}(-\int_x^{x'} dt^\mu \epsilon_{\mu\nu} \partial^\nu \Phi(t) - \alpha\Phi(x) - \beta\Phi(x'))} :N_{\alpha\beta} . \tag{24}$$

Because of the disappearance of the charged ("coloured") particles, the Schwinger model was from the beginning a model for quark confinement [60]. Since the gauge-field strength is proportional to Φ, Equ.(23), there is a mass gap $\Delta = e/\sqrt{\pi}$.

On the lattice the Schwinger model is not exactly solvable. However, approximations show interesting aspects of the "string breaking" described in the continuum by the exact solution [61].

I want to thank my colleagues P. Becher,
N. Kawamoto, H. Krasemann, M. Lüscher, G. Mack and
K. Symanzik for many helpful discussions.

REFERENCES

[1] H.Fritzsch, M.Gell-Mann, XVI.International Conference
    on High-Energy Physics, Chicago-Batavia (1972),
    Vol.2, p.135 (J.D.Jackson et al., eds., NAL (1972)).
    H.Fritzsch, M.Gell-Mann, H.Leutwyler, Phys.Lett.47B
    (1973) 365.
    S.Weinberg, Phys.Rev.Lett. 31 (1973) 494;
    Phys. Rev. D8 (1973) 4482.
    D.J.Gross, F.Wilczek, Phys.Rev. D8 (1973) 3633 and
    ibidem D9 (1974) 980.
    R.Dashen, Proceedings of the 1975 International
    Symposium on Lepton and Photon Interactions at
    High Energies, Stanford (1975), p.981 (W.I.Kirk,
    ed., SLAC (1975)).
[2] G.'t Hooft, Nucl. Phys. B33 (1971) 173;
    Nucl. Phys. B35 (1971) 167.
[3] K.G. Wilson, J.B.Kogut, Phys. Repts. 12C (1974) 75.
[4] M.Böhm, H.Joos, M.Krammer
    a) Nuovo Cim. 7A(1972) 21; Nucl.Phys.B51 (1973)397.
    b) in Recent Developments in Mathematical Physics
       (P.Urban ed.) (Springer Verlag, Wien and New
       York, 1973) p.p. 3-116
    c) TH-1949 CERN.
[5] E. Eichten, K. Gottfried, T. Kinoshita, J. Kogut,
    K.D. Lane, T.M. Yani, Phys.Rev. Lett. 34 (1975) 369.
[6] G. Bhanot, S. Rudaz, Phys. Lett. 78B (1978) 119.

[7] J.D. Bjorken, S.J. Brodsky, Phys. Rev. D1(1970)1416.
J.D. Bjorken, in "Current-Induced Reactions",
Springer Verlag Berlin, Heidelberg, New York
(1975) p.93.

[8] R.F. Schwitters et al., Phys.Rev.Lett. 35 (1975)
1320;
G.Hanson et al., Phys.Rev.Lett. 35 (1975) 1609;
G. Hanson, SLAC PUB 2118 (1978).
Ch. Berger et al., Phys. Lett. 78B (1978) 176;
Phys. Lett. 81B (1979) 410;
see also G. Flügge, this school.

[9] L. Susskind, Hamburg-Konferenz 1977, p.895.
J. Kogut, L. Susskind, Phys. Rev. D9 (1973) 2273.
G. Parisi, Phys. Rev. D11 (1975) 970.

[10] J.A. Swieca, Phys. Rev. D13 (1976) 312.
D. Buchholz, K. Fredenhagen (statement proven only
for abelian gauge fields!).

[11] G. Mack, DESY 77/58.
G. Mack, Phys.Lett. 78B (1978) 263.

[12] B. Schroer, G. Mack, private discussions.

[13] G.'t Hooft, Nucl. Phys. B138 (1978) 1.
G. Mack, Phys. Lett. 78B (1978) 263.

[14] K. Wilson, in New Developments in Quantum Field
Theory and Statistical Mechanics.

[15] K. Wilson, Phys. Rev. D10 (1974) 2445.
J.M. Drouffe, C. Itzykson, Phys. Rep. 38C (1978) 133.

[16] R.D. Parks, ed., "Superconductivity" (Marcel Dekker,
N.Y. (1969)), ch. 2.

[17] H.B. Nielsen, P.Oleson, Nucl. Phys. B61 (1973) 45.
B. Zumino, CERN preprint TH 1779 (1973).
Y. Nambu, Phys. Rev. D10 (1974) 4262.

[18] A.Chodos, R.L. Jaffe, K. Johnson, C.B. Thorn,
V.F. Weisskopf, Phys.Rev. D9 (1974) 3471.
C.G.Callan, R.F.Dashen, D.J.Gross, Phys.Lett. 78B
(1978) 307.

480

[19] J.Schwinger, Phys.Rev. 128 (1962) 2425;
      Theoretical Physics, Trieste Lectures, 1962,
      p.89, I.A.E.A., Vienna (1963).
      J.H. Lowenstein, J.A. Swieca, Ann.Phys.68(1971)172.
      See also: J.Wess, Lectures at the IV.Int. Univ.-
      Wochen für Kernphysik, Schladming 1965.
      P. Becher, H.Joos, DESY 77/43, Proceedings of the
      5th International Winter Meeting on Fundamental
      Physics (Candanchu 1977).
[20] H.A. Kramers, Rapports du 8^e Conseil Solvay (1948)
      p.241.(R.Stoops Brussels 1950).
[21] H.G. Loos, J. Math. Phys. 8 (1967) 2114.
      E. Lubkin, Ann. Phys. (NY) 23 (1963) 233.
      T.T. Wu, C.N. Yang, Phys. Rev. D12 (1975) 3845.
[22] G.'t Hooft, unpublished.
      D.J. Gross, F. Wilczek, Phys. Rev. Lett. 30 (1973),
      Phys. Rev. D8 (1973) 3633; D9 (1974) 980.
      H.D. Politzer, Phys. Rev. Lett. 30 (1973) 1346,
      H. Georgi, H.D. Politzer, Phys. Rev. D9 (1974) 416;
      D. Bailin, A. Love, and D. Nanopoulos, Nuov.Cim.
      Lett. 9 (1974) 501.
[23] K.G.Wilson, Phys.Rev.D10 (1974) 2445, Rev. Mod.
      Phys. 47 (1975) 773.
[24] W. Drechsler, M.E. Mayer, "Fibre Bundle Techniques
      in Gauge Theories", Lecture Notes in Physics, Vol.67
      (A. Böhm, J.D. Dollard, eds., Springer Berlin (1977)).
[25] Textbooks are:
      (a) E.Cartan: Differentialformen, Mannheim,
          Bibliograph.Institut 1974
      (b) I. Singer, J.A. Thorpe, Lecture Notes on
          Elementary Topology and Geometry. Berlin-
          Heidelberg-New York, Springer 1976.

[26] R.Balian, J.M. Drouffe, C.Itzkyson, Phys. Rev. D11
(1975) 2098.

T. Yoneya, Phys. Rev. D18 (1978) 1174.

A. Casher, DESY 78/43.

[27] M. Lüscher, Phys. Lett. 78B (1978) 465.

A. D'Adda, P. Di Vecchia and M. Lüscher, Nucl.
Phys. B146 (1978) 63 and DESY 78/75 (1978).

H. Eichenherr, Nucl. Phys. B146 (1978) 215.

V. Golo and A. Perelomov, ITEP preprints (1978).

[28] J.M. Drouffe, Phys. Rev. D18 (1978) 1174.

Textbooks on Algebraic Topology, f.i. P.S.
Aleksandrov, Combinatorial Topology (Gray Lock,
New York, 1956).

[29] Look for "De Rham Cohomology" in the textbook
I. Singer, J.A. Thorpe, Ref. [25].

[30] K. Wilson, Phys. Rev. D10 (1974) 2445.

A.M. Polyakov, Phys. Lett. 59B (1975) 82.

R. Balian, J.M. Drouffe, C. Itzykson, Phys. Rev.
D10 (1974) 3376, ibid. D11 2098 (1975) 2104.

[31] Superconductivity, Ed. R.D. Park (Marcel Bekker
Inc. NY 1969) Vol. II, ch.6 and 14.

[32] For an introduction: S. Coleman, Erice Lectures
1977.

[33] A. Belavin, A. Polyakov, A. Schwartz, Y. Tyupkin,
Phys. Lett. 59B (1975) 85.

[34] A.S. Schwartz, Physics Letters 67B(1977) 172.

Atiya et al., Proc. Nat. Acad. Scienc. 74 (1977)
2662.

R. Jackiw, C. Nohl, C. Rebbi, Phys. Rev.D15 (1977)
1642.

C.W. Bernard, N.H. Christ, A.H. Guth, C.J. Weinberg,
Phys. Rev.D16 (1977) 2967.

[35] G.'t Hooft, unpublished.

R. Jackiw et al. Ref.[34].

[36] E.F. Corrigan, D.B. Fairlie, S.Templeton,
P. Goddard, Nucl. Phys. B140 (1978) 31.

[37] A.M. Polyakov, Nucl. Phys. B120 (1977) 429.
R. Jackiw, C. Rebbi, Phys. Rev. Lett. 37 (1976)
172.
M. Creutz, T.N. Tudron, Phys. Rev. 16 (1976) 2978.

[38] Instanton configurations on the lattice are
discussed by J. Glimm, A. Jaffe, Comm. **Math.**
Phys. 56 (1977) 195.

[39] K. Wilson, Phys. Rev. D10 (1974) 2445.

[40] W. Fischler, CERN TH 2321 (1977).
Th. Appelquist, M. Dine, BNL 23511.

[41] For an early Review: G.F. Newell, E.W. Montroll,
Rev. Mod. Phys. 25 (1973) 353.
References [30].
M. Creutz, Rev. Mod. Phys. 50 (1978) 561.

[42] J.M. Drouffe, Phys. Rev. D18 (1978) 1174.

[43] K. Osterwalder, E. Seiler, Ann. Phys. (NY) 110
(1978) 440.
G.F. De Angelis, D. de Falco, F. Guerra, Lett.
Nuov. Cim. 19 (1977) 55.

[44] J. Glimm, A. Jaffe, Comm. Math. Phys. 51 (1976) 1.

[45] H.G. Dosch, V.F. Müller, Phys. Lett. 74B (1978) 241.

[46] A.A. Migdal, Zh. Eksp. Teor. Fiz. 69 (1975) 810,
69 (1975) 1477, (Sov. Phys. 42 (1975) 413, 42
(1975) 743).

[47] The critical dimension d for the Abelian gauge
group is discussed by T. Banks, R. Myerson,
J. Kogut, Nucl. Phys. B129 (1977) 493 and by
J. Glimm, A. Jaffe Ref.[38].

[48] L.D. Kadanoff, Physics 2 (1966) 263; Rev. mod.
Phys. 49 (1977) 267.
K.G.Wilson, Rev. Mod. Phys. 47 (1975) 773.

[49] A.A. Migdal, ref. [46].

B.E. Baaquie, SLAC-PUB 1964.

[50] B.M. McCoy, T.T. Wu, The Two-Dimensional Ising Model, Harvard University Press, Cambridge (Mass.) (1973), and literature quoted there.

[51] D. Isaacson, Comm. Math. Phys. 53 (1977) 257.

[52] A.M. Polyakov, Nucl. Phys. B120 (1977) 429.

[53] G.'t Hooft, Phys. Rev. D14 (1976) 3432,

Erratum: Phys. Rev. D18 (1978) 2199.

[54] C. Callan, R. Dashen, D. Gross, Phys. Rev. D17 (1978) 2717.

[55] V.A. Fateev, I.V. Frolov and A.S. Schwarz, ITEP preprint (1979). B. Berg, M. Lüscher, DESY 79/17.

[56] A.D'Adda, P. Di Vecchia, M. Lüscher, Nucl. Phys.B. G. Lazarides, DESY 79/01.

[57] Coty, F. Banks, S. Raby, S. Susskind, D.R. Jones, P.N. Scharbach, D.K. Sinclair, Phys. Rev. D15 (1977) 1111.

[58] M. Lüscher, Comm. Math. Phys. 54 (1977) 283.

[59] H. Suura, DESY 79/25.

S. Mandelstam, Phys. Rev. 175 (1968) 1580.

Y. Nambu, Phys. Lett. 80B (1979) 372.

T. Matsumoto, Univ. of Tokyo Preprint, UT-Komaba 78-10.

H. Leutwyler and J. Stern, Orsay preprint 1978 IPNO/TH 7.

A.M. Polyakov, Phys. Lett. 82B (1979) 247.

These authors consider dynamical equations based on string operators for 4-dimensional QCD and QED.

[60] A. Casher, J. Kogut, L. Susskind, Phys. Rev. Lett. 31 (1973) 792.

[61] P. Becher, H. Joos (in preparation).

Acta Physica Austriaca, Suppl. XXI, 485—558 (1979)
© by Springer-Verlag 1979

SOME CURRENT ISSUES IN GAUGE THEORIES[+]

by

E.A. PASCHOS[*][::]

Institut für Theoretische Physik III
Universität Dortmund, Postfach 500500
D-4600 Dortmund

CONTENTS

1. INTRODUCTION
2. NATURAL CONSERVATION OF FLAVOR
3. RESTRICTIONS ON THE MIXING ANGLES
4. NEUTRAL CURRENTS
    a. General Properties
    b. Neutrino Interactions
    c. Other Interactions

[+]Lectures given at the
XVIII. Internationale Universitätswochen für Kernphysik,
Schladming, Austria, February 28 — March 10, 1979.
[*]On leave of absence from Brookhaven National Laboratory
[::]Supported in part by Deutsche Forschungsgemeinschaft.

5. QCD PREDICTIONS FOR DEEP INELASTIC SCATTERING
   a. Renormalization of the Coupling Constants
   b. Scaling Corrections to the Structure Functions
   c. Experimental Comparisons
   d. Scaling Behaviour of the Quark-Fragmentation
     Functions
  ACKNOWLEDGEMENT
  REFERENCES
  FIGURE CAPTIONS

1. INTRODUCTION

The successes of gauge theories in explaining
many outstanding issues of the weak and electromagnetic
interactions created the optimism that a gauge theory
could also explain many problems of strong interactions as well. This led to the present picture where
there is a successful theory of the electroweak force
and a promising candidate for the theory of the strong
interactions. In these lectures I address two main
questions:
1) How good is the evidence in favor of SU(2) x U(1)?
2) How convincing is the evidence for QCD?

In attempting to answer these questions I discuss
several successful predictions of the theories and
also some unsettled problems and uncertainties
occuring in the determination of the parameters.

The subject matter is too large and a selection
was made. In the case of the weak and electromagnetic
interactions I discuss the conservation of flavor by
neutral couplings, the determination of the mixing
angles, and the status of neutral currents. This is
the subject matter of sections 2, 3 and 4. One interesting

result is the accurate determination of the Weinberg angle whose experimental value is in the proximity of the value predicted by the SU(5) grand-unified theory. The calculation of the mixing angle in SU(5) indicates that the angle is stable to higher-order corrections.

Section 5 covers the predictions of QCD for deep-inelastic scattering. The main quantitative tests of QCD deal with the renormalization of the coupling constants and scaling violations in deep-inelastic scattering. In general the predictions are in qualitative agreement with experiment. However, the comparisons are far from conclusive because the results are still sensitive functions of the underlying assumptions. Furthermore the predictions for deep inelastic scattering could be reproduced by alternative schemes. The tests so far are consistency checks. At this stage it is desirable to find additional tests of QCD which are qualitatively different from the predictions of the quark-parton model.

2. NATURAL CONSERVATION OF FLAVOR

After long activity in building unifying models of the weak and electromagnetic interactions it has become evident that the SU(2) x U(1) model [1] gives an accurate description of present day experiments. The choice of this group is the simplest one which still unifies the two forces. One of course can still contemplate larger groups for aesthetic or other reasons, but the main motivation resulting from conflicts with several experimental "anomalies" has now disappeared.

In such theories an immediate issue after the
selection of the group is the representation assignments
for the fermion, scalar, and other fields. Two requirements have been found which dictate the representation
of the fermion fields:

(i) Cancellation of the triangle anomalies, which
implies a relation between quark and lepton
fields;

(ii) natural conservation of flavor, which restricts
the weak T^2 and T_3 values for the quark and
lepton fields.

The cancellation of triangle anomalies [2] is necessary
in order to have a renormalizable theory. The Adler-
Bell-Jackiw [3] anomaly, Fig.1, requires that

$$\sum_i \tau_3 Q^2 = \frac{1}{2} \sum_i [Q_i^2 - (Q_i-1)^2] = \sum_{\text{all charges}} Q_i = 0 \quad (2.1)$$

with i being summed over all multiplets.

In addition another anomaly occurs when neutral
currents and gluons are present in the same theory as
shown in Fig.2. The couplings at the vertices of the
diagram are to an axial isovector current and two
gluons. The theories to be considered are anomaly-
free. This is evident by observing that the SU(4)
currents

$$A_\lambda^8 = \frac{1}{\sqrt{6}} [\bar{u}\gamma_\lambda\gamma_5 u + \bar{d}\gamma_\lambda\gamma_5 d - 2\bar{s}\gamma_\lambda\gamma_5 s] \quad (2.2)$$

and

$$A_\lambda^{15} = \frac{1}{\sqrt{12}} \; [\bar{u}\gamma_\lambda\gamma_5 u + \bar{d}\gamma_\lambda\gamma_5 d + \bar{s}\gamma_\lambda\gamma_5 s - 3\bar{c}\gamma_\lambda\gamma_5 c] \qquad (2.3)$$

are indeed anomaly-free. The cancellation of anomalies establishes a correspondence between leptons and quarks. In the models to be discussed they demand that the discovery of new quarks implies the existence of more leptons and vice versa. This requirement is consistent with the discovery of the b-quark and of the τ-lepton. It predicts the existence of the t-quark, but it does restrict the number of fundamental fields or their mixing angles.

Natural conservation of flavor is motivated by the very restrictive bounds for the absence of strangeness-changing and charm-changing neutral couplings. In order to make this statement quantitative consider the effective interaction

$$L_{eff} = \frac{G}{\sqrt{2}} \; \{\bar{s}\gamma^\mu(h_V^s + h_A^s\gamma_5)d \; \bar{s}\gamma_\mu(h_V^s + h_A^s\gamma_5)d$$

$$+ \; \bar{c}\gamma^\mu(h_V^c + h_A^c\gamma_5)u \; \bar{c}\gamma_\mu(h_V^c + h_A^c\gamma_5)u + \ldots \} \qquad (2.4)$$

where h_V^i and h_A^i are phenomenological coupling constants determined by experiments as shwon in Table 1. The large suppression of the strangeness-changing neutral couplings is well known from the K^0-$\bar{K}^0$ system and led to the Glashow-Iliopoulos-Maiani scheme [4]. The bounds on the charm-changing couplings are derived from observations on the mixing properties of the D^0-$\bar{D}^0$ system and it is worth reviewing it.

Consider the situation where a state D^o is produced at the initial time. As time goes on the state evolves into a superposition of $\bar{D}^o$ and D^o, the former contributing to abnormal decays. The net effect is that over long periods of time there are normal decays into

$$\ell^+ + \nu + \ldots \qquad \text{or} \qquad K^- + \ldots$$

and abnormal decays into

$$\ell^- + \bar{\nu} + \ldots \qquad \text{or} \qquad K^+ + \ldots \qquad .$$

Denoting by N^- the number of abnormal decays into a specific channel or sum of channels and by N^+ the normal decays into the corresponding channels, it was found [5] that

$$\rho_o \equiv \frac{N^-}{N^- + N^+} = \frac{1}{2} \frac{4(\Delta m)^2 + (\Gamma_L - \Gamma_S)^2}{4(\Delta m)^2 + (\Gamma_L + \Gamma_S)^2} \qquad (2.5)$$

where Δm is the mass difference between $D_L - D_S$ and Γ_L, Γ_S are the respective widths. Similarly, consider the production of a $D\bar{D}^o$ pair in the reaction

$$e^+ + e^- \to D^o + \bar{D}^o \qquad (2.6)$$

and study its time development. In the one-photon approximation the produced $D^o \bar{D}^o$ state has $C = -1$ and produces normal decays into $(K^+ K^- h_1)$, as well as abnormal decays into $(K^+ K^+ h_2)$ and $(K^- K^- h_3)$ where h_1, h_2,

h_3 are final states with total strangeness zero. Denoting the number of decays into each system by $N^{-+} +$ $+ N^{+-}$, N^{++} and N^{--}, respectively, it was found [6] that

$$\rho_1 \equiv \frac{N^{++} + N^{--}}{N^{+-} + N^{-+} + N^{++} + N^{--}} = \rho_0 \qquad (2.7)$$

ρ_0 being the same expression of masses and widths that occurs in Eq. (2.5). Relations (2.5) and (2.7) are general and they hold for other pairs of particles containing heavier quarks, like $B_d^o - \bar{B}_d^o$, $B_s^o - \bar{B}_s^o, \ldots$. If an experimental upper bound for ρ_1 is known, then it follows

$$\Delta m \leq (\frac{2\rho_1}{1-2\rho_1})^{1/2} \frac{(\Gamma_L + \Gamma_S)}{2} \qquad . \qquad (2.8)$$

Estimates of the widths give [7]

$$\Gamma_{11} = \frac{1}{2} (\Gamma_L + \Gamma_S) = \frac{G^2 M_D^5}{192\pi^3} g(\varepsilon) \qquad (2.9)$$

where $\varepsilon = M_s/M_c$ and

$$g(\varepsilon) = 1 - 8\varepsilon^2 - 24\varepsilon^4 \ln\varepsilon + 8\varepsilon^6 - \varepsilon^8 \qquad .$$

The mass difference is estimated from the effective Lagrangian as

$$M_{12} = \frac{G}{\sqrt{2}} (h_V^2 \text{ or } h_A^2) \; 2 \; f_D^2 m_D \qquad (2.10)$$

492

where h_V and h_A are assumed to be comparable and [8] the reduced matrix element is approximated by

$$<D^0|[\bar{c}\gamma_\mu(1-\gamma_5)u]^2|\bar{D}^0> = 4 \frac{(f_D M_D)^2}{2M_D} \quad . \tag{2.11}$$

For the D-decay coupling constant there are numerous estimates [9] using different approaches; a realistic value is f_D = 300 MeV. The above results together with the experimental bound [10]

$$\rho_1 \leq 15 \text{ \%} \qquad \text{at} \qquad 90\text{\% c.l.} \tag{2.12}$$

give the stringent bound [11] occuring in Table I. The estimate of the reduced matrix element can be improved by a calculation in the bag model [12]; however, the above estimate is conservative and suffices for these lectures. The reason that the bound on h_V or h_A is so restrictive is evident once it is realized that M_{12} is first-order weak while Γ_{11} is second-order weak.

It is of interest to determine what are also the experimental bounds for the mixing of flavor for the heavier quark systems like $\bar{b}d$, $\bar{b}s$, $\bar{t}u$, $\bar{t}c$ ... Some of them will be forthcoming in e^+e^- experiments and I will return to this topic later on.

In gauge theories [13,14] the large suppressions are incorporated by demanding that

i) Direct flavor-changing neutral couplings are suppressed to this level of accuracy. This is satisfied by choosing some of the coupling constants arbitrarily

small. What is more natural is to construct gauge
theories where direct flavor-changing neutral
couplings are absent.

ii) Even with the above choice flavor changing
effects can be induced by higher-order corrections
[13,14] that involve charged currents, diagram 3.
Higher-order effects are of $O(G)$ and contradict
the two entries in Table I. Such effects are
acceptable if they occur at the level $G\alpha \left(\frac{m_Q}{M_W}\right)^2$.

iii) If the symmetry breaking is through Higgs' mechanism,
then the resulting neutral couplings should preserve
flavor to the accuracy of Table I.

Couplings between quark pairs	Bound on h_A or h_V
$\bar{s}d$	2.5×10^{-4}
$\bar{c}u$	6.8×10^{-4}
$\bar{b}d$	?
$\bar{b}s$	?
$\bar{t}u$	?

Table I

Before one implements requirements (i) - (iii) into a
theory it is important to decide whether the suppression
in Table I should be limited to the strangeness- and

charm-changing neutral couplings or should be extended
to the bottom, top or other heavier quarks. I discuss
two cases.

Case A: All entries in Table I are of $O(10^{-4})$ or smaller.

Case B: The entries in Table I are of $O(\alpha \frac{m_Q^2}{m_W^2})$ where m_Q is
the mass of the quark.

The implementation of case (B) in SU(2) x U(1) theories
requires that all quarks of the same charge and same
chirality have the same [13,14] weak T^2 and T_3. This
can be implemented without adjusting any parameters of
the theory. It predicts that the large suppression of
flavor-changing couplings is a transient phenomenon,
which in general will not be true for heavy quarks whose
masses are comparable to the masses of intermediate
gauge bosons.

Case A is much harder to implement. In addition
to the requirements described in the previous paragraph
it also demands that some mixing angles, like those of
heavy to light quarks, must be small, or the existence
of new conservation laws [15]. There is also the
possibility of enlarging the group [16,17] and
selecting specific representations.

For SU(2) x U(1) models with quarks that carry
two distinct charges the above discussion requires
that all left-handed multiplets are doublets or
singlets. The experimental evidence obviously requires
left-handed doublets. Similarly all right-handed multiplets must have the same dimensionality. Right-handed
doublets could only couple light quarks to very heavy
quarks; but in this case the mixing angles must be

appropriately chosen so that the heavy states are not
yet produced in experiments and in addition they do not
induce large flavor-changing second-order corrections.
The simpler choice is to have right-handed singlets.
The fermion states are then assigned as follows

$$\begin{pmatrix} q_u \\ q_d \end{pmatrix}_L \quad ; \quad q_{uR} \; , \qquad q_{dR} \quad . \tag{2.13}$$

$$\begin{pmatrix} \nu_e \\ e^- \end{pmatrix}_L \quad ; \quad \bar{e}_R \; , \qquad ? \quad . \tag{2.14}$$

Here q_u can always be chosen to be quarks of definite
flavor and charge 2/3, while q_d are "unitary super-
positions" of quarks with charges -1/3 but different
flavors. This is the sequential model with the above
families of multiplets repeated many times. In the
sequential model the fermion multiplets couple only
to scalar doublets. Any number of doublets give masses
to the intermediate vector bosons which satisfy the
relation

$$\rho = \frac{M_W^2}{M_Z^2 \cos \theta_W} = 1 \quad . \tag{2.15}$$

This ratio is determined by neutral-current measurements
and has been established to be very close to one. Thus
there are compelling arguments in favor of the sequential
SU(2) x U(1) model arising from the natural conservation

of flavor and from measurements in neutral currents. But
even with this choice there are several unsolved problems.
For instance, in Eq.(2.14) there is no right-handed
neutrino and the direct correspondence between quarks
and leptons is broken. Then there is no understanding
or good reason for the large number of families like
(2.13) and (2.14). At the moment there is strong evidence
for at least three quark and three lepton families. In
this situation the masses and mixing angles of the quarks
are independent parameters whose number is alarmingly
large. In the next section I will describe restrictions
on the parameters and discuss some experiments which are
crucial for further development of the subject.

3. RESTRICTIONS ON THE MIXING ANGLES

Before discussing the bounds on the mixing angles
I outline the steps through which the mixing matrix is
obtained. The Lagrangian in the symmetry limit is

$$L = \bar{R} \not{D} R + \bar{L} \not{D} L + L_{YM}(W,A,Z) + L_\phi + h.c. \qquad (3.1)$$

where the covariant derivatives $D_\mu = \partial_\mu - ig\tau^i A^i_\mu$ and L
and R are fermion multiplets with left and right helicit-
ies belonging to representations of the group. L_{YM} is
the part containing the gauge fields and L_ϕ contains the
Higgs field. In the sequential model with one Higgs
doublet the Yukawa couplings are given

$$L_\phi = \sum_{i,j} C_{ij} (\bar{q}^u \bar{q}^d)_{iL} \begin{pmatrix} \phi^u \\ \phi^d \end{pmatrix} q^u_{jR} +$$

$$+ \sum_{i,j} C'_{ij} (\bar{q}^u \bar{q}^d)_{iL} \begin{pmatrix} \phi^u \\ \phi^d \end{pmatrix} q^d_{jR} + L_{YM}(\phi) - V(\phi) \quad . \quad (3.2)$$

C_{ij}, C'_{ij} are arbitrary constants and $\phi^u (\phi^d)$ are Higgs fields with upper and lower charges. The quark fields are still massless and unphysical. When the symmetry is spontaneously broken by giving non-zero expectation values to the neutral Higgs a fermion mass matrix is generated. This matrix is neither diagonal nor hermitean, but it can always be brought into diagonal form by unitary transformations acting to the left and to the right. Denote the mass matrices for up and down quarks by m^u and m^d, respectively; then the unitary transformations are defined by

$$U_L^+ \, m^u \, U_R = D^u$$

$$V_L^+ \, m^d \, V_R = D^R \qquad\qquad\qquad (3.3)$$

where D^u and D^R are diagonal matrices, whose elements correspond to quark masses. The physical quark states are defined as

$$u_L = U_L \, q^u$$

$$d_L = V_L \, q^d \qquad\qquad\qquad (3.4)$$

and the charged currents by

$$j_\mu^+ = \bar{q}_L^{-u} \gamma_\mu q_L^d = \bar{u}_L \gamma_\mu U_L^+ V_L d_L \qquad . \qquad (3.5)$$

The matrix

$$V = U_L^+ V_L \qquad (3.6)$$

is the generalized GIM matrix. For N flavors the N x N unitary matrix has N^2-parameters, which can be separated into $N(N-1)/2$ "angles" θ_i (the real parameters of $O(N)$) and $N(N+1)/2$ phases. A redefinition of the 2N-1 relative phases of u_L and d_L reduces the number of phases to $(N-1)(N-2)/2$. Thus for N = 3 there are 3 angles and one intrinsic phase which gives CP-violating effects. For N = 4 there are 6 angles and 3 phases. For six quarks the standard matrix was defined by Kobayashi and Maskawa [18]

$$V = \begin{pmatrix} c_1 & -s_1 c_3 & -s_1 s_3 \\ s_1 c_2 & c_1 c_2 c_3 - s_2 s_3 e^{i\delta} & c_1 c_2 s_3 + s_2 c_3 e^{i\delta} \\ s_1 s_2 & c_1 s_2 c_3 + c_2 s_3 e^{i\delta} & c_1 s_2 s_3 - c_2 c_3 e^{i\delta} \end{pmatrix} \qquad (3.7)$$

where $s_i = \sin \theta_i$ and $c_i = \cos \theta_i$. This matrix has the property that when all angles are chosen to be small and $s_i = O(\varepsilon)$ then some elements, V_{13}, are of second order in this parameter.

At this point the subject branches into two directions:

(α) Numerous attempts that try to relate the Cabbibo-type angles to the masses of the quarks and (β) realistic bounds imposed on the mixing angles. The general strategy in the first case is to impose a discret symmetry on the Yukawa couplings, which reduces the number of independent elements in the mass matrices. The results of this approach are relations between the eigenvalues and eigenfunctions of the mass matrix [19,20]; that is between the masses and the mixing angles. Realistic relations have been obtained in models based on the gauge group $SU(2)_L \times SU(2)_R \times U(1)$, but the good relations are lost when the gauge group is restricted to $SU(2) \times U(1)$. Here I will follow the second path and describe the known constraints on the mixing angles.

Estimates of the elements of V are made by drawing from several experiments. A general analysis of β-semi-leptonic- and K_{e3}-decays by Shrock and Wang [21] updated the analysis of the experimental data and included estimates of experimental errors as well as estimates of the theoretical uncertainties in the calculations. They obtained

$$\cos \theta_c = V_{11} = 0.973 \pm 0.0025 \qquad (3.8)$$

from β-decay,

$$V_{12} = 0.222 \pm 0.003 \qquad (3.9)$$

from hyperon decays and

$$V_{12} = 0.219 \pm 0.003 \qquad (3.10)$$

from K_{e3} decay. From these

$$\theta_1 = 13.17^\circ \tag{3.11}$$

and $c_3 = V_{12}/s_1 = 0.96$ or $s_3 = 0.28 \pm 0.22$. $\tag{3.12}$

These results within errors are consistent with earlier determinations [22]. To obtain restrictions of θ_2 it is necessary to consider higher-order corrections involved in the mixing of the $K^\circ - \bar{K}^\circ$ and $D^\circ - \bar{D}^\circ$ systems. I consider the general case for an off-diagonal mass-matrix element.

Consider a bound state consisting of a quark-antiquark pair of lower charge $M^\circ(q_i^d \bar{q}_j^d)$ and its anti-particle $\bar{M}^\circ(\bar{q}_i^d q_j^d)$. The mass-matrix element is given by

$$\langle \bar{M}^\circ | L_{eff} | M^\circ \rangle = \frac{G^2}{16\pi^2} \cdot$$

$$\text{Re}\{ \sum_\alpha m_\alpha^2 (V_{i\alpha} V_{j\alpha}^*)^2 + \sum_{\alpha \neq \beta} \frac{m_\alpha^2 m_\beta^2}{m_\beta^2 - m_\alpha^2} \ln(\frac{m_\beta}{m_\alpha})^2 (V_{i\alpha} V_{j\alpha}^* V_{i\beta} V_{j\beta}^*) \}.$$

$$\cdot \langle \bar{M}^\circ | \bar{q}_i \gamma_\mu (1+\gamma_5) q_j \ \bar{q}_i \gamma^\mu (1+\gamma_5) q_j | M^\circ \rangle \tag{3.13}$$

where m_i is the mass of the ith up-quark and V_{ij} are elements of the generalized GIM matrix. Combining this with Eq. (2.8)-(2.11) we obtain [23]

$$\sum_\alpha m_\alpha^2 (V_{i\alpha} V_{j\alpha})^2 \lesssim (\frac{2\rho_1}{1-2\rho_1})^{1/2} \frac{M_m^4}{12\pi f_m^2} \cdot \tag{3.14}$$

The application of such arguments to the $K^0 - \bar{K}^0$ system gave an accurate estimate [24,25] for the mass of the charm quark. Including the t-quark as well its contribution must be smaller in order for it not to spoil the good estimate. This gives [22]

$$s_2^2 \leq 0.17 \quad \text{for} \quad m_t \geq 18 \text{ GeV/c}^2 \, . \tag{3.15}$$

This argument can be generalized to even heavier quarks.

Transcribing the formalism to the $D^0-\bar{D}^0$ states together with the experimental bound on their mixing leads to a new constraint on the mixing angles. The new constraint is weaker than those occuring in Eq.(3.8) - (3.12).

In summary the conditions on the mixing angles are given in Table II.

Angle	$\sin \theta_i$	θ_i
$\theta_1 = \theta_c$	0.23 ± 0.01	$13.17 \pm 0.64^\circ$
θ_2	$0.1 < s_2 < 0.70$	$5.7 - 44.4^\circ$
θ_3	0.28 ± 0.22	$3.4 - 30.0^\circ$

Table II

The bounds on the angles lead to several predictions for states containing heavy quarks.

(α) The life-times of states containing b-quarks have been [26] estimated

$$10^{-14} \leq \tau_b \leq 10^{-11} \text{ sec.} \tag{3.16}$$

The experimental bound for long-lived states in the mass range of 4 to 10 (GeV/c^2) is

$$\tau_b < 5 \times 10^{-8} \text{ sec.}$$

(β) The decays of the b-quark follow mostly the sequence

$$b \rightarrow c \rightarrow s \tag{3.17}$$

leading to strange particles. By studying the charge topologies of the K's we can determine mixing parameters analogous to those occuring in Eq.(2.5) and (2.7). Similar studies can be made by observing the charges of the leptons in the final state. But now there is a confusion among the leptons arising from the two steps of the decay chain (3.17). Estimates of the decay distributions show that for $E_1 \geq 1.2$ GeV the majority of leptons come from the primary decay.

(γ) The small couplings of b to light quarks sets a bound for the production rate of such states by neutrinos or antineutrinos [22]. The fraction of the total cross section which contains a b-state is less than 1/2% which is roughly the level of the experimental measurements.

(δ) Arguments similar to those discussed in section B predict the mixing properties of the systems [27,28]

$$B_d(b\bar{d}) - \bar{B}_d(\bar{b}d)$$

and

$$B_s(b\bar{s}) - \bar{B}_s(\bar{b}s) \quad .$$

From the unitarity of the K-M matrix both mixings cannot be suppressed. In fact, for the B_s-$\bar{B}_s$ system substantial mixing is predicted, which is observable once B_s and $\bar{B}_s$ are produced in e^+e^--experiments [29]. In the sequential six-quark model there is no alternative. So the observation of such mixing will be a remarkable test for the estimates of higher-order effects. Even more dramatic would be the absence of such mixing at the 10% level, because then it implies that the sequential six-quark model as described is incomplete. One way out would be to enlarge the model to eight quarks where for several choices of the mixing angles the B_d-$\bar{B}_d$ and B_s-$\bar{B}_s$ mixing is eliminated.

4. NEUTRAL CURRENTS

a.General Properties

The subject of neutral currents is now well-founded and the determination of their coupling constants is becoming very precise. There are several excellent reviews of the subject matter [30-32] and what is intended in

504

this section is an up-to-date summary with emphasis on
the new results and new issues. The highlights over the
past year were

1) Observation of parity violation [33] in electron-
   deuteron experiments at SLAC.
2) Observation of parity violation [34] in atomic
   experiments.
3) Model-independent (almost) determinations [35,36,37]
   of the neutral-current couplings and of the Weinberg
   angle.
4) Revived interest on the axial - Baryon [38] (isoscalar) current.

The reactions that have been investigated are shown
schematically:

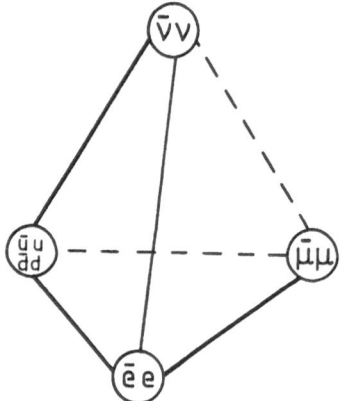

There are experimental results for most links and those
missing are related to others by ν-e universality, as
is the case with $\bar{\nu}\nu$-$\bar{\mu}\mu$ versus $\bar{\nu}\nu$-$\bar{e}e$,... . One of the
reactions $e^+e^- \rightarrow \mu^+\mu^-$ is expected to be observed with-
in a year at PETRA and PEP. The experiments are of
several types; some are with accelerator beams like

neutrinos and electrons, others in atoms and still others are colliding-ring experiments.

The first question on neutral currents concerned their existence. Once this question was settled a new issue arose concerning their space-time structure; are they V and A or Scalar (S), Pseudoscalar (P) and Tensor (T)? There are two measurements which show that the observed couplings are vector and axial. The y-distribut- ions in ν-experiments give the bounds

$$\frac{g(S,P)}{g(V-A) + g(V+A)} \quad = \quad -0.04 \pm 0.03 \quad \text{(BEBC)}$$

$$= \quad 0.02 \pm 0.07 \quad \text{(CDHS)} \tag{4.1}$$

where the quantity g(i) means the coupling of the ith type. In addition the observation of a parity-violating effect at SLAC implies that it is the interference of the electromagnetic current with the axial current, since (in the limit of neglecting the electron mass) scalar, pseudoscalar, and tensor interactions do not interfere with vector interactions. Thus the neutrino and electron interactions to be discussed here are mostly vector and axial.

b. Neutrino Interactions

The effective neutral-current interaction- Lagrangian is defined by

$$L_{INT} = \frac{G}{\sqrt{2}} \, \bar{\nu}\gamma^{\mu}(1+\gamma_5)\nu J_{\mu}^{z} \tag{4.2}$$

with the hadronic J_μ^z frequently given by one of two
equivalent forms. Both of the forms are expressible in
terms of the quark fields

$$J_\mu^z = \bar{u}\gamma_\mu \{u_L(1+\gamma_5) + u_R(1-\gamma_5)\}\, u +$$

$$+ \bar{d}\gamma_\mu \{d_L(1+\gamma_5) + d_R(1-\gamma_5)\}\, d +$$

$$+ \bar{s}\gamma_\mu \{s_L(1+\gamma_5) + s_R(1-\gamma_5)\}\, s +$$

$$+ \bar{c}c + \ldots \tag{4.3}$$

or alternatively

$$J_\mu^z = \frac{1}{2}\, \{\bar{u}\gamma_\mu (\alpha + \beta\gamma_5)u - \bar{d}\gamma_\mu (\alpha + \beta\gamma_5)d\} +$$

$$+ \frac{1}{2}\, \{\bar{u}\gamma_\mu (\gamma + \delta\gamma_5)u + \bar{d}\gamma_\mu (\gamma + \delta\gamma_5)d\} +$$

$$+ \bar{s}s + \bar{c}c + \ldots \qquad . \tag{4.4}$$

The constants α, β, γ, and δ are the coupling constants
of the vector-isovector, axial vector-isovector, vector-
isoscalar and axial vector-isoscalar currents, respectively.
Similarly, u_L, u_R, etc. are the coupling constants for
the left-handed up-quark current, right-handed up-quark
current, and so on. However, the use of these forms does
not require a quark picture since the occurance of the
quarks is merely a mnemonic for the transformation
properties of various parts of the total hadronic neutral
current. The chiral couplings $u_L,\ldots,d_R$ were introduced

by Bjorken [39] and the couplings $\alpha,...,\delta$ by Sakurai [40]. They are both included here since each of them is useful in specific calculations.

In the Weinberg-Salam model

$$u_L = \rho(\tfrac{1}{2} - \tfrac{2}{3} \sin^2\theta_w)$$

$$u_R = \rho(- \tfrac{2}{3} \sin^2\theta_w)$$

$$\text{with } \rho = \frac{M_W^2}{M_Z^2 \cos\theta_w} \quad .$$

$$d_L = \rho(- \tfrac{1}{2} + \tfrac{1}{3} \sin^2\theta_w)$$

$$d_R = \rho(\tfrac{1}{3} \sin^2\theta_w) \quad . \tag{4.5}$$

For any number of Higgs doublets $\rho = 1$, so this quantity tests the isospin assignment of the Higgs field. In addition, in this model $\delta = 0$; the baryon axial coupling vanishes, but there is still an isoscalar axial contribution arising from $\bar{s}s$, $\bar{c}c$ and other terms. Such quantities are of theoretical interest by themselves and I shall return to them later.

From general arguments the cross sections on isoscalar targets are given

$$\sigma_{NC} = \tfrac{1}{2} \{\alpha^2\sigma_V + \beta^2\sigma_A + \alpha\beta\sigma_I\} + s^0 \tag{4.6}$$

$$\bar{\sigma}_{NC} = \tfrac{1}{2} \{\alpha^2\sigma_V + \beta^2\sigma_A - \alpha\beta\sigma_I\} + \bar{s}^0 \tag{4.7}$$

where S^O, $\bar{S}^O$ are the contributions to the cross section from the isoscalar current only, σ_V is the contribution of the isovector vector current alone, σ_A receives contributions from isovector axial and σ_I is the interference term.

For charge currents on the other hand

$$\sigma_{CC} = \sigma_V + \sigma_A + \sigma_I \tag{4.8}$$

$$\bar{\sigma}_{CC} = \sigma_V + \sigma_A - \sigma_I \quad . \tag{4.9}$$

Then

$$\frac{1}{2}\alpha\beta = \frac{\sigma_{NC} - \bar{\sigma}_{NC}}{\sigma_{CC} - \bar{\sigma}_{CC}}$$

provided that $S = \bar{S}$. In the Weinberg-Salam model

$$\frac{1}{2}\alpha\beta = \frac{1}{2} - \sin^2\theta_W \quad , \tag{4.10}$$

which follows from general considerations [41] and holds for many processes. Inclusive reactions determine in this manner

$$\sin^2\theta_W = 0.24 \pm 0.02 \quad . \tag{4.11}$$

Alternatively, we can use the parton model to describe the detailed y-distributions (in units of $G^2 ME/\pi$)

$$\frac{d\sigma}{dy} = (u_L^2 + d_L^2)[Q + \bar{Q}(1-y)^2] + (u_R^2 + d_R^2)[\bar{Q} + Q(1-y)^2]$$

$$+ (s_L^2 + s_R^2) Q_S [1 + (1-y)^2] \qquad (4.12)$$

$$\frac{d\bar{\sigma}}{dy} = (u_R^2 + d_R^2) [Q + \bar{Q}(1-y)^2] + (u_L^2 + d_L^2) [\bar{Q} + Q(1-y)^2]$$

$$+ (s_L^2 + s_R^2) Q_S [1 + (1-y)^2] \qquad (4.13)$$

with

$$Q = \int x(u+d) dx; \quad \bar{Q} = \int x(\bar{u}+\bar{d}) dx$$

$$Q_S = \int x(s+\bar{s}) dx \quad . \qquad (4.14)$$

In the analysis the values for the ratios were

$$\frac{\bar{Q}}{Q} = 0.15 \quad \text{and} \quad \frac{Q_S}{Q} = 0.05$$

which are consistent [42] with the values

$$0.16 \pm 0.02 \text{ and } 0.025 \pm 0.01, \text{ respectively,}$$

reported recently by the CDHS group.

Separating the quantities $u_L^2 + d_L^2$ and $u_R^2 + d_R^2$ which occur in (4.12) and (4.13) gives [32] Table III.

The values in the Table include corrections which arise from

(i)   kinematic cuts,

(ii)   corrections due to the neutron excess in the
        target,

(iii)   a scalar structure function,

(iv)   strange sea-quark contribution,

(v)   corrections from asymptotic freedom.

Experimental Group	$u_L^2 + d_L^2$	$u_R^2 + d_R^2$
GGM	0.24 ± 0.05	0.06 ± 0.03
CITF	0.23 ± 0.04	0.08 ± 0.05
HPWF	0.29 ± 0.06	0.02 ± 0.06
CDHS	0.29 ± 0.02	0.03 ± 0.01
BEBC	0.33 ± 0.05	$0.02 \begin{array}{l}+\,0.04\\-\,0.02\end{array}$
World Average	0.28 ± 0.02	0.03 ± 0.01

TABLE III

Corrections from (i) and (iii) are well understood.
The entire corrections from the strange sea is less
than 4%, provided that $s_L^2 + s_R^2$ is not abnormally larger
than the other couplings. Corrections from scaling
violations can be estimated in terms of the violations
observed in charged-current reactions. Their effect is
at most one standard deviation [43] and the uncertainty
in the estimate much less.

The **values** in Table III together with Equation (4.5)
give two equations with ρ and $\sin^2\theta_w$ as unknowns. They
are solved pictorially in Fig. 4 giving precise values
of the parameters for SU(2) x U(1) models with any

arbitrary Higgs mechanism. Furthermore Table III indicates that data on deep-inelastic reactions restrict
the possible values of the left-handed coupling constants to a circular annulus in the u_L-d_L plane.
Similarly, u_R and d_R lie within a circular annulus in
the right-handed plane. The widths of the annuli depend
upon the confidence level and the present analysis uses
2 standard deviations. This more than accounts for uncertainties in the corrections (i)-(v). The overall sign
of the couplings is physically significant but cannot be
determined by available data. We make the arbitrary
choice $u_L > 0$. The allowed annular regions are shown
in Fig.5.

One can proceed to determine u_L^2, u_R^2, d_L^2 and d_R^2
separately from single-pion inclusive reactions

$$\nu + N \rightarrow \nu + \pi^i + x \tag{4.15}$$

as was done by Sehgal [44]. A unique determination of
the couplings then requires the removal of a four-fold
sign ambiguity still present in u_L, u_R, d_L, and d_R. This
was done by analyzing elastic and exclusive reactions
[35,45,46]. However, the final and unique solution has
unrealistically small errors because the errors do not
include (i) theoretical uncertainties some of which
have not been studied and others which cannot be
accurately estimated, and (ii) experimental uncertainties
resulting from the fact that the inclusive data were at
low energies. In the meanwhile it was realized that many
steps in the analyses can be eliminated by taking a
different approach [37].

In this approach we asked which regions of the domains in Fig. 4 are consistent with the elastic reaction, which is well understood theoretically. The matrix element for elastic scattering involves three form factors.

$$<p'|J_\mu^Z|p>=\overline{U(p')}[\gamma_\mu f_1(q^2)+\frac{i\sigma_{\mu\alpha}}{2M}q^\alpha f_2(q^2)+\gamma_\mu\gamma_5 g_1(q^2)]U(p) \ .$$

$$(4.16)$$

The form factors can be related at $q^2 = 0$ to corresponding quantities measured in elastic electron-nucleon and neutrino-nucleon reactions (except for the isoscalar axial term). In fact

$$G_E(0) = f_1(0) = \frac{1}{2}(\alpha + 3\gamma) \qquad\qquad (4.17)$$

$$G_M(0) = f_1(0) + f_2(0) = \frac{4.7}{2} (\alpha + 0.56\gamma) \qquad\qquad (4.18)$$

$$g_1(0) = \frac{1.23}{2} (\beta + 0.6\delta) \qquad . \qquad\qquad (4.19)$$

For the axial isovector current the charged-current relation

$$<p|A^+|n> = g_A(0)\bar{u}(p)\gamma_\mu\gamma_5 u(p) \ , \qquad\qquad (4.20)$$

with $g_A(0) = 1.23$, gives the coefficient in (2.20). The coefficient of δ is the ratio of the reduced matrix element for the axial isoscalar current to the matrix element for the axial isovector current. The value 0.6 is obtained by using SU(3) and η-PCAC. For the results

to be described this coefficient was varied from 0.4
to 0.6.

In the analysis the coupling constants were
restricted to lie within the regions of Fig.4; then a
lattice of points was selected in the four dimensional
space. For each point on this lattice the predicted
differential cross sections

$$\frac{d\sigma^\nu}{dq^2} \quad \text{and} \quad \frac{d\sigma^{\bar\nu}}{dq^2}$$

were compared with the experimental values. Chi-squared
maps are shown in Fig. 4 for M_A = 1.00 GeV/c^2. The best
values of the parameters lie in the vicinity of the
values predicted by the Weinberg-Salam model. The left-
handed couplings are accurately determined. For the
right-handed couplings a sizeable angular region is
still allowed. This solution remains stable when M_A is
changed by two standard deviations or when the matrix
element of the isoscalar-axial current is reduced by 30%.
Another solution is in gross (qualitative) disagreement
with Δ-production and was ignored. This is a convincing
determination of the parameters where the theoretical
uncertainties were precisely estimated. A possible uncertainty may still exist in the treatment of the isoscalar-axial term, which I discuss in the next paragraph.
But in spite of this point the above analysis still
provides a crucial confirmation of the theory, since
the form factors f_1 and f_2 are free of axial terms [38]
and they are found to agree with the theory. In summary
the values of the parameters under different assumptions
are given in Table IV.

	Model independent analysis		Glashow [47]	Weinberg-
	$m_A = 0.90$	$m_A = 1.00$	model	Salam
ρ	$1.03 \begin{smallmatrix} +0.07 \\ -0.22 \end{smallmatrix}$	$0.99 \begin{smallmatrix} +0.08 \\ -0.28 \end{smallmatrix}$	0.98 ± 0.05	1.00
δ	$0.03 \begin{smallmatrix} +0.14 \\ -0.40 \end{smallmatrix}$	$0.15 \begin{smallmatrix} +0.24 \\ -0.41 \end{smallmatrix}$	0.00	0.00
$\sin^2\theta_w$	0.20 to 0.30	0.22 to 0.30	0.23 ± 0.03	0.24 ± 0.02 inclusive 0.26 ± 0.04 elastic

TABLE IV

The isoscalar axial term of the neutral current requires special attention, because in Eq.(4.4) both the coupling δ and the matrix elements of the isoscalar operator

$$\bar{u}\gamma_\mu\gamma_5 u + \bar{d}\gamma_\mu\gamma_5 d \qquad (4.21)$$

are unknown. In Eq.(4.19) the coefficient of δ has the value determined by SU(3) and η-PCAC. Wolfenstein recently argued [38] that since the isoscalar-axial form factor can assume quite distinct values under different assumptions the corresponding matrix element should also be left as a free parameter to be determined by experiment. He observes that

$$<p|\partial^\mu A_\mu^8|p> = \frac{1}{\sqrt{6}}<p|(m_u+m_d)(\bar{u}\gamma_5 u + \bar{d}\gamma_5 d) - 3m_s\bar{s}\gamma_5 s|p>$$

$$= 2 M G_A^8 \bar{u}_p\gamma_5 u_p \qquad (4.22)$$

In the chiral SU(2) x SU(2) limit $m_u = m_d = 0$ and the first term vanishes. The second term also vanishes by Zweig's rule or Okubo's ansatz (the trace of the nonet tensor should be absent).

$$\text{Thus} \quad G_A^8 = 0 .$$

If on the other hand he applies η-PCAC for the evaluation of A_λ^8, he finds

$$2 M G_A^8 = \frac{g_{\eta NN} f_\eta m_\eta^2}{q^2 + m_\eta^2}\bigg|_{q^2 = 0} = g_{\eta NN} f_\eta . \qquad (4.23)$$

For SU(3) values of $g_{\eta NN}$ and f_η it gives the SU(6) value of

$$G_A^8 = \frac{3}{5} \frac{1}{2} G_A = .375 \qquad (4.24)$$

which is the value occuring in Eq.(4.19). So the issue is which value should be used? It is known from other estimates that current algebra gives an accurate representation of low-energy phenomena. The discrepancy with Eq.(4.22) is resolved by requiring that in the SU(2) x SU(2) limit the matrix elements also become singular. Thus it seems to me that a variation of G_A^8 around the central value of Eq.(4.24) is sufficient. A variation of the matrix element by ± 30% does not modify the couplings.

The isoscalar-axial term is interesting on a different account. In theories with $\delta = 0$ a finite value for δ arises in the deep-inelastic region from the residual terms of the triangle anomaly shown in Fig. 6. This is the same anomaly as discussed in Section 2. Collins, Wilczek, and Zee [48] point out that the effective contribution is

$$\Delta J_\mu^S = (\bar{u}\gamma_\nu\gamma_5 u + \bar{d}\gamma_\mu\gamma_5 d) \frac{1}{2} [\frac{\bar{g}^2(m_+^2)}{4\pi^2}][\frac{\bar{g}^2(m_-^2)}{4\pi^2}] \ln(\frac{m_+}{m_-}) \quad (4.25)$$

where m_+ and m_- are the masses of heavy quarks with positive and negative charges. This is an isospin-breaking term and it arises from the mass differences. Its strength is estimated to be 5% of the isovector-axial current. Mohapatra and Senjanovic [49] studied the possibility of obtaining a large isoscalar-axial

SUMMARY OF MEASUREMENTS FOR $\sin^2\theta_w$

WEIGHTED AVERAGE $\sin^2\theta_w = [0.235 \pm 0.015]$

$\nu_\mu + N \rightarrow \nu_\mu + \ldots$		0.24 ± 0.02
$\bar{\nu}_\mu + N \rightarrow \bar{\nu}_\mu + \ldots$		0.30 ± 0.1
$\nu_\mu + p \rightarrow \nu_\mu + p$		0.26 ± 0.09
$\bar{\nu}_\mu + N \rightarrow \bar{\nu}_\mu + N + \pi^0$		$0.15 \quad 0.52$
$\nu_\mu + N \rightarrow \nu_\mu + N + \pi^0$		0.22 ± 0.09
$e^- + d \rightarrow e^- + \ldots$		0.224 ± 0.020
$\bar{\nu}_\mu + e^- \rightarrow \bar{\nu}_\mu + e^-$		$0.30 \begin{smallmatrix} +0.10 \\ -0.30 \end{smallmatrix}$
$\nu_\mu + e^- \rightarrow \nu_\mu + e^-$		$0.2 \begin{smallmatrix} +0.09 \\ -0.06 \end{smallmatrix}$
$\bar{\nu}_e + e^- \rightarrow \bar{\nu}_e + e^-$		0.29 ± 0.05

.0   .2   .4

Table V

term in SU(2) x U(1) or SU(2)$_L$ x SU(2)$_R$ x U(1) models
from higher-order weak effects. They find the effects
to be very very small($\leq$ 1%). All these effects are
interesting but too small to the level of accuracy of
the previous determination.

Table V summarises [30] the values of the Weinberg-
Salam angle in the Weinberg-Salam model for different
experiments.

c. Other Interactions

(α) Electron-Nucleon Interactions

It is well known [50] that neutrino experiments
alone cannot determine whether the neutral current
interaction is parity violating or not. This motivated
a new class of experiments whose purpose is to measure
the electron-nucleon neutral-current interaction. The
first experiments were designed to look for parity
violation in atomic physics. Early experiments in
Bismuth [51,52] did not observe any parity violation
at the level predicted by the minimal theory. However,
the situation seems to have changed now when experiments
on Bismuth [53] and in Thallium [54] observe a parity
violation consistent with the theoretical expectations.

In addition parity violation was definitely observed in a second type of experiment [33] which studied
the inelastic scattering of polarized electrons on
deuterons, i.e.,

$$e(polarized) + d \rightarrow e' + x \qquad . \qquad\qquad (4.26)$$

The experiment measures the asymmetry

$$A = \frac{\sigma_R - \sigma_L}{\sigma_R - \sigma_L} \qquad (4.27)$$

where σ_R and σ_L are the cross sections for right-handed and left-handed polarized electrons. The observation of a non-vanishing asymmetry indicates that parity is violated. The dependence of the asymmetry on the scaling variables is

$$\frac{A(x,y,q^2)}{q^2} = a_1(x) + a_2(x) \frac{1-(1-y)^2}{1+(1-y)^2} . \qquad (4.28)$$

The functions $a_1(x)$ and $a_2(x)$ include the products of neutral-current couplings to electrons and nucleons. They also depend on the structure functions of the nucleons and were calculated originally in the quark-parton model. Subsequently several authors [55] studied the parton model assumptions and concluded that because of some fortuitous cancellations the dependence of the analysis on parton-model assumptions is minimal.

The asymmetry reported originally was at $y = 0.2$

$$\frac{A}{-q^2} = (-9.6 \pm 1.6) \times 10^{-5}/\text{GeV}^2 . \qquad (4.29)$$

This value is shown in Fig.7 together with the theoretical predictions for several Weinberg angles. In the meanwhile more data were taken in the range $0.14 \leq y \leq 0.40$. The new results [56] determine a_1 and a_2 of Eq.(4.28) separately:

$$a_1 = (-9.12 \pm 1.91) \times 10^{-5} \qquad\qquad (4.30)$$

$$a_2 = (2.10 \pm 6.89) \times 10^{-5} \qquad\qquad (4.31)$$

with a large uncertainty in a_2. The observed y-distribution is in agreement with the Weinberg-Salam model. The errors in (4.30) and (4.31) are correlated. Fig. 8 shows χ^2 contours at the 60% and 90% confidence levels. The Weinberg angle determined from this plot is

$$\sin^2\theta_w = 0.224 \pm 0.020 \quad .$$

This value was included in Table V and in the evaluation of the weighted average.

5. QCD PREDICTIONS FOR DEEP-INELASTIC SCATTERING

a. Renormalization of the Coupling Constants

Violations of scaling in deep-inelastic lepton-hadron scattering have been reported by several ex-periments. These reports study the quantitative deviations from scaling and provide a test of quantum field theories, where violations of scaling are predicted and are calculable. In such theories the coupling constants and other Green's functions vary with the external momenta in a way that is determined by the Callan-Symanzik equations [57]. For instance the coupling constant satisfies the equation

$$\mu \frac{dg(\mu)}{d\mu} = \beta(g(\mu)) = b_o g^3 + b_1 g^5 + \ldots \tag{5.1}$$

where μ is the renormalization point with units of mass and $\beta(g)$ is a function determined in perturbation theory. When β is negative the solution of (5.1) gives a $g(Q^2)$ that decreases to zero with increasing momentum. As a consequence the strong-interaction coupling constant becomes weaker at large momentum and perturbation theory becomes applicable. This is the case in quantum chromo-dynamics where b_o and b_1 have been calculated explicitly and b_o was found to be negative [58]

$$b_o = - \frac{1}{16\pi^2}(11-\frac{2}{3}N_f) \text{ and } b_1 = - \frac{1}{(16\pi^2)^2}(102 - \frac{38}{3}N_f). \tag{5.2}$$

Here N_f is the number of times that a generation of flavors, like Eq.(2.13), is repeated. The solution for the coupling constant including the two-loop effects is

$$a \ln \frac{g^2}{1+ag^2} + \frac{1}{g^2} = 2 b_o \ln\mu + c \tag{5.3}$$

with $a = b_1/b_o$ and c is a constant of integration. To lowest order Eq.(5.3) is rewritten as

$$\alpha_o(Q^2) = \frac{g^2}{4\pi} = \frac{12\pi}{(33-2f) \ln(Q^2/\Lambda^2)} \tag{5.4}$$

by redefining the constant of integration. Including two-loop effects gives

$$\alpha_s(Q^2) = \alpha_0(Q^2) \; [1 - \frac{b_1}{b_0} \; \frac{\ln \ln (Q/\Lambda)^2}{\ln (Q/\Lambda)^2}] \tag{5.5}$$

obtained from (5.3) and (5.4) by iteration. The corrections arising from the last term are less than 2% for $Q^2 = 100 \; (GeV)^2$ and $\Lambda^2 = 0.1 \; (GeV)^2$ and could be neglected for higher energies. The strong-coupling constant is directly related to the total electron-positron annihilation cross section and could provide a test of the theory. So far this comparison has not been decisive, because at low energies the cross section is dominated by resonances and at high energies the corrections from QCD are too small and are obscured by statistical errors and systematic uncertainties.

Similarly, it is possible to compute the asymptotic value of the vacuum polarization to an arbitrarily large order by applying the renormalization-group equation. The fine-structure constant (electric charge) satisfies an equation identical to (5.1) where the β-function is known from the three-loop computation of de Rafael and Rosner [59]

$$\beta(\alpha) = \frac{2\alpha}{3\pi} + \frac{\alpha^2}{2\pi^2} - \frac{121}{144} \; \frac{\alpha^3}{3} + O(\alpha^4) \quad . \tag{5.6}$$

It is observed that β is positive as $\alpha \to +0$. In contrast to the previous example the electromagnetic coupling constant increases with increasing momentum. In a

computation keeping only the first term in the beta
function   the solution is

$$\frac{1}{\alpha(Q^2)} = \frac{1}{\alpha(0)} \{1 - \frac{\alpha(0)}{3\pi} \ln(Q^2/\Lambda^2)\} \qquad . \qquad (5.7)$$

This expression was obtained some time ago by summing up
leading logarithms in QED and it exhibits the Landau
ghost-singularity.

Let us now derive a result promised earlier,
i.e. the computation of the mixing angle in grand
unified theories. I concentrate here on the SU(5)
model of Georgi and Glashow [60]. This theory aims
at the unification of the electroweak force based on
the group SU(2) x U(1), and the strong force, based
on the SU(3)-color group. It has the advantage of
replacing the three coupling constants by one, thus
enhancing the predictive power of the theory. The
predicted value of the mixing angle at asymptotic
energies is 3/8 and follows from group-theoretic
considerations. At present energies, however, this
value is substantially modified by the renormalization
of the coupling constants. What is needed is a
computation of

$$\sin^2\theta_w = \frac{e^2(Q^2)}{g_2^2(Q^2)} \qquad (5.8)$$

where the Q^2 is selected to be at the value of present-
day experiments. The change in the coupling constants
is governed by the equations

$$\mu \frac{dg_i}{d\mu} = b_i g_i^3 + \sum_{j=1}^{3} b_{ij} g_i^3 g_j^2 \tag{5.9}$$

with i = 1,2,3 refering to the groups U(1), SU(2) and SU(3), respectively. The summation in (5.9) includes two-loop contributions, which couple the three equations. The constants b_i, b_{ij} are calculated in perturbation theory. The solution for the Weinberg angle is

$$\sin^2\theta_w = \frac{1}{1+3c^2} \{1 + 2\alpha c^2 [\frac{1}{\alpha_s} + 4\pi a \ln \frac{\alpha_s}{\alpha_G}]\} \tag{5.10}$$

with a = $\frac{b_{33}}{b_3}$ and c^2 = 5/3. The first two terms are those obtained in the one-loop calculation. They depend on the fine-structure coupling constant, whose value is accurately known and also on the strong-coupling constant α_s which is not accurately known. The last term is new and arises from second-order terms. The quantity α_G is the coupling constant at the grand unification region. To this order it is sufficient to take

$$\alpha_G = 11 \ (1+3c^2) \frac{\alpha\alpha_s}{27\alpha_s + 2\alpha} \quad . \tag{5.11}$$

In presenting numerical results a range for α_s was chosen, since α_s is poorly known, and the fine-structure constant selected is $\alpha(10 \text{ GeV})$ = $\frac{1}{133}$. Table VI gives $\sin^2\theta_w$ and the mass for grand unification as a function of the α_s. It is noted that the values for

the mixing angle are slightly larger than those computed
by Georgi, Quinn and Weinberg [61]. The origin is two-
fold; a new value for α and the two-loop term in Eq.(5.6).
The values for $\ln(M/\mu)$ are substantially smaller than
those in ref.[61,62]. For the implication of these
results I refer to recent articles and the lecture by
J. Ellis at this meeting.

α_s	$\sin^2\theta_w$	$\log (M/\mu)$	τ_p in YR
0.05	0.250	9.70	Excluded experimen-tally
0.10	0.212	12.85	10^{27}
0.15	0.200	14.16	4×10^{28}
0.20	0.193	14.70	6×10^{30}
0.30	0.192	15.20	6×10^{32}
0.40	0.184	15.40	4×10^{33}
0.50	0.182	15.70	6×10^{34}

Table VI

There is one point that I wish to emphasize. The
calculation of the Weinberg angle is very stable and
for $\alpha_s > .20$ considerably smaller than the average experimental value presented in section 4. The values for

α_s determined from scaling violations are in general considerably larger $(0.3 \leq \alpha_s \leq 0.5)$. It is quite possible that a small value of α_s (< 0.10) cannot reproduce the scaling violations; thus a resolution of the discrepancy between the predicted and measured mixing angle becomes difficult.

b. Scaling Corrections to the Structure Functions

In direct analogy to the discussion of the previous section the moments of the deep inelastic structure functions are not constants, but functions of Q^2:

$$M_n^i \ (Q^2) = \int_0^1 dx \ x^{2n-1} \ F_i \ (x,Q^2) \quad . \tag{5.12}$$

The moments $M_N^i(Q^2)$ also satisfy a set of Callan-Symanzik equations

$$(\mu \frac{\partial}{\partial \mu} + \beta \frac{\partial}{\partial g} + 2\gamma_n) \ M_n^i \ (Q^2, \ g^2) = 0 \quad . \tag{5.13}$$

The β-function is the same one as before and the γ_n-functions are determined by low-energy calculations in field theory. Bjorken scaling is exact only when all $\gamma_n \equiv 0$; thus the origin of their name "anomalous dimensions". The moments of structure functions that are multiplicatively renormalizable satisfy

$$M_n^i(Q^2) = M_n^i(Q_0^2) \ [\log \ (Q^2/\Lambda^2)]^{-\gamma_n} \quad . \tag{5.14}$$

This implies that knowledge of a moment at Q_o^2 predicts its values at larger Q^2. It also predicts that

$$\frac{d \ln M_n^i(Q^2)}{d \ln M_{n+2}^i(Q^2)} = \frac{\gamma_n}{\gamma_{n+2}} \qquad (5.15)$$

and provides a clean test of the theory.

All structure functions are not multiplicatively renormalizable. This is the case when we consider the internal symmetry in greater detail. The amplitudes defined in the standard manner

$$W_{\mu\nu}^{ab}(p,q) = \int dx^4 e^{iq\cdot x} <p| [J_\mu^a(x), J_\nu^b(0)] |p> \qquad (5.16)$$

have a slightly more complicated structure on the internal indices a and b. In general, the matrix element of $J^a J^b$ between states can be classified by their transformation properties under the group. For SU(4) currents the short-distance behavior of the commutator is given by the Wilson expansion

$$[J^a(x), J^b(0)] = c_1(x) O_o(0) + c_2(x) O_3(0) + c_3(x) O_8(0)$$

$$+ c_4(x) O_{15}(0) + \ldots \qquad (5.17)$$

where c_i's are singular c-numbers and O_o, O_3, O_8, O_{15} are operators transforming like the 0th, 3rd, 8th and 15th component of the group. The dots indicate additional operators, but those present are the operators contributing to matrix elements between identical states. In terms of the quark fields the O_i's can be represented by

$$O = \bar{\psi} \, \lambda_i \, \psi$$

where λ_i's are the generators of the group and Lorentz indices have been suppressed. For the singlet operator O_o the matrix λ_o corresponds to the unit matrix. The currents A_8 and A_{15} in Eqs.(2.2) and (2.3) correspond to O_8 and O_{15}, respectively. Operators which transform like O_3, O_8 and O_{15} are non-singlets under SU(4), in contrast to O_o which is singlet. In gauge theories, in addition to the fermion singlet operator, there is another singlet constructed from the field tensor $F_{\mu\nu}^a$ as follows

$$O_{\mu,\mu_2\cdots\mu_N} = \sum_{s,\alpha} F_{\mu\alpha}^s \overset{\leftrightarrow}{\nabla}_{\mu_2} \cdots \overset{\leftrightarrow}{\nabla}_{\mu_{N-1}} F_{\mu_N\alpha}^s \tag{5.18}$$

with the α being a Lorentz- and s an internal-symmetry index. The total singlet operator is a superposition of these two operators each one having its own multiplicative renormalization constant.

The structure functions are also classified according to their transformation properties under the group. For instance

$$F_2^{\nu d}(x) = x[U((x,Q^2) + \bar{U} + D + \bar{D} + S + \bar{S} + C + \bar{C}] \tag{5.19}$$

is a singlet with U,D,... being the quark-distribution functions. $F_2^{ep} - F_2^{en}$ and F_3 are non-singlets.

The non-singlet structure functions are multi-plicatively renormalizable and the anomalous dimensions in QCD are given by

$$\gamma_S^{NS} = \frac{4}{(33-2f)} [1 - \frac{2}{N(N+1)} + 4 \sum_{j=2}^{N} \frac{1}{j}] . \qquad (5.20)$$

Consequently $F_3(x,Q^2)$ has a definite Q^2-development de-fined by (5.14) and (5.20). For the singlet-structure functions the situation is more complicated since the corrections can be of several types. In Fig.(9a) the correction arises from the radiation of a gluon by a quark. The amplitude in this case is proportional to the quark-distribution functions. In contrast the con-tribution from diagram (9b) involves the contribution from a gluon decaying into a quark-antiquark pair and is proportional to the distribution function of GLUONS. The Q^2-development of the singlet-structure function is a superposition of the quark- and the gluon-distribution functions. Let us denote by $M_i^n(Q^2,g^2)$ the nth moment of the singlet-distribution function with i = 1 denoting the quark distribution and i = 2 the gluon distribution. In the same notation denote the anomalous dimension by γ_{ij}^n where the indices i and j take the values 1,2 and correspond to the diagrams in Fig.10. The notation is such that γ_{11}^n is the anomalous dimension for the nth moment where both the external and internal lines are fermions, as in Fig.10a. Similarly γ_{12}^n corresponds to diagram (10b) where the external lines are gluons but the internal lines fermions. The Q^2-developments of the two singlet functions are coupled through the equation

$$[\mu\frac{\partial}{\partial\mu} +\beta(g)\frac{\partial}{\partial g}]M_i^n(Q^2,g^2)= \sum_{j=1,2} \gamma_{ij}^n(g) \, M_j^n(Q^2,g^2) . \qquad (5.21)$$

Denoting the eigenvalues of the anomalous dimensions by γ_+ and γ_- and the corresponding eigenfunctions by $F_+(Q^2)$ and $F_-(Q^2)$ the Q^2-development of a general moment is given by

$$\langle F(Q^2)\rangle_n = M_n(Q^2) = \langle F_{NS}(Q_o^2)\rangle_n \, e^{-\gamma_n^{NS} \cdot t} +$$

$$+ \langle F_+(Q_o^2)\rangle_n \, e^{-\gamma_+ \cdot t} + \langle F_-(Q_o^2)\rangle_n e^{-\gamma_- \cdot t} \qquad (5.22)$$

where

$$t = \ln\,[\ln(Q^2/\Lambda^2)/\ln(Q_o^2/\Lambda^2)].$$

The anomalous dimensions have been calculated several years ago [63] and a pedagogical account is to be found in ref.[64]. The next-order corrections have also been studied [65] and are important in comparison with ex-periments.

c. Experimental Comparisons

The scaling violations observed in electroproduction are in qualitative agreement with those expected in QCD. Fig. 11 shows the variations [66] of $F_2(x,Q^2)$ as a function of Q^2 observed in electron and muon experiments. At small x it increases with increasing Q^2 and at large x $\sim$ 1 it decreases. This is what is expected in QCD where the corrections arise from radiation of gluons. Their emission is the same, except for a normalization factor, as the emission of photons in the Weizsäcker-Williams approximation whose spectrum has a dx/x dependence and

thus produces an increase of $F_2(x,Q^2)$ at small x. Fig.12
shows the variation of $F_2(x,Q^2)$ for x^2 = 0.00909 as Q^2
increases.

Discussing the problem of deviations from exact
Bjorken scaling one must consider the ratio R = σ_L/σ_T.
This is the ratio of the absorption of a longitudinal
current by a nucleon divided by the corresponding transverse cross section. $R(x,Q^2)$ has been calculated [67] in
QCD and is expected to be decreasing with x and Q^2 and is
indeed very small for Q^2 > 5 and x > 0.5. Several experiments observed a sizeable R but the experimental
errors are also large [68]

$$R = 0.25 \pm 0.10 .$$

The kinematic ranges for the experiments do not overlap
and in several experiments Q^2 is not large enough to lie
comfortably within the scaling region. The neutrino experiments measure average values of R. An accurate value
was reported by the CDHS group [69]

$$R = 0.03 \pm 0.05 .$$

In this experiment $Q^2 \geq 5$ $(GeV)^2$ and the result is consistent with the theoretical predictions.

Other quantitative tests deal with the moments
of the structure functions. During the past year
neutrino experiments studied scaling violations of the
structure functions in terms of the quark-distribution
functions and of the fragmentation functions. The
general picture which emerges is consistent with the
QCD predictions, but one needs cautious words that the
present evidence is neither a confirmation nor a refusal

of QCD. This cautious presentation is necessary, because
there are several technical points, which have not been
investigated thoroughly. The moments of xF_3 have been
investigated by the BEBC [70] and the CDHS [71] groups.
Fig.13 shows log-log plots of various non-singlet moments
from the BEBC (ABCLOS)-group. The lines are predictions
from QCD. Fig. 14 shows similar curves with the CDHS data
also included. The errors for the CDHS data are indeed
small, but the lever arm is also smaller. There are two
features of the data in agreement with QCD: (α) the
dependence of the moments on log (Q^2/Λ^2) is as described
in Eq.(5.14). (β) The slopes of the curves are related
to the anomalous dimensions through Eq.(5.15). Table VII
gives ratios of the anomalous dimensions.

The CDHS slopes are slightly smaller than those of
BEBC. They are both consistent with the QCD prediction,
but the CDHS data could also be consistent with the
scalar-gluon theory. Scalar theories are not asymptotically
free and it is not evident that they should be treated on
the same footing with QCD. For scalar theories [72] it is
assumed that they have a fix-point in the neighborhood of
the origin (i.e. the coupling constant for large momenta
has a small value) and the anomalous dimensions are calculated. Recent papers [72] demonstrate that the CDHS
results can be fitted by a power law, i.e.

$$M_n(Q^2) = C_n(Q_o^2) \; (Q^2/Q_o^2)^{-dn} \qquad\qquad (5.23)$$

but this is perhaps a transient phenomenon due to the
limited range of Q^2. There are similar predictions and
measurements for the SU(4)-singlet-structure function,
which will not be discussed in these lectures.

Moment Ratio	Experiment		QCD Prediction	Scalar Gluon
	BEBC $q^2 = 1-100$	CDHS $q^2 = 6.5-75$		
γ_6/γ_4	$1.29 \pm .06$	1.13 ± 0.09	1.29	1.06
γ_5/γ_3	$1.50 \pm .08$	1.34 ± 0.12	1.456	1.12
γ_7/γ_3	$1.34 \pm .20$	–	1.760	1.16
γ_6/γ_3	–	1.38 ± 0.15	1.621	1.14

Table VII

d. Scaling Behaviour of the Quark-Fragmentation Functions

There is evidence that the quark-parton model, when extended to the inclusive particle reactions, gives an accurate representation of the data in inclusive electro-production, inclusive neutrino production, and inclusive reactions in e^+e^--annihilation. The model should also include scaling violations of the quark-distribution functions and the quark-fragmentation functions. The cross section is written in general as

$$\frac{d^2\sigma(Q^2)}{dxdz} = q(x,Q^2)F_{q\rightarrow h}(z,Q^2) \qquad (5.24)$$

where $q(x,Q^2)$ = quark-distribution function;

$F_{q\rightarrow h}(z,Q^2)$ = quark-fragmentation function of q becoming a hadron h;

$x = Q^2/2M\nu = B_j$-scaling variable;

$z = E_h/\nu$ = Feynman-scaling variable.

In (5.24) an overall normalization factor was omitted. Considering the double moments

$$\int_0^1 dx\ x^{N-1} \int_0^1 dz\ z^{N-1}\ \frac{d^2\sigma}{dxdz} = S_{NM}(Q^2) = Q_N(Q^2)F_M(Q^2) \quad (5.25)$$

the development in Q^2 of Q_N and F_M is expected to be the same as in Eq. (2.13)

$$Q_N(Q^2) \equiv \int_0^1 dx\ x^{N-1}\ q(x,Q^2) \sim Q_N(Q_0^2) \ln(Q^2/\Lambda_N^2)^{-\gamma_N} \qquad (5.26)$$

$$F_M(Q^2) = \int_0^1 dx\, x^{M-1} F_{q \to h}(x, Q^2) \sim F_M(Q_0^2) \ln(Q^2/\Lambda_M^2)^{-\gamma_M} \quad . \quad (5.27)$$

Note, however, that Q_N, F_M have their own anomalous dimensions and they also have distinct Λ-parameters. The simple forms of (5.26) and (5.27) assume that we consider non-singlet terms and there is an implicit factorization assumption. There are two predictions which follow in the above simple picture:

1) For a fixed N there is a power of $\log(Q^2/\Lambda^2)$ behaviour. For different values of M the slope of the moments in a log-log plot is again determined by QCD. The BEBC-group [73] has inclusive data, where it considers the differences of positive and negative particles in order to separate the non-singlet contribution. Fig.15 shows their preliminary and unpublished results for the moments of the fragmentation functions. The solid curves show the expectations of QCD.

2) In the simple picture described there is a factorization property

$$D_{NM}(Q^2) = \frac{S_{NM}(Q^2)}{Q_N(Q^2)} = F_M(Q^2) \quad .$$

$D_{NM}(Q^2)$ should be a function of Q^2 and M but does not depend on N. Fig. 15 shows the non-singlet combination of $D_{NM}^{h^+} - D_{NM}^{h^-}$. In Fig. 16 $N = 1$ and the moments are shown as functions of Q^2 and M. The curves are drawn to guide the reader. As N is changed factorization predicts that the experimental points must still agree with the dotted curves. This is clearly violated. The results on the quark-

fragmentation functions are still preliminary and they
depend critically on the analysis method since they are
not in the deep-inelastic region. For instance, the sum
total of the elastic events and the Δ-resonance comprise
50% of the total number of events. It is presented here
as a topic of further future investigations.

This brings me to the conclusions to be drawn
from studies of scaling violations in deep-inelastic
scattering. At the qualitative level there is agreement
between theory and experiment. At the quantitative level
the tests of QCD are not conclusive, because they depend
on several assumptions whose implications could be
significant.

1) Some data include   low-energy measurements and
contributions from elastic and quasi-elastic Δ-production
are still significant. All experiments use in one form or
another a low value of Q^2 in order to calculate the moments.
Thus corrections arising from the target mass are not
negligible. Part of this effect is included in using
Nachtmann moments [74], which decrease   the slopes of
Table VII by 10% and then the distinction between the
two types of theories is not definitive.

2) Second-order effects [75] on the coupling constants
are not always included. They can alter the ratio of
the moments substantially.

3) All experiments do not obtain the same effective
values of :

$$\Lambda = 0.47 \pm 0.01 \ (CDHS)$$
$$\Lambda = 0.73 \pm 0.03 \ (BEBC) \ .$$

Finally, there is the question of how critically is QCD
tested by just measuring the slopes of the moments? It
has been shown that a general empirical formula [76] for
the structure functions leads to upper and lower bounds
for the slope of the graph for log M_{n+1} versus log M_n.
Thus at this time we have several consistency checks and
other more critical checks should be coming in the future.
It is desirable to find tests of QCD which are qualitatively
different from the predictions of the quark-parton model.

ACKNOWLEDGEMENT

I wish to thank J. Ellis for discussion on scaling
violations and W. Plessas for helping in preparing the
manuscript.

REFERENCES

[1] S.Weinberg, Phys.Rev.Lett. 19 (1967) 264;
    A. Salam, in Elementary Particle Physics ed. by
    N. Svartholm (Stockholm 1967), p. 367.
[2] C. Bouchiat, I. Iliopoulos, and P. Meyer, Phys.
    Lett. 38B (1972) 519.
[3] S. Adler, Phys. Rev. 177 (1969) 2426;
    J. Bell and R.Jackiw, Nuovo Cim. 60 (1967) 47.
[4] S. Glashow, I. Iliopoulos, and L. Maiani, Phys.
    Rev. D2 (1970) 1285.
[5] A. Pais and S.B. Treiman, Phys. Rev. D12 (1975) 2744.
[6] L.B.Okun, V.I.Zakharov, and B.M.Pantecorvo, Nuovo
    Cim. Lett. 13 (1975) 218.
    M. Goldhaber and J.L.Rosner, Phys.Rev.D15 (1977) 1254.

[7] D.Fakirov and B.Stech, Nucl.Phys. B133 (1978) 315;
    N.Cabibbo and L.Maiani, Phys.Lett. 73B (1978) 418;
    M.B. Cavella, Orsay preprint, LPTHE 79/3 (1979).

[8] B.Lee, J.Primack and S.B.Treiman,Phys.Rev.D7(1973)510.

[9] R.N.Cahn and S.D.Ellis, Univ.of Michigan report,
    UM-HE 76-45 (1976);
    S.S.Gerstein and M.Khlopov, Serpukhov preprint,
    IHEP 76-73 (1976);
    V.N. Novikov et al., Phys. Rev. Lett. 38 (1977) 626,
    Err. ibid. 791.

[10] G.J.Feldman et al., Phys.Rev.Lett. 38 (1977) 1313.

[11] E.A.Paschos, Phys.Rev.Lett.39 (1977) 858; see also
    M. Gorn, Berlin preprint (1979).

[12] R.Shrock and S.B.Treiman, Princeton preprint (1978).

[13] S.Glashow and S.Weinberg, Phys.Rev. D15 (1977) 1958.

[14] E.A. Paschos, Phys. Rev. D15 (1977) 1966.

[15] E.Paige, E.A.Paschos,and T.L. Trueman, Phys.Rev.D15
    (1977) 3416.

[16] M.S.Chanowitz, J.Ellis,and M.K.Gaillard, Nucl. Phys.
    B128 (1977) 506.

[17] H.Georgi and A.Pais, Rockefeller preprint.

[18] M.Kobayashi and K.Maskawa, Progr.Theor.Phys.49
    (1973) 652.

[19] W.Kummer, Seminar at this meeting and Lecture at the
    Triangle Seminar on "Quarks and Gauge Fields",
    Matrafüred, Hungary (1978).

[20] S.Weinberg, in: Transactions of the New York Academy
    of Sciences, Series II, Vol.38 (1977);
    A.De Rujula, H.Georgi,and S.L. Glashow, Annals of
    Physics 109 (1977) 258;
    F.Wilczek and A.Zee, Phys. Lett. 70B (1977) 418;
    T. Hagiwara, T.Kitazoe, G.B.Mainland,and K.Tanaka,
    Phys. Lett. 76B (1978) 602;
    H.Fritzsch, Phys. Lett.70B (1977) 436, 73B (1978) 317;

T.Kitazoe and K.Tanaka, Phys. Rev. D18 (1978) 3476;
A.Ebrahim, Phys. Lett. 73B (1978) 181;
S. Pakvava and H. Sugawara, UH-511-240-77 and
additional articles in ref.[19].

[21] R.E. Shrock and L.L. Wang, Phys. Rev. Lett. 41
(1973) 1692.

[22] J.Ellis, M.K. Gaillard, D.Nanopoulos,and S. Rudaz,
Nucl. Phys. B131 (1977) 285.

[23] S.K.Bose and E.A. Paschos, to be published.

[24] M.K. Gaillard and B.Lee, Phys. Rev. D2 (1970) 1285.

[25] A.J. Vainshtein and I.B.Kriplovich, JETP Lett. 18
(1973) 83.

[26] H. Harari, SLAC-PUB-2234 (1978 and references therein).

[27] ref.[22];A.Ali and Z.Z. Aydin, DESY-preprint 78/11
(1978).

[28] M.K. Gaillard, Fermilab Conf. 78/64-Thy (1978).

[29] A.Ali, Z. Phys. C1 (1979) 25.

[30] C.Baltay, Invited paper presented at the XIX. Int.
Conference on High-Energy Physics, Tokyo (1978).

[31] J.J. Sakurai, Status of Neutral Currents, UCLA/78/
TEP/18 (1978).

[32] L.M. Sehgal, Status of Neutral Currents in Neutrino
Interactions, Invited Talk at the "Neutrino 78"
Int. Conf. Purdue Univ. (1978).

[33] C.Prescott et al., Phys. Lett. 77B (1978) 347.

[34] L.M. Barkov and M.C. Zolotorev, JETP Lett. 26
(1978) 379.
Conti et al., Phys. Rev. Lett. 42 (1979) 343.

[35] L.F. Abbot and R.M. Barnett, Phys. Rev. Lett. 40
(1978) 1303; Phys. Rev. D18 (1978) 3214.

[36] E.A. Paschos, Phys. Rev. D19 (1979) 83.

[37] M. Claudson et al., Phys. Rev. D19 (1979) 1373.

[38] L. Wolfenstein, Carnegie-Mellon preprint, COO-3066-
119 (1978).

540

[39] J.D.Bjorken, Proc. SLAC Summer Institute on Particle
     Physics, ed. M. Zipf (1976) p.1.

[40] J.J. Sakurai, CERN-TH-2099 (1976).

[41] E.A. Paschos and L. Wolfenstein, Phys. Rev. D7
     (1973) 91.

[42] J.G.H. de Groot et al., CERN-preprint (to be
     published in Zeitschrift für Physik).

[43] A.J. Buras and K.J.F. Gaermers, Phys. Lett. 71B
     (1977) 106.

[44] L.M. Sehgal, Phys. Lett. 71B (1977) 99.

[45] The elastic reaction was studied by C.H. Albright,
     C.Quigg, R.E.Shorck,and J.Smith, Phys. Rev. D14
     (1976) 1780;
     R.M. Barnett, Phys. Rev. D14 (1976) 2990;
     V.Barger and D.V.Nanopoulos, Phys. Lett. B63 (1976)
     168;
     P.Hung and J.J. Sakurai, Phys. Lett. 72B (1977) 208;
     G.Ecker, Phys. Lett. 72B (1978) 450;
     D.P. Sidhu and R.Langacker, Phys. Rev.Lett. 41 (1978)
     732;
     E.A. Paschos, Phys. Rev. D19 (1979) 83.

[46] The single-pion reactions were studied by
     C.H. Albright et al., Phys.Rev. D7 (1973) 2220;
     S.L. Adler, Phys. Rev. D9 (1974) 229;
     E.A. Monsay, Phys. Rev. Lett. 41 (1978) 728;
     ibid. 35.
     The above list is long but certainly incomplete.
     See also G. Rajasekaran and K.Sarma,Pramana Vol. 7
     (1976) 1944;
     Phys. Lett. 55B (1975) 201;
     M. Gourdin and X. Pham PAR LRTH 78/14.

[47] S.L. Glashow, Nucl. Phys. 22 (1961) 579.

[48] J. Collins, F.Wilczek,and A.Zee, Phys. Rev. D18
     (1978) 242.

[49] R.N.Mohapatra and G.Senjanovic, Maryland preprint
     79-49 (1979).

[50] See e.g. H.Fritzsch and P.Minkowski, Nucl. Phys.
     B103 (1976) 61,
     R.Mohapatra and D.P.Sidhu, Phys. Rev. Lett. 38
     (1977) 667.

[51] N.Fortson, Proceedings of Neutrinos 78, Purdue
     Univ. (1978).

[52] P.G.H. Sandars, Result presented at the RIGA Conf.
     (1978).

[53] ibid. 34.

[54] ibid. 34.

[55] J.D. Bjorken, Phys. Rev. D18 (1978) 3239;
     E.Derman, Rockefeller Univ. Report, 2232 B-158 (1978);
     H. Fritzsch, CERN-TH 2607 (1978);
     L. Wolfenstein, Carnegie-Mellon preprint, COO-3066-
     111 (1978);
     T. Rizzo, BNL-25837 (1979).

[56] I wish to thank Dr.M.Borghini for making their
     results available before publication.

[57] C.G. Callan, Phys. Rev. D2 (1970) 1541;
     K. Symanzik, Commun. Math. Phys. 18 (1970) 227.

[58] D.J. Gross and F.Wilczek, Phys. Rev. Lett. 30
     (1973) 1343; and H.D. Politzer, ibid. 1346; The
     calculation of the two loops was carried out by
     W.E. Caswell, Phys. Rev. Lett. 33 (1974) 244,
     D.R.T. Jones, Nucl. Phys. B75 (1974) 53.

[59] E.de Rafael and J.R. Rosner, Ann. of Phys. (N.Y.)
     82 (1974) 369.

[60] H.Georgi and S.L. Glashow, Phys. Rev. Lett. 32
     (1974) 438.

[61] H. Georgi, H.R. Quinn, and S.Weinberg, Phys. Rev.
     Lett. 33 (1974) 451.

542

[62] A.Buras, J.Ellis, M.K. Gaillard, and D.Nanopoulos,
Nucl. Phys. B135 (1978) 66.

[63] H.D. Politzer, Phys. Rev. 14 (1974) 129.
D.J. Gross and F.Wilczek, Phys. Rev. D8 (1973)
3633, D9 (1974) 980.

[64] M. Calvo, Phys. Rev. D15 (1977) 730.

[65] E.G. Floratos, D.A. Ross, and C.T. Sachrajda,
Nucl. Phys. B129 (1977) 66, B139 (1978) 545;
W.A. Bardeen, A.Buras, D.W.Duke, and T.Muta,
Phys. Rev. D18 (1978) 3998.

[66] The curves represent the variations of the electromagnetic structure functions. I wish to thank T.Kirk
for providing their data.

[67] H.D. Politzer, Phys. Reports, 14C (1974) 129;
W. Marciano, H. Pagels, Phys. Reports, 36C (1977)
137.

[68] R.Taylor, private communication.

[69] J.G.H. de Groot et al., Inclusive Interactions of
High-Energy Neutrinos and Antineutrinos in Iron.
To be published in Z. Physik C (1979).

[70] P.Bosetti et al., Nucl. Phys. B142 (1978) 1.

[71] J.G.N. de Groot et al., QCD Analysis for Charged-
Current Structure Functions (to be published in
Phys. Lett.).

[72] D.J. Gross, Les Houches Session XXVIII (1975) p.144;
M.Glück and E.Reya, Phys. Rev. D16 (1977) 3242;
E.Reya, DESY-preprint 79/09 (1979).

[73] D.H.Perkins, Rutherford preprint 2/79 (1979).

[74] O.Nachtmann, Nucl. Phys. B63 (1973) 237; B78 (1974)
455.

[75] ibid. 65.

[76] H.Harari, SLAC-PUB-2254 (1979).

FIGURE CAPTIONS

Fig. 1. Triangle anomaly with an axial current and two photons.

Fig. 2. Triangle anomaly with an axial current and two gluons.

Fig. 3. Feynman diagram for a flavor-changing transition.

Fig. 4. The mixing angle vs.ρ (ref.[32]).

Fig. 5. Constraints for neutral-current couplings from deep-inelastic and elastic scattering.

Fig. 6. Generation of an isoscalar-axial coupling through the triangle graph.

Fig. 7. The y-dependence of the asymmetry A in polarized electron scattering off deuterons in the minimal SU(2) x U(1) theory as a function of $\sin^2\theta_w$.

Fig. 8. χ^2 contours vs. a_1 and a_2.

Fig. 9. Gluon diagrams contributing to scaling violation.

Fig.10. Feynman diagrams contributing to the anomalous dimensions.

Fig.11. Scaling violations observed in electroproduction experiments.

Fig.12. Scaling violation [66] observed in $\nu w_2(x,Q^2)$ for small x as a function of Q^2.

Fig.13. BEBC [70] data on several moments $M_N(Q^2)$ of the F_3 structure function plotted logarithmically. The straight lines are predictions of QCD.

Fig.14. Several moments of the structure function F_3 with the CDHS [71] data also included.

Fig.15. Preliminary BEBC data [71] on the moments of non-singlet fragmentation functions [$D_u^{h^+}(M,N,Q^2)$ - $D_u^{h^-}(M,N,Q^2)$]. N denotes the moment of the quark-distribution function and the ratios correspond to moments of the fragmentation functions.

Fig. 16. Test of factorization. Double moments plotted as functions of Q^2. The notation is the same as in Fig. 15.

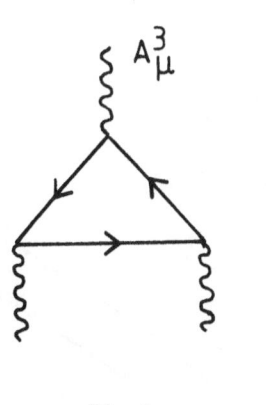

Fig.1

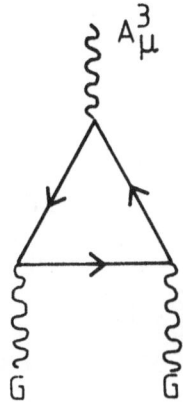

Fig. 2

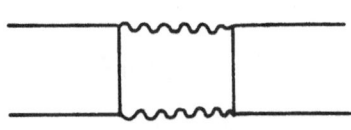

Fig. 3

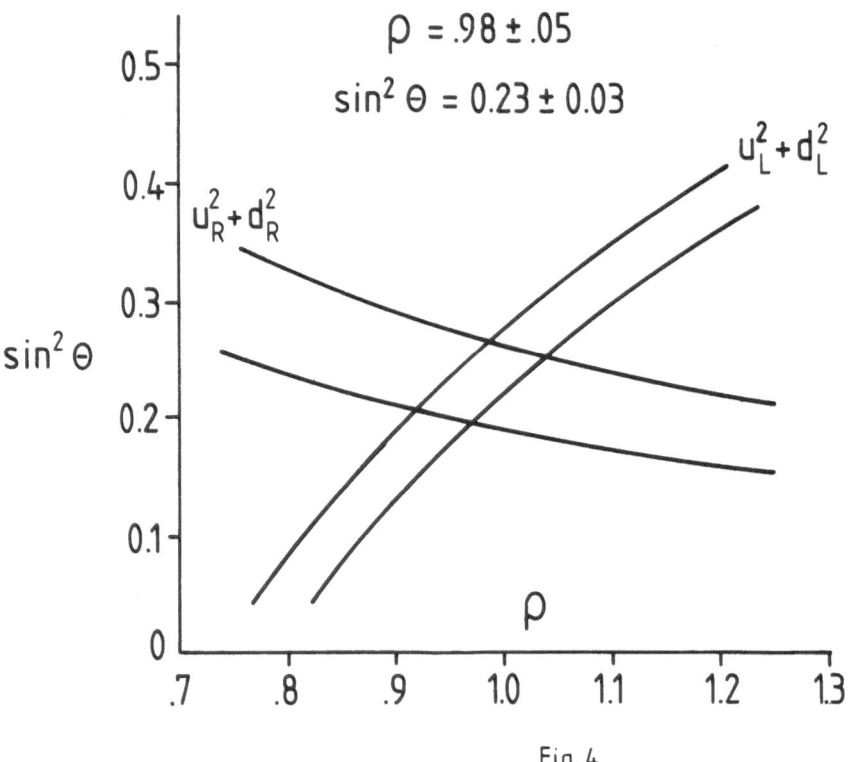

Fig. 4

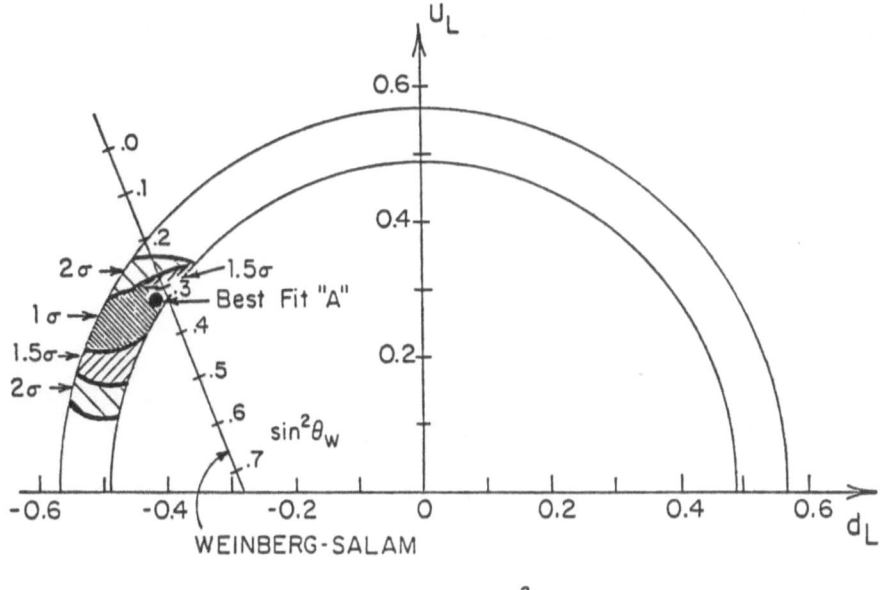

$M_A = 1.00$ GeV/c^2

Fig. 5a

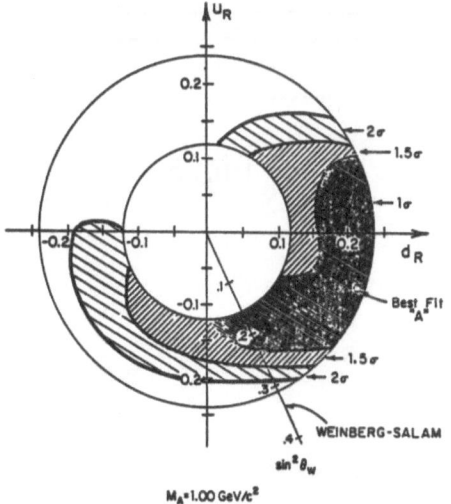

Fig. 5b

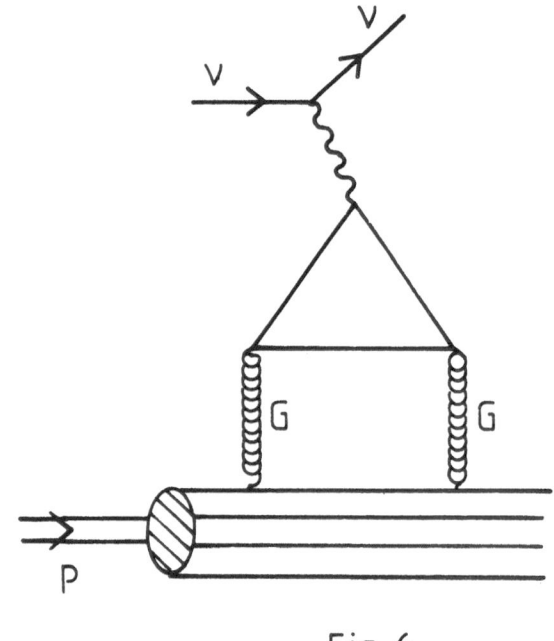

Fig. 6

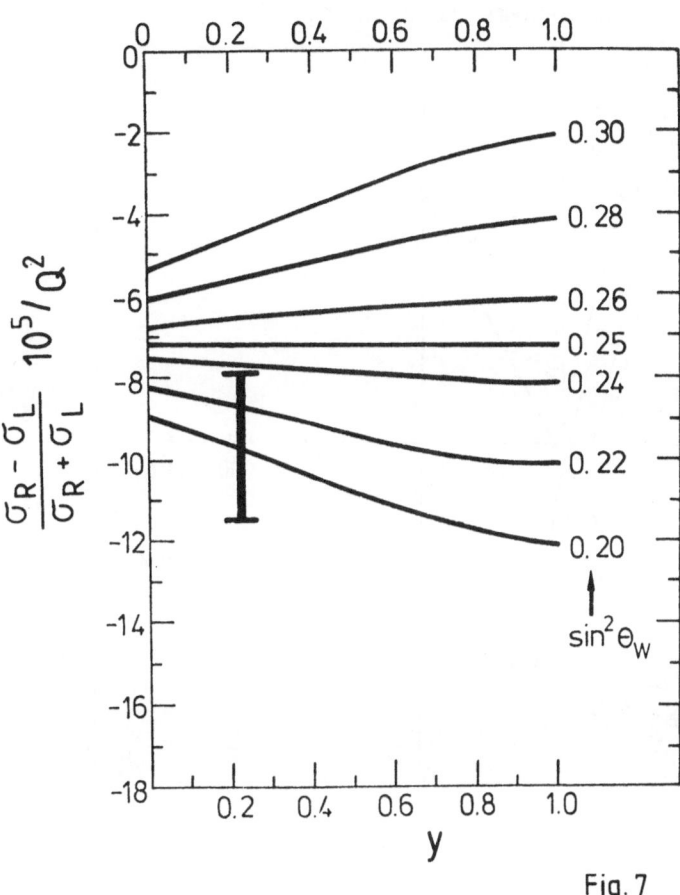

Fig. 7

550

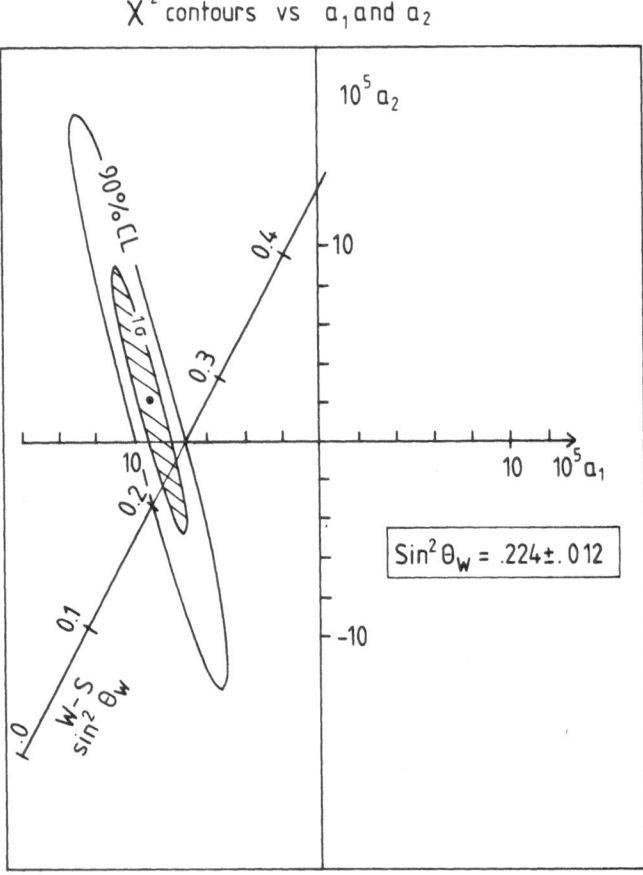

X^2 contours vs a_1 and a_2

Fig. 8

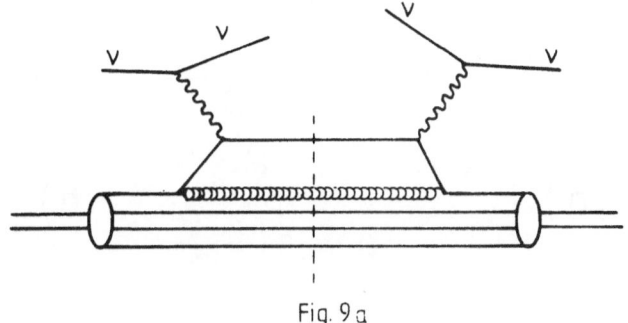

Fig. 9 a

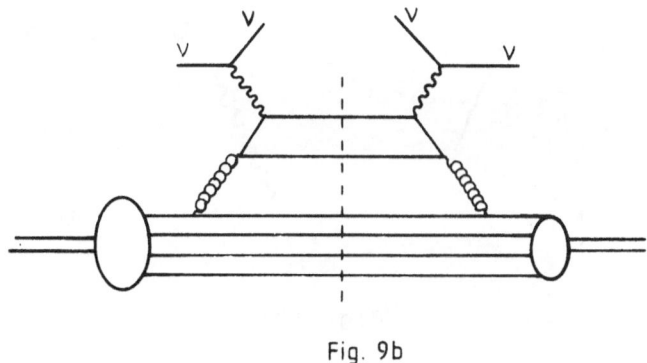

Fig. 9b

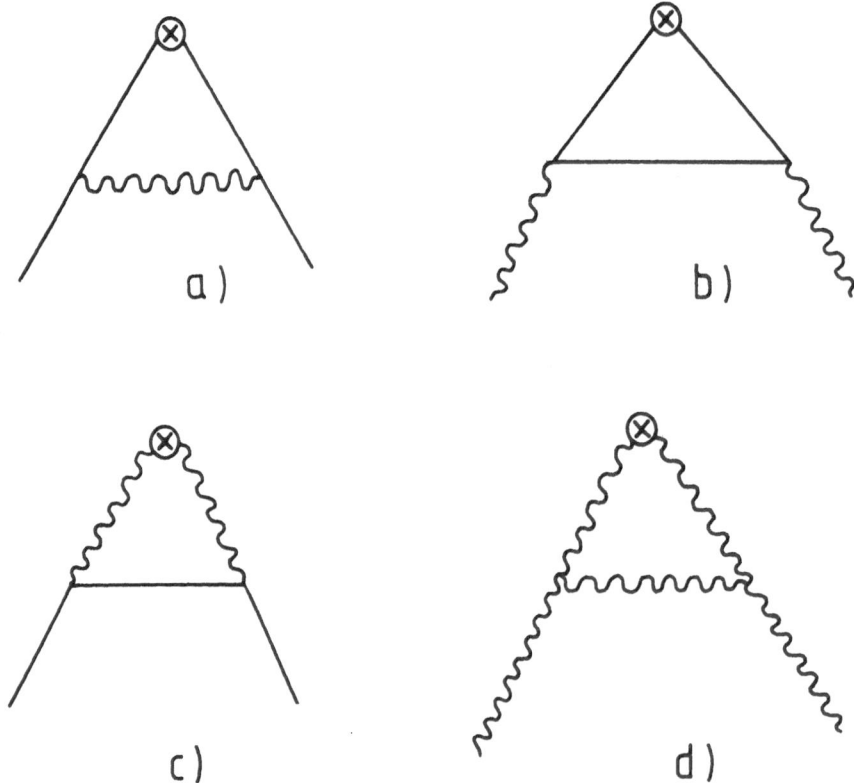

Fig. 10

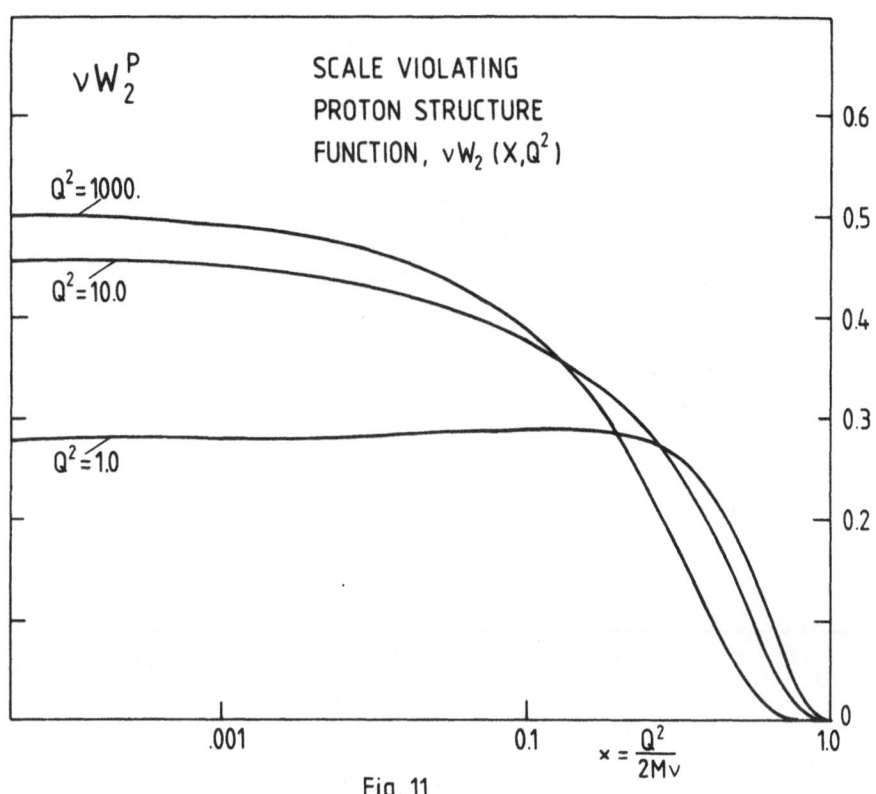

Fig. 11

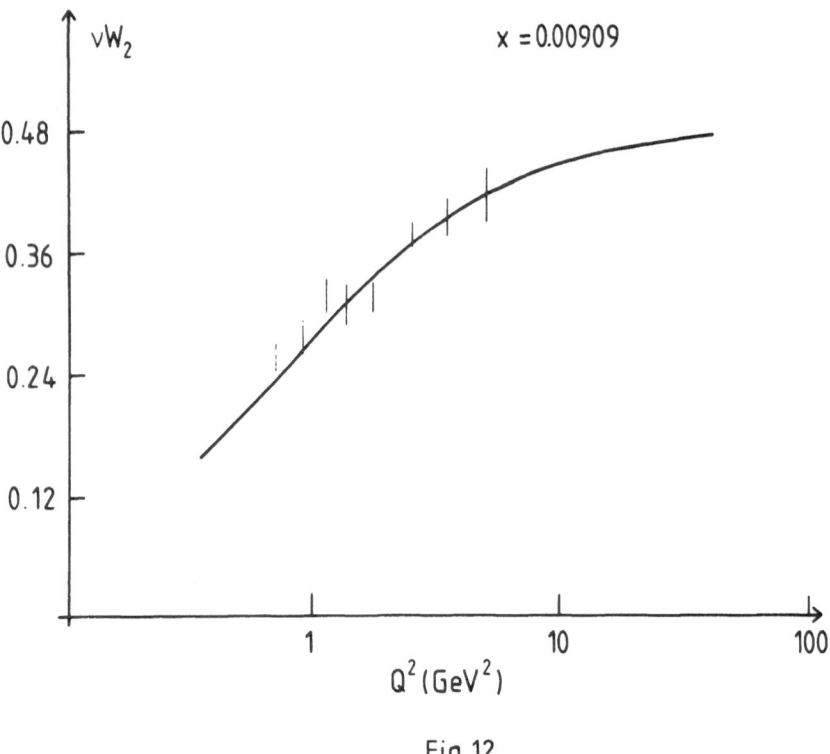

Fig. 12

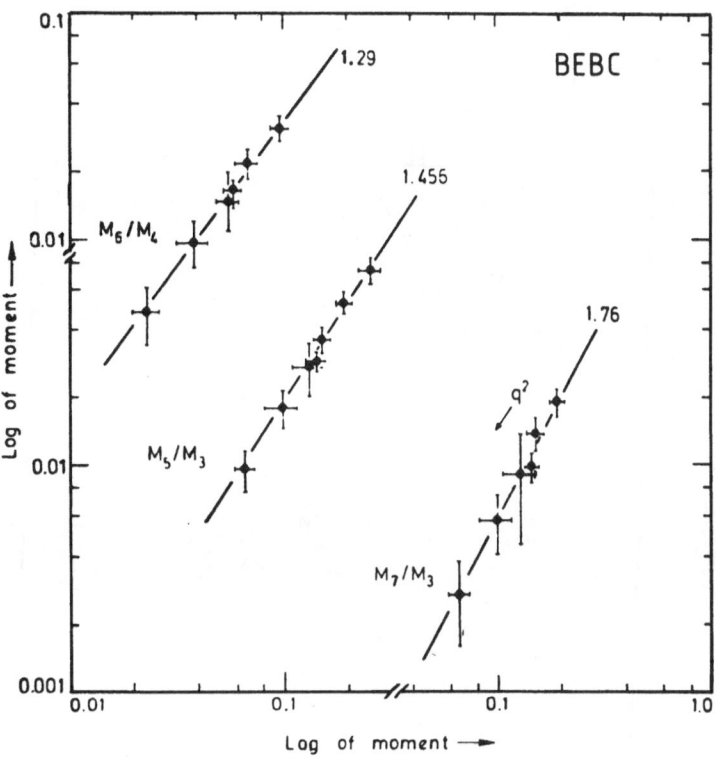

Fig. 13

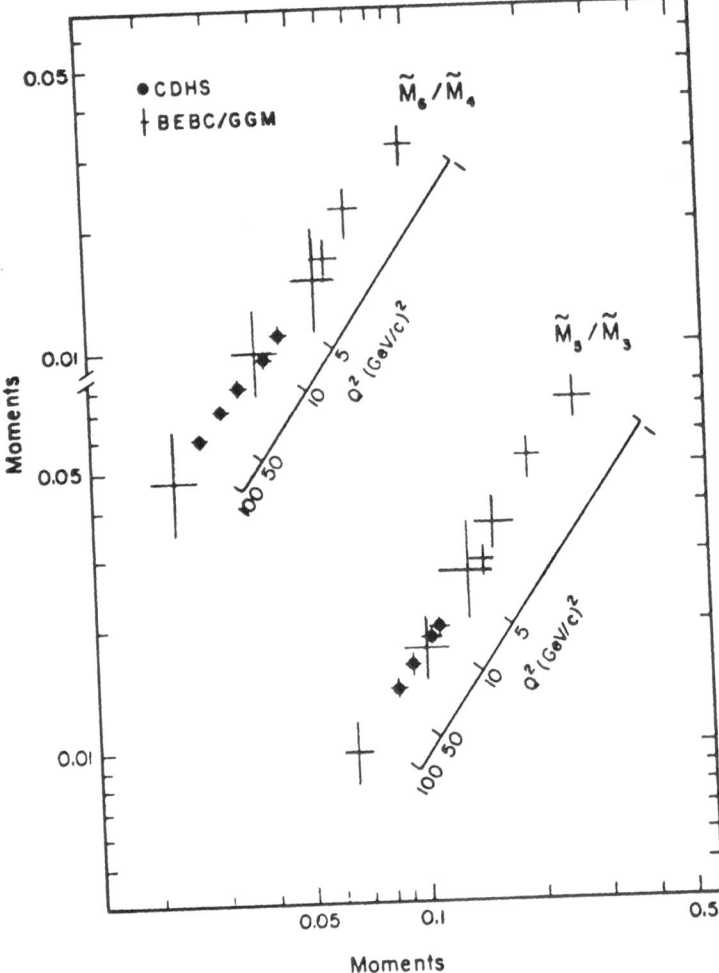

Fig. 14

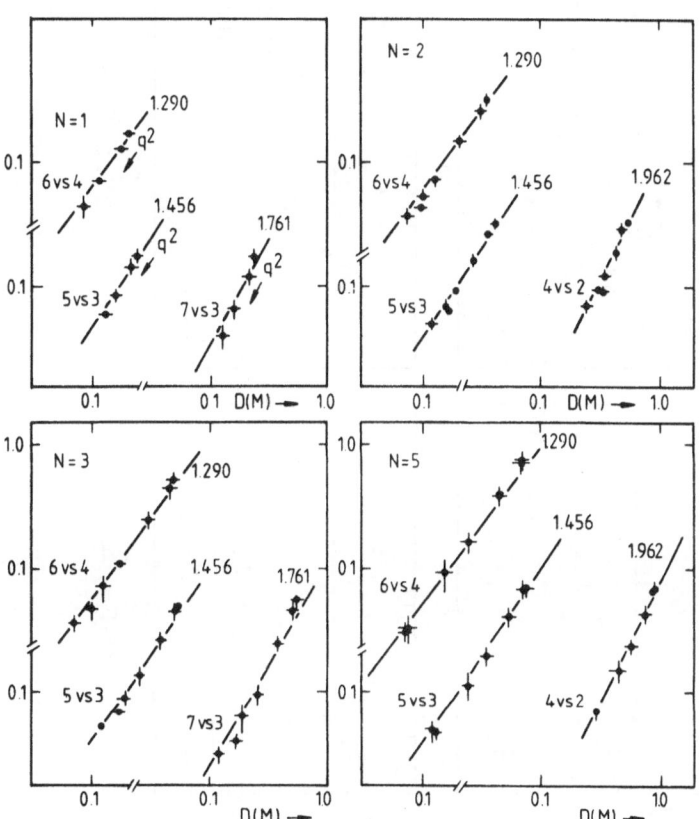

Fig. 15

558

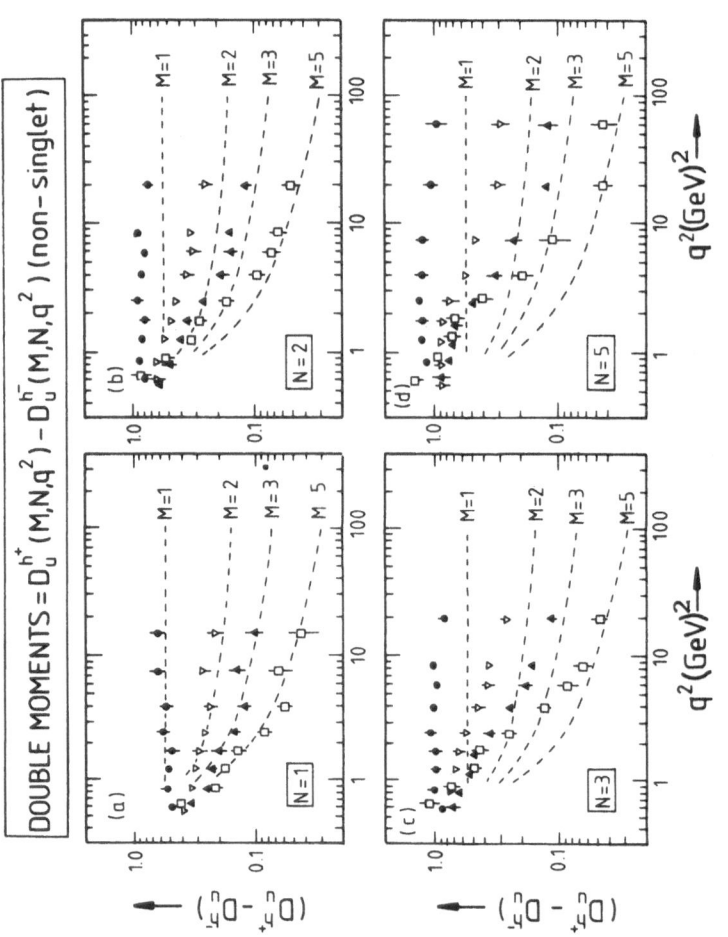

Fig. 16

Acta Physica Austriaca, Suppl. XXI, 559—618 (1979)
© by Springer-Verlag 1979

STRONG AND ELECTROMAGNETIC DECAYS

OF THE NEW PARTICLES[+]

by

H. GENZ

Institut für Theoretische Kernphysik

Universität Karlsruhe

Kaiserstraße 12, D-7500 Karlsruhe

CONTENTS

[+]Lectures given at the
XVIII. Internationale Universitätswochen für Kernphysik
Schladming, Austria, February 28 - March 10, 1979.

I. INTRODUCTION

Experimental information on decays [1-4] of the
$q\bar{q}$-particles with q = c,b,t... can provide simple,
direct and new tests of old-fashioned ideas such as
internal symmetries, meson mixings and dominance of
meson poles. It is the purpose of these lectures to work
out and review [5-9,24] some of these tests and to compare
them to experiment in the case of charm. For the old u-,
d- and s-quarks at low energies, the simple concepts of
symmetry, mixing and dominance (SYMIDO) work well and
thus these new tests are of much interest. From the
point of view of the old physics, this is obvious and -
if these tests are successful - SYMIDO should also
be of interest for the genuine QCD-type physics of the
new particles. SYMIDO - in that case a consequence of
hadron dynamics at low <u>and</u> high energies - would then
be an interesting property of QCD which might be understood in the perturbative limit at high energies. In any
case, SYMIDO would provide a compact description of
the data.

As compared to low energies, we have at high
energies additional information resulting from hard
gluon counting and asymptotic freedom. This must be
taken into account. It is also nice if we agree with
non-relativistic potentials, wave functions, overlap
integrals and all that.

According to QCD, mixing of mesons proceeds via
annihilation into vector gluons (e.g. ref.[10,11]). The
relevant processes are $q\bar{q} \to g...g \to q'\bar{q}'$ with a minimum
number of 2 gluons for pseudoscalar and tensor mesons
and of 3 gluons for vector mesons, as indicated in
figs. 1 and 2.

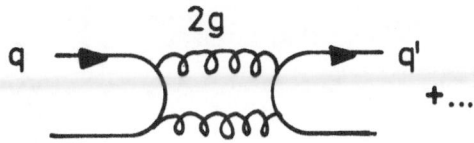

Fig. 1. Mixing of pseudoscalar and tensor mesons.

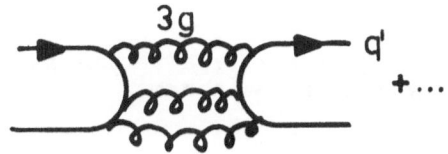

Fig.2. Mixing of vector mesons.

According to QCD, the vector mesons thus are "purer" than the pseudoscalars and the tensors and this has direct implications for processes such as $\psi \rightarrow \gamma\eta$ and $\psi \rightarrow \gamma f$ which proceed via mixing. For $\psi \rightarrow \gamma\eta$ the two diagrams of figs. 3 and 4 contribute and we expect from QCD-mixing that fig. 3 dominates over fig. 4. (e.g. ref.[12]).

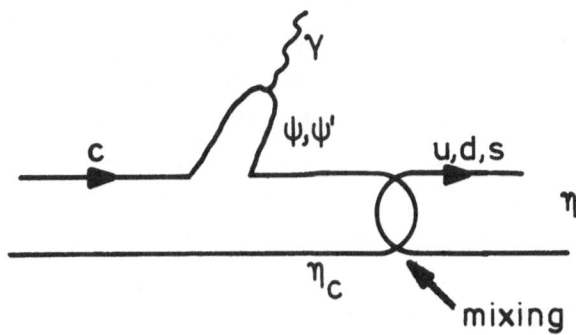

Fig.3. The decay $\psi \rightarrow \gamma\eta$ via pseudoscalar mixing and vector meson dominance.

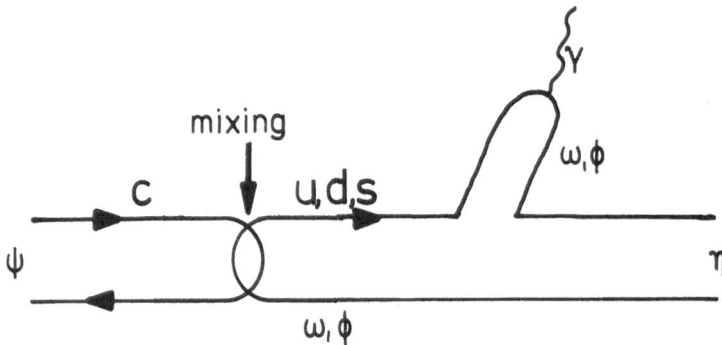

Fig. 4. The decay $\psi \to \gamma\eta$ via vector meson mixing and vector meson dominance.

Without the mixing concept, the relevant QCD-diagrams for the two processes are given in Fig. 5. One expects that mixing and computation of these diagrams - if possible - approximately yield the same results at least for the total rates.

We conclude that the η_c-η-mixing is mainly responsible for the decay and neglect the other diagrams. There is also direct experimental evidence that this is correct [1]. If the diagrams we neglect were the main source of the decay, vector meson dominance (VMD) would then imply that $\Gamma(\psi \to \gamma\eta)$ should at most be approximately 1% of $\Gamma(\psi \to \omega\eta)$ or $\Gamma(\psi \to \phi\eta)$. Experimentally, the branching ratios for $\psi \to \gamma\eta$ and $\psi \to \phi\eta$ are, however, equal (i.e. 0.08 ± 0.018% and 0.1 ± 0.06%, respectively). Thus there must be a larger source for $\psi \to \gamma\eta$, i.e. the diagrams with η_c-η-mixing.

For the decays [4] $\psi \to \gamma f$ (1.27) and $\psi \to \omega f$ (1.27), with branching ratios of 0.2 ± 0.07 % and 0.28 ± 0.11%, respectively, completely analogous

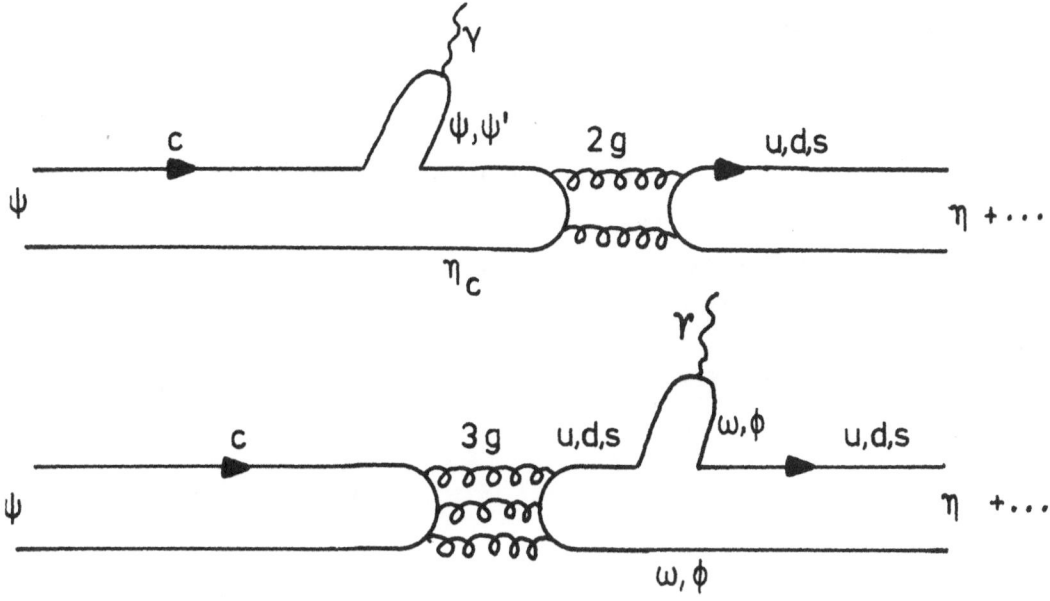

Fig. 5. The QCD-diagrams and vector meson dominance for
the decay $\psi \rightarrow \gamma\eta$.

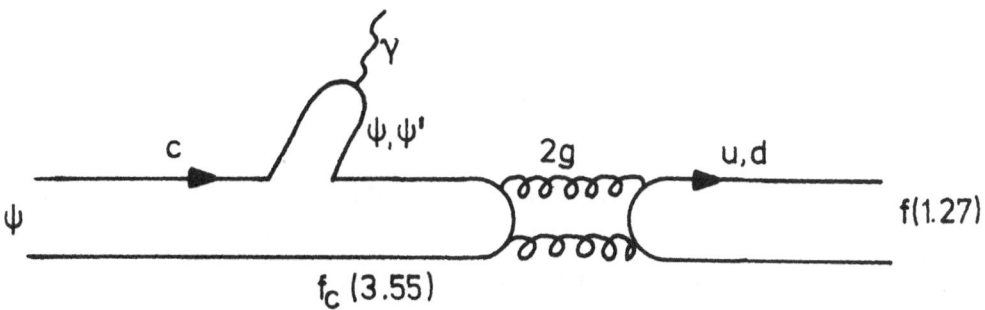

Fig.6. The largest QCD-diagrams and VMD for the process
$\psi \rightarrow \gamma f$.

considerations are valid. The decay $\psi \to \gamma f$ is described by the diagram in fig. 6 or by an equivalent mixing diagram. We have denoted the χ-state observed at 3.55 GeV by $f_c(3.55)$, assuming it is a 2^{++}-meson with $c\bar{c}$-content.

We use (e.g. ref.[5]) the decays $\psi \to \rho^o \pi^o$ and $\psi \to \gamma \pi^o$ to obtain a direct measure of VMD alone. Neglecting any radial excitations of ω and ρ, the only contributing diagrams - fig. 7 - are connected by VMD such that

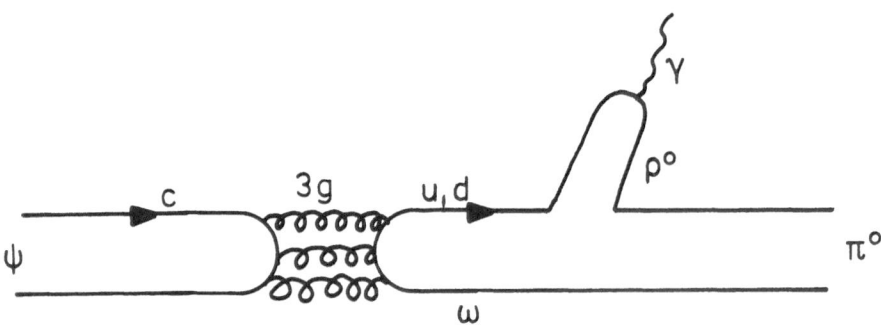

Fig. 7. The QCD-diagram for $\psi \to \gamma\pi$.

$$\Gamma(\psi \to \gamma\pi^o) = [\frac{q(\psi\to\gamma\pi^o)}{q(\psi\to\rho^o\pi^o)}]^3 \pi\alpha \frac{1}{\gamma_\rho^2} \Gamma(\psi \to \rho^o\pi^o) \qquad (1)$$

with $q(...)$ the decay momentum, $\alpha = \frac{1}{137}$ the fine structure constant and γ_V the coupling of the vector meson V to the photon normalized such that

$$\Gamma(V \rightarrow e^+e^-) = \frac{4\pi M_V \alpha^2}{12\gamma_V^2} \quad . \tag{2}$$

One finds $\gamma_\rho = 2.6$ from the experimental $\Gamma(\rho \rightarrow e^+e^-)$. The experimental values of $\Gamma(\psi \rightarrow \rho^\circ\pi^\circ)$ are

$$\Gamma(\psi \rightarrow \rho^\circ\pi^\circ) \quad = \quad \begin{cases} 0.39 \pm 0.1 \text{ \%} & \text{(DASP)} \\ 0.31 \pm 0.1 \text{ \%} & \text{(D-HD)} \\ 0.42 \pm 0.1 \text{ \%} & \text{(SLAC-LBL)} \end{cases} \tag{3}$$

and for $\Gamma(\psi \rightarrow \rho^\circ\pi^\circ)$ between 0.21 and 0.52 % we find as prediction from VMD

$$\Gamma^{VMD}(\psi \rightarrow \gamma\pi^\circ) = (0.6 - 1.4) \text{ eV} , \tag{4}$$

which must be compared to the experimental

$$\Gamma^{EXP}(\psi \rightarrow \gamma\pi^\circ) = (4.9 \pm 3.2) \text{ eV} . \tag{5}$$

Thus we expect to be able to predict the correct order of magnitude from VMD. Better agreement is, however, not excluded by eqs. (4) and (5).

Experimental results are taken from ref.[1], whenever possible. The original work is quoted there. Sections II, VI and VII have been taken from ref.[24].

II. MASS AND MIXING OF ISOSCALAR MESONS

In this section, we review with some changes, the work of Fritzsch and Jackson [11] on the mixing of pseudoscalar and vector mesons. This will later be applied to tensor mesons also.

Since deviations from ideal mixing are caused by the annihilation diagrams of figs. 1 and 2, one expects [11] the mass-squared matrix M of the neutral mesons to be

$$M \; = \; \begin{pmatrix} M_u^2+\lambda_{uu} & \lambda_{ud} & \lambda_{us} & \lambda_{uc} \\ \lambda_{ud} & M_d^2+\lambda_{dd} & \lambda_{ds} & \lambda_{dc} \\ \lambda_{us} & \lambda_{ds} & M_s^2+\lambda_{ss} & \lambda_{sc} \\ \lambda_{uc} & \lambda_{dc} & \lambda_{sc} & M_c^2+\lambda_{cc} \end{pmatrix}$$

(1)

in the basis $u\bar{u}$, $d\bar{d}$, $s\bar{s}$ and $c\bar{c}$.

The M_q^2 are the contributions of the direct diagrams to the meson masses and yield the total mass in the ideal mixing limit. Assuming isospin symmetry one has

$$\lambda_{uq} = \lambda_{dq} \quad \text{(any q)} ,$$

(2)

$$M_u^2 = M_{\pi^\pm}^2 = M_{\pi^o}^2 ,$$

(3)

$$M_s^2 = 2M_K^2 - M_\pi^2$$

(4)

and

$$M_c^2 = 2M_D^2 - M_\pi^2$$
$$= 2M_F^2 - 2M_K^2 + M_\pi^2 .$$

(5)

We assume the same formulas for vector mesons and tensor mesons. The π in these cases is replaced by ρ or A_2, respectively, etc.

We will assume - following an expectation expressed and successfully tested for the pseudoscalars in ref.[11] - that the $\lambda_{qq'}$ factorize as

$$\lambda_{qq'} = \pm \left| \sqrt{\lambda_{qq}\, \lambda_{q'q'}} \right| , \qquad (6)$$

where the sign does not depend on the quarks q or q'. It may however be different for different multiplets. We abbreviate by

$$\lambda_q \equiv \lambda_{qq} . \qquad (7)$$

Taking out the eigenstate $|\pi^0\rangle = \left| \dfrac{u\bar{u}-d\bar{d}}{\sqrt{2}} \right\rangle$, we now have the mass-squared matrix

$$M = \begin{pmatrix} M_u^2 + 2\lambda_u & \pm\sqrt{2\lambda_u\lambda_s} & \pm\sqrt{2\lambda_u\lambda_c} \\[2ex] \pm\sqrt{2\lambda_u\lambda_s} & M_s^2 + \lambda_s & \pm\sqrt{\lambda_s\lambda_c} \\[2ex] \pm\sqrt{2\lambda_u\lambda_c} & \pm\sqrt{\lambda_s\lambda_c} & M_c^2 + \lambda_c \end{pmatrix} \qquad (8)$$

in the basis $(u\bar{u}+d\bar{d})/\sqrt{2}$, $s\bar{s}$ and $c\bar{c}$. M is easily diagonalized by use of (e.g.) the invariants trace M, det M and det M^{-1} with the results

$$2\lambda_u = \frac{(M_\eta^2 - M_u^2)(M_{\eta'}^2 - M_u^2)(M_{\eta c}^2 - M_u^2)}{(M_s^2 - M_u^2)(M_c^2 - M_u^2)} ,$$

$$\lambda_s = \frac{(M_\eta^2 - M_s^2)(M_{\eta'}^2 - M_s^2)(M_{\eta c}^2 - M_s^2)}{(M_u^2 - M_s^2)(M_c^2 - M_s^2)} \quad ,$$

$$\lambda_c = \frac{(M_\eta^2 - M_c^2)(M_{\eta'}^2 - M_s^2)(M_{\eta c}^2 - M_c^2)}{(M_u^2 - M_c^2)(M_s^2 - M_c^2)} \tag{9}$$

and

$$|\mu^2> =$$

$$N(\mu^2)(\frac{\sqrt{2}|\lambda_u|}{\mu^2 - M_u^2}|\frac{\overline{uu} + \overline{dd}}{\sqrt{2}} > + \frac{\sqrt{|\lambda_s|}}{\mu^2 - M_s^2}|s\bar{s} > + \frac{\sqrt{|\lambda_c|}}{\mu^2 - M_c^2}|c\bar{c}>) \tag{10}$$

with $\mu^2 = M_\eta^2$, $M_{\eta'}^2$, $M_{\eta c}^2$ and $N(\mu^2)$ a normalization factor.
These formulas do not depend on the sign in eqs. (6).
We replace η by ω (or f), η' by ϕ (or f') and η_c by ψ
(or $f_c(3.55)$) to find the corresponding results for
vector mesons (or tensor mesons). We have checked that
the obvious generalization of eqs. (9) and (10) to any
number of flavours is indeed valid.

Eqs. (9) and (10) have some immediate consequences.
Since always $M_u^2 < M_s^2 < M_c^2$ it follows that for a given
multiplet either

$$\lambda_q > 0 \text{ together with } M_u^2 < M_{\eta,\omega}^2 < M_s^2 < M_{\eta',\phi}^2 < M_c^2 < M_{\eta_c,\psi}^2 \tag{11}$$

or

$$\lambda_q < 0 \text{ together with } M_f^2 < M_u^2 < M_{f'}^2, < M_s^2 < M_{f_c(3.55)}^2 < M_c^2 \quad . \tag{12}$$

Since the first alternative holds for pseudoscalar and
vector, the second for tensor mesons we have already
used the appropriate particle names. Secondly, the
relative signs of the various components of the physical
particles are fixed as

$$
\begin{array}{cccc}
 & (u\bar{u} + d\bar{d})/\sqrt{2} & s\bar{s} & c\bar{c} \\
|\eta,\omega> & + & - & - \\
|\eta',-\phi> & + & + & - \\
|\eta_c,-\psi> & + & + & +
\end{array}
\tag{13}
$$

for $\lambda_q > 0$ and

$$
\begin{array}{cccc}
 & (u\bar{u} + d\bar{d})/\sqrt{2} & s\bar{s} & c\bar{c} \\
|f> & + & + & + \\
|f'> & + & - & - \\
|f_c> & + & + & -
\end{array}
\tag{14}
$$

for $\lambda_q < 0$. Thus, an SU(3)-scalar admixture $\varepsilon(u\bar{u}+d\bar{d}+s\bar{s})$
to the $c\bar{c}$-particles η_c,ψ and f_c is consistent with the
signs derived here, whereas the 8th component of an
octet - i.e. $\varepsilon(u\bar{u}+d\bar{d}-2s\bar{s})$ - is inconsistent. This
agrees with the general features of the decays of the
$c\bar{c}$-particles. Furthermore, the $c\bar{c}$-contributions to η
and η' have the same sign leading to constructive
interference [13] in the isospin-violating decay [14]
$\psi' \rightarrow \psi\pi^o$. The predicted rate is then $0.09 < \Gamma(\psi'\rightarrow\psi\pi^o) <$
$< 0.18\%$ of $\Gamma(\psi' \rightarrow all)$ as compared to the upper experimental limit of 0.14% (compare previous editions
of ref.[1]).

III. MIXING OF PSEUDOSCALARS AND DECAYS

Since clearly $M_u = M_\pi < M_\eta$, we must have

$$M_s = \sqrt{2M_K^2 - M_\pi^2} < M_{\eta'},$$

which is the case. Thus $\lambda_q > 0$. We have $M_c \approx 3$ GeV and thus may compute λ_u and λ_s as $\lambda_u = 0.28$ GeV^2 and $\lambda_s = 0.17$ GeV^2. The factor $M_{\eta c} - M_c$ in λ_c is, however, essentially unknown since - for $M_{\eta c} = 2.82$ GeV - one finds $M_{\eta c} - M_c = 0.2$ GeV or 0.03 GeV if one computes M_c from the D-meson or the F-meson mass in eqs. (5), respectively. The sign computed this way is, however, in agreement with eq. (11). To proceed we may - following ref.[11] - use asymptotic freedom. To lowest order in QCD one expects

$$\lambda_{qq'} = c_{P,T} \kappa(M_q^2) \kappa(M_{q'}^2) \text{ for P and T} \tag{1}$$

and

$$\lambda_{qq'} = c_V (\kappa(M_q^2) \kappa(M_{q'}^2))^{\frac{3}{2}} \text{ for V} \tag{2}$$

with c independent of the quarks and $\kappa(M_q^2) = \dfrac{g_s^2(M_q^2)}{4\pi}$. The running strong coupling constant at M_q^2 is denoted by $g_s(M_q^2)$. At large M_q^2 and $M_{q'}^2$, we have

$$\kappa(M_q^2) = \frac{\kappa(M_{q'}^2)}{1 + \dfrac{25}{12\pi} \kappa(M_{q'}^2) \ln(\dfrac{M_q^2}{M_{q'}^2})}. \tag{3}$$

We follow ref.[11] to use this relation down at $M_s^2 = 0.7^2$ GeV^2 to compute

$$\frac{\lambda_c}{\lambda_s} = \left(\frac{\kappa \, (M_c^2 \approx 9 \; \text{GeV}^2)}{\kappa \, (M_s^2 = 0.49 \; \text{GeV}^2)} \right)^2 = 0.39 \tag{4}$$

and thus fix λ_c in terms of λ_s. The states $|\eta>$, $|\eta'>$ and $|\eta_c>$ are now completely determined. In particular, the $|\eta_c>$ is

$$|\eta_c> = |c\bar{c}> + 0.016 \, (|u\bar{u}> + |d\bar{d}> + 0.81 |s\bar{s}>) , \tag{5}$$

implying some SU(3)-violation. The η-η'-mixing angle turns out to be

$$\theta_{PS} = -12.3^{\circ} . \tag{6}$$

We may also define parameters ε_η and $\varepsilon_{\eta'}$ by writing

$$|\eta_c> = |c\bar{c}> + \varepsilon_\eta |\eta> + \varepsilon_{\eta'} |\eta'> \tag{7}$$

and find

$$\varepsilon_\eta = 0.82 * 10^{-2} , \tag{8}$$

$$\frac{\varepsilon_{\eta'}}{\varepsilon_\eta} = 3 \tag{9}$$

or

$$\varepsilon_{\eta'} = 2.6 * 10^{-2} . \tag{10}$$

The η-η'-mixing angle agrees with everything which is known about it. For example, from VMD or from PCAC with $f_\pi = f_\eta$ one derives

$$\frac{\Gamma(\eta \to 2\gamma)}{\Gamma(\pi^o \to 2\gamma)} = 18 \ (\frac{M_\eta}{M_{\pi^o}})^3 \ (\sqrt{\frac{3}{2}} \ \cos \theta_{PS} - 2\sqrt{3} \ \sin \theta_{PS})^2 \qquad (11)$$

for the theoretical ratios $\gamma_\rho/\gamma_\omega$ and γ_ρ/γ_ϕ. Experimentally, with about 20% error,

$$\frac{\Gamma(\eta \to 2\gamma)}{\Gamma(\pi^o \to 2\gamma)} = 42 \qquad (12)$$

corresponding to $\theta_{PS} = -7.8^o$. From $\theta_{PS} = -12.8^o, 56$ is obtained for this ratio.

We now discuss phenomenologically the decays of the new particles involving the pseudoscalar mixing parameters. We define coupling constants by effective Lagrangians and write for VVP (e.g. $\psi\psi\eta_c$) and V'VP (e.g. $\psi'\psi\eta_c$)-transitions

$$L(VVP) = \frac{G(VVP)}{2} \ \varepsilon_{\alpha\beta\gamma\delta} (\partial^\alpha V_a^\beta) \ (V_b^\gamma) \ (\partial^\delta P_c) d_{abc} \qquad (13)$$

and

$$L(V'VP) = G(V'VP) \ \varepsilon_{\alpha\beta\gamma\delta} (\partial^\alpha V_a'^\beta) \ (V_b^\gamma) \ (\partial^\delta P_c) d_{abc} . \qquad (14)$$

We have assumed, as we always will, that internal symmetries are defined by a Lagrangian without mass factors. This prescription is unique if only one amplitude contributes to the corresponding decays. If that is not the case we need further information (e.g. tensor mesons dominance). In eqs. (13) and (14), the Lorentz indices are $\alpha, \beta, \gamma, \delta$, and the SU(4)-indices are a,b,c.

Two-gluon diagrams such as the one in fig. 5 can be drawn for $\psi' \to \psi\eta$, $\psi \to \eta\gamma$, $\eta'\gamma$ such that

$$\Gamma(\psi \to \eta\gamma^{(\prime)}) = \varepsilon_\eta^2(\prime) \left[\frac{q(\psi \to \eta\gamma^{(\prime)})}{q(\psi \to \gamma\eta_c)}\right]^3 \Gamma(\psi \to \gamma\eta_c) \tag{15}$$

$$\Gamma(\psi' \to \gamma\eta) = \frac{(q(\psi' \to \psi\eta))^3}{6\pi} G^2(V'VP) \; \varepsilon_\eta^2 \tag{16}$$

and, writing out $\Gamma(\psi \to \gamma\eta_c)$ in eq. (15),

$$\Gamma(\psi \to \gamma\eta^{(\prime)}) = \frac{\alpha}{6}(q(\psi \to \gamma\eta^{(\prime)}))^3 \left(\frac{G(VVP)}{\gamma_\psi} + \frac{G(V'VP)}{\gamma_{\psi'}}\right)^3 \varepsilon_\eta^2(\prime) \; . \tag{17}$$

One knows that

$$\frac{4\pi}{\gamma_\psi^2} = 0.34 \text{ and } \frac{4\pi}{\gamma_{\psi'}^2} = 0.12 \; .$$

Eqs. (17) may now be used to obtain an experimental value for $(\varepsilon_{\eta'}/\varepsilon_\eta)^2$:

$$\frac{\Gamma(\psi \to \gamma\eta')}{\Gamma(\psi \to \gamma\eta)} = \left(\frac{M_\psi^2 - M_{\eta'}^2}{M_\psi^2 - M_\eta^2}\right)^3 (\frac{\varepsilon_{\eta'}}{\varepsilon_\eta})^2 \; . \tag{18}$$

The experimental widths are (ref.[1])

$$\frac{\Gamma(\psi \to \gamma\eta)}{\Gamma(\psi \to all)} = \begin{cases} (0.8 \pm 0.18) * 10^{-3} & \text{(DASP)} \\ (1.3 \pm 0.4) * 10^{-3} & \text{(Bartel et al.)} \end{cases} \tag{19}$$

and

$$\frac{\Gamma(\psi \rightarrow \gamma\eta')}{\Gamma(\psi \rightarrow all)} = \begin{cases} (2.2 \pm 1.7) * 10^{-3} & \text{(DASP)} \\ (2.4 \pm 0.7) * 10^{-3} & \text{(Bartel et al.).} \end{cases} \qquad (20)$$

The ratio with the smallest errors is

$$\frac{\Gamma_{Bartel\ et.al.}(\psi \rightarrow \gamma\eta')}{\Gamma_{DASP}(\psi \rightarrow \gamma\eta)} = 3 \pm 1.1 \qquad (21)$$

yielding

$$\left|\frac{\varepsilon_\eta'}{\varepsilon_\eta}\right| = 1.9 \pm 0.7 , \qquad (22)$$

to be compared with the value 3 obtained for this from mixing. The agreement is sufficiently good. With the new value

$$\frac{\Gamma(\psi \rightarrow \gamma\eta')}{\Gamma(\psi \rightarrow \gamma\eta)} = 4.8 \pm 1.2 \qquad (21a)$$

from the crystal ball [15] one obtains

$$\left|\frac{\varepsilon_\eta'}{\varepsilon_\eta}\right| = 2.4 \pm 0.8 \qquad (22a)$$

in even better agreement with the theoretical value. From the width formula in eq. (16), the experimental $\Gamma(\psi' \rightarrow all)$ and

$$\frac{\Gamma(\psi' \rightarrow \psi\eta)}{\Gamma(\psi' \rightarrow all)} = \begin{cases} 0.065 \pm 0.026 & \text{(DASP)} \\ 0.075 \pm 0.014 & \text{(SLAC-LBL)} \end{cases} \qquad (23)$$

we obtain approximately

$$|\varepsilon_\eta G(V'VP)| = 0.2 \text{ GeV}^{-1} \ . \tag{24}$$

To determine $|\varepsilon_\eta|$, we combine this result with eq.(17) for the η and the experimental value [1]

$$\Gamma(\psi \to \gamma\eta) = \begin{cases} 0.054 \text{ keV} & \text{(DASP)} \\ 0.087 \text{ keV} & \text{(Bartel et.al.)} \end{cases} \tag{25}$$

and obtain $|G(VVP)\varepsilon_\eta|$. Internal symmetries are then used to compute $|G(VVP)|$ from the old mesons. To do so, we first note that in eq.(17) the $|\varepsilon_\eta G(V'VP)|$ from eq.(24) alone, without the VVP-term, would predict 1.7 keV for $\Gamma(\psi \to \gamma\eta)$, which is much too large. Thus, the VVP-term is there and practically cancels the V'VP-term. Consequently,

$$\text{sign } G(VVP)/\gamma_\psi = -\text{sign } G(V'VP)/\gamma_{\psi'}, \tag{26}$$

and one finds

$$|G(VVP)\varepsilon_\eta| = \begin{cases} 0.143 \text{ GeV}^{-1} & \text{(DASP)} \\ 0.157 \text{ GeV}^{-1} & \text{(Bartel et.al.).} \end{cases} \tag{27}$$

Now, due to internal symmetries assumed in writing $L_I(VVP)$, we may obtain G(VVP) from [16,17] $\omega \to \rho\pi \to 3\pi$ or from six measured photonic decays of the old mesons [17,18] ($\pi^0 \to 2\gamma$, $\rho \to \gamma\pi$, $\omega \to \gamma\pi$, $K^{*0} \to \gamma K^0$, $\eta \to 2\gamma$ and $\phi \to \gamma\eta$). One finds as extreme values $|G(VVP)| = 17 \text{ GeV}^{-1}$ from $\omega \to \rho\pi \to 3\pi$ and $|G(VVP)| = 9 \text{ GeV}^{-1}$ from $K^{*0} \to \gamma K^0$. To free ourselves from uncertainties resulting from the photon mass extrapolation, we use the value of $|G(VVP)|$ from $\rho \to 3\pi$ and find

$$|\varepsilon_\eta| = \begin{cases} 0.9 * 10^{-2} & \text{(DASP} \\ 1.2 * 10^{-2} & \text{(Bartel et.al.)} \end{cases} \tag{28}$$

to be compared with $|\varepsilon_\eta| = 0.82 * 10^{-2}$ from the mass matrix. We also get

$$|G(V'VP)| \stackrel{\sim}{\sim} 20 \text{ GeV .} \tag{29}$$

Consistency of the present approach requires existence of a pseudoscalar meson at approximately 3 GeV with $c\bar{c}$-content, which we call η_c and which may or may not be the controversial - it is not found in the crystal ball experiment - $\eta_c(2.82)$. We will therefore not assume it to be at precisely 2.82 GeV and thus cannot obtain a number for $\Gamma(\psi \to \gamma\eta_c)$ from

$$\Gamma(\psi \to \gamma\eta_c) = \frac{1}{\varepsilon_\eta^2} \left(\frac{q(\psi \to \gamma\eta_c)}{q(\psi \to \gamma\eta)}\right)^3 * \Gamma(\psi \to \gamma\eta) , \tag{30}$$

which strongly depends on the $\psi\gamma\eta_c$-phase space. This relation amounts approximately to

$$\Gamma(\psi \to \gamma\eta_c) = \left(\frac{q(\psi \to \gamma\eta_c)}{q(\psi \to \gamma\eta_c(2.82)}\right)^3 * (3 \text{ keV) .} \tag{31}$$

There is, however, a number of results on the η_c which only weakly depend on its precise mass. These include the phase space of $\psi'\gamma\eta_c$ and of $\eta_c\gamma\gamma$. For easy comparison with the literature, we continue to use $M_{\eta_c} = 2.82$ in predictions of this kind. Turning to $\psi' \to \gamma\eta$, $\gamma\eta'$, $\gamma\eta_c$ we can improve the upper experimental limits

$$\Gamma(\psi' \to \gamma\eta) < 0.091 \text{ keV} \tag{32}$$

and

$$\Gamma(\psi' \to \gamma\eta') < 0.25 \text{ keV} \tag{33}$$

by almost an order of magnitude if we incorporate that only 5% of the ψ'-decays are unknown.

We need the $\psi'\psi'\eta_c$-coupling constant and define it in analogy to eq. (13) via a $G(V'V'P)$; the symmetry implied by the corresponding $L_I(V'V'P)$ is however not used here. The widths of $\psi' \to \gamma\eta$, $\gamma\eta'$ can be written as

$$\Gamma(\psi' \to \gamma\eta^{(1)}) = \frac{\alpha}{6} (\sigma(\psi' \to \gamma\eta^{(1)})^3$$

$$\tag{34}$$

$$(\frac{G(V'V'P)}{\gamma_{\psi'}} + \frac{G(V'VP)}{\gamma_\psi})^2 \; \varepsilon_\eta^2{(1)} \quad .$$

From eqs. (29) and (28) the V'VP-contribution can be computed with the result $\Gamma(\psi' \to \gamma\eta^{(1)}) = 7.7 \text{ keV}$ (19 keV), much above the experimental upper limit. Thus, also the V'V'P-term is there and almost cancels the V'VP-term!

$$\frac{G(V'V'P)}{\gamma_{\psi'}} \approx - \frac{G(V'VP)}{\gamma_\psi} \quad , \tag{35}$$

yielding

$$G(V'V'P) \approx - \frac{\gamma_{\psi'}}{\gamma_\psi} G(V'VP) \approx -(28-40) \text{ GeV}^{-1} \quad . \tag{36}$$

Since evidently

$$\Gamma(\psi' \rightarrow \gamma\eta_c) = (\frac{q(\psi' \rightarrow \gamma\eta_c)}{q(\psi' \rightarrow \gamma\eta)})^3 \frac{1}{\epsilon_\eta^2}\Gamma(\psi' \rightarrow \gamma\eta) = 763 * \Gamma(\psi' \rightarrow \gamma\eta)$$

(37)

we get

$$\Gamma(\psi' \rightarrow \gamma\eta_c) < 70 \text{ keV}$$

(38)

from the experimental limit on $\Gamma(\psi' \rightarrow \gamma\eta)$. Due to eq. (37) or even due to

$$\Gamma(\psi' \rightarrow \gamma\eta) + \Gamma(\psi' \rightarrow \gamma\eta') + \Gamma(\psi' \rightarrow \gamma\eta_c)$$

(39)

$$= (1 + 2.51 + 763) \Gamma(\psi' \rightarrow \gamma\eta)$$

we furthermore get the improved limits

$$\Gamma(\psi' \rightarrow \gamma\eta) < 0.014 \text{ keV}$$

(40)

and

$$\Gamma(\psi' \rightarrow \gamma\eta') < 0.035 \text{ keV},$$

(41)

since the sum in eq. (39) must be below the 11 keV un-known ψ'-decays. If less than almost all unknown ψ'-decays are $\psi' \rightarrow \gamma\eta_c$, these limits are reduced even further.

Finally for this chapter, we also obtain a prediction for $\eta_c \rightarrow 2\gamma$. Incorporating the relative signs obtained above, the widths formula is

$$\Gamma_{\eta_c \to 2\gamma} = \frac{\alpha^2 \pi M_{\eta_c}^3}{32} \left(\frac{1}{\gamma_\psi} \left\{ \left| \frac{G(VVP)}{\gamma_\psi} \right| - \left| \frac{G(VV'P)}{\gamma_{\psi'}} \right| \right\} \right.$$

$$\left. \pm \frac{1}{\gamma_{\psi'}} \left\{ \left| \frac{G(V'V'P)}{\gamma_{\psi'}} \right| - \left| \frac{G(VV'P)}{\gamma_\psi} \right| \right\} \right)^2 . \tag{42}$$

We use eqs.(17) and (34) to obtain, including both signs,

$$0.3 \text{ keV} < \Gamma(\eta_c \to 2\gamma) < 0.7 \text{ keV}. \tag{43}$$

IV. MIXING OF VECTOR MESONS AND DECAYS

We first perform the phenomenological analysis of the decays and then compare to the mass-squared matrix mixing formalism. Internal symmetry for the VPP-vertex is defined by the Lagrangian.

$$L_I(VPP) = \frac{G(VPP)}{2} V_a^\mu (P_b \overleftrightarrow{\partial}_\mu P_c) f_{abc} \tag{1}$$

without mass factors. It yields the width formula

$$\Gamma(V \to P_A P_B) = \frac{|G(VPP) f_{ABV}|^2}{6\pi M_V^2} (\sigma(V \to P_A P_B))^3 , \tag{2}$$

which is compared in Table 1 to known decays. The agreement is good with

$$G(VVP) = 6 . \tag{3}$$

(VMD implies $G(VPP) = 2\gamma_\rho = 5.2.$).

To discuss decays of $c\bar{c}$-particles via vector meson mixing, we define ε_ω and ε_ϕ as

$$|\psi> = |-c\bar{c}> + \varepsilon_\omega |\frac{u\bar{u}+d\bar{d}}{\sqrt{2}}> + \varepsilon_\psi |-s\bar{s}> \tag{4}$$

and obtain

$$|\varepsilon_\omega + \sqrt{2} \ \varepsilon_\phi| < 2 * 10^{-4} \tag{5}$$

from the experimental

$$\Gamma(\psi \to K^+K^-) < 10^{-4} \ \Gamma(\psi \to \text{all}) . \tag{6}$$

To gain further information on ε_ω and ε_ϕ, we use the vertex VVP. The reactions $\psi \to \pi\rho$, $KK^*(.892)$, $\phi\eta$ and $\phi\eta'$ proceed via mixing the ψ with ϕ and ω; the width are

$$\Gamma(\psi \to VP) = \frac{(G(VVP))^2}{12\pi} \ (q(\psi \to VP))^3 \tag{7}$$

$$* \ |\varepsilon_\omega d_{\omega VP} + \varepsilon_\phi d_{\phi VP}|^2 .$$

From $G(VVP) = 17$ GeV^{-1} (i.e. from $\omega \to 3\pi$) and the experimental $\Gamma(\psi \to \pi^\pm\rho) = 0.8\%$ we get

$$|\varepsilon_\omega| = 10^{-4} , \tag{8}$$

whereas $\Gamma(\psi \to K^{\pm}K^{*}(.892)) = 0.35\%$ implies

$$|\varepsilon_{\omega} - \sqrt{2}\ \varepsilon_{\phi}| = 1.5 * 10^{-4} \ . \tag{9}$$

The width for $\psi \to \phi n^{(\prime)}$ is determined by

$$\Gamma(\psi \to \phi n^{(\prime)}) = \frac{(G(VVP))^2}{12\pi}(q(\psi \to \phi n^{(\prime)}))^3 \varepsilon_{\phi}^2 |<n^{(\prime)}|+s\bar{s}>_p|^2 |d_{\phi\phi}|^2 \tag{10}$$

with $<n^{(\prime)}|+s\bar{s}>_p = -0.68$ (0.74) for our $\theta_{PS} = -12.3^{\circ}$. The experimental numbers $\Gamma(\psi \to \phi n) = 0.1\%$ and $\Gamma(\psi \to \phi n') < 0.13\%$ thus imply

$$|\varepsilon_{\phi}| = 0.6 * 10^{-4} \tag{11}$$

and

$$|\varepsilon_{\phi}| < 7.7 * 10^{-5} \ . \tag{12}$$

Thus we predict

$$\Gamma(\psi \to \phi n') = 0.8 \ \% \ , \tag{13}$$

near to the experimental upper limit. To get consistency of eqs. (8), (9) and (11), ε_{ω} and ε_{ϕ} must have opposite signs, in agreement with expectation based on the mass-squared matrix. We now obtain

$$|\varepsilon_{\omega}| = 10^{-4} \ , \qquad \frac{\varepsilon_{\phi}}{\varepsilon_{\omega}} = -0.6 \tag{14}$$

from eqs. (8) and (11) and thus

$$\left| \varepsilon_\omega - \sqrt{2}\ \varepsilon_\phi \right| = 1.8 * 10^{-4} \tag{15}$$

and (compare eq.(5))

$$\left| \varepsilon_\omega + \sqrt{2}\ \varepsilon_\phi \right| = 0.15 * 10^{-4}\ . \tag{16}$$

Eqs. (9) and (15) check this type of mixing and are sufficiently consistent. The value in eq.(16) depends on a cancellation and if - rather than eqs. (14) - eq.(9) would be exact, we would get $\varepsilon_\phi/\varepsilon_\omega = -0.5$ and $\left| \varepsilon_\omega + \sqrt{2}\ \varepsilon_\phi \right| = 0.3 * 10^{-4}$. At any rate, it is predicted that $\underline{\Gamma(\psi \rightarrow K^+K^-)\ \text{is more than an order of magnitude below}}$ $\underline{\text{the present upper experimental limit.}}$ The ψ-meson contains an admixture of old quarks which is almost SU(3)-symmetric

$$\left| \psi \right> = -\left| c\bar{c} \right> - \left| u\bar{u} + d\bar{d} + 0.85\ s\bar{s} \right> * 0.7 * 10^{-4}\ . \tag{17}$$

To compute ε_ω and ε_ϕ from the mass-squared matrix, we use

$$M_u = M_\rho\ ,\quad M_s = \sqrt{2M_K^2 - M_\rho^2}$$

and

$$M_c = \begin{cases} \sqrt{2M_D^2 - M_\rho^2} = 2.73\ \text{GeV} \\[2mm] \sqrt{2M_F^2 - 2M_{K^*}^2 + M_\rho^2} = 2.86\ \text{GeV}\ . \end{cases} \tag{18}$$

Now, all the λ_q contain unstable factors, namely

$$M_\omega - M_u = 0.0066\ \text{GeV}\ ,$$
$$M_\phi - M_s = 0.025\ \ \text{GeV} \tag{19}$$

and

$$
M_\psi - M_c = \begin{cases} 0.37 & \text{GeV} \\ 0.24 & \text{GeV} \end{cases} . \tag{20}
$$

We have seen that the signs of these factors must agree. It is irrelevant for what follows if all of them are negative or positive. We assume

$$
M_u < M_\omega < M_s < M_\phi < M_c < M_\psi , \tag{21}
$$

such that all the λ_q are positive. Due to the 3-gluon annihilation of the vector mesons, we find to lowest order

$$
\frac{\lambda_q}{\lambda_{q'}} = (\frac{\kappa(M_q^2)}{\kappa(M_{q'}^2)})^3 . \tag{22}
$$

Ref.[11] has used the asymptotic freedom formula to compute the $\kappa(M_q^2)$ from $\kappa(M_c^2)$. One then finds $\kappa(M_s^2) = 0.27$ as compared to $\kappa(M_\phi^2) = 0.47$ from $\Gamma(\phi \to$ nonstrange hadron$)/\Gamma(\phi \to e^+e^-)$. Deriving $\kappa(M_u^2)$ via asymptotic freedom from $\kappa(M_s^2)$ one gets

$$
\frac{\varepsilon_\phi}{\varepsilon_\omega} = -0.74 \sqrt{\frac{\lambda_s}{\lambda_u}} = -0.65 \text{ for } \kappa(M_s^2) = 0.27 , \tag{23}
$$

whereas -0.56 results this way from $\kappa(M_s^2) = 0.47$. The agreement with eq.(14) is surprisingly good.

For $|\varepsilon_\omega|$ one finds

$$|\varepsilon_\omega| = 0.048 \sqrt{\frac{\lambda_c}{\lambda_s}} |tg(\theta_{ideal} - \theta_V)| , \qquad (24)$$

with θ_V the vector meson mixing angle. We may obtain θ_V
from $\Gamma(\phi \to \gamma\pi^o)/\Gamma(\omega \to \gamma\pi^o)$ or from $\Gamma(\phi \to \rho\pi \to 3\pi)/$
$\Gamma(\omega \to \rho\pi \to 3\pi)$ and find - taking into account also the
signs in eq. (2.14) - $\theta_V = 38.2^o$. The $|\varepsilon_\omega|$ obtained from
this is

$$|\varepsilon_\omega| = 15 * 10^{-4} , \qquad (25)$$

whereas $|\varepsilon_\omega| = 6 * 10^{-4}$ from $\kappa(M_\phi^2) = 0.47$. This is far
away from the "experimental" $|\varepsilon_\omega| = 10^{-4}$ since it would
predict 30% rather .8% for $\Gamma(\psi \to \pi^\pm\rho)$. On the other
hand, from the "experimental" $|\varepsilon_\omega|$ we derive $\theta_V = 35.5^o$.
Since the two ratios we have used to determine θ_V are
proportional to $A(\theta_V) = |\cos\theta_V - \sqrt{2} \sin\theta_V|^2$ we have
$A(\theta_V = 35.5^o)/A(\theta_V = 38.2^o) = 1/155$, which violently
disagrees with experiment. Thus, the use of the
asymptotic freedom formula in eq.(22) is not justified
for the ω and ϕ vector mesons even though it correctly
yields $\varepsilon_\phi/\varepsilon_\omega$.

V. TENSOR MESON DOMINANCE FOR ANGULAR DISTRIBUTIONS
AND OZI-ALLOWED PROCESSES

Tensor meson dominance (TMD) is analogous to vector
meson dominance. Perhaps TMD is more fundamental than
VMD since, basically, TMD is the theory of the universal
coupling of the graviton to hadrons and is therefore
related to general relativity [19]. We are concerned

here with phenomenological applications of these ideas
and extend the work of - mainly - B. Renner [20,21] to
the new particles. As we will see, this yields a rich
phenomenology.

In any relativistic local quantum field theory
there exists a local conserved

$$\partial^\mu T_{\mu\nu}(x) = 0 \qquad (1)$$

and symmetric

$$T_{\mu\nu}(x) = T_{\nu\mu}(x) \qquad (2)$$

operator $T_{\mu\nu}(x)$, the symmetric energy-momentum tensor.
We will need two properties of $T_{\mu\nu}(x)$, namely that the

$$P_\mu = \int d^3x \, T_{o\mu}(x) \qquad (3)$$

are the energy-momentum-operators and the

$$M_{\mu\nu} = \int d^3x \, (x_\mu T_{o\nu}(x) - x_\nu T_{o\mu}(x) \qquad (4)$$

are the rotation and boost-operators.

In Lagrangian field theories, operators with these
properties are e.g. the Belinfante tensor and the new
and improved energy momentum tensor [22]. We shall
assume that-just as these tensors in an effective
Lagrangian theory-$T_{\mu\nu}$ is diagonal in the physical
single-hadron states. We allow $T_{\mu\nu}$ to connect a single
hadron to a continuum state. It is however assumed
that $|<z|T_{\mu\nu}|z'>|$ is small as compared to $|<z|T_{\mu\nu}|z>|$

with $|z>$ and $|z'>$ states of different single hadrons.
In particular, we assume

$$|<\psi|T_{\mu\nu}|\psi'>| \ << \ |<\psi|T_{\mu\nu}|\psi>|, |<\psi'|T_{\mu\nu}|\psi'>| \ . \qquad (5)$$

In a nonrelativistic picture with $T_{\mu\nu}$ given by the QCD-
Lagrangian, this can be checked by working out the
matrix elements of the active part of $T_{\mu\nu}$, i.e. of

$$T_{\mu\nu} = \frac{i}{4} : \bar{c} \ (\gamma_\mu\partial_\nu + \gamma_\nu\partial_\mu)) \ c : \ . \qquad (6)$$

The nondiagonal matrix elements differ from the diagonal
ones by an overlap integral and thus are smaller. It
should be stressed that our assumption of dominance of
the diagonal single-particle matrix elements of $T_{\mu\nu}$
does not necessarily follow from the Lagrangian of QCD
since this quark-gluon-Lagrangian does not necessarily
imply an effective $T_{\mu\nu}$ which is diagonal in the physical
hadron field. Our assumption follows if the graviton
has mainly diagonal couplings on physical hadrons. This
is plausible since the graviton couples to the mass and
thus a non-diagonal coupling would be analogous to a
photon coupling connecting states of different charges.
We propose to test our assumption below and find good
agreement with experiment.

In what follows we first describe in detail the
"classical" application of TMD to $f\pi\pi$ and fNN by
B.Renner [20]. We then apply the same technique to more
complicated kinematics and the new particles [7-9,21,23,
24].

The matrix element of $T_{\mu\nu}$ between pions contains

two form factors, $F_1(t)$ and $F_2(t)$:

$$<\pi^o(p)|T_{\mu\nu}(x)|\pi^o(q)> \tag{7}$$

$$= \{(p+q)_\mu(p+q)_\nu F_1(t)+((p-q)_\mu(p-q)_\nu-tg_{\mu\nu})F_2(t)\}e^{i(p-q)x}.$$

The form of the matrix element follows from the properties of $T_{\mu\nu}$ listed above and Lorentz covariance. In order to use

$$H|\pi^o(p)> = p^o|\pi^o(p)> , \tag{8}$$

we integrate the time-time-component of eq.(7) over x and find

$$F_1(t = 0) = \tfrac{1}{2} . \tag{9}$$

We might also try to exploit

$$M_3|\pi^o(\vec{0})> = 0 . \tag{10}$$

This however yields no additional result such that F_2 is not restricted. We will use $T_{\mu\nu}$ to describe tensor mesons and thus the physical matrix element involves $\varepsilon^{\mu\nu}(p-q)T_{\mu\nu}$. Due to $\varepsilon^\mu_\mu = 0$, $\varepsilon^\mu(k)k_\mu = 0$ there is no contribution of the unknown F_2 to physical matrix elements of tensor mesons.

The tensor mesons are (almost) ideally mixed and thus only the f together with its radial excitations can couple to $T_{\mu\nu}$ and the pions. Neglecting radial excitations and assuming an unsubtracted dispersion relation for $F_1(t)$, we dominate $F_1(t)$ by the f(1.271)-

meson pole:

$$F_1(t) = \frac{g_f G_{f\pi\pi}}{M_f^2 - t} \quad .$$
(11)

The asymptotically normalized f-meson field is $g_f^{-1} T_{\mu\nu}$, i.e.

$$\langle 0 | T_{\mu\nu} | f(1.271) \rangle = g_f \varepsilon_{\mu\nu} \quad .$$
(12)

The $f\pi\pi$-coupling constant is $G_{f\pi\pi}$ such that

$$\Gamma(f \to \pi^o \pi^o) = \frac{2(g(f \to \pi^o \pi^o))^5}{15\pi \, M_f^2} \, G_{f\pi^o\pi^o}^2 \quad .$$
(13)

From eq.(9) we read off

$$\frac{1}{2} M_f^2 = g_f \, G_{f\pi\pi} \quad .$$
(14)

Thus we may derive $|g_f|$ from the experimental

$$\Gamma_{exp.}(f \to \pi^o \pi^o) = \frac{0.146}{3} \text{ GeV}$$

as

$$|g_f| = 0.179 \text{ Gev}^3 \quad .$$
(15)

This value will be used to predict other matrix elements of $T_{\mu\nu}$. In particular, the nucleon matrix element of $T_{\mu\nu}$ is

$$\langle N(\lambda', p') | T_{\mu\nu} | N(\lambda, p) \rangle =$$

$$= \bar{u}_{\lambda'}(p') \; \{ \frac{\gamma_\mu (p+p')_\nu + (p+p')_\nu \gamma_\mu}{2} \; G_1(t)$$

(16)

$$+ \; \frac{(p+p')_\mu (p+p')_\nu}{4} G_2(t) \; + \; ((p'-p)_\mu (p'-p)_\nu - t g_{\mu\nu}) \, G_3(t) \} u_\lambda(p) ,$$

with $\lambda^{(l)}$ the 3rd component of the nucleon spin. From

$$H|N(\lambda,p)> \; = \; p_o |N(\lambda,p)>$$

(17)

and

$$M_3 |N(\lambda = \tfrac{1}{2}, \; \vec{p} = 0)> \; = \; \tfrac{1}{2} |N(\lambda = \tfrac{1}{2}), \; \vec{p} = 0)>$$

one finds

$$G_1(0) \; = \; \tfrac{1}{2}$$

(18)

and

$$G_2(0) \; = \; 0 \; .$$

(19)

As for the π-matrix element, we may also here assume f-meson dominance and find for the coupling constant from eqs.(18) and (19)

$$\frac{g_f}{M_f^2} \; G_{1,fNN} \; = \; \tfrac{1}{2}$$

(20)

and

$$G_{2,fNN} \; = \; 0 \; .$$

(21)

With $|g_f|$ from eq.(15), we obtain the predictions [20]

$$G_{1,fNN} = 8.7 \pm 0.5 \tag{22}$$

and

$$G_{2,fNN} = 0 . \tag{23}$$

This can be compared to the latest "experimental" results [25]

$$G_{1,fNN} = 21 \pm 2 \tag{24}$$

and

$$G_{2,fNN} = 10.8 \pm 16.4 . \tag{25}$$

There is a discrepancy in G_1 whereas the prediction for G_2 agrees with experiment within the errors. We mention that the "experimental" results in eqs.(24) and (25) are the result of an involved analysis. The discrepancy may also be due to the low masses.

We now turn to decays involving particles with intrinsic charm. The χ at 3.55 GeV is interpreted as the ground state 2^{++} $c\bar{c}$-tensor meson which we call $f_c(3.55)$. It is observed in

$$\psi'(3.7) \rightarrow \gamma\, f_c(3.55) \tag{26}$$

and has

$$f_c(3.55) \rightarrow \gamma\, \psi(3.4) \tag{27}$$

as one of its decay channels. We next derive the angular

distribution of decays of these types, i.e.

$$V \to \gamma T, \quad T \to \gamma V \tag{28}$$

from TMD. In case of

$$\psi(3.1) \to \gamma f(1.271)$$

the angular distribution is nontrivial and has experimentally [4] been derived by the Pluto group. The agreement
between experiment and TMD is excellent in this case.

The decay mechanism of $\psi \to \gamma f$ is given in fig.6.
Obviously, also the diagrams responsible for the OZI-
allowed decays $\psi' \to \gamma f_c$ and $f_c \to \gamma \psi$ can be read off
that figure. Using $T_{\mu\nu}$ as interpolating field for the
tensor mesons, we neglect the nondiagonal couplings of
2^{++}-mesons as compared to the diagonal ones and use VMD
to interpret the decays $V \to \gamma T$ and $T \to \gamma V$ as

$\psi \to \gamma f$:
$$\psi \to \psi f_c \to \psi(q^2 = 0)f \to \gamma f \ ,$$
$\psi' \to \gamma f_c$:
$$\psi' \to \psi' f_c \to \psi'(q^2 = 0)f_c \tag{29}$$

and

$f_c \to \gamma\psi$:
$$f_c \to \psi(q^2 = 0)\psi \to \gamma\psi \quad .$$

In any case, we need the matrix element [21]

$$<V(\sigma,p)|\varepsilon^{*}_{\mu\nu} T^{\mu\nu}|V(\tau,\underline{q})> \quad (M^2_T - t)$$

$$= G_1(t)(\varepsilon^{*}_{p}\cdot\varepsilon_{q})((p+\underline{q})\cdot\varepsilon_T\cdot(p+\underline{q}))$$

$$+ G_2(t)(\varepsilon^{*}_{p}\cdot\underline{q})(\varepsilon_{q}\cdot p)((p+\underline{q})\cdot\varepsilon_T\cdot(p+\underline{q})) \tag{30}$$

$$+ 2G_3(t)\{(\varepsilon^{*}_{p}\cdot\underline{q})(\varepsilon_{q}\cdot\varepsilon_T\cdot(p+\underline{q}))+(p\leftrightarrow q)\}$$

$$-2G_4(t)\varepsilon^{*}_{p}\cdot\varepsilon_T\cdot\varepsilon_{q} \quad .$$

(The matrix element of $T_{\mu\nu}$ without the factor $\varepsilon^{\mu\nu}_T$ has 6 independent form factors.) Using TMD, i.e.

$$H|V(\sigma,p)> = p_o|V(\sigma,p)> \tag{31}$$

and

$$M_3|V(\sigma,\vec{p} = 0)> = \sigma|V(\sigma,\vec{p} = 0)>$$

one derives

$$G_1(0) = -\tfrac{1}{2} M^2_T \tag{33}$$

and

$$G_3(0) = - G_1(0) . \tag{34}$$

We assume unsubtracted dispersion relations for $G_1(t)$ and $G_3(t)$ and approximate the t-dependence by the appropriate tensor meson. Since the scale is set by the charmed masses due to the OZI-rule, this probably is a reasonable approximation. At the point $t = M^2_T$, $p^2 = M^2_V$, $q^2 = 0$ gauge invariance requires the amplitude to vanish if $\varepsilon_\mu(q)$ is replaced by q_μ. This yields

$$G_2(t = M_T^2, \ p^2 = M_V^2, \ q^2 = 0) = 0 \tag{35}$$

and

$$G_4(t = M_T^2, \ p^2 = M_V^2, \ q^2 = 0) = M_T^2(p \cdot q) . \tag{36}$$

At $p^2 = q^2 = 0$, $t = M_T^2$ the __same__ results follow from gauge invariance. Thus, all OZI-allowed processes are completely determined by TMD in terms of the one parameter g_T defined by

$$\langle \Omega | T_{\mu\nu} | T \rangle = g_T \ \varepsilon_{\mu\nu} . \tag{37}$$

Introducing the helicity amplitude B_s and A_s for $V \rightarrow \gamma T$ and $T \rightarrow \gamma V$, respectively, with $s = 0,1,2$, the helicity of the T, one predicts

$$|B_0| = |A_0| = \frac{(2\pi)^4 M_V^2}{\sqrt{6}|g_T|} \ \frac{e}{2\gamma_V} \ |M_V^2 - M_T^2| , \tag{38}$$

$$x_B = x_A = \sqrt{3} \ \frac{M_T}{M_V} \tag{39}$$

and

$$y_B = y_A = \sqrt{6} \ (\frac{M_T}{M_V})^2 . \tag{40}$$

We have defined [26]

$$x_B = \frac{B_1}{B_0} \; , \quad x_A = \frac{A_1}{A_0} \; , \quad y_B = \frac{B_2}{B_0} \; \text{and} \; y_A = \frac{A_2}{A_2} \; . \tag{41}$$

The width of $\psi' \to \gamma f_c$ is

$$\Gamma(\psi' \to \gamma f_c) = \frac{\alpha}{144 g_{f_c}^2 \gamma_{\psi'}^2} \; M_{\psi'} (M_{\psi'}^2 - M_{f_c}^2)^3 (1 + x_B^2 + y_B^2) . \tag{42}$$

Experimentally,

$$\Gamma(\psi' \to \gamma f_c) = 16 \text{ keV} . \tag{43}$$

We may compute g_{f_c} in complete analogy to the convential computation of γ_V from the wave function of the f_c. Namely, the active part of $T_{\mu\nu}$ in eq.(37) now is $\frac{i}{4} : \bar{c}(\gamma_\mu \partial_\nu + \gamma_\nu \partial_\mu) c :$ and thus one finds

$$|g_{f_c}|^2 = 9 \frac{M_{f_c}}{\pi} |\phi'(0)|^2 \tag{44}$$

with $\phi'(0)$ the derivative of the f_c-wave function at the origin. Using

$$|\phi'(0)|^2 = 0.09 \tag{45}$$

from the work of Gatto [27] et al., we may compute $|g_{f_c}|$ and in turn $\Gamma(\psi' \to \gamma f_c)$ and find 16 keV for this in agreement with the experimental result in eq.(43). Other potential models [28,29] yield predictions which are up to a factor two bigger. In any case, TMD and potential models predict the correct order of magnitude. We shall use internal symmetries to compute g_{f_c} from g_f in a later chapter.

Obviously, many more predictions along these lines are possible. We mention

$$\Gamma(f_c \to \gamma\psi) = 600 \text{ keV} \tag{46}$$

which, together with the measured branchings of 27% (ref.[1]) and 16% (ref.[2]), which are not completely in agreement, yields

$$\Gamma(f_c \to \text{all}) = (2.2 - 4.3) \text{ MeV} . \tag{47}$$

Taking the f_b-T-mass splitting and the $\phi'_{f_b}(0)$ from ref.[32] we also predict

$$\Gamma_{f_b \to \gamma T} = 134 \text{ keV} . \tag{48}$$

For OZI-forbidden processes $V \to \gamma T$, $T \to \gamma V$, only angular distributions, i.e. the ratios x and y in eqs.(39) and (40) of the helicity amplitudes are predicted independent of the mixing factors. Experimentally, $\psi \to \gamma f$ has been analysed with the result [4]

$$x_{\psi \to \gamma f} = 0.6 \pm 0.3, \quad y_{\psi \to \gamma f} = 0.3 {}^{+ 0.6}_{- 1.6} \tag{49}$$

in comparison with our prediction

$$x_{\psi \to \gamma f} = 0.71, \quad y_{\psi \to \gamma f} = 0.41 . \tag{50}$$

The experimental results are compared to various theoretical possibilities in fig. 8. The QCD-point [33] practically agrees with our result. It is obvious that

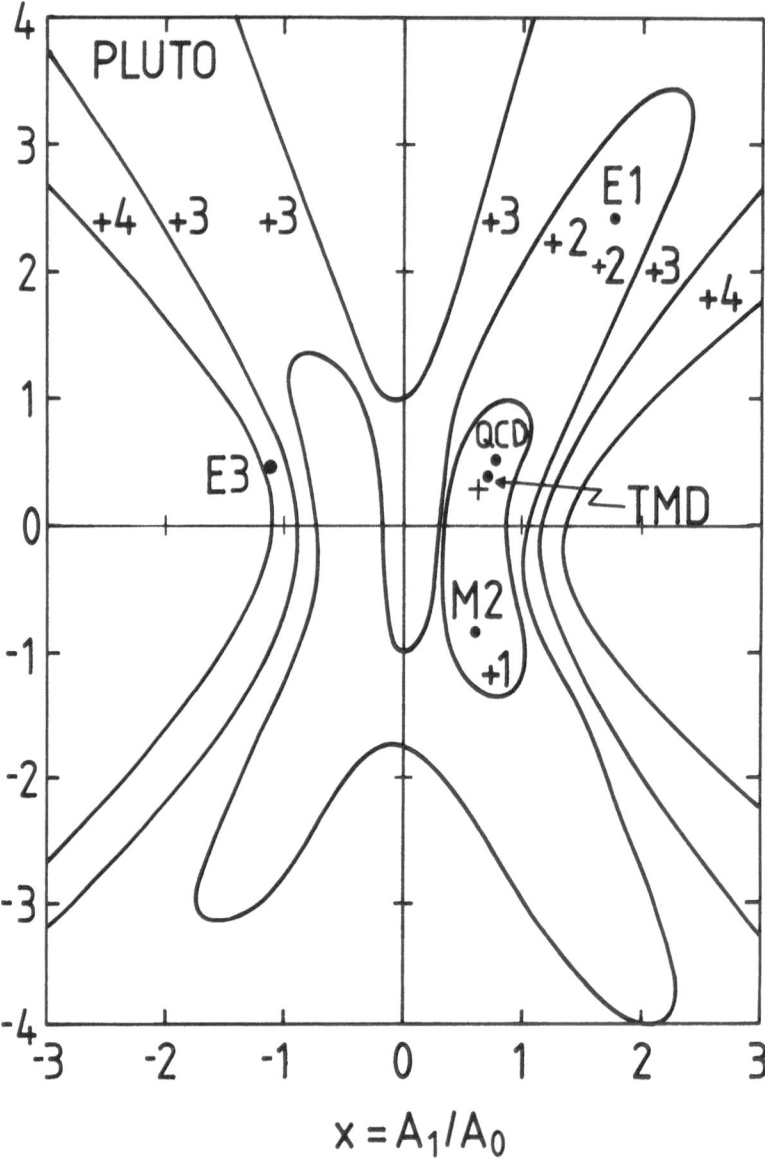

Fig. 8. The angular distributions of the decay
$\psi \to \gamma f$. The χ^2 for fits with given values of
x and y are shown. The cross is the point
of minimal χ^2. The result of refs.[33] (QCD)
and[7](TMD) are also indicated. - The fig.
has been taken from ref.[4].

eqs. (39) and (40) also predict the angular distributions of many further decays such as e.g. $\psi \to \gamma f'$, $T'(10.01) \to$ $\to \gamma f_c$ etc. Angular distributions such as e.g. $\psi' \to \gamma f_c$ are trivially nearly El due to the small mass difference of ψ' and f_c.

TMD also allows us to predict [9] the angular distributions of jets resulting from the f_b-decays. Since the f_b probably lies below $(\bar{b}q)(b\bar{q})$-threshold, the prominent decay should be in two gluons g which fragment into hadron jets j. Including production of f_b, the process under consideration is [30-32]

$$T'(10.01) \to \gamma f_b \to \gamma gg \to \gamma jj \ . \tag{51}$$

The jets should have the angular distribution of the gluons. The angular distributions of this decay chain are determined by the matrix element $\langle T'|f_{b\mu\nu}\varepsilon^{\mu\nu}|T'\rangle$ and $\langle g(p^2 = 0)|f_{b\mu\nu}\varepsilon^{\mu\nu}|g(q^2 = 0)\rangle$. For the former, we have already derived the x and y; from kinematical reasons the result is almost an El-transition. For the second, we can use our amplitudes at $M_V^2 = p^2 =$ $= q^2 = 0$ since the gluons have mass zero and are gauge-invariantly coupled in QCD. This implies that only the states of the f_b with helicities ± 2 are coupled to the gluons. The same holds in perturbative QCD. Compare also ref. [34].

We can compare the gg-decay to the jets resulting from the decay of the f_b into $q\bar{q}$ via mixing:

$$T'(10.01) \to \gamma f_b \to \gamma q\bar{q} \to \gamma jj \ . \tag{52}$$

Obviously, the ± 2-helicity states of the f_b decouple
from $q\bar{q}$, leading to $y_A = 0$. No result on $f_b \to q\bar{q} \to jj$
is known in perturbative QCD. We may use our results
on the fNN-coupling to obtain

$$G_2^{f_b qq} = 0 \; . \tag{53}$$

Translating into helicity amplitudes, one finds

$$x_A \equiv (\frac{A_1}{A_o})_{f_b q\bar{q}} = \sqrt{\frac{3}{8}} \; \frac{M_{f_b}}{M_q} \xrightarrow[q \to 0]{} \infty \; . \tag{54}$$

Thus we predict a dependence of the jet angular distributions on the mass of the parent quark. If one identifies
the quantum numbers of leading particles of a jet with
those of this quark, the prediction in eq.(54) can be
tested. This clearly goes beyond perturbative QCD.

VI. TMD AND INTERNAL SYMMETRIES

In extending SU(3) to SU(n) with $n > 3$, the problem
of mass factors in matrix elements is particularly severe.
We advocate the point of view that effective decay
Lagrangians should not contain any such factors. As we
have said, this prescription is unique if there is only
one amplitude. For tensor mesons, a formulation of TMD
in analogy to VMD leads to the same results.

In VMD, one effectively writes for the e.m.
current J_μ

$$J_\mu = \sum_V \frac{M_V^2}{2\gamma_V} \phi_\mu^V \quad , \tag{1}$$

with ϕ_μ^V asymptotically normalized vector meson fields,

$$<\Omega|\phi_\mu^V|v'> \sqrt{} = \varepsilon_\mu \delta_{VV'} \tag{2}$$

and γ_V^{-1} subject to SU(n) for the V of any given multiplet:

$$\frac{1}{\gamma_V} = (\text{const.}) \; \text{Trace} \; (\Omega M_V) \quad , \tag{3}$$

with Ω = diagonal $(\frac{2}{3}, -\frac{1}{3}, -\frac{1}{3}, \frac{2}{3}, -\frac{1}{3}, \ldots)$ the quark charge matrix and M_V the matrix of the vector meson V in the quark basis, $|V> \sim |\bar{q} M_V q>$. The vanishing of the matrix element in eq.(2) for $V \neq V'$ is an extra assumption. This works well for ρ, ω and ϕ. For the ψ it predicts $\gamma_\psi^{-1} = 0.37$ whereas experimentally $\gamma_\psi^{-1} = 0.16$. The order of magnitude is predicted all right.

We next show that for the VPP matrix element, VMD as defined above and our decay Lagrangian

$$L_I = \frac{G(VPP)}{2} \; V_a^\mu \; (P_b \overset{\leftrightarrow}{\partial}_\mu P_c) f_{abc} \tag{4}$$

without mass factors are consistent. (We use $V_a^\mu \equiv \phi_a^\mu$.) From L_I we get

$$<P_\alpha(p) | (M_V^2 - t) \phi_\mu^V | P_\beta(q) > \sqrt{}$$

$$\tag{5}$$

$$= -(p+q)_\mu \; G(VPP) f_{\alpha V \beta}^*$$

such that

$$<P_\alpha(p)|J_\mu|P_\beta(q)> \Big|_{(p-q)^2 = t = 0} \qquad \cdot \sqrt{} \cdot \tag{6}$$

$$= -(p+q)_\mu \; G(VPP) \sum_V f_{\alpha^*V\beta} \frac{1}{\gamma_V} e^{i(p-q)x} \; .$$

Constructing $Q = \int d^3x J_0(x)$ and using $\Omega|\text{Charge } q> = q|\text{Charge } q>$ we thus obtain from $<P_\alpha|Q|P_\beta> = q_\alpha <P_\alpha|P_\beta>$ the consistency requirement

$$g_\alpha^{-1} \sum_V f_{\alpha^*V\alpha} \frac{1}{\gamma_V} = g_\beta^{-1} \sum_V f_{\beta^*V\beta} \frac{1}{\gamma_{V'}} \; , \tag{7}$$

which is fulfilled if the γ_V satisfy eq.(3). We also get

$$G(VPP) = 2\gamma_\rho \; . \tag{8}$$

Thus the choice of mass factors in VMD and eq.(4) is consistent.

We are interested in TMD and write in analogy to VMD the effective relation:

$$T_{\mu\nu} = \sum_T \frac{M_T^2}{2\chi_T} \phi_{\mu\nu}^T \tag{9}$$

for the spin-2-part of $T_{\mu\nu}$ with

$$<\Omega|\phi_{\mu\nu}^T|T'> \sqrt{} = \varepsilon_{\mu\nu}\delta_{TT'} \tag{10}$$

and

$$\frac{1}{X_T} = (\text{const.}) \ \text{Trace} \ M_T \ . \tag{11}$$

The g_T defined before is

$$g_T = \frac{M_T^2}{2X_T} \ .$$

The SU(4)-symmetric decay Lagrangian for $T \rightarrow PP$ is

$$L_I = \frac{G(TPP)}{2} \ \phi^{a\mu\nu}(P_b \overleftrightarrow{\partial}_\mu \overleftrightarrow{\partial}_\nu P_c)d_{abc} \tag{13}$$

leading to

$$<P_\alpha(p)|(M_T^2 - t)\phi_{\mu\nu}^T|P_\beta(q)> \ \sqrt{}$$

$$\tag{14}$$

$$= -G(TPP)(p+q)_\mu(p+q)_\nu d_{\alpha^*T\beta} \ .$$

At $t = 0$ the diagonal matrix elements of $T_{\mu\nu}$ are therefore

$$<P_\alpha|T_{\mu\nu}|P_\alpha> \ = \ -G(TPP)(p+q)_\mu(p+q)_\nu \sum_T d_{\alpha^*T\alpha} \frac{1}{2X_T} \tag{15}$$

such that we must have, using again $H = \int d^3x \ T_{oo}(x)$,

$$\sum_T d_{\alpha \ T\alpha^*} \frac{1}{2X_T} = \sum_T d_{\beta \ T\beta^*} \frac{1}{2X_T} \ , \tag{16}$$

which is obviously correct for X_T fulfilling eq.(11).
The relation between $G(TPP)$ and X_T which we also could
obtain in analogy to eq.(8), is obviously useless since

graviton scattering experiments are impossible. We conclude that internal symmetry for $T_{\mu\nu}$ as formulated in eqs.(9) and (11) is consistent with the absence of mass factors in the Lagrangian of eq.(13). We thus assume internal symmetry for TMD in this form.

For g_f and g_{f_c} this reads

$$\frac{g_f}{M_f^2} = \sqrt{2} \; \frac{g_{f_c}}{M_{f_c}^2} \; , \tag{17}$$

which with $|g_f| = 0.18 \text{ GeV}^3$ from $f \to \pi\pi$ and $|g_{f_c}| = = 3 \sqrt{M_{f_c}/\pi}|\phi'(0)| = 0.96 \text{ GeV}^3$ from $\Gamma(\psi' \to \gamma f_c)$ or ref.[27] is fulfilled with

$$\frac{|g_f|}{M_f^2} = \sqrt{2} \; \frac{|g_{f_c}|}{M_{f_c}^2} = 0.11 \text{ GeV} \; . \tag{18}$$

Thus at least the correct order of magnitude is predicted by our choice of mass factors.

In order to test SU(3) as assumed in the $(T \to PP)$-decay Lagrangian, we write the resulting width formula as

$$\Gamma(T \to P_A P_B) = \frac{4(G(TPP))^2}{15\pi \; M_T^2}(q(T \to P_A P_B))^5 |d_{T^*P_A P_B}|^2 \tag{19}$$

and compare to experiment in table 2. The agreement is good.

VII. MIXING OF TENSOR MESONS AND DECAYS IN TMD

The λ_q for the tensor mesons are all negative since

$$M_f = 1.271 \text{ GeV} < M_u = M_{A_2} = 1.312 \text{ GeV} < M_{f'} = 1.516 \text{ GeV} <$$

$$< M_s = \sqrt{2M_{K^{**}}^2 - M_{A_2}^2} = 1.546 \text{ GeV} < M_{f_c} = 3.55 \text{ GeV} , \qquad (1)$$

such that we also need to have $M_{f_c} < M_c$. We do not know $M_{D^{**}}$ and $M_{F^{**}}$ to check the accuracy of the M_q. Assuming an error of $\pm 3\%$ we get $M_u = 1.31 \pm 0.04$ GeV and $M_s = 1.546 \pm 0.05$ GeV such that the factors $M_f - M_u = -0.04$ GeV and $M_{f'} - M_s = -0.03$ in λ_u and λ_s are essentially un-known. Thus λ_u, λ_s and λ_c cannot be computed from the tensor meson masses. We may, however, rather safely assume asymptotic freedom for $M_{\text{Tensor}}^2 > 1.3^2$ GeV^2. We obtain

$$\frac{\lambda_c}{\lambda_u} = 0.58, \quad \frac{\lambda_c}{\lambda_s} = 0.63 \text{ and } \frac{\lambda_s}{\lambda_u} = 0.9 \qquad (2)$$

for the ratios of the λ_q. Approximately, the λ_q can be expressed by the unstable factors as (units: GeV)

$$\lambda_u = 1.1 \, (M_f - M_u) ,$$

$$\qquad (3)$$

$$\lambda_s = 3.5 \, (M_{f'} - M_s)$$

and

$$\lambda_c = 7.3 \, (M_{f_c} - M_c)$$

such that, with N_T normalization constants, the meson

states are

$$\frac{1}{N_f}|f> = \frac{0.56}{\sqrt{M_f-M_u}}|\frac{u\bar{u}+d\bar{d}}{\sqrt{2}}>+2.26\sqrt{M_{f'}-M_s}|s\bar{s}>+0.25\sqrt{M_{f_c}-M_c}|c\bar{c}>$$

$$\frac{1}{N_{f'}}|f'> = 2.5\sqrt{M_f-M_u}|\frac{u\bar{u}+d\bar{d}}{\sqrt{2}}>-\frac{0.61}{\sqrt{M_{f'}-M_s}}|s\bar{s}>-0.26\sqrt{M_{f_c}-M_c}|c\bar{c}>$$

and   (4)

$$\frac{1}{N_{f_c}}|f_c> = 0.13\sqrt{M_f-M_u}|\frac{u\bar{u}+d\bar{d}}{\sqrt{2}}>+0.18\sqrt{M_{f'}-M_s}|s\bar{s}>-\frac{0.38}{\sqrt{M_{f_c}-M_c}}|c\bar{c}> .$$

Defining ε_f and $\varepsilon_{f'}$ by

$$|f_c> = \varepsilon_f |\frac{u\bar{u}+d\bar{d}}{\sqrt{2}}> + \varepsilon_{f'}|-s\bar{s}> + |-c\bar{c}> ,$$ (5)

we may compute the ratio

$$\frac{\varepsilon_{f'}}{\varepsilon_f} = \frac{-0.18}{0.13} \sqrt{\frac{M_{f'}-M_s}{M_f-M_u}} = -\frac{3}{4}\frac{\lambda_s}{\lambda_u} = -0.72 ,$$

which is stable.

The states in eq.(4) are obviously all determined
in terms of one remaining parameter, e.g. ε_f, which cannot be computed from asymptotic freedom and the meson
masses. We next discuss tensor meson decays and note
first that ε_f is given by the f-f'-mixing angle θ_T as

$$\varepsilon_f = 0.047 \ tg(\theta_{id.} - \theta_T) \ . \tag{6}$$

Since θ_T is experimentally only known from $f \to \pi\pi/f \to K\bar{K}$, not from $f' \to \pi\pi/f' \to K\bar{K}$, the error of $\theta_{id} - \theta_T$ is unfortunately very big and eq.(6) can only be used for an approximate check. We may however use the experimental results [1]

$$\Gamma(f_c \to \pi^+\pi^- + K^+K^-) = 2.9 * 10^{-3} \ \Gamma(f_c \to all) \ , \tag{7}$$

$$\Gamma(\psi \to \gamma f) = 2 * 10^{-3} \ \Gamma(\psi \to all) \tag{8}$$

and

$$\Gamma(\psi \to \gamma f') < 0.34 * 10^{-3} \ \Gamma(\psi \to all) \tag{9}$$

together with the TMD-result

$$\Gamma(f_c \to all) = (2.2 - 4.4) \ MeV \tag{10}$$

such that

$$\Gamma(f_c \to \pi^+\pi^- + K^+K^-) = (6.4 - 12.8) \ keV \tag{11}$$

to determine $\varepsilon_{f'}$, compare our result on $\varepsilon_{f'}/\varepsilon_f$ to experiment and check the consistency of the mixing formalism for decays.

The decay $f_c \to \pi^\circ\pi^\circ$ proceeds via the diagram in fig. 9. To apply TMD, we first note that the decay amplitude $f \to \pi^\circ\pi^\circ$ is given by

$$< \pi^\circ(p) | (M_f^2 - t) \ \frac{1}{g_f} \ T_{\mu\nu} \varepsilon^{\mu\mu}(p-q) | \pi^\circ(q)> \ \sqrt{\ }$$

$$= \frac{M_f^2}{2g_f} \ (p+q)_\mu (p+q)_\nu \varepsilon^{\mu\nu}(p-q) \ . \tag{12}$$

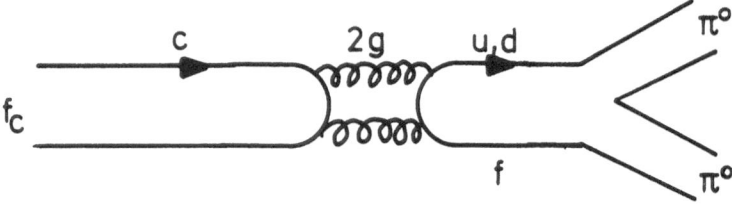

Fig.9. The decay $f_c \to \pi^o \pi^o$ proceeds via f_c-f-mixing.

Due to the saturation assumption, except for kinematical factors, this matrix element is a constant independent of t. At $t = M_{f_c}^2$ we have

$$T_{f_c \to \pi^o \pi^o} = \varepsilon_f \, T_{f \to \pi^o \pi^o}(t = M_{f_c}^2) \tag{13}$$

such that

$$\Gamma(f_c \to \pi^o \pi^o) = \frac{2\,(q(f_c \to \pi\pi))^5}{15\pi \, M_{f_c}^2} \, \frac{M_f^2}{(2g_f)^2} \, \varepsilon_f^2 \quad . \tag{14}$$

The same formula incidentally follows from $L_I(TPP)$. The same way one derives $\Gamma(f_c \to K^+ K^-)$ such that

$$\Gamma(f_c \to \pi^+ \pi^- + K^+ K^-) = \frac{1}{15\pi \, M_{f_c}^2} \, \frac{M_f^2}{g_f^2} \, (q(f_c \to \pi\pi))^5.$$

$$\cdot \left(\varepsilon_f^2 + \frac{1}{2} \left(\frac{q(f_c \to K\bar{K})}{q(f_c \to \pi\pi)}\right)^5 \left(\frac{\varepsilon_f}{\sqrt{2}} - \varepsilon_{f'}\right)^2\right) \quad . \tag{15}$$

We have used SU(3) for $g_f/g_{f'}$, and find from experiment

$$|\epsilon_f| \sqrt{1+0.21(1-\sqrt{2}\,\frac{\epsilon_{f'}}{\epsilon_f})^2} = (1.4-2.5) * 10^{-3} . \tag{16}$$

We next consider $\psi \to \gamma f$, $\gamma f'$. Due to the large number of amplitudes SU(4) alone yields no definite prediction. We consider the decay chains $\psi \to \psi f_c \to \gamma f_c \to \gamma f$, $\gamma f'$ in TMD and thus need to know the $(\psi \to \psi f_c)$-matrix element
$$\frac{1}{g_{fc}} < \Psi(p) | (M_f^2-t) T_{\mu\nu} | \psi(q) > \epsilon^{\mu\nu}(p-q) \text{ at } p = M_\psi^2, q^2 = 0$$
and $q^2 = 0$ and $t = M_f^2$. At this point, $G_1,...,G_4$ have already been derived from TMD. Including mixing of f_c and $f^{(1)}$, we therefore find

$$\Gamma(\psi \to \gamma f^{(1)}) = \frac{\alpha}{144\gamma_\psi^2} (\frac{1}{M_f(1)})^4 (\frac{M_{f_c}^2}{g_{f_c}})^2 M_\psi (M_\psi^2-M_f^2(1))^3 \cdot$$

$$(1 + x^2 + y^2) \epsilon_f^2(1) \tag{17}$$

with

$$x = \sqrt{3}\,\frac{M_f(1)}{M_\psi} , \quad y = \sqrt{6}\,(\frac{M_f(1)}{M_\psi})^2 . \tag{18}$$

The experimental results in eqs. (8) and (9) now directly imply

$$|\epsilon_f| = 7.6 * 10^{-4} \tag{19}$$

and

$$|\epsilon_{f'}| < 4.6 * 10^{-4} \tag{20}$$

such that

$$|\varepsilon_{f'}/\varepsilon_f| < 0.6 \ . \tag{21}$$

We repeat

$$|\varepsilon_f| \ \sqrt{1+0.21\,(1-\sqrt{2}\ \frac{\varepsilon_{f'}}{\varepsilon_f})^2} = (1.4-2.5) * 10^{-3} \ . \tag{22}$$

The value

$$\frac{\varepsilon_{f'}}{\varepsilon_f} = -0.72 \ , \tag{23}$$

which has been derived from the mixing formalism, approximately agrees with the upper limit in eq.(21). We conclude that $\psi \rightarrow \gamma f'$ cannot be much below its present experimental upper limit in eq.(9).With $\varepsilon_{f'} = 0$, eq.(19) would imply $.81 * 10^{-3}$ for the square root in eq.(22). To get better agreement $\varepsilon_{f'}/\varepsilon_f$ must be negative (or bigger than the experimental upper limit) in agreement with theoretical expectation. With the values in eqs.(23) and (19) we compute

$$|\varepsilon_f| \ \sqrt{1+0.21\,(1-\sqrt{2}\ \frac{\varepsilon_{f'}}{\varepsilon_f})^2} = 1.04 * 10^{-3} \ , \tag{24}$$

whereas the experimental upper limit in eq.(21) yields as upper limit 10^{-2} for this. Thus the agreement of TMD and the mixing formalism with experiment is good if the bigger branching of $f_c \rightarrow \gamma\psi$ (27%, ref.[1]) is correct whereas the smaller value (16%, ref.[2]) leads to a discrepancy of about a factor 2 in eq.(21). We also note that the ε_f in eq.(19) yields

$$|\theta_T - \theta_{id.}| = 0.926^o , \tag{25}$$

which is in agreement with the value in table 2.

As an example for further possible tests we work out $f_c \to$ proton $\overline{\text{proton}}$. Since we have, at any value of t,

$$< p(q)|\epsilon^{\mu\nu}(p-q)\frac{1}{g_f} T_{\mu\nu}(M_f^2-t)|p(q)> \sqrt{} \tag{26}$$

$$= \bar{u}(p)\epsilon^{\mu\nu}(p-q)\gamma_\mu(p+q)_\mu(p-q)_\nu u(q) \frac{M_f^2}{2g_f} ,$$

we find from the matrix element at $t = M_{f_c}$ and mixing the width formula

$$\Gamma(f_c \to p\bar{p}) = \frac{(g(f_c \to p\bar{p}))^3}{60\pi} \frac{M_{f_c}^2}{g_{f_c}})^2(3 + \frac{8M_p^2}{M_{f_c}^2})\epsilon_f^2 , \tag{27}$$

such that

$$\Gamma(f_c \to p\bar{p}) = 0.02 \text{ keV} , \tag{28}$$

which is smaller than any observed branching of the f_c.

We $\underline{\text{cannot}}$ predict $f_c \to \rho\rho$, $K^*\bar{K}^*$,... and $\psi \to f\omega$, $f'\phi$,..., since this requires extrapolation away from the gauge invariant point $q^2 = 0$. Away from that G_4 is not known. This also holds for angular distributions.

Finally,we look at the vertex TVP in SU(4). This will enable us to compute $\Gamma(\psi \to K^{\pm}K^*(1.420))$ in terms of quantities such as $\Gamma(A_2 \to \pi\rho)$. There is only one amplitude involved in this D-wave-decay and thus our prescription for constructing an SU(4)-symmetric Lagrangian is unique:

$$L_I = G(TVP) \partial^\rho \phi_a^{\mu\nu} \partial_\nu \partial^\tau V_b^\sigma P_c \varepsilon_{\rho\mu\tau\sigma} f_{abc} \quad , \tag{29}$$

such that the width formulas are

$$\Gamma(T \to VP) = (q(T \to VP))^5 \frac{(G(TVP))^2 |f_{T^* VP}|^2}{40\pi} \tag{30}$$

and

$$\Gamma(V \to TP) = (\frac{M_V}{M_T})^2 (q(V \to TP))^2 \frac{(G(TVP))^2 |f_{V^* TP}|^2}{24\pi} \quad . \tag{31}$$

In Table 4, we have compared eq.(30) for the SU(3)-particles to experiment and find good agreement with

$$\frac{(G(TVP))^2}{40\pi} = 3 * 10^{-12} \, \text{MeV}^{-4} \quad . \tag{32}$$

With this value, we may compute $\Gamma(\psi \to K^\pm K^{**})$, which proceeds via $\psi\omega$- and $\psi\phi$-mixing and thus contains a factor $|\varepsilon_\omega + \sqrt{2}\varepsilon_\phi|^2$. From the experimental upper limit

$$\Gamma(\psi \to K^\pm K^{**}) < 3.3 * 10^{-3} \, \Gamma(\psi \to \text{all}) \tag{33}$$

we find

$$|\varepsilon_\omega + \sqrt{2}\,\varepsilon_\phi| < 2 * 10^{-4} \quad . \tag{34}$$

With our values $|\varepsilon_\omega| = 1.2 * 10^{-4}$ and $\varepsilon_\phi = -0.3\,\varepsilon_\omega$ we predict

$$\Gamma(\psi \to K^\pm K^{**}) = 1.5 * 10^{-4} \, \Gamma(\psi \to \text{all}) \quad . \tag{35}$$

VIII. CONCLUSIONS

We have tested internal symmetries, particle mixing, and dominance of meson poles in the strong and electromagnetic decays of particles with implicit charm and have found good agreement with experiment. The predictions for the Υ- and higher families will be given elsewhere [24].

ACKNOWLEDGEMENTS

The author would like to thank K. Bongardt and W. Gampp for discussions and help. The final version of this report was partly written during a visit of the author to DESY. He would like to thank the DESY theory group, and in particular H. Joos, for their kind hospitality.

APPENDIX

Comparing TMD to Potential Models

Nonrelativistic potential models predict the widths of $\psi' \rightarrow \gamma f_c$, $f_c \rightarrow \gamma\psi$, $f_c \rightarrow \gamma\gamma$ and $f_c \rightarrow$ all involving the f_c. Out of these, the first two widths are also predicted by TMD. Without any theoretical speculation we compare in this appendix the corresponding width formulas of TMD to those of the harmonic oscillator potential model of ref.[29]. We have chosen this particular model since it is extremely simple and since ref.[29] contains a collection of all the formulas we require. For the two

E1-transitions $\psi' \to \gamma f_c$ (i.e. $2^3S_1 \to 2^3P_2$) and $f_c \to \gamma\psi$ (i.e. $2^3P_2 \to 1^3S_1$) the widths are in the nonrelativistic case

$$\Gamma(\psi' \to \gamma f_c) = \frac{5\alpha}{54} \frac{(M_{\psi'}^2 - M_{f_c}^2)^3}{M_{\psi'}^3} (e_Q)^2 |{<}2p|r|2s{>}|^2 \qquad (1)$$

and

$$\Gamma(f_c \to \gamma\psi) = \frac{\alpha}{18} \frac{(M_{f_c}^2 - M_{\psi}^2)^3}{M_{f_c}^3} (e_Q)^2 |{<}1s|r|2p{>}|^2 . \qquad (2)$$

The charge of the quark is $e_Q = \frac{2}{3}$ and the matrix element $<2p|r|2s>$ and $<2s|r|2p>$ depend on the special model. The corresponding TMD-formulas are

$$\Gamma(\psi' \to \gamma f_c) = \frac{\alpha\pi}{1296} \frac{1}{\gamma_{\psi'}^2} \frac{1}{|R'_{2p}(o)|^2} \frac{M_{\psi'}}{M_{f_c}} (M_{\psi'}^2 - M_{f_c}^2)^3 .$$

$$\cdot (1 + 3 (\frac{M_{f_c}}{M_{\psi'}})^2 + 6 (\frac{M_{f_c}}{M_{\psi'}})^4) \qquad (3)$$

and

$$\Gamma(f_c \to \gamma\psi) = \frac{\alpha\pi}{2160} \frac{1}{\gamma_{\psi}^2} \frac{1}{|R'_{2p}(o)|^2} (\frac{M_{\psi}}{M_{f_c}})^4 (M_{f_c}^2 - M_{\psi}^2)^3$$

$$\cdot (1 + 3 (\frac{M_{f_c}}{M_{\psi}})^2 + 6 (\frac{M_{f_c}}{M_{\psi}})^4) . \qquad (4)$$

We also know

$$\frac{1}{\gamma_V^2} = \frac{12}{M_V^3 \pi} (e_Q)^2 |R_V(0)|^2 \tag{5}$$

with the radial wave functions given by (ξ is a numerical parameter)

$$R_\psi(r) \equiv R_{1s}(r) = \sqrt{\frac{4\xi^3}{\sqrt{\pi}}} \, e^{-\xi^2 \frac{r^2}{2}} \, , \tag{6}$$

$$R_{f_c}(r) \equiv R_{2p}(r) = \sqrt{\frac{8\xi^3}{3\sqrt{\pi}}} \, e^{-\xi^2 \frac{r^2}{2}}$$

and

$$R_{\psi'}(r) \equiv R_{2s}(r) = \sqrt{\frac{6\xi^3}{\sqrt{\pi}}} \, (1-\tfrac{2}{3}\xi^2 r^2) \, e^{-\xi^2 \frac{r^2}{2}} \, .$$

With these we have

$$|R'_{2p}(0)|^2 = \frac{8\xi^5}{3\sqrt{\pi}} \, ,$$

$$\frac{1}{\gamma_\psi^2} = \frac{12}{M_\psi^3 \pi} (e_Q)^2 \frac{4}{\sqrt{\pi}} \xi^3 \, , \tag{7}$$

$$\frac{1}{\gamma_{\psi'}^2} = \frac{12}{M_{\psi'}^3 \pi} (e_Q)^2 \frac{6}{\sqrt{\pi}} \xi^3$$

and

$$|\langle 2s|r|2p\rangle|^2 = \xi^{-2}, \quad |\langle 2p|r|1s\rangle|^2 = \tfrac{3}{2}\xi^{-2} \, . \tag{8}$$

Introducing these expressions in the above formulas of TMD and QCD, and equating Γ_{TMD} with Γ_{QCD}, the ξ-dependence drops and one finds equality if the mass formulas

$$\frac{5}{54} = \frac{1}{48} \frac{M_{\psi'}}{M_{f_c}} \left(1 + 3\left(\frac{M_{f_c}}{M_{\psi'}}\right)^2 + 6\left(\frac{M_{f_c}}{M_{\psi'}}\right)^4\right) \tag{9}$$

(for $\psi' \to \gamma f_c$) and

$$1 = \frac{1}{10} \frac{M_{\psi}}{M_{f_c}} \left(1 + 3\left(\frac{M_{f_c}}{M_{\psi}}\right)^2 + 6\left(\frac{M_{f_c}}{M_{\psi}}\right)^4\right) \tag{10}$$

(for $f_c \to \gamma\psi$) hold. Numerically, with the experimental masses, there is approximate agreement (i.e. 0.093 = = 0.13 and 1 = 1.3 for $\psi' \to \gamma f$ and $f_c \to \gamma\psi$, respectively). Thus TMD and the simple potential model of ref.[29] are approximately equivalent in case of these two decays. More refined models approximately also agree with this. The reader might compare his favorite model to TMD by himself.

Decay	Γ (GeV)	G^2 computed from Γ
$\rho^0 \to \pi^+\pi^-$	0.15	36
$\phi \to K^+K^-$	0.0019	35.8
$\phi \to K^0\bar{K}^0$	0.00144	40.8
$K^{*+} \to K^+\pi^0$		
$+ K^0\pi^+$	0.050	42.1

Table 1. SU(3) for V $\to$ PP.

Decay	Γ (MeV)	$(G(\text{TVP}))^2/40\pi$ 10^{12} MeV^4
$A_2 \to \rho\pi$	72	3.1
$K^{**} \to K^*\pi$	27	2.6
$K^{**} \to \rho K$	6.6	2.4
$K^{**} \to \omega K$	3.7	4.4
$f' \to K^*K^+$	13	
$+ \bar{K}^*K^-$	Prediction from 3 (average value)	

Table 3. SU(3) for T $\to$ VP.

Decay	Γ (MeV)	$\frac{4G^2}{15\pi}$ $\times 10^7 \mathrm{MeV}^2$	Remark
$A_2 \to K\bar{K}$	4.8	11	
$A_2 \to \eta'\pi$	<10.2	<21	$\theta_p = -10.2^{\circ}$
$A_2 \to \eta\pi$	14.4	12	$\theta_p = -10.2^{\circ}$
$f \to \pi\pi$	145	18	yields $\theta_T = 34.2^{\circ}$
$f \to K\bar{K}$	5.58	(18)	
$f' \to K\bar{K}$	<65	<30	with $\theta_T = 34.2^{\circ}$. Partial width is "dominant"[2]. Including the prediction on the channel $f' \to K^*\bar{K}$ reduces G^2 by 20% (Table 3).
$f' \to \pi\pi$	seen	./.	$\frac{\Gamma(f' \to \pi\pi)}{\Gamma(f' \to KK)} < 5 * 10^{-4}$ prediction
$K^{**} \to K\pi$	49	14	
$K^{**} \to K\eta$	2.5 ± 2.5		Prediction: $\frac{\Gamma(K^{**} \to K\eta)}{\Gamma(K^{**} \to K\pi)} = 10^{-3}$

Table 2. SU(3) for $T \to PP$.

REFERENCES

[1] B. Wiik and G. Wolf, DESY 78/23.

[2] Particle Data Group, Phys.Lett. 75B (1978) 1.

[3] W. Tannenbaum et al., Phys. Rev. 17 (1978) 1731.

[4] G. Alexander et al., Phys. Lett.76B (1978)652.

[5] R. Heimann, Nuovo Cimento 39A (1977) 461.

[6] E. Kazi, G.Kramer,and D.H. Schiller, DESY 76/18 and
    references quoted there.

[7] W. Gampp and H.Genz, Phys. Lett. 76B (1978) 319.

[8] W. Gampp and H.Genz, Phys.Lett. 79B (1978) 267.

[9] W. Gampp and H.Genz, Z. Physik C1 (1979) 199.

[10] H.Fritzsch and P.Minkowski, Nuovo Cimento 30 (1975)
     393.

[11] H.Fritzsch and J.D. Jackson, Phys.Lett. 66B (1977)
     365.

[12] H. Harari, Phys. Lett. 60B (1975) 172.

[13] H. Genz, Lett. Nuovo Cimento 21 (1978) 270.

[14] G. Segre and J.Weyers, Phys.Lett. 62B (1976) 91.

[15] Talks by E.Bloom on the crystal ball experiment,
     April 1979.

[16] M. Gell-Mann, D.Sharp,and W.G.Wagner, Phys.Rev.Lett.8
     (1962) 261.

[17] H. Genz and C.B. Lang, Nuovo Cimento 40A (1977) 313.

[18] P.J. O'Donnell, Phys.Rev. Lett. 36 (1976) 177;
     T.Barnes, Phys.Lett. 63B (1976) 65 and references
     quoted there.

[19] E.g.: J.Wess in: Scale and Conformal Symmetry in
     Hadron Physics, ed. R. Gatto (Wiley, New York, 1973)
     and references quoted there.

[20] B. Renner, Phys. Lett. 33B (1970) 599.

[21] B. Renner, Nucl. Phys. B30 (1971) 634.

[22] E.g.: G.Källén in: Handbuch der Physik, ed. by S.Flügge,

618

Vol. 5 (Berlin, Göttingen, Heidelberg, 1975),
p. 169;
C.G. Callan, S.Coleman, and R.Jackiw, Ann. Phys.
59 (1970) 42.

[23] P. Budini and G. Calucci, Lett. Nuovo Cimento 1 (1971) 1071.

[24] K. Bongardt, W. Gampp, and H. Genz, Karlsruhe preprint, in preparation.

[25] F. Kaiser, Dissertation, Karlsruhe (1977).

[26] G. Karl, S. Meshkov, and J.L. Rosner, Phys. Rev.D13 (1976) 1203.

[27] R. Barbieri, R. Gatto, and R. Kögerler, Phys.Lett. 60B (1976) 183.
R. Barbieri, R. Gatto, R. Kögerler, and Z. Kunszt, Phys. Lett. 57B (1975) 455; Nucl. Phys. B105 (1976) 125.

[28] E. Eichten et al., Phys.Rev.Lett. 34 (1975) 369.
K. Gottfried and K.M. Lane, private communication to J.D. Jackson, ref.[29].

[29] J.D. Jackson, LBL-5500 (1976).

[30] M.Krammer and H.Krasemann, Phys.Lett. 73B (1978) 58.

[31] A.de Rujula et al., Nucl.Phys. B138 (1978) 387.

[32] H.Krasemann, DESY 78/46.

[33] M.Krammer, Phys. Lett. 74B (1978) 361.

[34] P.Grassberger, Acta Phys. Austr. Suppl. XV (1976) 571.

Acta Physica Austriaca, Suppl. XXI, 619 (1979)
© by Springer-Verlag 1979

GRAND UNIFIED THEORIES

by

J. ELLIS
CERN
CH-1211 Genève 23

Since the topics discussed in these lectures are covered by other recent publications of the author, we agreed with the author's decision not to publish lecture notes in the proceedings.

Acta Physica Austriaca, Suppl. XXI, 621—661 (1979)
© by Springer-Verlag 1979

THE QUARK RECOMBINATION MECHANISM IN MULTI-HADRON PRODUCTION[+]

by

L. Van HOVE

CERN

CH-1211 Genève 23

ABSTRACT

The lectures discuss the role of quarks and gluons in high energy hadron collisions, in particular in the dominant class of such collisions which are those of non-diffractive low-p_T type. The main progress made recently in this field concerns the secondaries produced in the fragmentation regions of the incident hadrons. The observed similarity in the x-distributions of incident valence quarks and secondary mesons at medium and large x is well accounted for by the quark recombination model. Studying the case of pp collisions,

---

[+]Lectures given at the
XVIII. Internationale Universitätswochen für Kernphysik,Schladming,Austria,February 28-March 10,1979.

we explicitly exhibit how this similarity comes about
and find on the basis of recent ISR data that, contrary
to earlier claims, only a modest enhancement of the
parton sea is needed to explain the absolute magnitude
of the observed cross section. We also find that inclusive meson production at large x is best accounted
for by additive contributions of recombination and
triple-Regge mechanism. Starting from the meson production data, we then show that by using the hyperon
production rate as additional input, one can estimate
the probabilities for the various ways in which the
three valence quarks of an incident proton recombine
with each other and/or whith sea partons to produce
secondaries in the fragmentation region.

CONTENTS

1. INTRODUCTION

The aim of these lectures is to discuss certain aspects of low-p_T reactions of hadrons in terms of quarks and gluons. The general problem of what happens to quarks and gluons in such collisions is under intensive investigation along several lines. I shall talk about a possible picture on which I have worked with S. Pokorski since 1974 and which underwent new developments since 1977 due to the work of Ochs, Das and Hwa, Teper, and others. The work reported here was done in collaboration with Pokorski and Kalinowski.

The picture says the following. At very high energy, non-diffractive collisions of low-p_T type are characterized by excitation of the set of gluonic constituents of the incident hadrons, the valence quarks tending to fly through with retention of their original fraction x of incident momentum. These valence quarks hadronize by recombining with each other and/or with additional low-x partons from the parton sea (q, $\bar{q}$ and gluons); they produce in this way the outgoing hadrons in the fragmentation regions (regions of medium and high x of either sign). As to the hadrons produced in the central region (low-x region), they mainly result from hadronization of the excited gluon sea.

It should be noted that the quark behaviour postulated in the recombination model is distinct from the assumptions made for quasi-free quark behaviour in deep inelastic processes. The latter processes are characterized by short times, whereas

quark recombination is a soft process involving longer
time intervals. Hence, we cannot say a priori that the
quark recombination picture is bound to be right, just
as one was entitled to question the assumption of quasi-
free behaviour of quarks in deep inelastic processes
before the experimental evidence imposed itself. Experiment has to decide on the validity of the quark
recombination picture and, if qualitative success is
encountered, it will reveal the quantitative features
of recombination. As to Quantum Chromodynamics, it may
be quite compatible with the quark recombination model,
but this compatibility may be hard to establish since
in QCD soft processes require a non-perturbative
approach which is still lacking.

2. QUARK RECOMBINATION (GENERALITIES)

In the picture here discussed, the basic mechanism
for production of hadrons in the fragmentation regions
is a <u>recombination mechanism</u> involving the valence
quarks and sea partons of the same incident hadron[+].
In the case of an incident proton, it can operate in
three ways:

i) <u>Recombination of a single valence quark:</u> Consider
the inclusive reaction

$$p + hadron \rightarrow M + ... \tag{1}$$

producing a meson M in the proton fragmentation
region. M is assumed to originate through recombina-

[+]Recombination of sea partons from different incident
hadrons gives hadronic systems of a mass growing with
collision energy (except when all the recombining
partons have $x \rightarrow 0$).

tion (some authors prefer the word fusion) of a
valence quark q of p, which retains its original
x-value, with a sea antiquark $\bar{q}$ of low x, the x-
value of M being approximately

$$x_M \simeq x_q + x_{\bar{q}} \simeq x_q \ . \tag{2}$$

This relates the x-distribution of the meson M to
the x-distribution of the valence quark q in the
incident proton. A single valence quark could also
recombine with two sea quarks to give a baryon.

ii) <u>Recombination of two valence quarks:</u> Consider, for
example, the reaction

$$p + hadron \rightarrow Y + \ldots \tag{3}$$

producing a hyperon Y in the proton fragmentation
region. Y is assumed to originate:

(a) either through recombination of two valence
quarks q, q' of p with a strange quark s of
the sea, in which case

$$x_Y \simeq x_q + x_{q'} + x_s \simeq x_q + x_{q'} \ , \tag{4}$$

(b) or as in i) through recombination of one valence
quark q of p with two sea quarks s, q", in which
case

$$x_Y \simeq x_q + x_s + x_{q''} \simeq x_q \ .$$

iii) <u>Recombination of all three valence quarks:</u> The
reaction

$$p + \text{hadron} \rightarrow p + \ldots \tag{5}$$

in the proton fragmentation region can proceed as in i) or ii), or through recombination of the three valence quarks q, q', q" of the incident proton giving for the outgoing proton the x-value

$$x_p \simeq x_q + x_{q'} + x_{q''} \, .$$

The recombination mechanisms ii) and iii) involve the <u>joint</u> probability distribution $g(x_1, x_2, x_3)$ of the valence quarks in the proton. The recombination mechanism iii) was proposed in 1974 for the "leading particle" reaction (5) by S. Pokorski and the present author, on the basis of the observation that the same average value ~ 0.5 is found experimentally for $<x_p>$ and $<x_q + x_{q'} + x_{q''}>$, the former from leading particle measurements in pp collisions, the latter from deep inelastic lepton-nucleon collisions [1]. Possible forms for the function $g(x_1, x_2, x_3)$ were discussed in this context [2].

The recent upsurge of interest in the recombination model started with an observation by Ochs [3] to the effect that the x-distribution of pions in the fragmentation region of an incident proton is similar to that of the valence quarks which they share with this proton. The first model calculation for recombination was published by Das and Hwa [4]. Refs.[3] and [4] concern the recombination mechanism i) for meson production. The most extensive proton-proton collision data available for this type of analysis are the inclusive meson distributions measured at the CERN Intersecting Storage Rings, especially by the CERN-Holland-Lancaster-

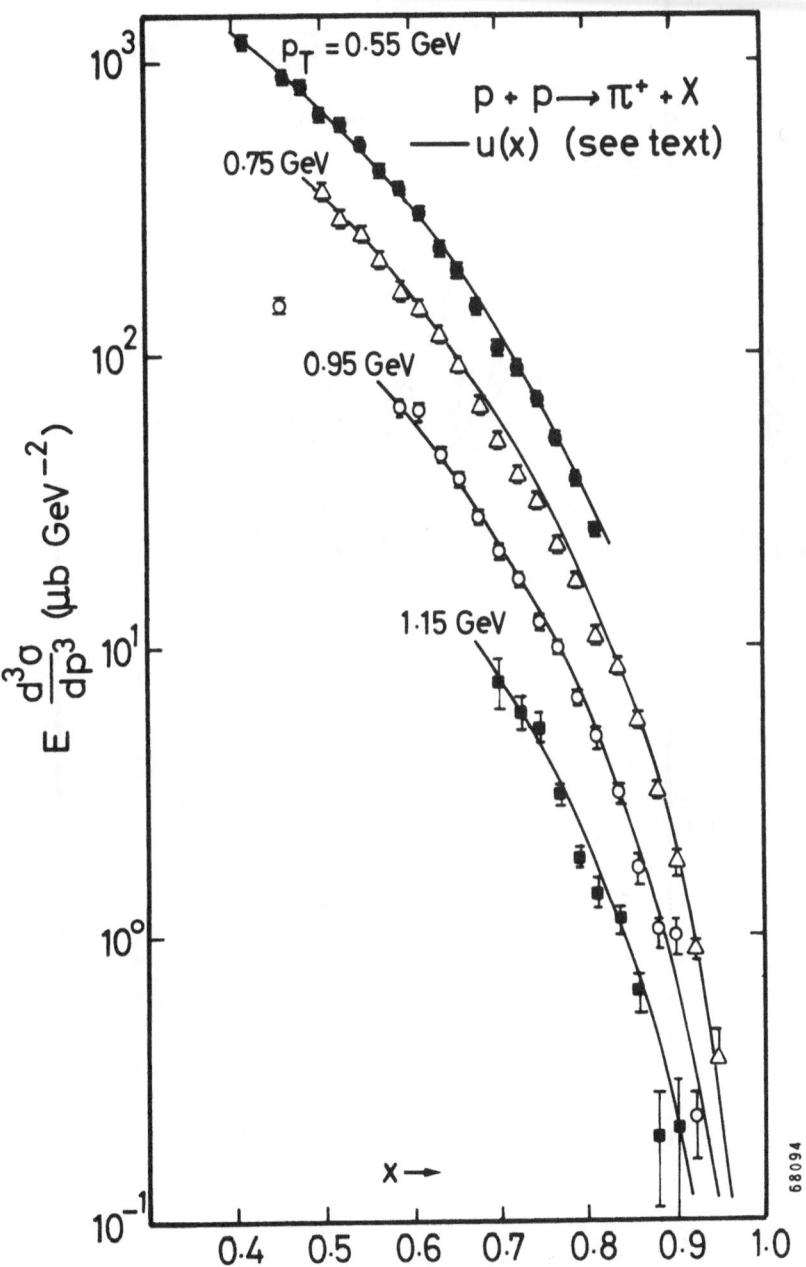

Fig. 1. Inclusive π+ production cross section at the
CERN ISR, for centre-of-mass energy 45 GeV.
Data from J. Singh et al.[5].

628

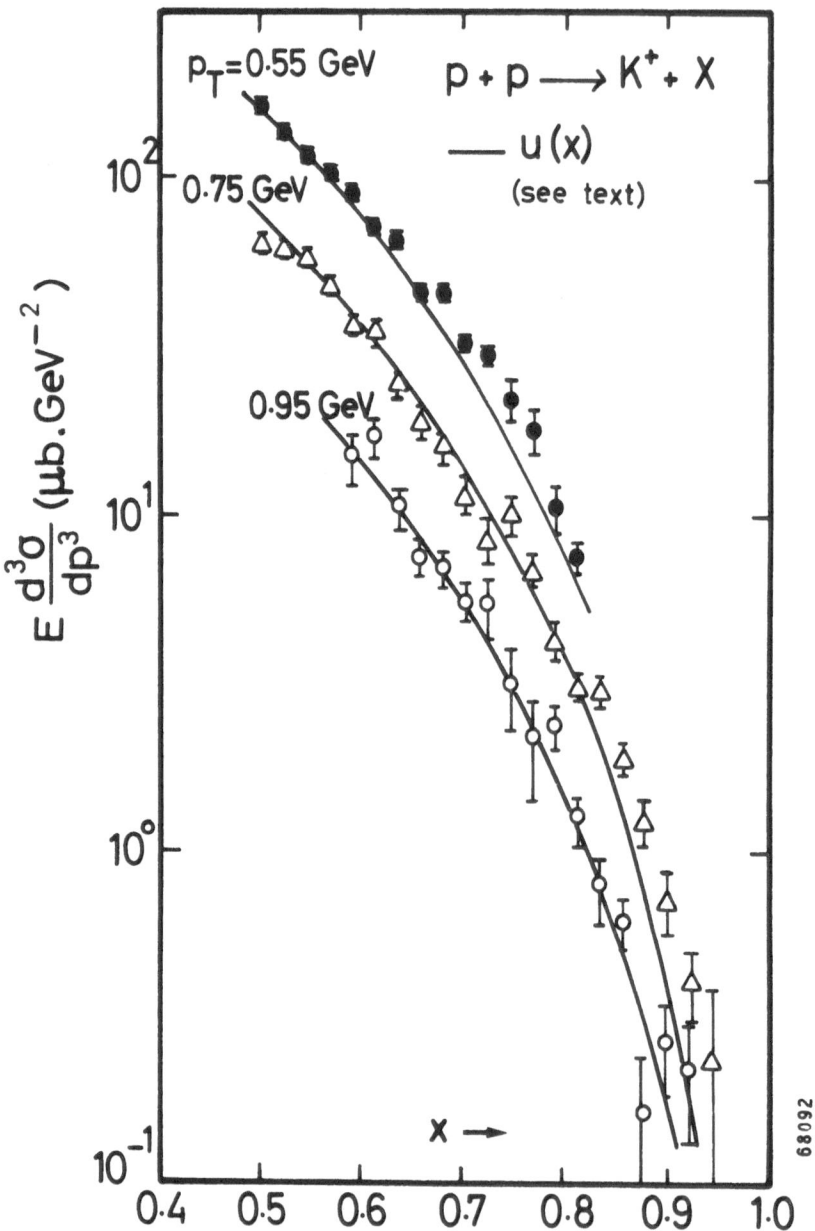

Fig. 2. Same as Fig. 1 for K$^+$ production.

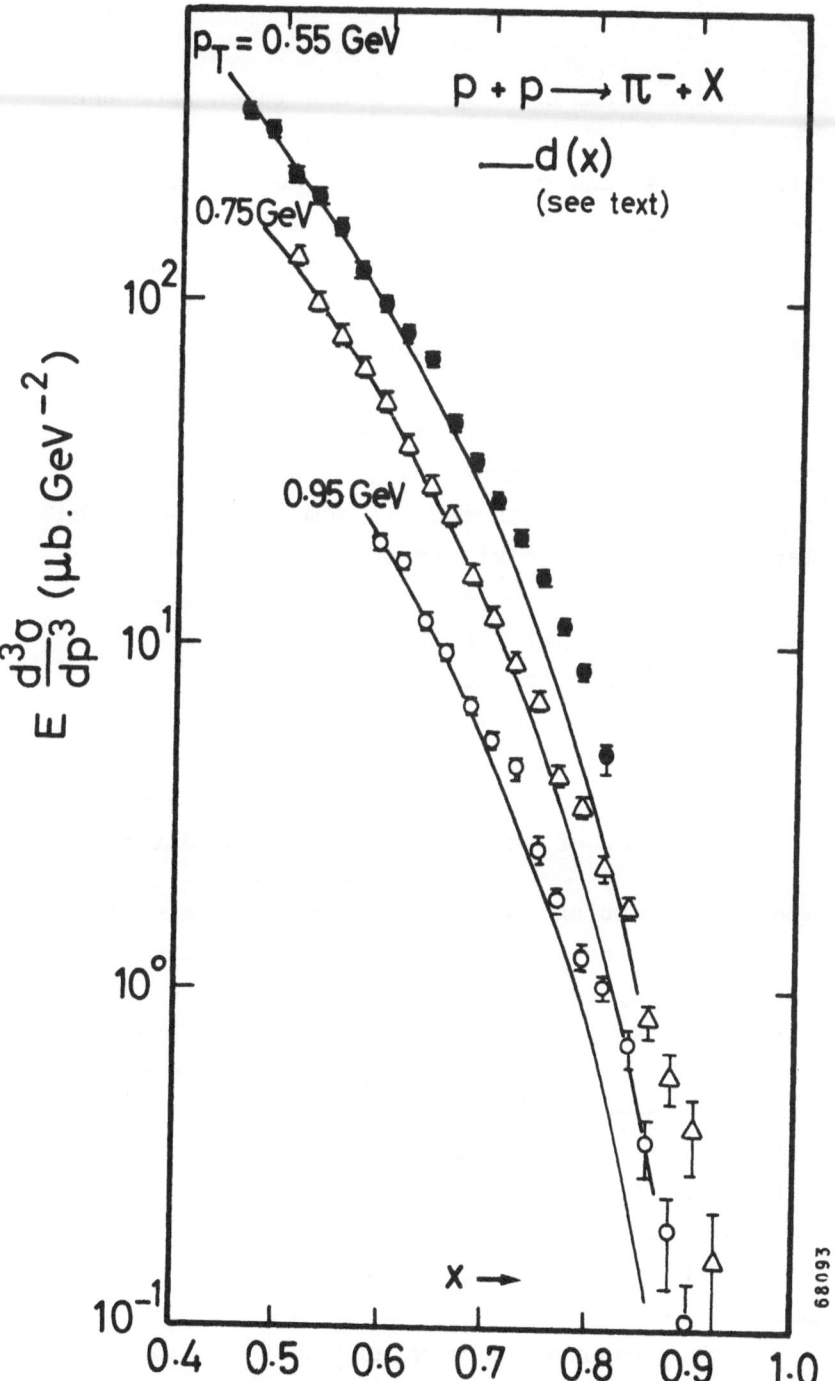

Fig.3. Same as Fig.1 for π^- production.

Manchester Collaboration. Figs. 1-3 are taken from the latest publication of these data by Singh et al.[5]. The centre-of-mass energy is $s^{1/2}$ = 45 GeV and the x region covered is x $\gtrsim$ 0.4. Figs. 1 and 2 show that, up to normalization, the x-distributions of π^+ and K^+ mesons (which can be formed by recombination of an up quark of the proton with a sea antiquark) are strikingly similar to the up quark distribution u(x) derived from the SLAC data on electron-nucleon deep inelastic scattering [6], the agreement being particularly good for π^+. The π^- distribution and its comparison with the down quark distribution d(x) are shown on Fig. 3, where the agreement is good up to x $\simeq$ 0.7, the π^- spectrum being slightly above d(x) for larger x. Note that the u(x) and d(x) used for the comparison refer to the region of relatively low momentum transfers (low Q^2) studied in the SLAC experiments, where Bjorken scaling was nevertheless found to hold approximately. We come back to the question of scaling violation in section 7.

The K^- production spectrum is very different from the $\pi^\pm$ and K^+ spectra. The invariant cross section $E\, d^3\sigma/dp^3$ drops rapidly, about as $(1-x)^6$, and is only 2.5% of the K^+ cross section at x = 0.7. Note that the exponent is slightly lower than the one predicted by the dimensional-counting rule, which is 7 or 9 depending on whether one assumes single quark or no quark interchange in the collision. The recombination model here discussed makes no specific prediction for K^- production because this meson has no valence quark in common with the incident proton. This holds also for the K^{*0} or φ resonances of which the K^- could be a decay product. One might consider K^-, K^{*0} or φ production as a recombination of two sea quarks, and various authors have done this. We believe that such a picture is doubtful for sizable x of the meson produced, and that sea gluons must play an important role by contributing a considerable amount of additional momentum.

As shown by Singh et al., none of the $\pi^{\pm}$ and K^{+} spectra follows a simple $(1-x)^{N}$ law, in contradistinction with the indications given by cruder data. This means that applications of the dimensional-counting rule of Brodsky, Farrar and Matveev et al. [7] can only have indicative value. Another interesting result of Ref.[5] is that for π^{+} and K^{+} a triple-Regge fit works, but only for $x \gtrsim 0.7$, and in this interval it gives the expected Regge trajectories with the correct slope (the N_{α} trajectory for π^{+} and a superposition of the Λ and Σ_{δ} trajectories for K^{+}). For π^{-} a triple-Regge fit again works for $x \gtrsim 0.7$, but in this case it gives the N_{α} trajectory instead of the expected Δ_{δ} one. We note that there is no theoretical reason against quark recombination and triple-Regge mechanism being both at work, and it seems that the contribution of the latter becomes apparent at large x only. This matter is studied further in section 5.

3. RECOMBINATION OF A VALENCE QUARK WITH A
SEA ANTIQUARK (EQUATIONS)

The most striking fact mentioned above is the similarity between the invariant meson cross section of Figs. 1-3 and the valence quark probability distributions $u(x)$, $d(x)$. How this similarity can be

understood in the recombination model will now be discussed in some detail for process (1)

$$p + hadron \rightarrow M + \ldots \quad .$$

For the inclusive cross section integrated over transverse momenta the basic formula is

$$\frac{1}{\sigma} \frac{d\sigma}{dk_M} = \int f(k_p, k_v, k_s) \, r(k_v, k_s) \, \delta(k_M - k_v - k_s) \, dk_v \, dk_s \qquad (6)$$

where momenta are taken in the centre-of-mass frame and
only their longitudinal components are written down,
k_p for the incident proton, k_v and k_s for two of its
constituents (a valence quark q_v and a sea antiquark $\bar{q}_s$),
and k_M for the meson M produced by the recombination of
q_v and $\bar{q}_s$. All these longitudinal momenta are taken to
be positive.

The function f is the joint distribution function
of k_v and k_s for given k_p. The recombination probability
is denoted by the function r multiplied by the delta
function for momentum conservation. If the recombination
would involve additional partons, for example, gluons,
their momenta would have to be included in f, r and the
delta function. σ is the relevant integrated cross
section (total inelastic cross section excluding
diffraction dissociation, $\sigma \simeq 28$ mb for pp collisions).

We now go over to the x-variables

$$k_M = x_M k_p, \qquad k_v = x_v k_p, \qquad k_s = x_s k_p \quad . \qquad (7)$$

For large k_p, the combination $k_p^2 f$ is expected to become a function of x_v and x_s with only a slow dependence on k_p reflecting the increasing excitation of the parton sea as k_p increases. We shall write

$$k_p^2 f(k_p, k_v, k_s) = F(x_v, x_s) . \tag{8}$$

As to $r(k_v, k_s)$, it is not expected to depend on k_p. This makes it invariant for a longitudinal Lorentz boost, so that for large momenta it depends only on the ratio of k_v and k_s. We therefore write

$$r(k_v, k_s) = R(x_s/x_v) . \tag{9}$$

With (8) and (9) the cross section formula becomes

$$\frac{1}{\sigma} \frac{d\sigma}{dx_M} = \int_0^1 \int_0^1 F(x_v, x_s) R(x_s/x_v) \delta(x_M - x_v - x_s) dx_v\, dx_s . \tag{10}$$

The single valence quark distribution $G_v(x)$ is related to F by

$$G_v(x) = n_v \int_0^1 F(x, x_s) dx_s \Big/ \int_0^1 \int_0^1 F(x', x_s) dx'dx_s \tag{11}$$

with n_v the number of valence quarks of type q_v in the proton ($n_v = 2$ for u quarks, $n_v = 1$ for d). Depending on whether q_v is up or down, $G_v(x)$ is the function $u(x)$ or $d(x)$ compared with the data in Figs.1-3. We recall that the structure functions νW_2 of deep inelastic lepton-nucleon scattering are linear combinations of the x $G_V(x)$ with the squares of the quark charges as coefficients.

The distribution of sea partons is believed to be concentrated at small x and to become singular as x^{-1} for x → 0, this singularity being needed to reproduce the diffractive low-x limit of deep inelastic lepton-nucleon scattering. We therefore expect our function $F(x_V, x_S)$ to be concentrated at small x_S, say $x_S \lesssim \varepsilon \simeq 0.1$, and to grow as x_S^{-1} for x_S → ∞. For eq.(11), this means that the ratio must be calculated first for the integrals taken from a positive x_0 to 1, the limit x_0 → 0 being taken afterwards. The situation is different for eq.(10), which would give an infinite cross section unless the recombination function has a zero at $x_S = 0$. By symmetry it should also have a zero at $x_V = 0$. We shall write

$$R(x_S/x_V) = x_S x_V (x_S + x_V)^{-2} \, \rho(x_S/x_V) \tag{12}$$

and assume $\rho(0)$ finite. Das and Hwa [4] adopt this form with constant ρ, justifying it by applying the dimensional-counting rule to R:

$$R \propto (1 - x_V) \quad \text{for} \quad x_V + x_S = 1, \quad x_V \to 1 \, .$$

With (12) the cross section is given by

$$x_M \frac{d\sigma}{dx_M} = \sigma \int x_S F(x_M - x_S, x_S) \, \rho\left(\frac{x_S}{x_M - x_S}\right) \frac{x_M - x_S}{x_M} \, dx_S \, . \tag{13}$$

For small ε and not too small x_M, we approximate $(x_M - x_S)/x_M$ by one and $\rho(x_S(x_M - x_S)^{-1})$ by $\rho(0)$ over the interval

$0 < x_s \lesssim \varepsilon$ where the sea contributes. The invariant cross section formula then reduces to the simple form

$$x_M \frac{d\sigma}{dx_M} = \sigma\rho(0) \tilde{G}_v(x_M) \tag{14}$$

with

$$\tilde{G}_v(x) = \int_0^x F(x-x_s,x_s) \, x_s \, dx_s \quad . \tag{15}$$

Comparison of eqs. (11) and (15) suggests that $\tilde{G}_v(x)$ is likely to be similar but not identical in shape to the valence quark distribution $G_v(x)$, as found experimentally (Figs.1-3). The extent of the similarity can be discussed as follows. For $x_v \lesssim 1 - \varepsilon$, one can assume factorization of F, i.e.

$$F(x_v,x_s) = G_v(x_v) G_s(x_s) \quad . \tag{16}$$

$\tilde{G}_v$ is then given by the convolution

$$\tilde{G}_v(x) = \int_0^x G_v(x-x') G_s(x')x' \, dx' \quad . \tag{17}$$

For $x > \varepsilon$ and small ε one gets the simple approximation

$$\tilde{G}_v(x) \simeq G_v(x) \int_0^1 G_s(x')x' \, dx' \tag{18}$$

i.e. proportionality of $\tilde{G}$ and G for $\varepsilon < x < 1-\varepsilon$. The situation is different for $x \lesssim \varepsilon$, where one must know the low-x behaviour of $G_v(x)$, which is believed to be

$$G_v(x) \propto x^{-\alpha} \quad \text{with} \quad \alpha \sim 0.5 \quad . \tag{19}$$

Eqs. (17) and (14) then give the small-x behaviour

$$\overset{\curvearrowright}{G}_V(x) \propto x^{1-\alpha}, \quad \frac{d\sigma}{dx_M} \propto x_M^{-\alpha} \ . \tag{20}$$

No meaningful comparison with data can be made here. For $x_V \gtrsim 1-\varepsilon$ the constraint $x_V + x_s \leq 1$ prevents the validity of the factorization (16) and again $\overset{\curvearrowright}{G}_V$ is no longer expected to follow the shape of G_V, as seems to be indicated by the data.

Eq. (18) can also be written

$$\overset{\curvearrowright}{G}_V(x) = \bar{x}_s \ G_V(x) \tag{21}$$

where

$$\bar{x}_s = \int_0^1 G_s(x') \ x' \ dx' \tag{22}$$

is the average momentum fraction carried by the sea antiquark $\bar{q}_s$ involved in the recombination. Under the assumptions for which (18) holds, in particular,

$$\varepsilon < x < 1-\varepsilon, \quad \varepsilon \ll 1 \tag{23}$$

the cross section formula (14) becomes

$$x_M \frac{d\sigma}{dx_M} = \sigma \ \bar{x}_s \ \rho(0) \ G_V(x_M) \ . \tag{24}$$

We shall use it in this approximate form for our further discussion.

4. RECOMBINATION OF A VALENCE QUARK WITH A SEA ANTIQUARK
(COMPARISON WITH DATA)

To compare with data, one needs the invariant cross section $E \, d^3\sigma/dp^3$ integrated over the transverse momentum p_T. The ISR measurements of Refs.[5] and [8] cover the range $0.35 < p_T < 1.45$ GeV. For π^+ and K^+ the data are fitted in Ref.[5] to expressions of the form:

i) $\exp(-B \, p_T)$, ii) $\exp(-C \, p_T^2)$. (25)

Within the experimental errors, the variation of B or C with x is found to be negligible for $x > 0.5$. Using those fits, the value 28 mb for σ and the $G_v(x) = u(x)$ function from deep inelastic electron-nucleon scattering [6] one can estimate $\bar{x}_s \rho(0)$. We have worked at $x = 0.7$ and we have adopted an estimate of 0.16 for $u(0.7)$, a rather high value which takes into account that low Q^2 are relevant to our case (see section 6). We then find

$$P_{u\bar{d}\to\pi^+} \equiv \bar{x}_s \rho(0) \approx 0.06 - 0.09 \qquad (26)$$

for π^+ production ($u\bar{d}$ recombination), the lower (upper) value corresponding to the form ii) (i)) of p_T dependence. While we have taken a rather high $u(x)$, the cross section value used contains some π^+ produced by diffraction dissociation or resulting from resonance decay, so that (26) is probably a reasonable estimate. For K^+ production ($u\bar{s}$ recombination), the value of $\bar{x}_s \rho(0)$ is found from table 9 of Ref.[5] to be $\sim 25\%$ of the value for π^+ production. Note that in the model the quantity (26) has a simple physical meaning. It is the probability $P_{u\bar{d}\to\pi^+}$

that a valence u quark recombines with a sea $\bar{d}$ antiquark to produce a π^+.

We can now discuss the size of (26), and in particular whether it requires a strong enhancement of the sea, as was found to be necessary by Das and Hwa [4], whose approach differs from ours by the fact that they adopted for F the following factorized ansatz

$$F(x_v, x_s) = \beta \; G_v(x_v) \; G_s(x_s) \; (1-x_v-x_s) \qquad (27)$$

β being a constant, G_v and G_s the valence and sea quark distributions, and $(1-x_v-x_s)$ a phase space factor introduced ad hoc. This enhancement question is discussed in detail in a recent paper by Teper [9].

The value of $\rho(0)$ is basically unknown. If the function $\rho(x_s/x_v)$ in eq. (12) is taken to be a constant ρ, one must have $\rho \leq 4$ to satisfy the requirement that the recombination probability cannot exceed one, $R \leq 1$, a requirement noted by Teper. Das and Hwa adopted $\rho = 6$ on the basis of an alleged sum rule, eq.(6) of Ref.[4], which is not compelling as far as we can see; in view of $R \leq 1$ it could only hold if $R \equiv 1$. Of course, $\rho(x_s/x_v)$ need not be a constant, and therefore $\rho(0)$ can be larger than 4, but probably not by a big factor. Adopting $\rho(0) \simeq 4$ we deduce for the average momentum fraction of the sea $\bar{d}$ antiquark involved in the recombination

$$\bar{x}_s \simeq 0.015 - 0.02 \qquad (28)$$

The total momentum fraction carried by all sea quarks and antiquarks in the nucleon is estimated from the SLAC deep inelastic scattering data to be $\bar{x}_{sea} \simeq 0.05$. It is natural to assume that it is shared equally between u, $\bar{u}$, d, $\bar{d}$, with a small additional fraction for s, $\bar{s}$ (this fraction is estimated to be 11% if we use our above determination that K^+ production is $\sim$ 25% of π^+ production, assuming the same $\rho(0)$ for both cases). The share of $\bar{d}$ in $\bar{x}_{sea}$ is therefore $x_{\bar{d}} \simeq 0.01$, and we conclude from (28) that the sea enhancement required is only a factor 1.5 - 2, instead of the factor 4 resulting from Das and Hwa's analysis (see Teper's discussion, Ref.[9]).

We believe that our result is reasonable and that the high value of Das and Hwa resulted from their use of eq.(27), which has the serious defect of not satisfying the consistency condition (11). Following Teper, it must be pointed out, however, that other suppression factors are ignored in going from $\bar{x}_{sea} \simeq 0.05$ to $\bar{x}_d \simeq 0.01$, namely a factor 9 for colour and a factor 4 for spin. Teper proposes that the necessary matching of colour and spin of u and $\bar{d}$ to recombine into π^+ should be achieved thanks to the sea gluons (presumably through capture or emission).

The comparisons of Figs. 1-3 between meson spectra and valence quark distributions, which represent a nice success for the recombination picture, assume that the pions and kaons are directly produced particles and are not decay products of resonances. As shown by Teper, this may be true to a reasonable approximation in the x range considered in Figs. 1-3 ($x \gtrsim 0.5$), but direct measurements of meson resonance production are needed, not only to verify this point, but also because such resonances will themselves be produced by the same quark recombination mechanism. They will provide further tests

of the model and allow it to be used at lower x values.
The importance of this question will also become
apparent in the next section.

5. TRIPLE-REGGE MECHANISM

We now expand on the remarks at the end of section
2. Figs. 4 and 5 from Ref.[5] give the triple-Regge fits
at large x for $\pi^{\pm}$, K^{+} production and the comparison with
the relevant Regge trajectories. The variables used are
the familiar ones, $M^2 = s(1-x)$ being the square of the
missing mass and t the relativistic square of the energy-
momentum transfer. We recall that the triple-Regge formula
is

$$E \frac{d^3\sigma}{dp^3} = \frac{s}{\pi} \frac{d^2\sigma}{dtdM^2} = \beta(t) \ (1-x)^{1-2\alpha(t)} \quad .$$

The data for pp $\rightarrow \pi^- X$ are particularly interesting. Fig.4
shows a clear break at x $\simeq$ 0.7 (already visible in Fig.3),
and Fig.5 shows poor agreement with the Δ trajectory. The
explanation we propose is that valence quark recombination
and triple-Regge mechanism contribute additively:

$$E \frac{d^3\sigma}{dp^3} = (E \frac{d^3\sigma}{dp^3})_{Recomb.} + \beta(t) \ (1-x)^{1-2\alpha(t)}$$

the latter becoming important only at large x. Fig. 6
shows that the π^- data are easily accounted for by this
additive model. The Regge trajectory used is the Δ one,

$$\alpha_\Delta(t) = 0.13 + 0.90 \ t$$

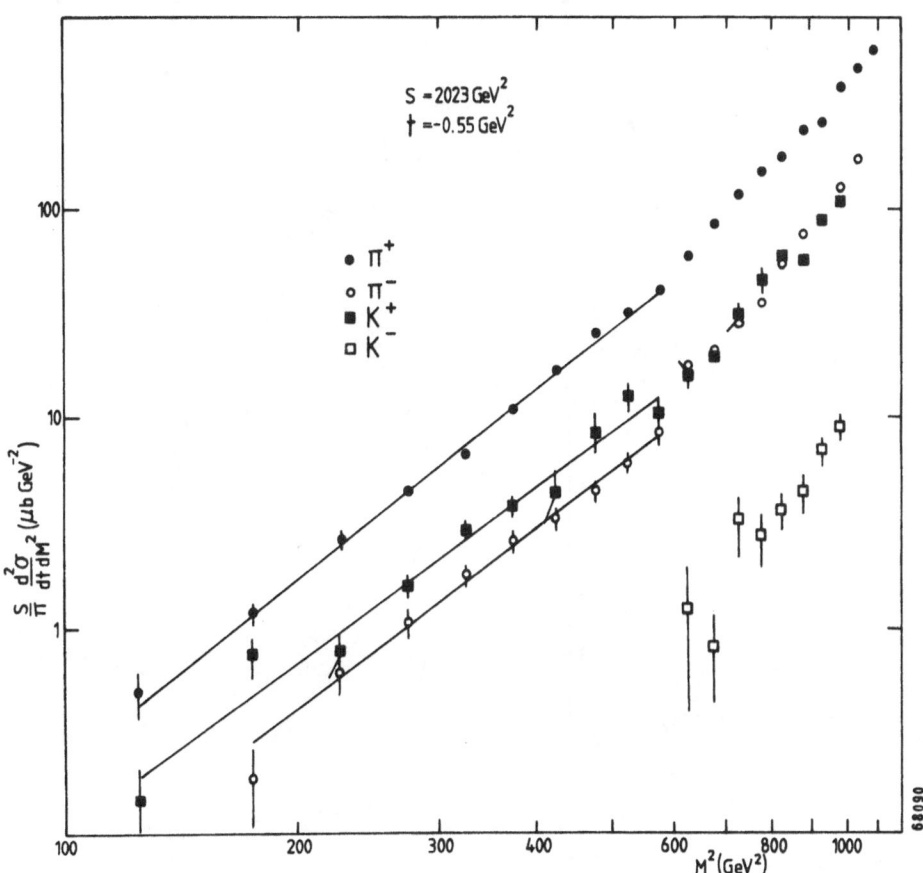

Fig. 4. Comparison of the data of Figs. 1,2,3 at
t = 0.55 GeV^2 with triple-Regge fits, from
Ref.[5].

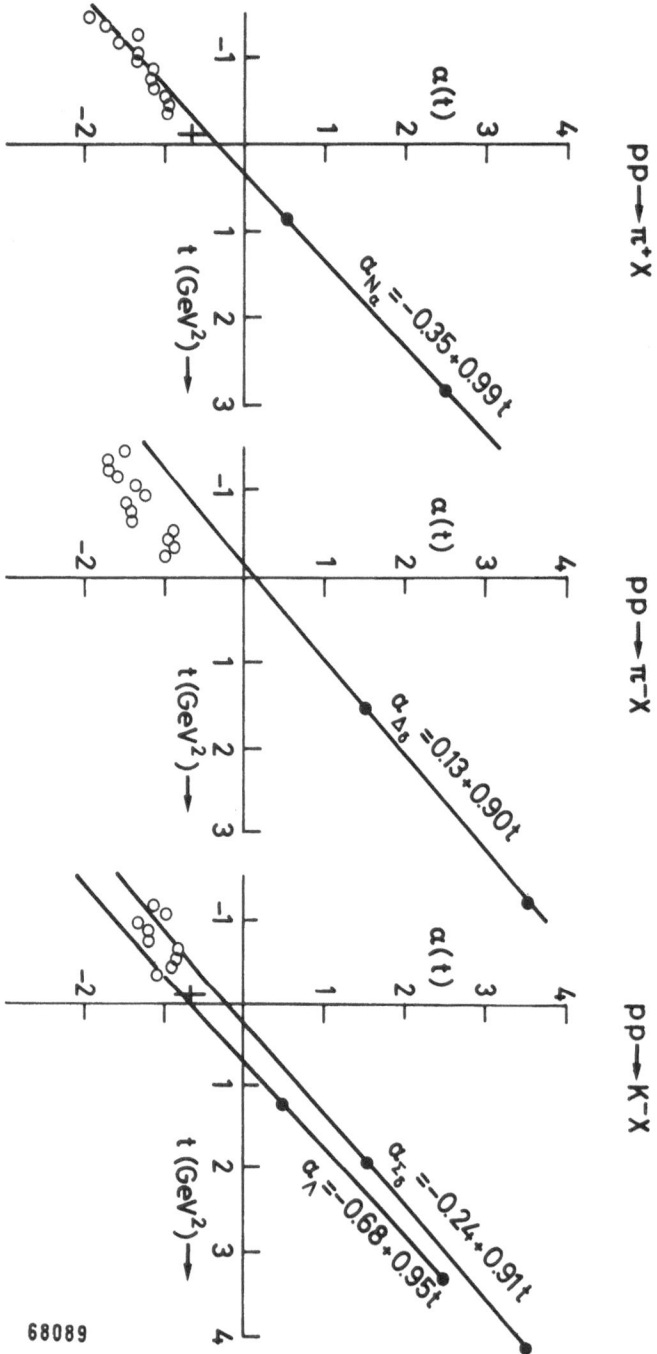

Fig.5. Comparison of triple-Regge fits with expected
Regge trajectories, from Ref.[5].

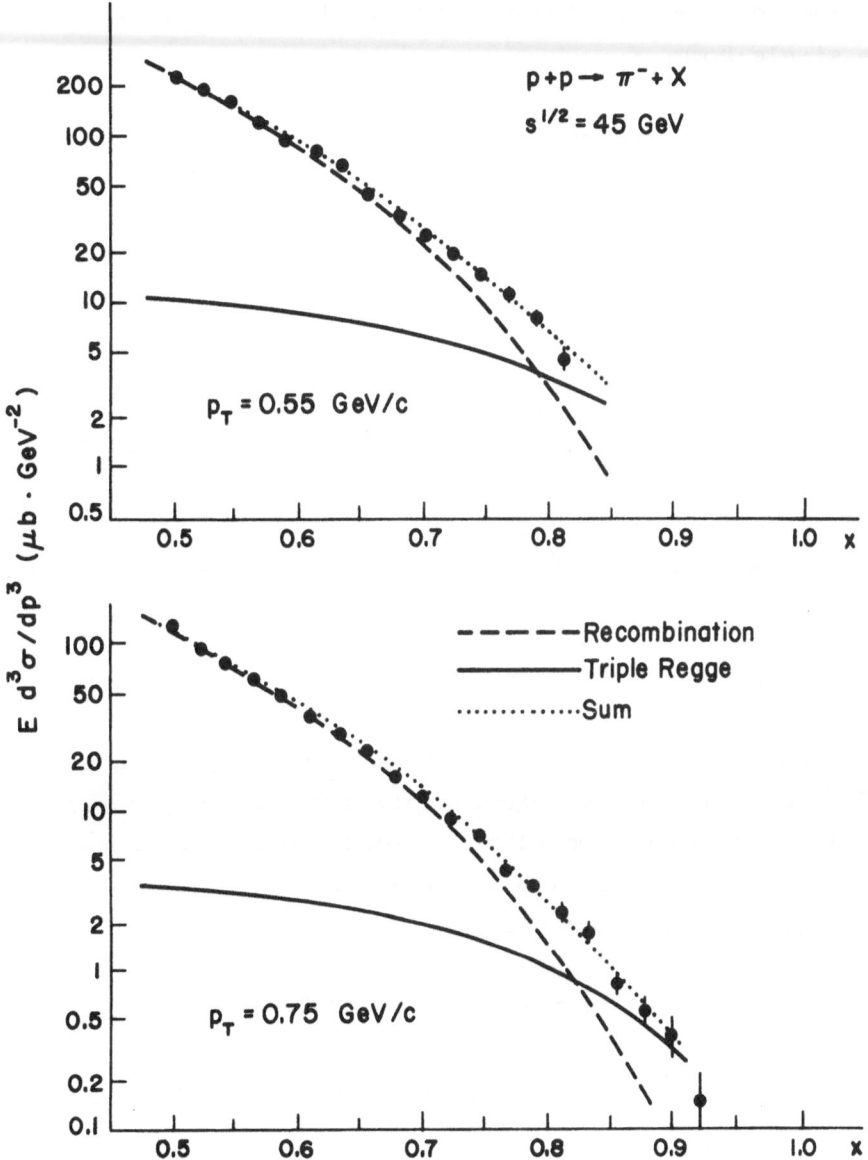

Fig.6. Fit of inclusive π^- production at p_T = 0.55 and
0.75 GeV/c with a sum (dotted line) of a triple-
Regge term (full line) and a recombination term
(dashed line). Data from Ref.[5].

and the residue function is found to be described by
the simple form

$$\beta_\Delta(t) = 27 \, e^t \quad \mu b \, GeV^{-2}$$

in the range $-0.55 \lesssim t \leq -0.2 \, GeV^2$. At smaller t
$(t \lesssim -0.6) \, \beta_\Delta(t)$ is found to flatten off. For the recombination term we have taken

$$(E \frac{d^3\sigma}{dp^3})_{Recomb.} = \gamma(p_T) \, d(x)$$

with the coefficient $\gamma(p_T)$ reduced by 10% compared with
the fits of Fig. 3. One has $\gamma(0.55 \, GeV) \simeq 2\gamma(0.75 \, GeV)$.

Fig. 5 shows that, contrary to the case of π^-
production, the π^+ and K^+ production data are in good
agreement with the corresponding Regge trajectories.
The reason is that, in these cases, both triple-Regge
and recombination mechanisms imply similar x dependences
for large x. Thus recombination leads to a dependence as
$u(x) \propto (1-x)^3$, while with the trajectories given in
Fig. 5 the Regge exponents become at $t = -0.55 \, GeV^2$

$$1 - 2\alpha(-0.55) = 2.8 \text{ for } \pi^+, \quad 2.5 \text{ and } 3.4 \text{ for } K^+.$$

In contrast we have for π^- production $d(x) \propto (1-x)^{4.5}$
and

$$1 - 2\alpha_\Delta(-0.55) = 1.7,$$

i.e. a much steeper decrease for the recombination than
for the Regge term. The large-x dominance of the Regge
term will even be stronger at $p_T = 0$, where t becomes
positive.

Another case of interest is $\pi^- p \to \pi^o X$, experimentally studied in Ref.[10] at large x and interpreted in terms of the triple-Regge mechanism. The authors of Ref.[10] find quite good agreement, except for the fact that the trajectory $\alpha(t)$ derived from the triple-Regge fit for $0.71 \leq x \leq 0.98$ falls somewhat below the expected ρ-trajectory, whereas there is agreement for $0.81 \leq x \leq 0.98$. This can again be explained by addition of Regge and re-combination terms. The latter is expected to decrease as 1-x, while the former has the Regge exponent

$$1 - 2\alpha_\rho(t) \simeq 0 \qquad \text{for} \qquad t \simeq 0.$$

We end this section by pointing out that additivity of recombination and triple-Regge mechanisms is physically reasonable, because the mechanisms are different when expressed in terms of partons. The triple-Regge case amounts to the substitution of an incident valence (anti)quark of low x by another one (in $\pi^- p \to \pi^o X$ for example), or of two incident valence quarks of low x by a low-x antiquark (in pp $\to \pi X$ for example), but the incident valence quark flying through need not have its x-value close to the one of the outgoing hadron, as is the case in the recombination mechanism.

6. PROBABILITIES FOR SINGLE, DOUBLE, AND TRIPLE VALENCE QUARK RECOMBINATION

Returning to the recombination model, we now make use of the probability (26) for $u\bar{d} \to \pi^+$ recombination in order to estimate the probability $P_{u \to M}$ that a given valence u quark recombines with any sea antiquark to give any meson including the meson resonances. Such an estimate was made by Teper [9], with the result

$$P_{u \to M} \simeq (3.4 - 4.5) \; P_{u\bar{d} \to \pi^+} \qquad . \tag{29}$$

We take the same range of values for $P_{d \to M}$. We have
checked that at $x = 0.7$ this is compatible with the
π^- production data of Ref.[5] and the value of the
valence d quark distribution function. Combining (26)
and (28) we find

$$P_1 = P_{u \to M} = P_{d \to M} \simeq (3.4 - 4.5) \cdot (0.06 - 0.09)$$

$$\simeq 0.2 - 0.4 \tag{30}$$

P_1 is the probability that a given valence quark recombines with any sea antiquark. In the framework of
the model (sections 1 and 2), $1-P_1$ is then the
probability for a given valence quark to recombine
with quarks and form a baryon. The above estimates
hold for $x \gtrsim 0.5$ and for recombination of one valence
quark. We do not know what happens at small x, nor what
correlations may exist between the behaviour of the
three valence quarks. But if we adopt the estimate (29)
for all x we can attempt to guess the probabilities for
the various types of recombinations listed in section 2.
This will now be done.

The probability $1-P_1$ is composed of three contributions

$$1 - P_1 = P_1' + P_2 + P_3 \tag{31}$$

corresponding to the three ways in which a given quark

can recombine to form a baryon, either with two sea quarks (probability P_1'), or with one sea and one valence quark (probability P_2), or with the two remaining valence quarks (probability P_3). These contributions can be related to the probabilities for the overall fate of the set of three valence quarks of an incident nucleon, namely the probabilities A_1 that they all recombine separately with sea partons, A_2 that two of them recombine together with a sea quark to form a baryon, and A_3 that all three recombine together to form a nucleon. The relations are

$$P_1 + P_1' = A_1 + \frac{1}{3} A_2$$

$$P_2 = \frac{2}{3} A_2 \qquad\qquad (32)$$

$$P_3 = A_3$$

and (31) is equivalent to $A_1 + A_2 + A_3 = 1$. All probabilities A_i, P_i and P_i' refer to non-diffractive collisions.

In addition to our above estimate of P_1, one can make use of the production rate of hyperons to obtain information on the size of the A_i. Let 2B be the mean number of hyperons produced in the fragmentation regions of a non-diffractive pp collision. The recombination model gives for the hyperon production per proton

$$B = A_2[\xi\eta + (1-\xi)\eta'\xi' + \xi\eta'(1-\xi') + 2\xi\eta'\xi']$$

$$+ A_1 [3\eta^2\eta'\xi' + 3\eta\eta'^2 (2\xi'(1-\xi') + 2\xi'^2)$$

$$+ \eta'^3 (3\xi'(1-\xi')^2 + 6\xi'^2(1-\xi') + 3\xi'^3)]$$

$$+ A_1 \eta^3 [1-(1-\xi)^3]$$

$$= 3\eta'\xi'A_1 + (\xi + \xi'\eta')A_2 + A_1\eta^3[1-(1-\xi)^3] . \tag{33}$$

ξ is the probability for a sea quark to be strange and $1-\xi' = (1-\xi)^2$ the probability for two sea quarks to be both non-strange. We have also defined

$$\eta = 1 - \eta' = P_1/(P_1 + P_1') . \tag{34}$$

A valence quark which recombines with sea partons only has probability η to do so with an antiquark, and η' to do so with two sea quarks. We assume η to be the same irrespective of what happens to the other valence quarks. The various terms of eq. (33) were first written so as to correspond to the various ways in which 1, 2 or 3 hyperons can be formed by recombination of the valence quarks with 1 to 6 sea quarks, or by recombination of the 3 surplus sea quarks available when all 3 valence quarks have recombined with sea antiquarks into mesons (term in $A_1\eta^3$, which was not considered before but should be included since it is likely to populate the fragmentation region). We have assumed these various recombinations to be uncorrelated.

Using (32) and (34) we rewrite (33) as

$$B = \xi A_2 + 3\xi'P_1' + A_1\eta^3 [1-(1-\xi)^3] . \tag{35}$$

To first order in ξ this simplifies to

$$b = B/\xi = A_2 + 6P_1' + 3A_1\eta^3 \tag{36}$$

a formula which becomes exact if in counting hyperons one gives weight 2 and 3 to hyperons of strangeness -2 and -3 respectively. These equations combined with the fact that P_1, P_1', P_2 and P_3 are all positive, impose interesting constraints on b. We shall use the simpler form (36). For given P_1, P_1' one finds that the minimum of b is obtained for $A_2 = 0$ and has the value

$$b_{min} = 3(P_1 + P_1')(\eta^3 - 2\eta + 2) = 3P_1\eta^{-1}(\eta^3 - 2\eta + 2) . \tag{37}$$

The maximum of b at fixed P_1, P_1' is obtained for $A_1 = 0$ if $P_1 + P_1' \leq \frac{1}{3}$, and for $A_1 + A_2 = 1$ if $\frac{1}{3} \leq P_1 + P_1' \leq 1$. Its value is

$$b_{max} = \begin{cases} 3(P_1+P_1')(3-2\eta)=3P_1+9P_1' & \text{for } P_1+P_1'\leq\frac{1}{3} \\ \frac{3}{2}(1-\eta^3)+\frac{3}{2}(P_1+P_1')(3\eta^3-4\eta+3) & \text{for } \frac{1}{3}\leq P_1+P_1'\leq 1 \end{cases} \tag{38}$$

Notice that $b = b_{min} = b_{max}$ for $P_1' = 0$ as well as for $P_1 + P_1' = 1$. Fig. 7 illustrates the shape of the allowed domain of b versus P_1, P_1'.

We now discuss the values of our parameters for pp collisions. From the K^+/π^+ production ratio of 0.25 we deduce

$$\xi = 0.25/(2 + 0.25) = 0.11 .$$

The best experimental data to estimate B are those from Blobel et al.[11], who give for pp collisions at 24 GeV

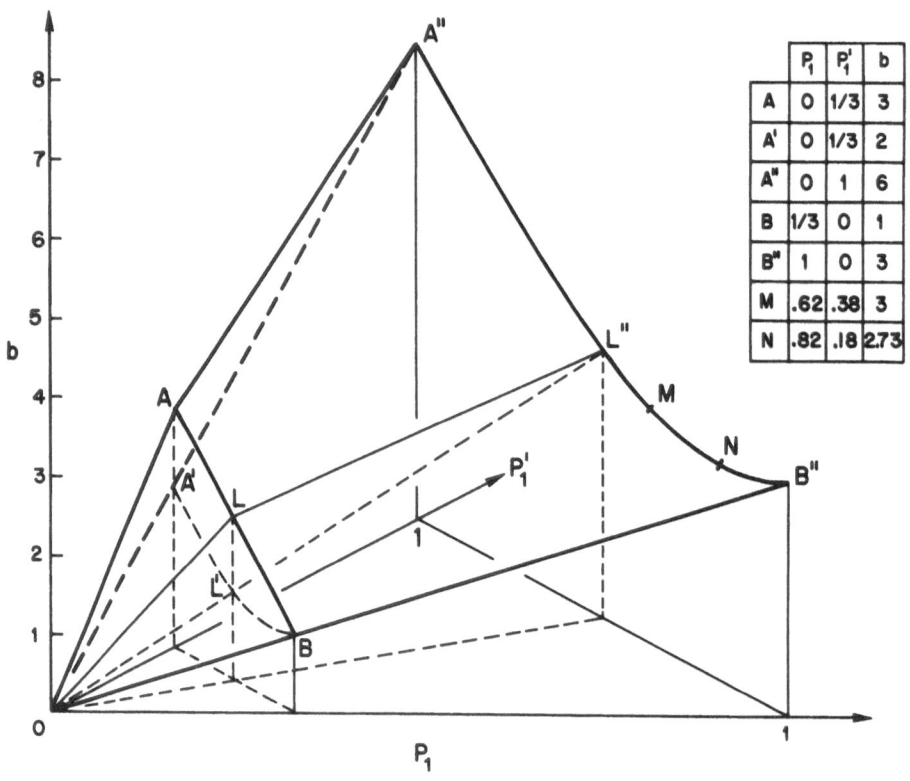

	P_1	P_1'	b
A	0	1/3	3
A'	0	1/3	2
A"	0	1	6
B	1/3	0	1
B"	1	0	3
M	.62	.38	3
N	.82	.18	2.73

Fig.7. Allowed domain of b as function of P_1, P_1'. The domain
is bounded by the surfaces OAB, AA"B"B and OA"B".
The straight lines OL, LL" and OL" show how these
surfaces are constructed. The curve A"B" has the
equations

$$P_1 + P_1' = 1, \qquad b = 3(P_1^3 - 2P_1 + 2) .$$

Its minimum is in the point N with $P_1 = (2/3)^{1/2} \approx 0.82$.
The table gives the co-ordinates of various points
of the figure.

$$2B_\Lambda = 0.057, \qquad 2B_{\Sigma^+} = 0.028, \qquad 2B_{\Sigma^-} = 0.0009 .$$

The Λ production rate includes Λ from Σ^0 decays. These data are at sufficiently low energy to assume that most hyperons result from beam and target particle fragmentation, but still high enough not to expect a large change of the fragmentation cross sections as one goes to higher energies. The data of Ref.[11] include diffraction dissociation, but this is approximately corrected for by the fact that the values (39) are obtained by dividing the production cross section by the full inelastic cross section. The Ξ production is not given in Ref.[11], but for pp collisions at 19 GeV, Alpgard et al.[12] give $2B_{\Xi^-} \simeq 0.001$ with a large error. Adding up we obtain $B = 0.048$. Hyperon production data at Fermilab and ISR energies are too incomplete to determine B reliably, but they suggest a somewhat larger multiplicity. Values in the range

$$b = B/\xi \simeq 0.6 - 0.75 \qquad\qquad (39)$$

may therefore be expected and shall be adopted for our discussion. Since P_1 is in the range $0.2 - 0.4$ and the function $b_{min}/3P_1$ of eq.(37) has the shape shown in Fig. 8, we find that our constraints limit P_1 to low values, $P_1 \simeq 0.2 - 0.25$, and η to values close to one, $\eta \gtrsim 0.8$ or $P_1' \lesssim 0.04$.

We can now estimate ranges of values for the probabilities A_1, A_2, A_3. If $P_1' = 0$, i.e. $\eta = 1$, we have

$$b = 3P_1 = 3A_1 + A_2 \qquad\qquad (P_1' = 0) \qquad\qquad (40)$$

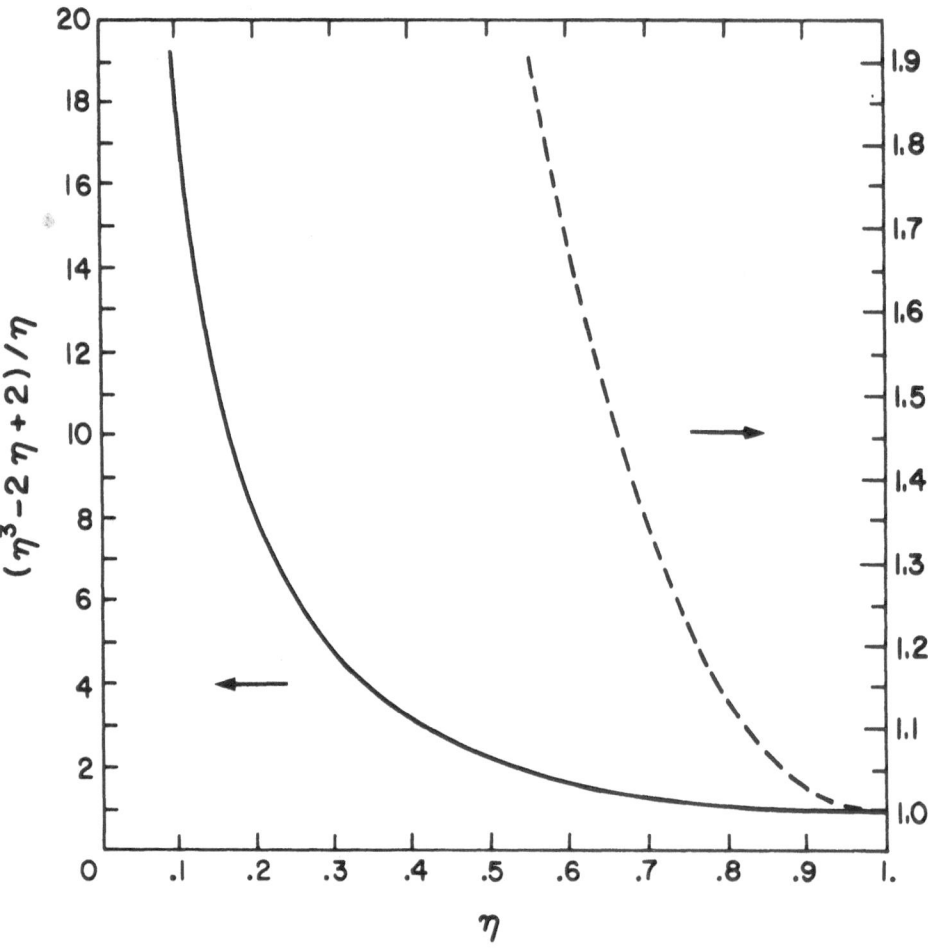

Fig.8. The shape of the function $b_{min}/3P_1$ of eq. (37).

as the only constraints beyond positivity. Hence

$$A_1 \leq P_1, \quad A_2 \leq 3P_1, \quad 1 - 3P_1 \leq A_3 \leq 1 - P_1 . \qquad (41)$$

Taking $P_1 \approx 0.2 - 0.25$ this gives

$$A_1 \lesssim 0.25, \quad A_2 \lesssim 0.75, \quad 0.25 \lesssim A_3 \lesssim 0.8 . \qquad (42)$$

If $P_1' > 0$, i.e. $\eta < 1$, the A_i are determined in terms of P_1, P_1' and b:

$$A_1 = (P_1 + 3P_1' - \tfrac{1}{3}b) / (1 - \eta^3)$$

$$A_2 = 3(P_1 + P_1' - A_1) \qquad\qquad (43)$$

$$A_3 = 1 - 3(P_1 + P_1') + 2A_1 .$$

They can be rewritten as

$$A_1 = P_1 + P_1' - \tfrac{1}{3}(b - b_{min}) / (1 - \eta^3)$$

$$A_2 = (b - b_{min}) / (1 - \eta^3)$$

$$A_3 = 1 - P_1 - P_1' - \tfrac{2}{3}(b - b_{min}) / (1 - \eta^3) .$$

As long as $P_1 + P_1' \leq \tfrac{1}{3}$, b_{max} is given by the first equation (38) and corresponds to

$$A_1 = 0, \quad A_2 = 3(P_1 + P_1'), \quad A_3 = 1 - 3(P_1 + P_1') .$$

As we have seen in connection with eq.(39), unless P_1 is

close to the lower limit of its range ($P_1 \simeq 0.2$) and P_1' close to zero, b is expected to be close to b_{min}, which implies

$$A_1 \simeq P_1 + P_1', \qquad A_2 \simeq 0, \qquad A_3 \simeq 1 - P_1 - P_1' .$$

Closer estimates can be obtained if we make more special but still plausible probabilistic assumptions on the relative sizes of P_1, P_1', P_2 and P_3. Consider a given valence quark of the proton and call p_v its probability to recombine with another given valence quark, and p_s its probability to recombine with a given sea quark or antiquark. We neglect the dependences of these probabilities on momenta and flavours, as well as possible correlations. We then have

$$P_1 = n_s\, p_s$$

$$P_1' = \frac{1}{2}\, n_s (n_s - 1)\, p_s^2 \qquad\qquad (44)$$

$$P_2 = 2\, n_s\, p_s\, p_v$$

$$P_3 = p_v^2$$

where n_s is the effective number of sea quarks (or anti-quarks) available for recombination. Hence

$$P_2^2 = 4 P_1^2\, P_3 \qquad\qquad (45)$$

$$P_1' = c\, P_1^2, \qquad c = (n_s - 1)/(2n_s) . \qquad\qquad (46)$$

Taking $1 \leq n_s \leq \infty$ as range for n_s, we have for c

$$0 \leq c \leq 0.5 . \qquad\qquad (47)$$

Using $P_3 = 1 - P_1 - P_1' - P_2$, we solve eq.(45) for P_2

$$P_2 = 2P_1(1-P_1-P_1'+P_1^2)^{1/2} - 2P_1^2 = 2P_1(1-P_1+(1-c)P_1^2)^{1/2} - 2P_1^2. \quad (48)$$

For given P_1, P_1' the values of the A_i and of b are then all fixed. We give a few examples for $c = 0$ and $c = 0.5$ (the values for $c = 0.5$ being given between brackets).

P_1	b	A_1	A_2	A_3
0.2	0.6(0.72)	0.06(0.08)	0.43(0.42)	0.51(0.50)
0.3	0.9(1.12)	0.12(0.17)	0.53(0.51)	0.35(0.32)
0.4	1.2(1.53)	0.21(0.31)	0.57(0.51)	0.22(0.18)

Keeping in mind eq.(39) we conclude that under our more specific assumptions we are again led to a low value for P_1, $P_1 \simeq 0.2$, and to $P_1' = c\,P_1^2 \ll P_1$. They imply small A_1, and A_2 of the same order of magnitude as A_3 (whereas the early considerations of Refs.[1] and [2] assumed dominance of A_3). Since P_1' is small, A_2-type processes are mostly those where, out of the three valence quarks of the incident proton, two recombine with a sea quark q_s to give a baryon b and the third with a sea antiquark $\bar{q}_s'$ to give a meson m, where b and m may be resonances but the b-m system is not. Hence the usefulness of meson-baryon correlation studies in the proton fragmentation region of non-diffractive collisions. The recombination model relates them to correlations among the proton valence quarks.

Under our assumptions, we find A_3 to be considerable, probably $\gtrsim 0.4$. This is in contradiction with the

two-quark-chain model of inelastic collisions (as obtained by cutting the cylindrical Pomeron structure of
the dual topological unitarization scheme), which
implies A_3 = 0. Our findings are in favour of schemes
with an important gluon exchange contribution to the
Pomeron. They are compatible with complete dominance
of gluon exchange in hadronic collisions at very high
energy.

7. DISCUSSION

   Our analysis suggests that lower values of P_1 are
more likely than higher ones. As already mentioned in
section 4, P_1 and $\bar{x}_s \rho(0)$ may be somewhat smaller than
the estimates in eqs. (30) and (26), due to diffraction
dissociation, resonance decay, and the fact that the
value of $G_v(x) = u(x)$ to be used when applying eq.(24)
to π^+ production may have to be taken somewhat larger
than the one given at the same x by the SLAC electron-
nucleon scattering data. The reason for this latter expectation is that the recombination process here considered involves low momentum transfers, and the
violation of Bjorken scaling in lepton-nucleon scattering is such that u(x) at large x increases for decreasing momentum transfer (decreasing Q^2). Physically, the
valence quark takes along more of its gluon cloud in
the recombination process than in high Q^2 lepton-nucleon
scattering. At x $\gtrsim$ 0.5, the effects of diffraction
dissociation, resonance decay and Bjorken scaling
violation should not be very large, however, if the
similarity in shape of $(x^{-1} d\sigma/dx)_{\pi^+}$ and u(x) is to be
explained by the recombination model. In particular,

we doubt that they could completely eliminate the need
for an enhancement of the parton sea, although at ISR
energies the enhancement may be somewhat less than the
factor 1.5 - 2 mentioned in section 4, and much less
than the factor required by Das and Hwa [4].

Our whole discussion has been limited to the
region $x \gtrsim 0.5$. To extend it to lower x one has to face
the problem of resonances, mainly mesonic[+], but also
baryonic. Accurate work will become possible when more
data will be available on resonance production in the
fragmentation region. Upper limits for the resonance
multiplicities can be estimated in terms of the
probabilities P_1, P_1', A_2 and A_3. They are rather small,
which is encouraging for the experimental work. It would
be advisable to study the resonance contents of the
fragmentation region by imposing a cut-off of perhaps
$x \gtrsim 0.2$ on the particles considered, also when they are
resonance decay products, so as to avoid contamination
from decays of the resonances produced in the central
region. The study of strange particle production should
be very useful, as illustrated already by the information
we were able to extract above from very limited data.
Also the study of non-resonant systems of two or three
particles is instructive, as shown by Lockman et al.
for the production of multi-pion systems in the fragmentation region at the ISR [14].

In these lectures, the equations of the quark recombination model have only been given for single valence
quark recombination with a sea antiquark and for cross
sections integrated over transverse momenta. Extension
to the various other recombination processes can be

[+]For a study of meson resonance production in the recombination model see R.G. Roberts et al.[13].

carried out to test the model and exploit all the information to be extracted from the experimental study of the fragmentation regions. Also the extension to incident mesons is likely to become very interesting, all the more so that it should be, with the Drell-Yan process, one of the very few ways of getting information on the quark-gluon structure of mesons[+].

8. OTHER PRODUCTION PROCESSES AND OTHER MODELS

We add a few brief remarks on other cases of hadron production. The quark recombination model proposes an answer to the question of what happens in low-p_T non-diffractive collisions to the valence quarks of the incident hadrons. It extends naturally to processes involving leptons and to high-p_T hadronic collisions; it can then be applied to those incident valence quarks which undergo no high-p_T process in the reaction. In deep inelastic lepton-nucleon scattering, for example, the model can be applied to the target fragmentation region and will permit detailed comparisons with nucleon fragmentation in low-p_T hadronic reactions. Similar applications can be found in Drell-Yan processes. Much work is being done in this direction [16].

Coming back to the purely hadronic reactions, we note that the quark recombination model covers only the fragmentation region. In the spirit of the model, most of the centrally produced particles, i.e. at very high energy the majority of the secondaries, must originate entirely from the sea. They may be produced through hadronization of the excited gluon sea, as would be

---

[+]For an application to incident pions, see R.C.Hwa and R.G. Roberts [15].

expected if the Pomeron is gluon-exchange dominated.

As mentioned above, the two-quark-chain model of
central particle production deduced from dual topological
schemes is of another nature and, if dominant, would
require $A_3 = 0$. A comparison of this model with central
particle production data has recently been worked out
by Capella et al.[17]. For the fragmentation regions,
they reject the recombination model and are led to adopt
a model where a diquark carrying about 95% of the incident
proton momentum fragments into the observed secondaries.
This type of parton fragmentation model had been proposed
and studied by B. Andersson et al.[18]. While it can
account for the data, it does not explain the similarities shown by Figs. 1-3 between inclusive meson spectra
and valence quark distributions, similarities which give
at present the main justification for the recombination
model[+].
Note added in proof: on section 6 see also [20].

REFERENCES

[1] S. Pokorski and L. Van Hove, Acta Phys. Pol. B5
    (1974) 229.
[2] L. Van Hove and S. Pokorski, Nucl. Phys. B86 (1975)
    243;
    L. Van Hove, Acta Phys. Pol. B7 (1976) 339.
[3] W. Ochs, Nucl. Phys. B118 (1977) 397; see also
    H. Goldberg, Nucl. Phys. B44 (1972) 149.

[+]Another approach to hadron fragmentation in the dual
topological scheme has recently been proposed by
G.Cohen-Tannoudji et al.[19]. It leads to results
which are rather close to those of the recombination
model.

[4] K.P. Das and R.C.Hwa, Phys. Lett. 68B (1977) 459;
see also D.W. Duke and F.E. Taylor, Phys.Rev.D17
(1978) 1788; and C.B. Chiu et al., Fast Meson
Production and the Recombination Model, University
of Texas at Austin, Preprint ORO 356 (October 1978).

[5] J.Singh et al. (CHLM Collaboration), Nucl. Phys.
B140 (1978) 189.

[6] G.Miller et al., Phys.Rev. D5 (1972) 528;
A.Bodek et al., Phys.Rev. Lett. 30 (1973) 1087 and
Phys.Lett. 51B (1974) 419.

[7] S.Brodsky and G.Farrar, Phys.Rev.Lett. 31 (1973) 1153;
V.A.Matveev et al., Nuovo Cimento Lett. 1 (1973) 719.

[8] M.G.Albrow et al., Nucl.Phys. B73 (1974) 40.

[9] M.J.Teper, Applying the Parton Model to the Fast
Hadrons at Low p_T, Rutherford Laboratory, Preprint
RL-78-022/A, March 1978.

[10] A.V.Barnes et al., Phys.Rev.Lett. 41 (1978) 1260.

[11] V. Blobel et al., Nucl. Phys. B69 (1974) 454.

[12] K.Alpgard et al., Nucl. Phys. B103 (1976) 234.

[13] R.G. Roberts, R.C.Hwa and S.Matsuda, Parton Recombination Model Including Resonance Production, Rutherford Laboratory, Preprint RL-78-040 (May 1978).

[14] W. Lockman et al., Phys.Rev. Lett. 41 (1978) 680.

[15] R.C.Hwa and R.G.Roberts, Z. Physik, C1 (1979)
81.

[16] T.A. DeGrand and H.I.Miettinen, Phys.Rev.Lett. 40
(1978) 612;
J.Ranft, Phys.Rev. D18 (1978) 1491;
T.A.DeGrand, Hadronic Fragmentation Initiated by
Pointlike Probes, SLAC, Preprint SLAC-PUB-2182
(August 1978), submitted to Phys.Rev.D.

[17] A.Capella et al., Jets in Small-p_T Hadronic
Collisions, Universality of Quark Fragmentation,
and Rising Rapidity Plateaus, Orsay, Preprint
LPTE 78/30 (October 1978).

[18] B.Andersson, G.Gustafson and C.P.Peterson,
Phys. Lett. 69B (1977) 221.

[19] G.Cohen-Tannoudji et al., A Quark-Parton Model
from Dual Unitarization, CEN-Saclay, Preprint
DPh-T/78/88 (October 1978), submitted to Phys.
Rev. D.

[20] J. Kalinowski, S. Pokorski and L. Van Hove,
Z. Physik, to appear.

Acta Physica Austriaca, Suppl. XXI, 663—707 (1979)
© by Springer-Verlag 1979

TESTS OF QCD IN TWO-PHOTON PROCESSES
IN VERY HIGH ENERGY e^+e^- COLLISIONS$^+$

by

K. KAJANTIE

Research Institute for Theor. Physics

Univ. of Helsinki, Siltavuorenpenger 20 C

SF-00170 Helsinki, Finland

ABSTRACT

The $1\gamma^*$ process $e^+e^- \to \gamma^* \to \bar{q}q \to$ hadrons has a fundamental status in QCD. At energies in the $\sqrt{s} = 100$ GeV range another fundamental process, $e^+e^- \to e^+e^-\gamma\gamma \to e^+e^-\bar{q}q \to$ $\to e^+e^-$ hadrons, becomes experimentally accessible. These lectures describe what QCD predicts for these two-photon processes.

$^+$Lectures given at the
XVIII. Internationale Universitätswochen für Kernphysik
Schladming, Austria, February 28 - March 10, 1979.

CONTENTS

1. INTRODUCTION

A great number of QCD predictions for e^+e^- annihila-
tion in the $1\gamma^*$ approximation ($e^+e^- \to \gamma^* \to$ various quark
and gluon states) have been worked out. One class of them
are tests on jet production properties. Here an important
assumption is that on factorizability of the problem to a
hard and a soft part: the former can be calculated with
perturbative field theory techniques, the latter (the
fragmentation of a produced quark or gluon to hadrons)
is at present unsolved in QCD. A second class of tests
refers to properties of various produced bound states of
heavy quarks.

The cross section of $1\gamma^*$ events is proportional
to 87 nb/s which is about 4.4 pb (or 1 event/hour at

luminosity = 10^{32} cm^{-2}s^{-1}) at LEP energies (70 + 70 GeV
e^+e^- collisions). The smallness of this cross section
is due to the propagator $1/s$. A way of increasing the
cross section is to effectively reduce the value s. This
can take place through the emission of various hard quanta.
In these lectures, I shall discuss one example of this:
the two-photon mechanism [1-3]. In this the e^+ and e^- both
radiate an almost real photon which then collides with total
energy squared $\hat{s} = x_1 x_2 s$ (x_1 and x_2 are the longitudinal
momentum fractions of the photons). If x_1 and x_2 are very
small or the photons very soft, s can be brought down to
threshold $4m_e^2$ and total cross section of these events is
very large, of the order of $\alpha^2 \log^3 (s/m_e^2)/m_e^2$. However,
the bulk of these events is totally uninteresting (they
go down the beam pipe). The interesting ones have $\hat{s}$ so
large that jet effects are visible.

It may seem silly to build a storage ring for
10^3MSfr and then be happy when some process effectively
looses a significant fraction of this dearly acquired
energy. What probably will happen is that the genuinely
new and exciting phenomena will be observed via the $1\gamma^*$
mechanism; the small cross section just is the price one
has to pay. However, the higher order processes with
large cross sections are also there and cause very interesting jet effects [4-6].

From a technical point of view, a treatment of jet
effects in the $1\gamma^*$ approximation amounts to a discussion
of the basic diagram (2 back-to-back jets)

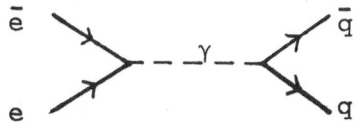

and the higher order corrections to it (3 and more jets):

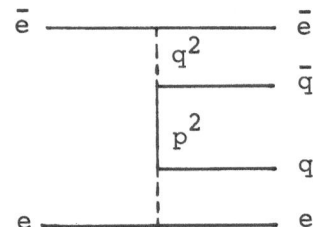

etc.

In the 2γ case the basic diagram is similarly

e̅ ——————— e̅

q̅

q̅

q^2

p^2

q

e ——————— e

Here one has to distinguish between the following cases

1. q^2 and p^2 both soft, i.e. constant when s → ∞. This is where the bulk of the cross section lies and the final state consists of particles nearly parallel with the initial particles. No hard process is involved and one cannot say anything more about this in QCD.

2. q^2 soft and p^2 hard, i.e. proportional to s when s→∞. The final state now contains two non-back-to-back quark jets at large transverse momentum (see fig. 8 below) taking only a part of the total energy (and, in addition, the nearly forward and backward electron and positron).

3. q^2 hard and p^2 soft. This case, in fact, corresponds to deep inelastic scattering on an electron: $e^+e^- \to e^+X$ or, if the final state e^- is also detected, one can interpret it as deep inelastic scattering on a (slightly off-shell) photon. Now the final state contains one large q_T and one small q_T jet.

The main aim of these lectures is to in detail discuss

the cases 2. and 3. above, together with the QCD
corrections to the basic two-photon diagram. We shall
see that QCD makes definite predictions about the
angular and q_T distributions of produced jets and
scattered electrons which should be easily testable
at LEP energies.

Physically, what happens in case 2. above is that
the electron and positron both radiate a nearly on-shell
and parallel photon which then collide and via the hard
process $\gamma\gamma \rightarrow q\bar{q}$ produce the final state quark jets which
are constrained to have large q_T so as to make the ex-
changed quark hard. To begin with, I shall derive the
formula which permits one to relate the cross section on
the $\gamma\gamma$ level to the e^+e^- level. As the formula has a wide
range of applicability I shall phrase it in more general
terms than really needed here.

2. HARD SCATTERING FORMULA

Assuming that the factorization to an uncalculable
soft part (probability $f(x_1)$ of finding a parton p_1 in
the hadron h_1) and a calculable hard part (cross section
$d\hat{\sigma}/d\hat{t}$ for $p_1 + p_2 \rightarrow q + k$) is valid we shall first derive
a simple and useful formula (eq. 2.6-7) [7-8] for the
cross section of the two-jet process $h_1 + h_2 \rightarrow q + k + X$
(see fig. 1 for notation). The most convenient jet
variables are its mass M, rapidity y and transverse
momentum q_T. The resulting formula can be applied to the
following cases in increasing (decreasing) order of
theoretical (experimental) complexity

 - the two-jet process $h_1 + h_2 \rightarrow q + k + X$, where
  the two final state partons are really identified

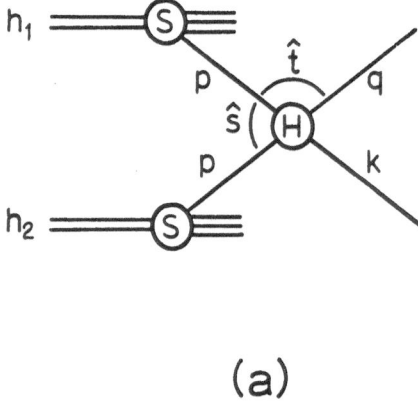

(a)

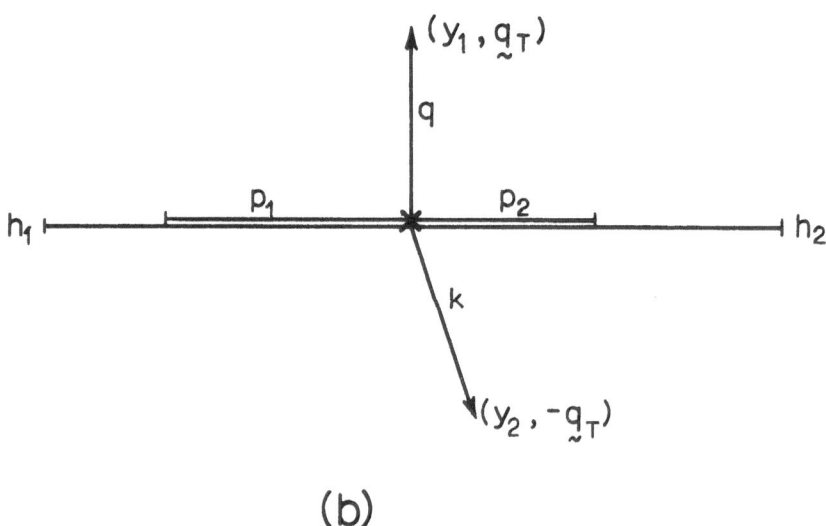

(b)

Fig.1. (a) a hard scattering process
(b) the same in momentum space

as jets. This is hardly possible today but, with
increasing total energy, will soon become the
primary measurement

- the one-jet process $h_1 + h_2 \to q + X$ by integrating
  over (the rapidity of) the other unobserved jet.
  This is particularly simple if q is a heavy virtual
  photon (a muon pair)
- hadron production reactions $h_1 + h_2 \to h + X$ by
  introducing a further soft element in the calculation:
  the materialization probability of a parton to a
  hadron.

The calculation proceeds as follows. Firstly, on the
parton level one has, using the standard normalisation.

$$\hat{\sigma}(p_1 + p_2 \to q + k) = \frac{1}{2\hat{s}} \frac{1}{(2\pi)^2} \int \frac{d^3q}{2q_o} \frac{d^3k}{2k_o} \delta^4(p_1 + p_2 - q - k) \sum |M|^2 \tag{2.1}$$

Equivalently,

$$q_o k_o \frac{d\hat{\sigma}}{d^3q\, d^3k} = \frac{1}{32\pi^2 \hat{s}} \sum |M|^2\, \delta^4(p_1 + p_2 - q - k)$$

$$= \frac{\hat{s}}{2\pi} \frac{d\hat{\sigma}}{d\hat{t}} \delta^4(p_1 + p_2 - q - k) \ . \tag{2.2}$$

Since it is more convenient to catalogue hard scattering
processes in terms of $d\hat{\sigma}/d\hat{t}$ than of $\sum |M|^2$, the former has
been introduced in (2.2). Since (2.2) is written in in-
variant form one can write on the hadronic level for the
invariant cross section of $h_1 + h_2 \to q + k + X$ the formula

$$q_o k_o \frac{d\hat{\sigma}}{d^3q\, d^3k} = \int_o^1 dx_1 dx_2 f(x_1) f(x_2) \frac{\hat{s}}{2\pi} \frac{d\hat{\sigma}}{d\hat{t}} \delta^4(p_1 + p_2 - q - k) \tag{2.3}$$

The 4 δ-functions can then be used to integrate over x_1, x_2, and k_T. In the hadron-hadron CMS (Fig.1(b)) the four-vectors have the following components ($k^2 = 0$ is assumed for simplicity):

$$p_1 = x_1 \tfrac{1}{2} \sqrt{s}(1,\underline{0},1)$$

$$p_2 = x_2 \tfrac{1}{2} \sqrt{s}(1,\underline{0},-1)$$

$$(2.4)$$

$$q = (q_o, \underline{q}_T, q_L), \qquad q^2 = M^2$$

$$k = (k_o, \underline{k}_T, k_L), \qquad k^2 = 0$$

which are related to rapidity by

$$q_o \pm q_L = \sqrt{M^2+q_T^2}\; e^{\pm y_1} \equiv \tfrac{1}{2}\sqrt{s}\; \bar{x}_T\; e^{\pm y_1}$$

$$k_o \pm k_L = q_T\; e^{\pm y_2} \equiv \tfrac{1}{2}\sqrt{s}\; x_T\; e^{\pm y_2}$$

$$(2.5)$$

$$\bar{x}_T = \frac{2}{\sqrt{s}}\sqrt{M^2+q_T^2} = \sqrt{x_T^2+4\tau}\; .$$

Using $d^3k/k_o = dy_2\, d^2k_T$ and $\delta(a)\,\delta(b) = 2\delta(a+b)\,\delta(a-b)$

we have

$$q_o\,\frac{d\sigma}{d^3q\, dy_2} = \int d^2k_T \int_o^1 dx_1 dx_2\, f(x_1)\, f(x_2)\frac{\hat{s}}{2\pi}\frac{d\hat{\sigma}}{d\hat{t}}\delta^2(\underline{q}_T + \underline{k}_T)\; \cdot$$

$$\cdot\; 2\delta(x_1\sqrt{s}-(q_o+q_L)-(k_o+k_L))\,\delta(x_2\sqrt{s}-(q_o-q_L)-(k_o-k_L))$$

which gives the final formula

$$\frac{d\sigma}{dy_1 d^2 q_T dy_2} h_1 h_2 \to qkX = x_1 x_2 f(x_1) f(x_2) \frac{1}{\pi} \frac{d\hat{\sigma}}{d\hat{t}}$$ (2.6)

$$x_1 = \frac{1}{2}\bar{x}_T \, e^{y_1} + \frac{1}{2}x_T \, e^{y_2}, \quad x_2 = \frac{1}{2}\bar{x}_T \, e^{-y_1} + \frac{1}{2}x_T \, e^{-y_2} .$$ (2.7)

Physically, measuring the rapidities y_1 and y_2 and transverse momentum q_T of the produced jets fixes the momentum
fractions x_1 and x_2 of the colliding partons (fig.1(b) and
eq.(2.7)). The overall cross section is then given by a
probability factor $x_1 x_2 f(x_1) f(x_2)$ (where $x_1 x_2$ arises from
$\hat{s} = x_1 x_2 s$ in (2.2)) times the hard cross section. In
practice, there are several hard processes one has to
sum over.

The subprocess invariants are often needed ($Y=y_1-y_2$):

$$\hat{s} = x_1 x_2 s = M^2 + \frac{1}{2}s \, x_T(x_T+\bar{x}_T \cosh Y) \quad M \equiv 0 \quad \frac{1}{2}sx_T^2(1+\cosh Y)$$

$$\hat{t} = M^2 - \frac{1}{4}s \, \bar{x}_T(\bar{x}_T+x_T \, e^{-Y}) \quad M \equiv 0 \quad -\frac{1}{4}sx_T^2(1+e^{-Y})$$ (2.8)

$$\hat{u} = M^2 - \frac{1}{4}s \, \bar{x}_T(\bar{x}_T+x_T e^{Y}) \quad M \equiv 0 \quad -\frac{1}{4}sx_T^2(1+e^{Y}) .$$

If the other jet k is unobserved, eq.(2.6) has to be
integrated over y_2. On the $x_1 x_2$ plane this amounts to
an integration over the curve

$$x_1 x_2 + \tau = \frac{1}{2}\bar{x}_T(x_1 e^{-Y_1}+x_2 e^{Y_1}) ,$$ (2.9)

obtained by eliminating y_2 from eqs.(2.7). Examples of

these curves are shown in fig. 2.

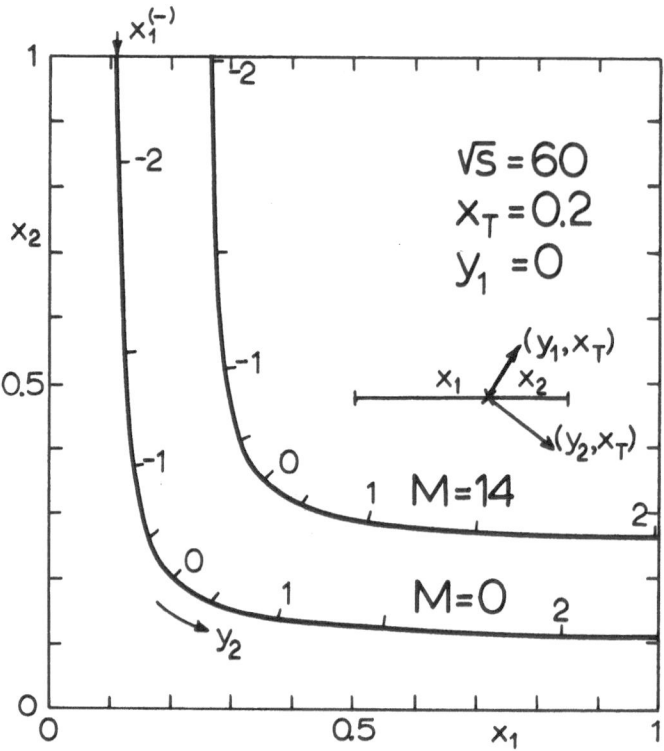

Fig. 2. Examples of the values of x_1 and x_2 (longitudinal momentum fractions of the colliding constituents) probed by a process leading to two final state jets with kinematic variables (y_1, x_T) and (y_2, x_T) (eq.(2.7)). The numbers on the curves give values of y_2. M is the invariant mass of jet 1.

The limits of y_2 are given by $x_i \leq 1$:

$$\log \frac{x_T}{2-\bar{x}_T e^{-y_1}} = y_2^{(-)} \leq y_2 \leq y_2^{(+)} = \log \frac{2-\bar{x}_T e^{y_1}}{x_T} \qquad (2.10)$$

and the cross section of the single jet process $h_1 + h_2 \rightarrow$ $\rightarrow$ q+X is given by

$$\frac{d\sigma}{dy_1 d^2 q_T} = \int_{y_2^{(-)}}^{y_2^{(+)}} dy_2 \; x_1 x_2 f(x_1) f(x_2) \frac{1}{\pi} \frac{d\hat{\sigma}}{d\hat{t}} \qquad . \qquad (2.11)$$

Alternatively, eq. (2.9) gives

$$\int_{y_2^{(-)}}^{y_2^{(+)}} dy_2 = \int_{x_1^{(-)}}^{1} \frac{dx_1}{x_1 - \frac{1}{2}\bar{x}_T e^{y_1}} \qquad (2.12)$$

where

$$x_1^{(-)} = \frac{\frac{1}{2}\bar{x}_T e^{y_1 - \tau}}{1 - \frac{1}{2}\bar{x}_T e^{-y_1}} \quad \overset{M \; \equiv \; 0}{\equiv} \quad \frac{\frac{1}{2}\bar{x}_T e^{y_1}}{1 - \frac{1}{2}\bar{x}_T e^{-y_1}} \qquad . \qquad (2.13)$$

When using x_1 as a variable one has to be careful near the lower limit $x_1^{(-)}$ (curve almost vertical in fig. 2). It may be necessary to divide the range of x_1 in two parts:

$$x_1^{(-)} \le x_1 \le \sqrt{\tau}_+ \, e^{y_1} \equiv \tfrac{1}{2}(\bar{x}_T + x_T) e^{y_1} \qquad (2.14)$$

$$\sqrt{\tau}_+ \, e^{y_1} \le x_1 \le 1$$

where

$$(x_1, x_2) = (\sqrt{\tau}_+ \, e^{y_1}, \ \sqrt{\tau}_+ \, e^{-y_1})$$

is that point in which the two jets have the same
rapidity $(y_1 = y_2)$. In the former part one can then
use x_2 instead of x_1 as a variable with the result:

$$\int\limits_{x_1^{(-)}}^{1} \frac{dx_1 \, F(x_1, x_2(x_1))}{x_1 - \tfrac{1}{2}\bar{x}_T e^{y_1}} = \int\limits_{\sqrt{\tau}_+ e^{y_1}}^{1} \frac{dx_1 \, F(x_1, x_2(x_1))}{x_1 - \tfrac{1}{2}\bar{x}_T e^{y_1}} +$$

$$+ \int\limits_{\sqrt{\tau}_+ e^{-y_1}}^{1} \frac{dx_2 \, F(x_1(x_2), x_2)}{x_2 - \tfrac{1}{2}\bar{x}_T e^{-y_1}} . \qquad (2.15)$$

Note the interpretation: in the limit $x_T \to 0$ the first
term here corresponds to emission of parton q almost
parallel to p_1 ($\hat{t}$ small), the second to k parallel to
p_1 ($\hat{u}$ small).

Remarks:

1. Eq. (2.11) in the simpler case $M = 0$ is often given
the form [9]

$$E \frac{d\sigma^{AB \to cX}}{d^3 p} = \int\limits_{o}^{1} dx_1 dx_2 f_{a/A}(x_1) f_{b/B}(x_2) \frac{d\hat{\sigma}^{ab \to cX}}{d\hat{t}} \frac{\hat{s}}{\pi} \delta(\hat{s}+\hat{t}+\hat{u})$$

$$(2.16)$$

in obvious notation. Integrating over the δ function gives a formula analogous to (2.11-12), but the interpretation of the remaining integral and the explicitness of eq.(2.6-7) is lost.

2. If M = O it is possible to replace rapidities by angles through

$$\cosh y = 1/\sin \Theta, \quad \sinh y = 1/\tan \Theta, \quad \tanh y = \cos \Theta,$$

$$e^y = \cot\frac{\Theta}{2} \qquad (2.17)$$

so that (eq.(2.7))

$$x_1 = \frac{1}{2}x_T(\cot\frac{\Theta_1}{2} + \cot\frac{\Theta_2}{2}); \quad x_2 = \frac{1}{2}x_T(\tan\frac{\Theta_1}{2} + \tan\frac{\Theta_2}{2}) . \qquad (2.18)$$

Numerically, y = 0,1,2,3 correspond to $\Theta = 90^{\circ}$, $40^{\circ}.4$, $15^{\circ}.4$, $5^{\circ}.7$.

3. An interesting special case of eq.(2.6) is that in which either of the incident particles is pointlike, $f(x_1) = \delta(1-x_1)$, say. This includes the standard deep inelastic scattering (cf.fig. 3 for $e^+e^- \to e^+X$).

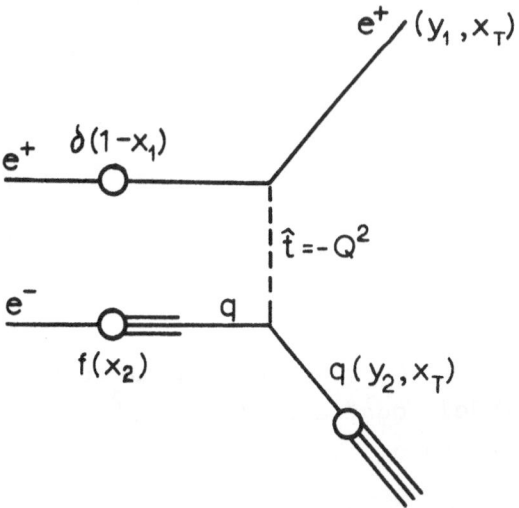

Fig.3. Variables and structure functions for deep inelastic scattering on an electron: $e^+e^- \to e^+qX$.

Using the constraint $x_1 = \frac{1}{2}x_T(e^{y_1} + e^{y_2}) = 1$ one can integrate over either y_2 or y_1. Eliminating y_2 gives the distribution of the scattered pointlike particle:

$$\frac{d\sigma}{dy_1 d^2 q_T} = \frac{x_2 f(x_2)}{1 - \frac{1}{2}x_T e^{y_1}} \frac{1}{\pi} \frac{d\hat{\sigma}}{d\hat{t}} \tag{2.19}$$

where

$$x_2 = \frac{\frac{1}{2}x_T e^{-y_1}}{1 - \frac{1}{2}x_T e^{y_1}} \quad , \quad 1 - x_2 = \frac{1 - x_T \cosh y_1}{1 - \frac{1}{2}x_T e^{y_1}} \geq 0$$

$$\hat{s} = x_2 s, \quad \hat{t} = -s \frac{1}{2} x_T e^{-y_1}, \quad \hat{u} = -s \frac{\left(\frac{1}{2}x_T\right)^2}{1 - \frac{1}{2}x_T e^{y_1}} \quad . \tag{2.20}$$

Similarly, eliminating y_1 gives the distribution of the recoil jet:

$$\frac{d\sigma}{dy_2 d^2 q_T} = \frac{x_2 f(x_2)}{1 - \frac{1}{2}x_T e^{y_2}} \frac{1}{\pi} \frac{d\hat{\sigma}}{d\hat{t}} \tag{2.21}$$

where

$$x_2 = \frac{\frac{1}{2}x_T e^{-y_2}}{1 - \frac{1}{2}x_T e^{y_2}} \tag{2.22}$$

$$\hat{s} = x_2 s, \quad \hat{t} = -s \frac{\left(\frac{1}{2}x_T\right)^2}{1 - \frac{1}{2}x_T e^{y_2}}, \quad \hat{u} = -s \frac{1}{2}x_T e^{-y_2} \quad .$$

These formulas, of course, are equivalent to the standard naive parton model equations for $d\sigma/dQ^2 d\nu$, now expressed in CMS rapidity and q_T. In fact,

$$Q^2 = s\tfrac{1}{2}x_T e^{-y_1} = s\frac{(\tfrac{1}{2}x_T)^2}{1-\tfrac{1}{2}x_T e^{y_2}}$$

$$(2.23)$$

$$2m_e \nu/s = 1-\tfrac{1}{2}x_T e^{y_1} = \tfrac{1}{2}x_T e^{y_2}$$

satisfying $Q^2/2m_e\nu = x_2$ and $\partial(Q^2,\nu)/\partial(y_1,q_T^2) = \nu/(1-\tfrac{1}{2}x_T e^{y_1})$, $\partial(Q^2,\nu)/\partial(y_2,q_T^2) = \nu/(1-\tfrac{1}{2}x_T e^{y_2})$. Curves of constant y_1 and y_2 (or Θ_1, Θ_2) on Q^2,ν plane are shown in fig. 4.

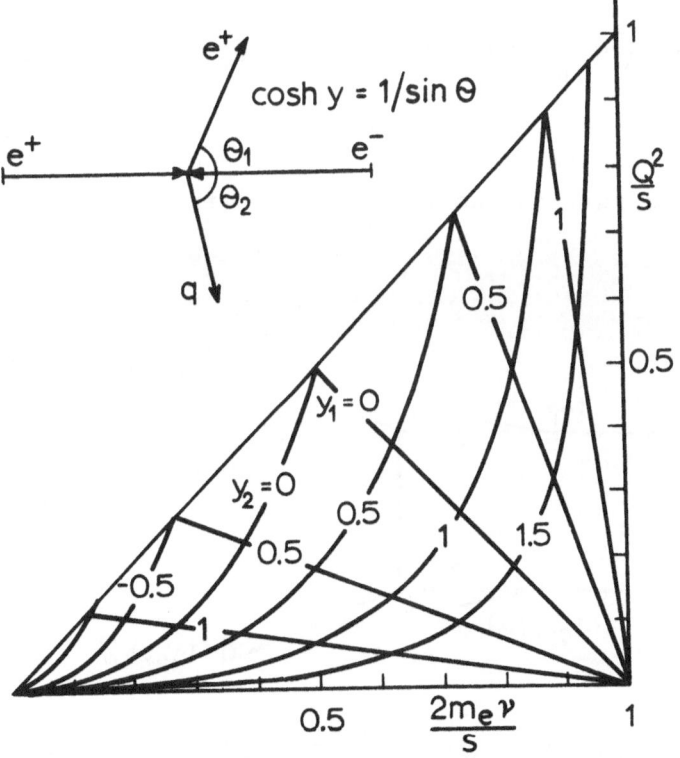

Fig.4. Relation between the standard deep inelastic scattering variables Q^2 and ν and the scattered electron rapidity y_1 and the recoil quark rapidity y_2 (note that $\cosh y = 1/\sin \Theta$).

3. TWO-JET EVENTS IN e^+e^- COLLISIONS PRODUCED
BY THE TWO-PHOTON MECHANISM

Present knowledge about QCD does not permit one to calculate the distribution function f(x) in (2.6) for a hadron. However, there is a simple and potentially important case in which f(x) is calculable: photon (electron) content of the electron (photon). For these one has, in the manner of the classic Weizsäcker-Williams approximation and in leading log approximation [10-11],

$$f_{\gamma/e}(x) = \frac{\alpha}{2\pi} \log \frac{s}{4m_e^2} \frac{1+(1-x)^2}{x}$$

$$f_{e/\gamma}(x) = \frac{\alpha}{2\pi} \log \frac{s}{4m_e^2} [x^2+(1-x)^2]$$

(3.1)

and eq.(2.6) can be used to evaluate the cross section for $e\bar{e} \rightarrow q\bar{q}X$ arising from the 2γ diagram in fig. 5.

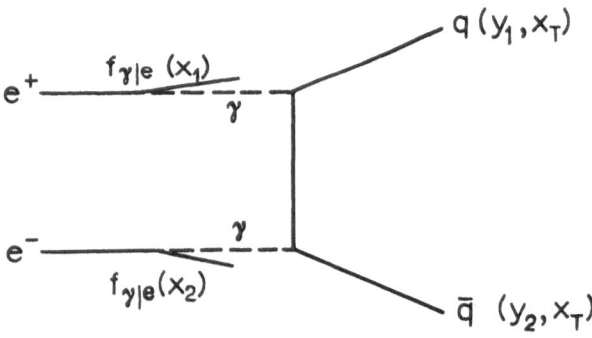

Fig. 5. Lowest-order diagram for $e^+e^- \rightarrow$
$\rightarrow \gamma\gamma X \rightarrow q\bar{q}X$.

One just needs

$$\frac{1}{e_q^4}\frac{d\hat{\sigma}^{\gamma\gamma\rightarrow q\bar{q}}}{d\hat{t}} = \frac{2\pi\alpha^2}{\hat{s}^2}\frac{\hat{t}^2+\hat{u}^2}{\hat{t}\hat{u}} = \frac{\pi\alpha^2}{q_T^4}\frac{\cosh Y}{(1+\cosh Y)^2} \qquad (3.2)$$

to obtain the final result [5-6] (a factor 2 comes from the fact that q and $\bar{q}$ jets are not distinguished between)

$$\frac{d\sigma^{e\bar{e}\rightarrow q\bar{q}X}}{dy_1 d^2q_T dy_2} = R_{\gamma\gamma}(\frac{\alpha}{2\pi}\log\frac{s}{4m_e^2})^2 \frac{\alpha^2}{q_T^4}\frac{2\cosh Y}{(1+\cosh Y)^2} \cdot$$

$$\cdot [1+(1-x_1)^2][1+(1-x_2)^2] \qquad (3.3)$$

where $Y = y_1 - y_2$ and

$$x_1 = \tfrac{1}{2}x_T(e^{y_1}+e^{y_2}), \quad x_2 = \tfrac{1}{2}x_T(e^{-y_1}+e^{-y_2}) , \qquad (3.4)$$

$$R_{\gamma\gamma} = 3\sum e_q^4 = \frac{34}{27} \text{ (for 4 flavours) .} \qquad (3.5)$$

Note that it is essential to have q_T^2 large in order that the exchanged quark line in fig. 5 be hard.

A detailed discussion of eq. (3.3) is given in ref. [6], containing consecutive integrations over y_2, y_1 and q_T. Numerical values of the single jet cross section at LEP energies are given in figs. 6-7.

Geometrically, the 2γ, 2 jet events described by (3.3) are (in general) non-back-to-back and take only a part of the total energy $(\hat{s}/s = \tfrac{1}{2}x_T^2(1+\cosh Y))$, in contrast

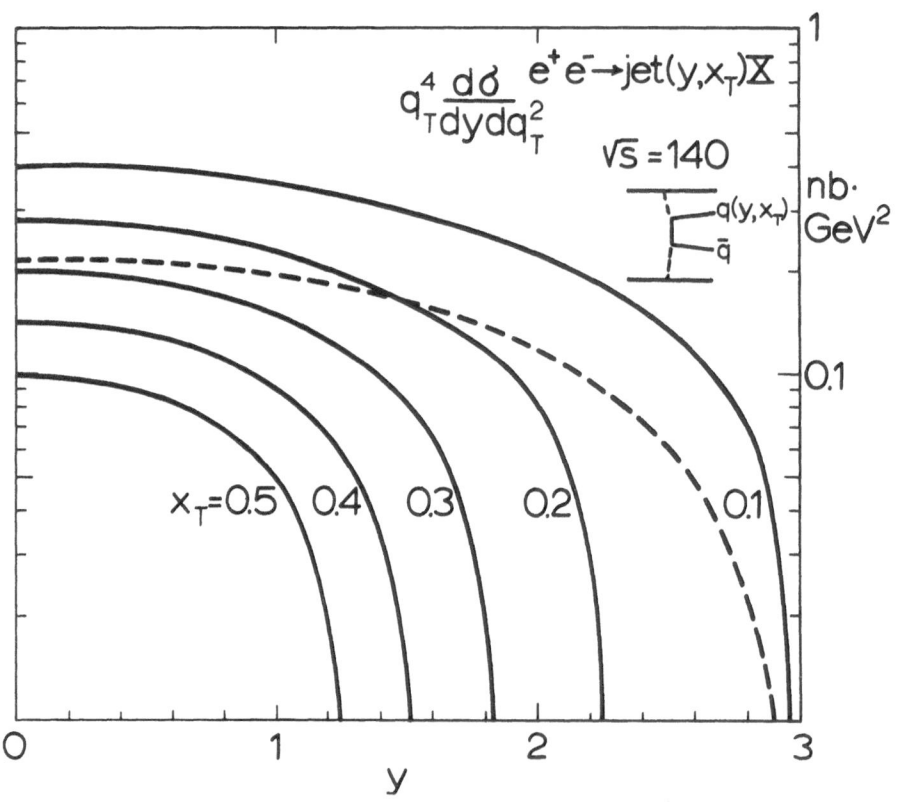

Fig. 6. Predicted rapidity-scaled transverse momentum
distribution for the single jet cross section
from $e^+e^- \to \gamma\gamma X \to q\bar{q} X$ at LEP energy. The
dashed curve gives the same quantity at $x_T=0.1$
for the recoil jet from deep inelastic electron
scattering (fig.3 and eq.(5.8)), symmetrised
by $y \to -y$ (scattered e^+ or e^- undetected).

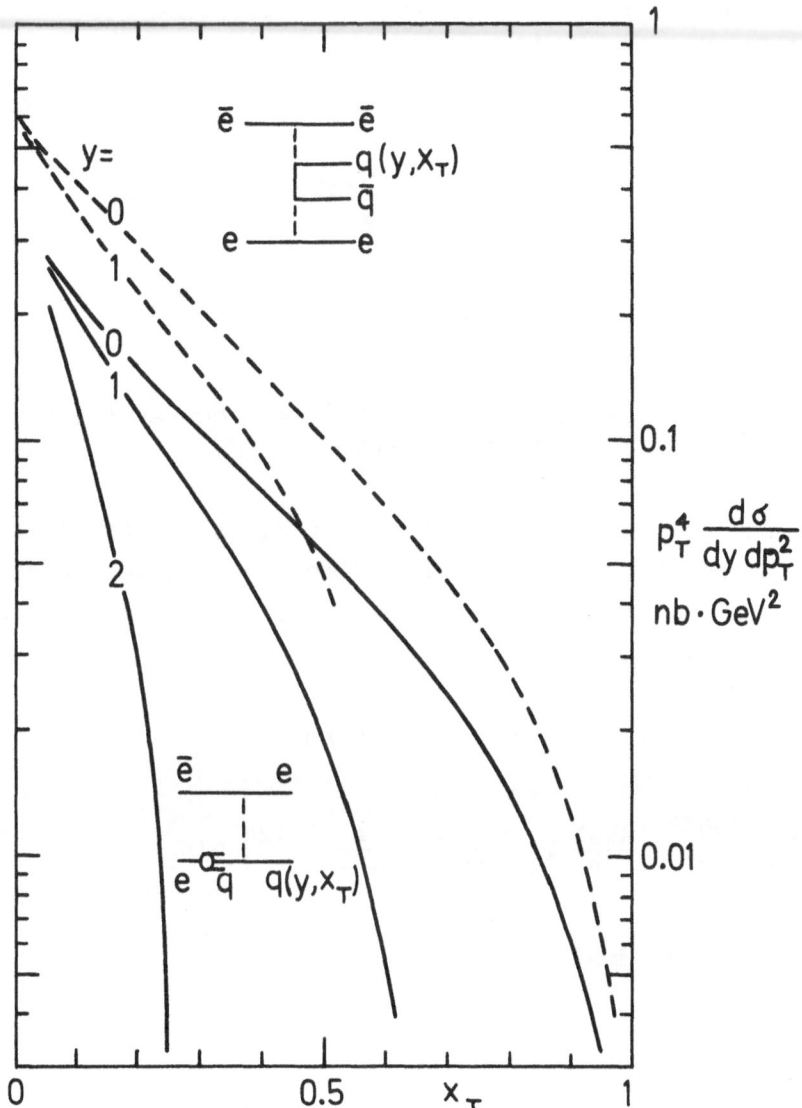

Fig.7. The same quantity as in fig. 6 plotted against
x_T at fixed y. The solid curves refer to deep
inelastic scattering, dashed curves are calculated
on the basis of eq.(3.3) integrated over y_2
(ref.[6]).

682

to the usual $1\gamma^*$, 2 jet events (fig.8).

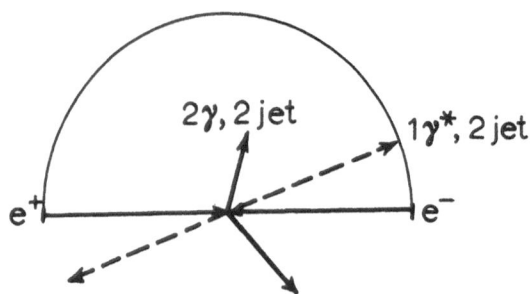

Fig.8.Appearance of $1\gamma^*$ jets compared with that of $\gamma\gamma$ jets.

However, their rate is larger if x_T is not close to 1. One thus has two competing effects:events get less interesting but their rate grows when $x_T \to 0$. The relation between 2γ and $1\gamma^*$ events is particularly clear if we integrate (3.3) over the two rapidities y_1 and y_2 (see refs.[5-6]) and write the result in the form

$$\frac{d\sigma^{e\bar{e}\to q\bar{q}X}}{dx_T} = \frac{4\pi\alpha^2}{3s} R_{\gamma\gamma} (\frac{\alpha}{2\pi} \log \frac{s}{4m_e^2})^2 \frac{128}{x_T^3} D(x_T) \tag{3.6}$$

where, for small x_T

$$D(x_T) = \log \frac{2}{x_T} - \frac{13}{12} + O(x_T) . \tag{3.7}$$

Thus the scale of the 2γ cross section is determined by the $1\gamma^*$ cross section $R_\gamma*4\pi\alpha^2/3s = R_\gamma*0.0044$ nb, which is firstly reduced by

$$(\frac{\alpha}{2\pi} \log \frac{s}{4m_e^2})^2 = (0.0275)^2 \tag{3.8}$$

(reflecting the rarity of photon emission by electrons) but secondly increased by the factor $128 \log(1/x_T)/x_T^3$ (reflecting the gain obtained by reducing the subprocess total energy $\hat{s}$). If $\sqrt{s}$ is so large that true jet events correspond to small values of x_T, the latter factor wins easily. At LEP energies $x_T = 0.1$ corresponds to $q_T = 7$ GeV which should give a clear jet signal. The latter factor is then $\sim 3 \cdot 10^5 \sim (0.0018)^{-2}$, winning clearly the photon emission factor (3.6). At PETRA energies so large an x_T is needed that the 2γ, 2 jet events are only marginally identifiable.

Summarising: one thus has an interesting and rather reliable (if one believes in perturbative QCD) prediction for LEP: non-back-to-back 2γ, 2 jet events will be copiously produced. The quantitative properties of these jets will form a good check on the quark propagator in fig. 5. From the experimental point of view it is a great simplification that there is no need to detect the forward and backward fast electrons: in fact, the above calculation assumes that they are undetected. Note also that the magnitude of the cross section depends on $R_{\gamma\gamma}$ and can be used to test the group theoretical structure of quarks.

However, one also has to keep a reservation in mind: the interesting $\gamma\gamma$ phenomena have a large cross section at the expense of lost energy and the truly novel phenomena probably will be observed via the $1\gamma^*$ diagram.

4. QCD CORRECTIONS: THE DISTRIBUTION OF QUARKS AND GLUONS IN A PHOTON AND AN ELECTRON

A well-known result about $1\gamma^* e^+ e^- \rightarrow$ hadrons is that

$$\sigma(e\bar{e} \rightarrow \gamma^* \rightarrow q\bar{q} \rightarrow hadrons) \equiv \frac{4\pi\alpha^2}{3s} \ 3\sum e_q^2 [1+\frac{4}{3} \ \frac{3\alpha_s(s)}{4\pi} +...] \quad (4.1)$$

where the second 4/3 is a color factor, $3\alpha_s/4\pi$ was cal-culated by Jost and Lüttinger in 1950 in QED (for the diagrams in fig.9a) and

$$\alpha_s(s) = \frac{1}{b \ \log\frac{s}{\Lambda^2}} \ , \quad b = \frac{33-2n_f}{12\pi} \ , \ \Lambda \approx 0.5 \ GeV \ . \quad (4.2)$$

One may now ask how the result (3.3), which corresponds to the first term in the brackets of (4.1), is modified by QCD corrections (fig. 9b). Note that in the $\gamma\gamma$ case one has to have s and q_T^2 large (x_T fixed) to avoid the infrared sensitive quark pole. The answer [4] is inter-esting: the correction term does not depend on $\alpha_s(s)$ but is a calculable scaling function of x_T and y. This result will here be motivated by deriving the quark and gluon content of a photon and an electron [12].

The calculation of the correction is based on the leading logarithm approximation [13]. The dominant con-tribution comes from emission of nearly parallel quanta, the last one participating in a hard collision (see fig.11c below). When summing over leading logs the following equations about consecutive longitudinal distributions are useful: if $f_{a/b}(x)$ and $f_{b/c}(x)$ are known, then

$$f_{a/c}(x) = \int_0^1 dy \ dz\delta(x-yz) f_{a/b}(y) f_{b/c}(z) \equiv f_{a/b}(x) \otimes f_{b/c}(x)$$

$$= \int_x^1 \frac{dy}{y} \ f_{a/b}(y) f_{b/c}(x/y) \ . \quad (4.3)$$

The δ function in this convolution is decoupled by using moments:

$$f_{a/c}^{(n)} = f_{a/b}^{(n)} \cdot f_{b/c}^{(n)} \tag{4.4}$$

$$f^{(n)} = \int_0^1 dx\, x^{n-1} f(x) \ . \tag{4.5}$$

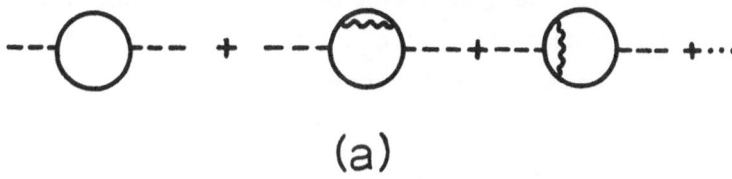

(a)

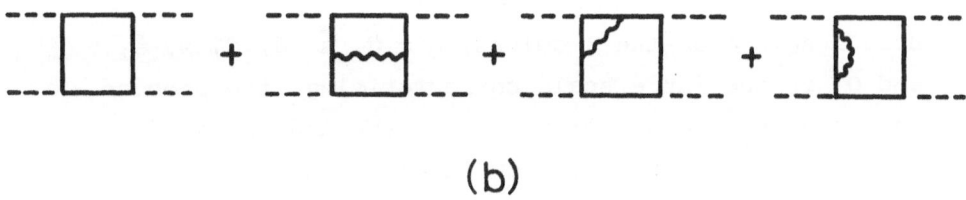

(b)

Fig.9. QCD corrections to (a) $1\gamma^*$
(b) $\gamma\gamma$ total cross sections.

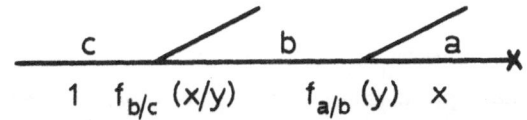

Fig.10. Consecutive longitudinal decays; the
cross refers to a hard process
initiated by constituent a.

Some simple convolutions needed in the following are

$$\frac{1}{x} \otimes 1 = \frac{1-x}{x} \qquad\qquad \frac{1}{x} \otimes \frac{1}{x} = \frac{1}{x} \log \frac{1}{x}$$

$$\frac{1+(1-x)^2}{x} \otimes 1 = \frac{1}{x}[2-2x \log\frac{1}{x} -x-x^2] \; , \; [] \xrightarrow[x\to 1]{} 1-x \qquad (4.6)$$

$$\frac{1+(1-x)^2}{x} \otimes \frac{1}{x} = \frac{1}{x} [2\log\frac{1}{x} - \frac{3}{2} + 2x-\frac{1}{2}x^2], \; [] \xrightarrow[x\to 1]{} 1-x \; .$$

Consider first the quark content of a photon. The photon couples to quarks through e_q and, to lowest order in α_s, one obtains from eq.(3.1) that

$$f_{q/\gamma}^{Born}(x) = 3 \frac{\alpha}{2\pi} e_q^2 \log \frac{Q^2}{4m_q^2} [x^2 + (1-x)^2] \; . \qquad (4.7)$$

Here 3 sums over the 3 colors, $q = u, \bar{u}, d, \bar{d}, s, \bar{s},\ldots$ and Q^2 is the large scale characterising the process initiated by the q. To appreciate the meaning of (4.7) it is useful to remember how it arises in QED [10-11]. The rate of producing large angle electrons in $\gamma e^- \to$ $\to e^+ e^- e^-$ is obtained by convoluting $f_{e/\gamma}(x)$ with the large angle cross section for $e^+ e^- \to e^+ e^-$ (fig. 11b):

$$\sigma_{\gamma e^-}(E) = \int_0^1 dx \; f_{e/\gamma}(x) \sigma_{e^+ e^-}(xE) \; .$$

Now, of course, the m_q in (4.7) is to be regarded as a phenomenological parameter or a field theoretic sub-traction point.

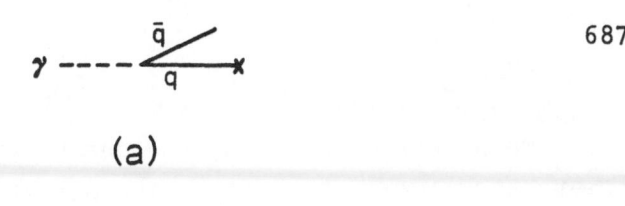

(a)

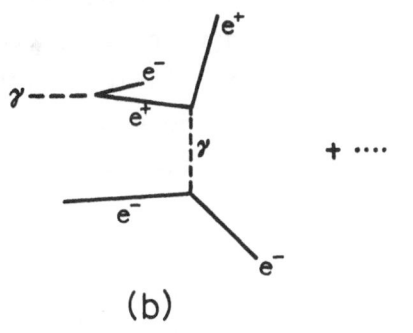

(b)

$$\sum \text{----} \diagup\!\!\!\diagdown\!\!\diagdown\!\!\diagup\!\!\diagdown\!\!\diagdown\!\!\diagup\!\!\!\text{×}$$

$$= \text{----}\diagup \quad p^2 \quad \text{(}\!\text{IIII}\!\text{)} \quad Q^2 \quad \text{×}$$

$$\frac{3\alpha}{2\pi}e_q^2[x^2+(1-x)^2] \quad f_{q'/q}(x,Q^2,p^2)$$

(c)

$$\frac{\mu^2 \;\text{(}\!\!\text{O}\!\!\text{)}\; Q^2}{q_n(Q^2,\mu^2)}\!\times = \frac{\mu^2}{q_n(\mu^2,\mu^2)}\!\times + \frac{\mu^2\;\text{(}\!\!\text{O}\!\!\text{)}\;p^2 \quad Q^2}{q_n(p^2,\mu^2)\; A_n^{qq}}\!\times$$

$$\alpha_s(p^2)$$

(d)

Fig.11. (a) γ transforms to a q which initiates a hard
process (cross)
(b) example of a hard process initiated by an e^+
coming from a γ
(c) sum over all possible consecutive decays before
the hard process
(d) interpretation of the integral equation (4.8).

To higher order in α_s the quark can emit further quarks and gluons (fig. 11c). The effect of emissions is, in leading log approximation [13], summed by an integral equation (fig. 11d) [14-16] which reads (written, for simplicity in the valence quark approximation, i.e., neglecting the flavor changing $q \to q \to q'$ transitions)

$$q_n(Q^2,\mu^2) = q_n(\mu^2,\mu^2) + A_n^{qq} \int_{\mu^2}^{Q^2} \frac{dp^2}{p^2} \frac{\alpha_s(p^2)}{2\pi} q_n(p^2,\mu^2) \qquad (4.8)$$

where $q_n(Q^2,\mu^2)$ is the nth moment of the quark distribution $q(x,Q^2,\mu^2)$ ($\equiv f_{q/q}(x,Q^2,\mu^2)$) renormalised so that its value at $q^2 = \mu^2$, $q(x,\mu^2,\mu^2)$ ($= \delta(1-x)$, for example) is known and $A_n^{qq} = \frac{4}{3}(-2\sum_j^n \frac{1}{j} + \frac{1}{n(n+1)} - \frac{1}{2})$. Eq. (4.8) contains an integration over the mass of the intermediate quark and a convolution over longitudinal momenta (cf.(4.3-5)). Eq.(4.8) basically formulates the renormalisation condition stating that one obtains the same result with (the last term) or without (LHS) gluon emission. Its solution is well known:

$$q_n(Q^2,\mu^2) = q_n(\mu^2,\mu^2) \left[\frac{\alpha_s(Q^2)}{\alpha_s(\mu^2)} \right]^{-\frac{A_n^{qq}}{2\pi b}} . \qquad (4.9)$$

Using (4.9) it is now simple to evaluate the quark content of a photon (fig. 11c):

$$f_{q/\gamma}^n(Q^2) = \frac{3\alpha}{2\pi} e_q^2 a_n \int_{\mu^2}^{Q^2} \frac{dp^2}{p^2} q_n(Q^2,p^2) \qquad (4.10)$$

where

$$a_n = \int_0^1 dx \, x^{n-1}[x^2+(1-x)^2] = \frac{1}{n} - \frac{2}{n+1} + \frac{2}{n+2} \qquad (4.11)$$

The simplicity of the result is due to the fact that
the dependence of q_n on Q^2 and p^2 is known and the
integration over p^2 is of the form

$$Q^2 \int_{\mu^2} \frac{dp^2}{p^2} [\frac{\alpha_s(Q^2)}{\alpha_s(p^2)}]^d = \frac{\log \frac{Q^2}{\Lambda^2}}{1+d} \qquad . \tag{4.12}$$

In the general case eq.(4.8) is replaced by a 2 x 2
matrix equation for q_n and g_n (gluon moments) and the
solution for $f^n_{q'/q}(Q^2,\mu^2)$ contains 3 terms of the type
(4.9) corresponding to A^{qq}_n and the two eigenvalues of
the anomalous moment matrix. Inserting to (4.10) and
using (4.12) one obtains the general solution

$$f^n_{q/\gamma}(Q^2) = \frac{3\alpha}{2\pi}\log\frac{Q^2}{\Lambda^2} \; a_n[\frac{e^2_q - <e^2_q>}{1+d^{qq}_n} + <e^2_q> \frac{1+d^{gg}_n}{(1+d^+_n)(1+d^-_n)}] \tag{4.13}$$

$$\approx \frac{3\alpha}{2\pi}\log\frac{Q^2}{\Lambda^2} \; e^2_q \frac{a_n}{1+d^{qq}_n} \quad \text{(valence approximation)} \tag{4.14}$$

where $<e^2_q> = 5/18$ for 4 quarks and $d^{qq}_n = -A^{qq}_n/2\pi b$, d^{gg}_n
and $d^\pm_n$ are standard quantities given in many places (see,
for instance[17] eqs.(4)-(5)). Similarly,

$$f^n_{g/\gamma}(Q^2) = \frac{3\alpha}{2\pi} \log \frac{Q^2}{\Lambda^2}<e^2_q> \frac{a_n d^{gg}_n(-2n_f)}{(1+d^+_n)(1+d^-_n)} \qquad . \tag{4.15}$$

The functions $xf_{q/\gamma}(x)$ and $xf_{g/\gamma}(x)$ obtained by numerically inverting [18] eq. (4.5) with (4.13-15) are shown
in figs. 12-13.

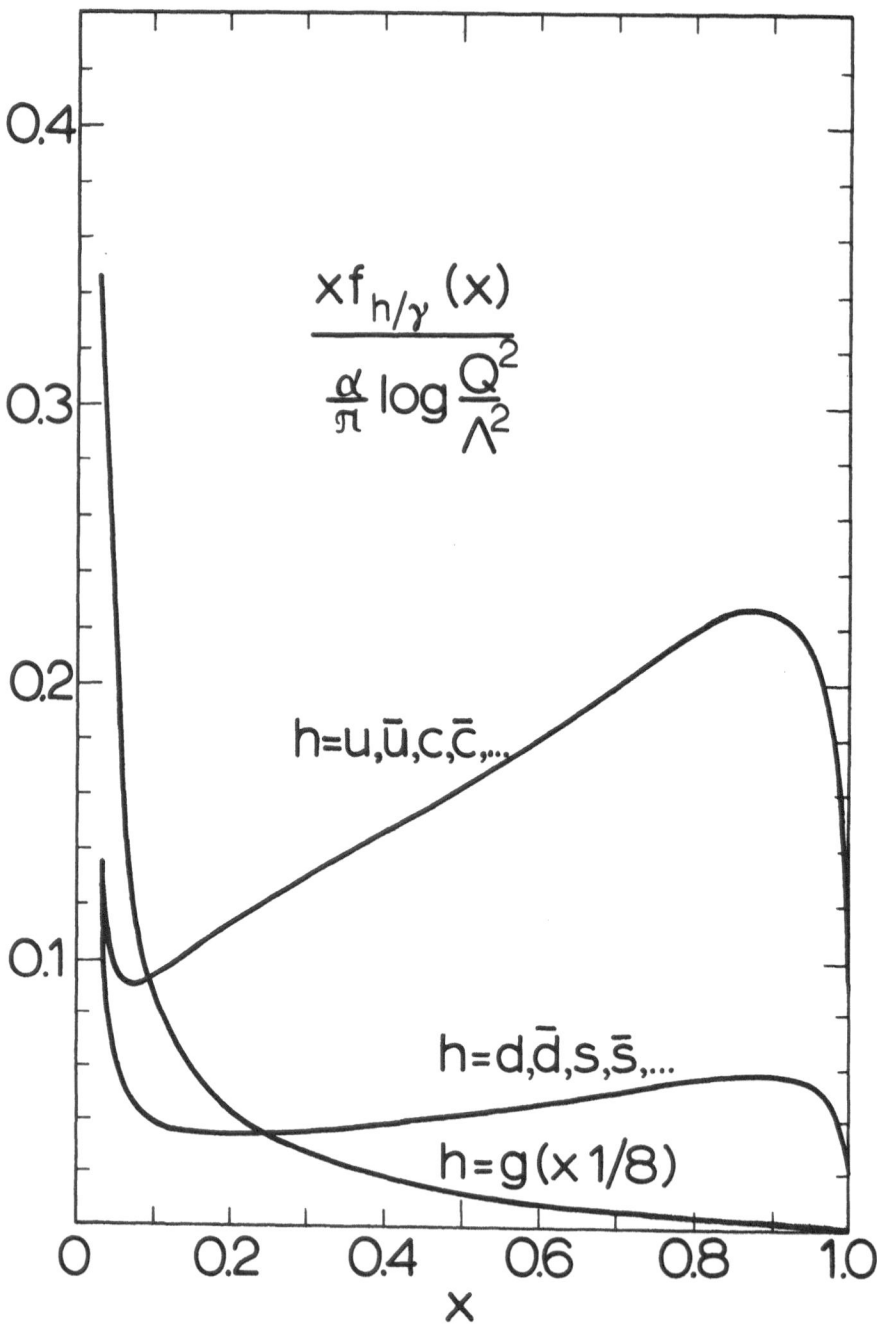

$$\frac{x f_{h/\gamma}(x)}{\frac{\alpha}{\pi} \log \frac{Q^2}{\Lambda^2}}$$

$h = u, \bar{u}, c, \bar{c}, ...$

$h = d, \bar{d}, s, \bar{s}, ...$

$h = g(\times 1/8)$

x

Fig. 12. The distribution function of quarks and gluons
calculated on the basis of eq. (4.13).

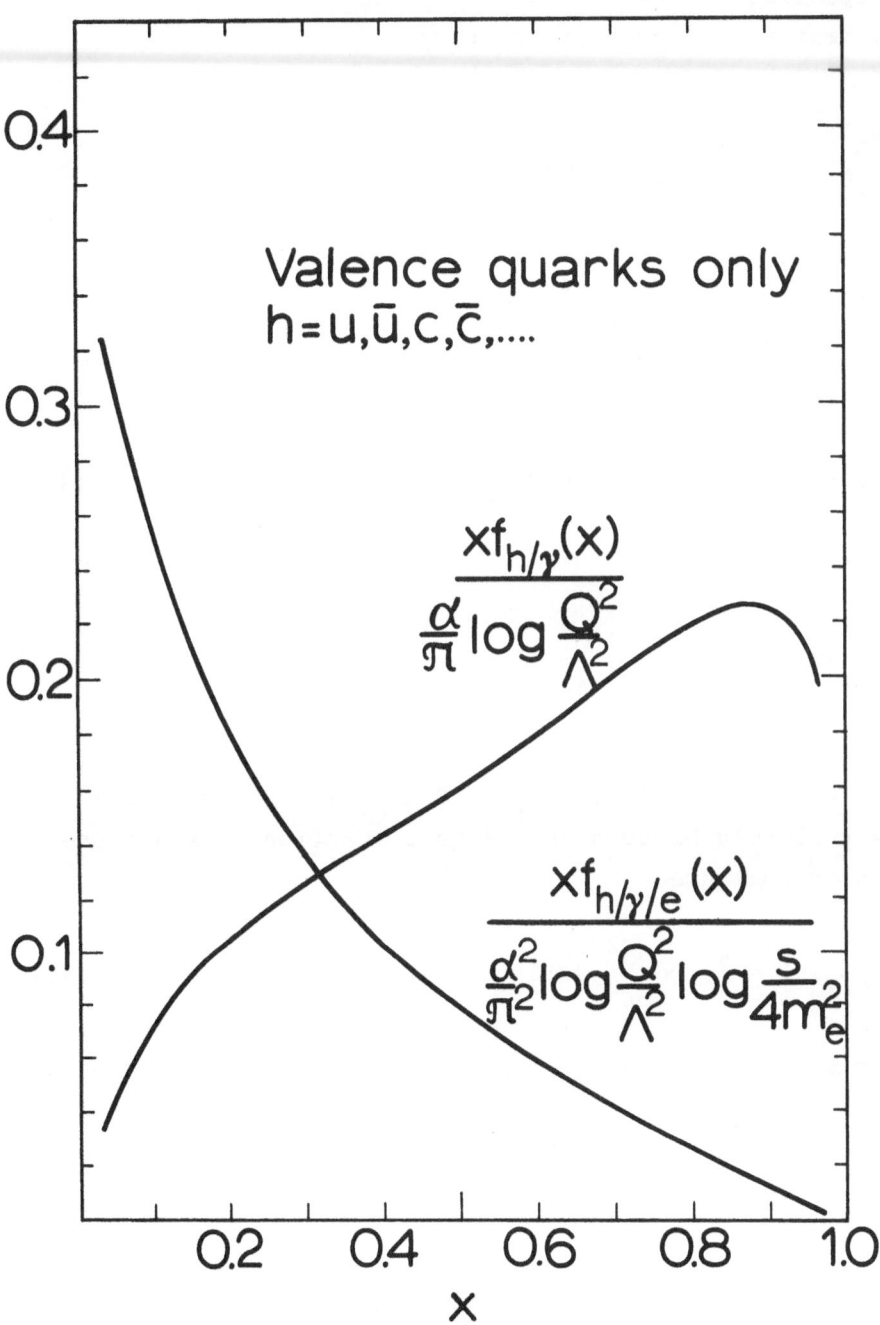

Fig. 13. The distribution function of $e_q = \frac{2}{3}$ quarks in a photon and an electron calculated on the basis of valence quark approximation (eqs.(4.14) and (4.22)).

Properties of the result:

1. Near $x \to 1$ one obtains using

$$\frac{a_n}{1+d_n^{qq}} \xrightarrow[n\to\infty]{} \frac{1}{n} \frac{1}{A \log n - B + 1} \tag{4.16}$$

$$A = \frac{16}{33-2n_f} \qquad B = \frac{4(3-4\gamma^{Euler})}{33-2n_f}$$

and the relation

$$f(1 - \tfrac{1}{n}) = (n-1)f_{n-1} - O(\tfrac{1}{n^2}) \tag{4.17}$$

that

$$f_{q/\gamma}(x) \xrightarrow[x\to1]{} \frac{3\alpha}{2\pi} \log \frac{Q^2}{\Lambda^2} e_q^2 \frac{1}{1-B+A \log\frac{1}{1-x}} \quad . \tag{4.18}$$

2. In the range $0.05 \lesssim x \lesssim 0.9$ $xf_{q/\gamma}(x)$ for $e_q = 2/3$ very closely behaves as a linear function of x and one can approximate

$$xf_{q/\gamma}(x) = \frac{\alpha}{\pi} \log \frac{Q^2}{\Lambda^2} 0.16 \ (x + \tfrac{1}{2}), \ e_q = \tfrac{2}{3} \quad . \tag{4.19}$$

For x near 1 the $e_q = -1/3$ contribution is smaller by a factor 4; for small x the sea gets more important and the ratio of $e_q = 2/3$ to $e_q = -1/3$ goes to 1. For the gluon content a similar analytic approximation is

$$xf_{g/\gamma}(x) = \frac{\alpha}{\pi} \log \frac{Q^2}{\Lambda^2} 0.088 \ \frac{1-x}{x} \quad . \tag{4.20}$$

The approximations (4.19-20) are useful for rough estimates.

3. For $x \to 0$ the calculated $f_{q/\gamma}(x)$ grows like an inverse power of x. In fact, the leading behaviour for $x \to 0$ is determined by the rightmost singularity of the moments in the complex n plane. For the moments in eqs. (4.13) and (4.15) a closer inspection shows that the leading singularity is a pole caused by the vanishing of $1 + d_n^+$, which takes place at $n \approx 1.60$ (remember that $d_n^+ \to -36/25/(n-1)$ when $n \to 1$ and that $d_2^+ = 0$). Evaluating the residue at this point one obtains for $x \to 0$

$$\sum_{q=1}^{2n_f} xf_{q/\gamma}(x) \to \frac{\alpha}{\pi} \log Q^2 \ 2n_f \cdot 0.0072 \ x^{-0.6}$$

$$xf_{g/\gamma}(x) \to \frac{\alpha}{\pi} \log Q^2 \ 2n_f \cdot 0.0210 \ x^{-0.6} \ .$$

Comparing with the numerically inverted value in figs. 12-13 shows that the method of ref.[19] overestimates the correct result. Note also that the convergence to the above $x \to 0$ limits is non-uniform in x.

4. A good check on the normalisations of eqs.(4.13) and (4.15) is obtained by evaluating the momentum carried by the quarks and gluons separately. This is given by the $n = 2$ moment. In the Born approximation all the momentum is carried by the quarks:

$$\sum_{1}^{2n_f} {}^f \langle x \rangle_q^{Born} = \frac{\alpha}{\pi} \log Q^2 \ 2n_f \cdot \frac{5}{36}$$

while inclusion of QCD correction gives

$$\sum_{1}^{2n_f} <x>_q = \frac{\alpha}{\pi} \log Q^2 \ 2n_f \cdot \frac{5}{36} \cdot \frac{99}{131}$$

$$<x>_g = \frac{\alpha}{\pi} \log Q^2 \ 2n_f \cdot \frac{5}{36} \cdot \frac{32}{131} \ .$$

Momentum is thus shared among quarks and gluons with the ratio 99/32 (independent of Q^2 and n_f). For a hadron at $Q^2 \to \infty$ the corresponding ratio is considerably smaller, 3/4. The reason for the difference is, of course, precisely the very hard valence quark component in a photon.

5. In the region $x \to 0$ one should also include the in-calculable vector meson contribution. Physically, this enters since in the calculation of the $\gamma \to q\bar{q}$ transition above (eq.(4.7) and fig.11a) no interaction between the emitted q and $\bar{q}$ has been included. These terms do not get a $\log Q^2$ from the angular integration and they are thus negligible in leading log approximation. However, one knows that physically there is a case of strong final state interaction: that leading to vector meson formation. It will not contain a $\log Q^2$ but may still be large.

The vector meson contribution may be estimated using vector dominance. One associates the coupling constant e/f_V ($\alpha = e^2/4\pi$) with the γV transition and calculates f_V from $\Gamma(V \to e\bar{e}) = 4\pi\alpha^2 m_V/3f_V^2$. Then, using experimental values

$$f_{q/V/\gamma}(x) = \sum_V \frac{3\Gamma(V \to e\bar{e})}{\alpha m_V} f_{q/V}(x)$$

$$\approx \frac{\alpha}{\pi}(1.52f_{q/\rho^0} + 0.17f_{q/\omega} + 0.22f_{q/\varphi} + 0.27f_{q/\psi}+...).$$

It is obvious that only ρ^0 need be included for a rough
estimate. To estimate f_{q/ρ^0} one may use one's experience
with $f_{q/p}$ and $f_{q/\pi}$. However, one is now faced with the
difficulty that the QCD and the vector dominance contributions need in no way be additive. It is quite
possible that the QCD calculation contains a sizable
fraction of the valence quark (u, $\bar{u}$, d, $\bar{d}$) contribution
to f_{q/ρ^0}. Let us then estimate

$$f_{q/\rho^0}(x) = 2<x>_q \frac{1-x}{x} \qquad q = u,\ \bar{u},\ d,\ \bar{d},....$$

where $<x>_q \leq 0.25$, the upper limit corresponding to
glue taking 50% and u, $\bar{u}$ 25% each of the total momentum.
Then

$$xf_{q/V/\gamma}(x) \approx \frac{\alpha}{\pi} 3<x>_q (1-x) \tag{4.21}$$

which has to be compared with (4.19) (with log $Q^2/\Lambda^2 = 4-5$).
Due to the double counting problem it is hard to make a
definite statement, but at small x (x $\lesssim$ 0.1) the vector
meson contribution is of the same order as the QCD part.
The most favourable situation would be if the QCD part
effectively simulated the vector meson part at all
reasonable Q^2.

To conclude this section, the quark and gluon
content of an electron is now calculable through the
convolution

$$f_{q/\gamma/e}(x) = f_{q/\gamma}(x) \otimes f_{\gamma/e}(x) \tag{4.22}$$

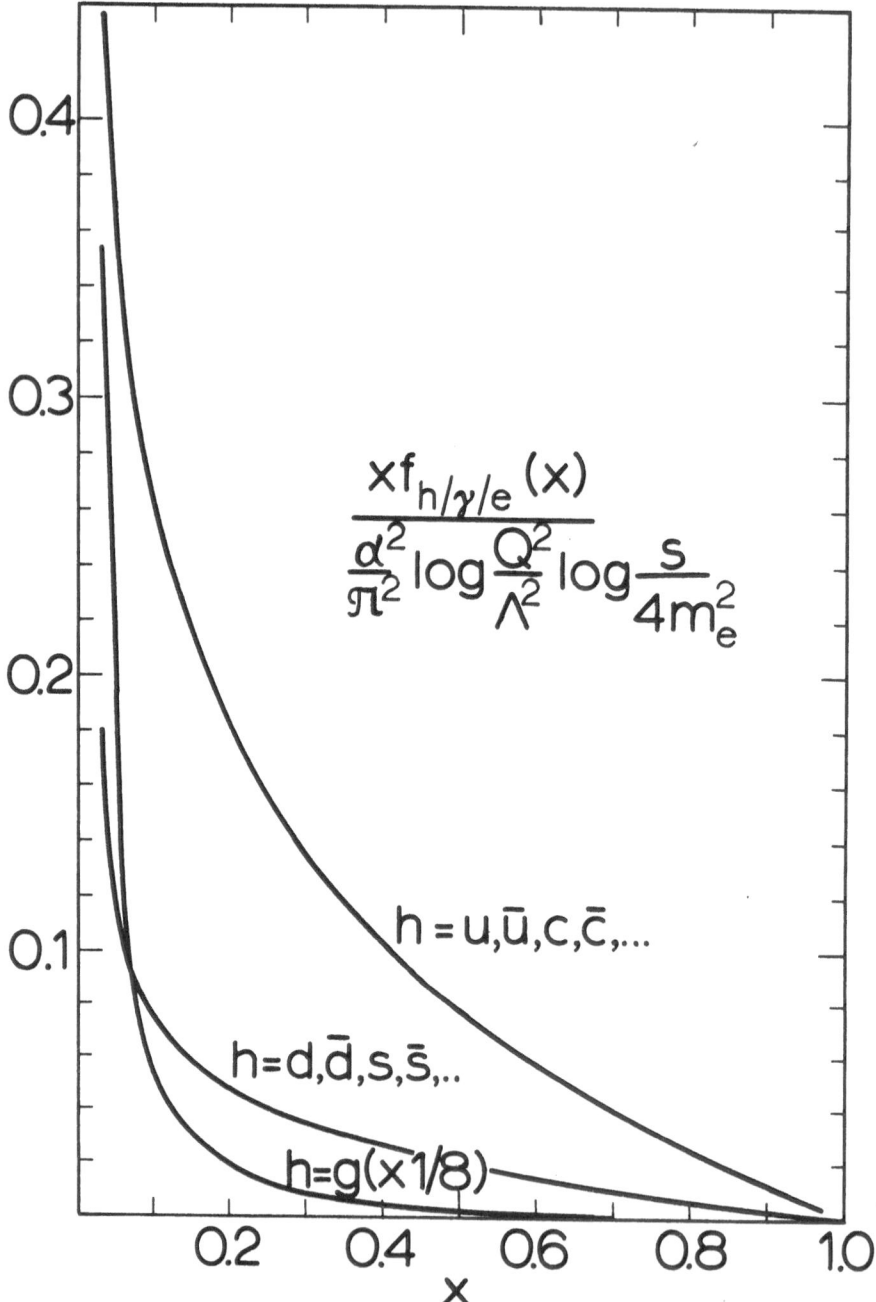

$$\frac{x f_{h/\gamma/e}(x)}{\frac{\alpha^2}{\pi^2} \log \frac{Q^2}{\Lambda^2} \log \frac{s}{4m_e^2}}$$

$h = u, \bar{u}, c, \bar{c}, ...$

$h = d, \bar{d}, s, \bar{s}, ..$

$h = g (\times 1/8)$

Fig. 14. The distribution function of quarks and gluons in an electron calculated on the basis of eqs. (4.13) and (4.22).

The result calculated by multiplying moments and inverting is shown in fig. 14. Using the approximation
(4.19) and the convolutions (4.6) one obtains the
analytic approximation

$$xf_{q/\gamma/e}(x) \stackrel{\sim}{=} \frac{\alpha^2}{\pi^2} \log \frac{Q^2}{\Lambda^2} \log \frac{s}{4m_e^2} Q(x) \qquad e_q = \frac{2}{3} \qquad (4.23)$$

$$Q(x) = 0.08[(1-2x) \log \frac{1}{x} + \frac{5}{4}(1-x^2)] \qquad [] \xrightarrow[x \to 1]{} \frac{3}{2}(1-x) .$$

For the gluons the approximate form is

$$xf_{g/\gamma/e}(x) = \frac{\alpha^2}{\pi^2} \log \frac{Q^2}{\Lambda^2} \log \frac{s}{4m_e^2} 0.048 \, G(x) \qquad (4.24)$$

$$G(x) = \frac{4}{3x} - 2 \log \frac{1}{x} - \frac{1}{2} -x + \frac{1}{6} x^2 \xrightarrow[x \to 1]{} \frac{1}{2}(1-x)^2 .$$

When applying these formulas in quantitative calculations
it will also be assumed that for $e_q = -1/3$ the distribution is 1/4 of that in (4.23). This is an underestimate
(cf. fig. 12).

5. MEASUREMENTS TESTING THE QCD PREDICTION FOR QUARK
AND GLUON DISTRIBUTION IN A PHOTON AND AN ELECTRON

A. $e^+e^- \to$ (2 large q_T jets) + (1 or 2 small q_T jets)

With the structure functions $f_{q/e}(x,Q^2)$ and
$f_{g/e}(x,Q^2)$ derived in the previous section and the hard
scattering formula (2.6) one can now calculate the cross

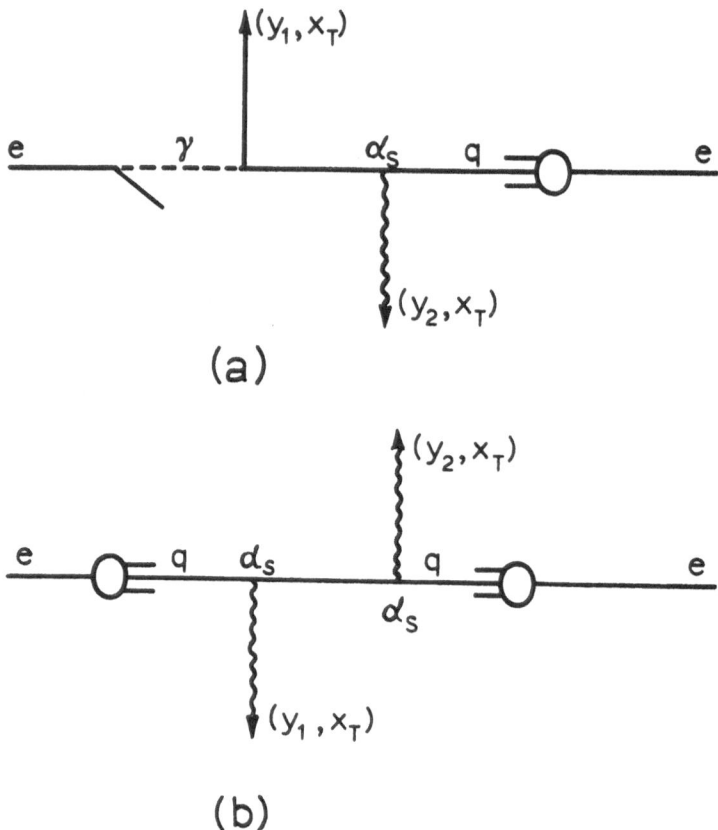

Fig.15. (a) a 3 jet diagram
(b) a 4 jet diagram

sections for the processes shown in fig. 15. Here the
materialization of the electron to a quark produces a
number of nearly parallel quarks and gluons (fig.11c)
which are seen as a small q_T hadron jet.

Consider the process in fig. 15a. We have

$$\frac{d\sigma^{\gamma q \to gq}}{dy_1 d^2q_T dy_2} = \sum_q x_1 x_2 f_{\gamma/e}^{(x_1)} f_{q/e}^{(x_2, q_T^2)} \frac{1}{\pi} \frac{d\hat{\sigma}^{\gamma q \to qg}}{d\hat{t}} \qquad (5.1)$$

where

$$\frac{d\hat{\sigma}^{\gamma q \to qg}}{d\hat{t}} = \frac{4}{3} \frac{2\pi e_q^2 \alpha \alpha_s}{\hat{s}^2} (\frac{\hat{s}}{-\hat{t}} + \frac{-\hat{t}}{\hat{s}}) . \qquad (5.2)$$

One observes that the product of the α_s in $d\hat{\sigma}/d\hat{t}$ and the
log Q^2 in $f_{q/e}$ (since the mass of the exchanged quark in
fig. 15 is $\hat{t} \approx -q_T^2$ (eq.2.8) the large scale is most
naturally $Q = q_T$) is just a constant $= 1/b$. The same
clearly holds for the 4 jet events in fig. 15b. This
is the result [4] referred to in the previous section:
QCD corrections to leading log accuracy to the lowest
order diagram (fig.5) are finite and scaling.

Inserting to eqs. (5.1-2) the structure functions
(eqs.(3.1) and (4.23))and invariants (eq.(2.8))one
obtains

$$\frac{d\sigma^{\gamma q \to gq}}{dy_1 d^2q_T dy_2} = R_{\gamma\gamma} (\frac{\alpha}{2\pi} \log \frac{s}{4m_e^2})^2 \frac{\alpha^2}{q_T^4} [1+(1-x_1)^2]\frac{2}{\pi b} Q(x_2) .$$

$$\cdot \frac{1}{(1+\cosh Y)^2}[1+e^Y + \frac{1}{1+e^Y}] \tag{5.3}$$

where $x_1 = \frac{1}{2}x_T(e^{Y_1} + e^{Y_2})$, $x_2 = \frac{1}{2}x_T(e^{-Y_1} + e^{-Y_2})$, $Y = y_1 - y_2$. This is to be compared with the two-jet cross section (3.3). The ratio is

$$\frac{d\sigma^{\gamma q \to gq}}{d\sigma^{\gamma\gamma \to q\bar{q}}} = \frac{Q(x_2)}{1+(1-x_2)^2} \frac{2}{\pi b} R(Y) \tag{5.4}$$

where $Q(x_2)$ can be directly read from fig.14, $R(Y)$ is the ratio of the Y dependent parts of (3.3) and (5.3), and $R(0) = 5/4$.

To get the complete 3 jet cross section one has to add to the process in fig.15a the same interchanging the electron and positron (this amounts to symmetrizing (5.3) with respect to $y \to -y$) and the two 3 jet processes caused by the electron turning to a gluon. The $\gamma g \to q\bar{q}$ processes can be neglected at large x_T, since $f_{g/\gamma}(x)$ decreases faster by a factor $1-x$ than $f_{q/\gamma}(x)$ when $x \to 1$. However, at $x_T \lesssim 0.1$ they are as important as the $\gamma q \to gq$ processes. Let us estimate the ratio of the latter 3 jet processes to 2 jet processes at $y_1 = y_2 = 0$ (both large q_T jets produced at $90°$) so that $x_2 = x_T$. Then

$$\frac{d\sigma^{3jet}}{d\sigma^{2jet}}\Bigg|_{y_1=y_2=0} = \frac{12}{5} \frac{Q(x_T)}{1+(1-x_T)^2} \approx \frac{1}{\pi}(1-x_T) \tag{5.5}$$

where $Q(x_T)$ (eq.(4.23)) can be read from fig.14.

Similarly $d\sigma^{4jet}/d\sigma^{2jet} \sim \frac{1}{\pi^2} (1-x_T)^2$. For detailed
numerical curves at $y = 0$, see ref.[5]. Note, however,
that while the 2 jet cross section peaks at $y = 0$, the
3 jet and 4 jet cross sections may have a peak at
larger y, due to the fast increase of $Q(x)$ and $G(x)$
at small x. A comprehensive discussion of all 3 and
4 jet processes can be found in ref.[20].

B. Deep inelastic scattering on an electron: $e^+e^- \to$
   $\to$ (1 large q_T jet) + (1 small q_T jet).

The standard and most direct way of testing the
QCD predictions for $f_{q/e}(x)$, the quark distribution in
an electron, would, of course, be a measurement of the
deep inelastic reaction $e^+e^- \to e^+X$ (fig.3) by only
detecting the final e^+. Normally these measurements are
carried out in the target rest frame; now we are automatically in the overall CMS and the kinematics is
slightly different. This has already been discussed
in section 2 and the formulas given there can be directly
applied to give quantitative predictions for deep
inelastic scattering experiments at LEP energies. A
direct measurement of the photon structure function
$f_{q/\gamma}(x)$ would necessitate the detection of the final
state e^-, too.

To use the formulas (2.19-22) one again first
needs the hard scattering (eq$\to$eq) cross section:

$$\frac{1}{e_q^2} \frac{d\hat{\sigma}^{eq \to eq}}{d\hat{t}} = \frac{4\pi\alpha^2}{\hat{t}^2} \frac{\hat{s}^2+\hat{u}^2}{2\hat{s}^2} = \frac{2\pi\alpha^2}{q_T^4} (\frac{1}{2}x_T e^{y_1})^2 [1+(\frac{1}{2}x_T e^{y_1})^2] \quad (5.6)$$

$$= \frac{2\pi\alpha^2}{q_T^4} (1-\frac{1}{2}x_T e^{y_2})^2 [1+(1-\frac{1}{2}x_T e^{y_2})^2] \quad (5.7)$$

where y_1 and y_2 are the electron and recoil jet rapidities (fig.3). Inserting to (2.19-22) one directly obtains for $\bar{e}e \to \bar{e}X$ (i=1) or $\bar{e}e \to \bar{q}X$ (i=2):

$$\frac{d\sigma}{dy_i d^2 q_T} = R_{\gamma\gamma} \left(\frac{\alpha}{2\pi}\log\frac{s}{4m_e^2}\right)^2 \frac{\alpha^2}{q_T^4} \frac{\log(q_T^2/\Lambda^2)}{\log(s/4m_e^2)} J_i(y_i,x_T) \qquad (5.8)$$

where

$$J_1(y_1,x_T) = Q(x_2) \frac{3x_T^2 e^{2y_1}(1+\frac{1}{4}x_T^2 e^{2y_1})}{1-\frac{1}{2}x_T e^{y_1}} \qquad (5.9)$$

$$J_2(y_2,x_T) = Q(x_2)\ 12(1-\frac{1}{2}x_T e^{y_2}) \cdot$$

$$\cdot [1+(1-\frac{1}{2}x_T e^{y_2})^2]^2 \qquad (5.10)$$

$$x_2 = \frac{\frac{1}{2}x_T e^{-y_i}}{1-\frac{1}{2}x_T e^{y_i}} \qquad i = 1,2\ . \qquad (5.11)$$

Numerical values of (5.8) are given in figs. 16 and 6. Some properties of the result are as follows:

1. The electron tends to go forward and essentially all the cross section is near $y_1 = y_1^{max}$ (cosh $y_1^{max} = 1/x_T$). Physically this is due to the fact that it is in the electron forward direction that $-\hat{t}$ is as small as possible (of the order of q_T^2); for $y_1 \approx 0, -\hat{t} \approx q_T\sqrt{s}$. The recoil quark jet, on the other hand has a broad angular distribution. To measure the electron structure

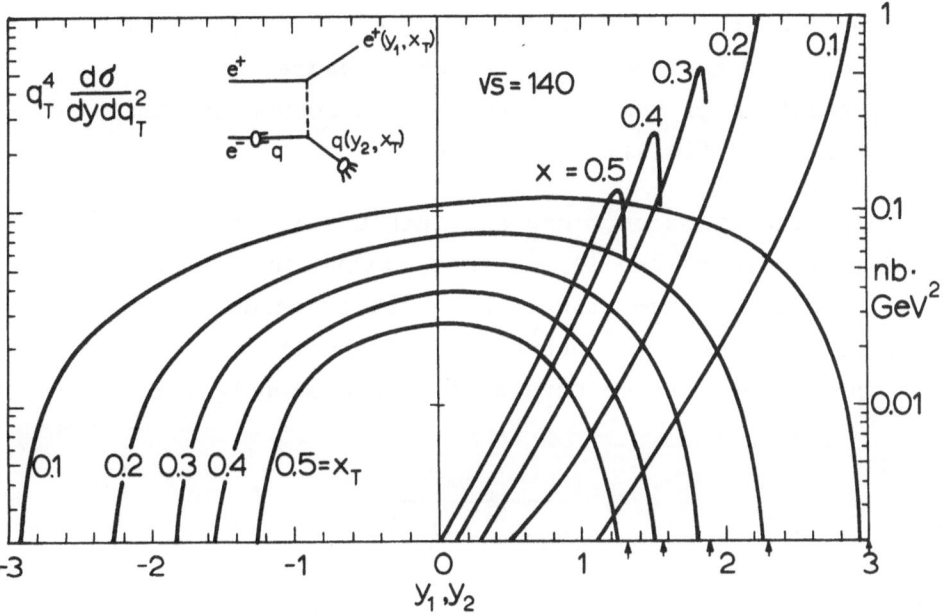

Fig.16. The rapidity - x_T - distribution of scattered electrons (rapidly rising curves at right) and produced jets in deep inelastic scattering on an electron at LEP energy ($\sqrt{s}$ = 140) calculated from eqs.(5.8-11). The arrows denote values of y_{max} corresponding to the value of x_T on the nearest curve.

function the most straightforward procedure is to observe the scattered e^+; this is difficult near the
forward direction. Equivalent information can be
collected by observing the scattered quark; this is
difficult because of soft effects in the transition
q → hadrons.

2. According to fig. 6 the magnitude of the single jet
production cross section is not much less than that
coming from the two-photon process $e\bar{e}$ → $q\bar{q}X$ (fig. 5).
However, now the accompanying jet is caused by
the evolution of the quark from the electron and thus
has small q_T (fig.3).

3. That the cross sections are rather similar is not
unexpected since one is, in fact, computing on the basis
of a single diagram: first the quark is hard and the
photon soft, then the quark is soft and the photon hard.
In the latter case on obtains a suppression factor

$$\frac{1}{\alpha_s(q_T^2) \log \frac{s}{4m_e^2}}$$

which, for relevant values of q_T and s, is about 1/4.
Asymptotically, in leading log approximation at fixed
x_T, it goes to 1.

6. SUMMARY

These lectures have discussed a new source of jet effects in e^+e^- collisions: jets arising from the $\gamma\gamma$ mechanism. QCD predicts a rich structure of phenomena and it will be interesting to see how well these predictions agree with future experimental tests.

Concerning the experimental verification of the predictions one may note the following points:

1. A detection of the fast forward and backward e^+ and e^- is not necessary provided that the jets can be identified with sufficient accuracy. More work has to be done on what these jet events look like after the jets have fragmented to hadrons.

2. If the forward and backward e^+ and e^- are detected much more information is, of course, gained. This will probably require so large scattering angles that the counting rate is considerably reduced.

3. PETRA energies are only marginally sufficient to test the QCD predictions; reliable tests are possible only at LEP energies.

Further important work on the photon structure functions has been carried out in refs.[21] (mass effects), [22] (independent diagrammatic derivation of the result in [4]) and [23] (derivation of second order asymptotic freedom effects).

REFERENCES

[1]   S. Brodsky, T.Kinoshita and H. Terazawa, Phys.
      Rev. D4 (1971) 1532.

[2]   H. Terazawa, Rev. Mod. Phys. 45 (1973) 615.

[3]   V. Budnev, I. Ginzburg, G. Meledin and V. Serbo,
      Physics Reports 15C (1975) 182. This also contains
      an extensive list of references to the pre-1975
      work.

[4]   C. Llewellyn Smith, Phys. Lett. 79B (1978) 83.

[5]   S. Brodsky, T. de Grand, J. Gunion and J. Weis,
      Phys. Rev. Lett. 41 (1978) 672, SLAC preprint
      SLAC-PUB-2199, 1978.

[6]   K. Kajantie, Physica Scripta 29 (1979) 230.

[7]   J. Bjorken, Phys. Rev. D8 (1973) 4098.

[8]   S. Ellis and M. Kislinger, Phys. Rev. D9 (1974) 2027.

[9]   D. Sivers, S.J. Brodsky and R. Blankenbecler, Phys.
      Reports 23C (1976) 1.

[10] V.N. Baier, V.S. Fadin and V.A.Khoze, Nucl.Phys.
      B65 (1973) 381.

[11] M.S. Chen and P.Zerwas, Phys.Rev. D12 (1975) 187.

[12] E.Witten, Nucl.Phys. B120 (1977) 189.

[13] V.Gribov and L.Lipatov, Yadernaya Fizika 15 (1972)
      781, 1218.

[14] L. Lipatov, Yadernaya Fizika 20 (1974) 181.

[15] Yu. Dokshitzer, D.Djakonov and S. Trojan, Hard
      processes in QCD (in Russian), Proceedings of the
      13th winter school of Leningrad institute of
      nuclear physics, Leningrad 1978.

[16] G.Altarelli and G.Parisi, Nucl. Phys. B125 (1977)
      298.

[17] C.Llewellyn Smith and S.Wolfram, Nucl. Phys.
     B138 (1978) 333.
[18] A.B. Govorkov, Sov. J. Part. Nucl. 8 (1978) 431.
[19] F.J. Yndurain, Phys. Lett. 74B (1978) 68.
[20] K. Kajantie and Risto Raitio, Univ. of Helsinki
     preprint HU-TFT-79-13, 1979.
[21] C.T. Hill and G.G. Ross, Nucl. Phys.B148 (1979) 373
[22] W.R.Frazer and J.F.Gunion, San Diego preprint
     UCSD-10P10-199 (1978).
[23] W.A. Bardeen and A.J. Buras, Fermilab preprint
     78/91-THY (1978).

Acta Physica Austriaca, Suppl. XXI, 709—716 (1979)

SUMMARY[+]

by

H. PIETSCHMANN

Institut für Theoretische Physik
Univ. Wien, Boltzmanngasse 5, A-1090 Wien

When - during the first days of this school -
Professor Mitter asked me to summarize Schladming's
Winter school again, I accepted with great pleasure.
It is indeed a challenge to try to put into perspective
particle physics of our days for it is at the same time
easy and difficult. It is easy and satisfying because
of the tremendous steps forward we have taken during the
last years; it is difficult and frustrating because we
can easily overshoot our goal by taking for granted what
is only a first hint. Thus there are two positions which
I try to avoid: One is the euphoric position which claims
that we have now solved all the problems and only small
obstacles are in the way to a final "unified theory" or

[+]Summary given at the
XVIII. Internationale Universitätswochen für Kernphysik,
Schladming, Austria, February 28 - March 10, 1979.

"world formula". The other one is the ever-pessimistic
position which keeps pointing out that we have achieved
nothing because there are alternative explanations for
everything.

We ask a lot of questions and Nature says mostly
"No", sometimes "Perhaps" but never "Yes", said Einstein,
pointing out that we can never be absolutely sure that the
theory is correct, we have to be content with an overwhelming support by experiments. Such is the position of
Gauge theories (the Salam-Weinberg model) in the realm of
weak and electromagnetic interactions. It has become the
basis for all other extensions which try to incorporate
more phenomena. I am, of course, referring to Quantum
Chromodynamics which is modelled after the so successful
electro-weak theory.

But in this connection we are suddenly faced with
a big puzzle! As the title of this school indicates,
leptons and quarks are the truly fundamental particles.
This is fine for leptons, but quarks apparently do not
exist! It seems to be the fundamental contradiction in
our scheme. Thus we can divide the lectures into two
strictly opposing opinions and I shall indicate this by
dividing space on the pages.

There is a set of lectures
which simply assumes the
existence of quarks taking
them as the basis for
computations and predictions.
For example, Paschos showed
how the phenomenology of
neutral currents can be

On the other hand, Morpurgo
talked about the experiments
trying to find quarks. He
pointed to the ingenius
ideas which have been ex-
plored with no success.
But Fairbanks finally found
one. The trouble, however, is

understood on the basis of quark currents. Kajantie told us how we are going to do hadron physics with lepton machines once we reach high enough energies. He too assumed quarks as the basis for his computations. Van Hove showed how useful quarks are in determining the best experiments in purely hadronic collisions and how they help to interpret the results. Margarete Krammer (and also Genz) culminated this game with Quarkonia, the possibility to understand and predict heavy narrow meson states which are such a wonderful support for this picture.

that Morpurgo's own experiment is not too different from Fairbank's and he found none. While this is not necessarily a contra-indication, I am reminded to similar situations in the history of physics. When Le Verrier, after successfully predicting (simultaneously with Adams) the planet Uranus, also predicted a new planet between Mercury and the sun in order to account for the perihelian motion of Mercury, this planet was also found several times. But we know it does not exist. Anyway, the overwhelming experimental evidence is against quarks.

Thus we can summarize all these lectures by saying: ANYONE WHO STILL DOES NOT BELIEVE IN QUARKS IS A COMPLETE FOOL!

Thus we can summarize Morpurgo's lecture by saying: ANYONE WHO STILL BELIEVES IN QUARKS IS A COMPLETE FOOL!

> If we now try to unify the two sides,
> the conclusion can only be:
> WHATEVER YOU BELIEVE, YOU ARE A COMPLETE
> FOOL!

It is quite obvious, that we cannot live with this situation, the contradiction must be eliminated!

The first try is a compromise: Quarks exist as fundamental particles, but they are confined and can never be detected experimentally. Joos showed us how this can be done mathematically; the basis for the idea is Quantum Chromo Dynamics.

When atoms were analyzed successfully, chemistry became a branch of physics. Is it all too surprising when, on the assumption that particles ARE successfully analyzed into quarks, Chan Hong-Mo showed how notions of chemistry can be useful to particle physics? Finally, being euphoric about the assumed success, one can develop dreams: Ellis talked about GUT, grand unified theories. It is great to see how far we can already develop such a scheme and it is not harmful, because as a balance we had the very fine exposition of Flügge on experiments. This, I think, is absolutely necessary and important: to be reminded on the basis of everything we are doing in physics: measurement. And to see, how tedious these measurements are and the amount of ingenuity and skill which goes into them in order to derive results on those quantities which are important to rule out((or corroborate) another theoretical scheme.

Well, I have said that confinement is a compromise.

The question is, is it a good or a foul compromise? This question can only be answered by history, i.e. in the future. But we can speculate and compare it with past experience. For a good compromise will stay with us and be improved to general satisfaction. A foul compromise will not last long - it will throw us back into the mad situation with the open contradiction unless we develop new concepts to achieve a synthesis.

Let me say right away that I believe we have not yet reached a synthesis, that the compromise is still rather foul. But - as I said - this is a personal opinion which I have no way to prove. I can only support it by historical analogies.

We all know, that the ptolemeic system of planetary motion ran into a difficult situation when more and more epicycles had to be added for a good description of the orbits. The Kopernican revolution, which put the sun into the center, was certainly a compromise, though - with circles as planetary orbits - it was less accurate. The synthesis was found through Kepler's laws based on Newton's unified description of mechanical processes.

We all know that Thomson's plum pudding model for the atom ran into difficulties when Rutherford discovered the nucleus. His model showed a number of contradictions.

THE ATOM IS STABLE	THE ATOM IS HIGHLY UNSTABLE
THE ATOM IS SPHERICAL	THE ATOM IS A SMALL PLANETARY SYSTEM
RADIATION SHOWS A DISCRETE SPECTRUM	RADIATION SHOULD FOLLOW A CONTINUOUS SPECTRUM

Bohr's model certainly was a compromise which eliminated some of the contradictions. But the synthesis was reached only through Quantum Mechanics.

We all know that the quark model ran into difficulties when quarks could not be found. Confinement certainly is a compromise, but is it the final answer?

Let us - once more - look into the history of our science. If a compromise is foul, the contradiction cannot be kept out of the theory forever, it keeps bouncing back. The outstanding example is the history of the concept of the atom. It dates back to the old Greeks. The underlying contradiction has most explicitly been pronounced by Zeno from Elea about 5 centuries B.C. Zeno - in his apories (contradictions with no way out) - stated flatly: "The flying arrow cannot reach its target". This is, of course, in contradiction to experience. But Zeno had a proof! Before the arrow reaches its target, it has to pass through the middle point of the distance. Once it is there, it has to pass the middle point of the remaining distance before it reaches its target and so on, proving the statement. In modern language, we would say that the series 2^{-n} TENDS to zero, but never REACHES it. Many scientists believe, that modern calculus can SOLVE this problem. It cannot! It can overcome it by a technical trick. Let us quote from the famous mathematician Courant: "Since the time of Zeno and his paradoxes all attempts at an exact mathematical formulation of the **intuitive** physical or metaphysical notion of continuous motion have failed... The intuitive idea of a continuum and a continuous

flow is completely natural. But we cannot appeal to it,
when we want to clarify a mathematical situation; there
remains a gap between the intuitive idea and the
mathematical formulation which should describe the
scientifically important elements of our intuition in
precise terms.... Zenos paradoxes point to this gap."
Courant then points out how Cauchy eliminated the
paradox by a trick, as far as mathematical analysis
is concerned. The paradox clearly stems from the
contradiction between continuum and discrete points.
We cannot decide which of the two contradicting statements is right and which is wrong: "A continuum can
never be made up of discrete points" and "In order to
handle the continuum formally, it must be divisible
into mathematical points."

For physics, the contradiction was eliminated
(not solved!) by Demokrit from Abdera about 400 B.C.
He simply postulated the atom as indivisible in spite
of its extended nature. Thus the atom represented both
continuous and discrete properties, the contradiction
was - so to speak - buried within the atom and all was
fine as long as the question as to the nature of the
atom was not raised. However, this question could no
longer be suppressed, when J.J. Thomson discovered the
electron in 1897. The dichotomy discrete - continous
was beautifully unveiled in his plum pudding model,
where the positive dough represented the continuum and
the electrons were discrete. But Rutherford upset this
model and tried to construct an atom purely from discrete
particles. The contradiction reappeared in the form
stated before. As I have said, Bohr found a compromise,
in his model, but the final synthesis was achieved by
Quantum Mechanics. The electron is neither discrete nor

continuous, neither a particle nor a wave, because it
is both! It was Niels Bohr, who dared to leave the
contradiction in the physical picture of the microcosm,
the man who often said: "The contrary of a correct
statement is a wrong statement, but the contrary of a
deep truth may again be a deep truth."

The struggle to handle the dichotomy discrete -
continuous ended in a peaceful coexistence of
contradicting notions: waves and particles. Parallel
to this struggle we endeavour to find out what an
elementary particle is. The atom of Demokrit was the
element which buried the contradiction discrete -
continuous, as we have seen. When the question about
the structure of the atom came up, we went to the
deeper level of elementary particles, which again
carry the contradiction in the concept of dualism.
But we go a step further and ask the question about
the structure of hadrons and now we face a new
contradiction which we must distinguish from the old
discrete - continuous problem (which can rest as long
as we are satisfied with Quantum Mechanics). The new
contradiction is the one we have started out with:
Quarks are constituents of hadrons, but they do not
exist as particles.

At this point my own view becomes cloudy and
blurred and I have to stop my wise talk. All I can do
is to express my feeling that confinement is - analogous
to the Bohr model - the first compromise and that we are
waiting for a breakthrough to a completely new picture.
But this is a foul way to end such a fine meeting, thus
I will instead make a prediction: Before 1990 there will
be either a fascinating breakthrough or physics will become very boring.